W9-AWY-626

AN INTRODUCTION TO HUMAN GENETICS

AN INTRODUCTION TO HUMAN GENETICS

THIRD EDITION

H. ELDON SUTTON
The University of Texas at Austin

1980 Saunders College
Philadelphia

Saunders College
West Washington Square
Philadelphia, PA 19105

Library of Congress Cataloging in Publication Data

Sutton, Harry Eldon, 1927-
An introduction to human genetics.

Bibliography.
Includes index.
1. Human genetics. I. Title.
QH431.S95 1980 573.2′1 79-25443
ISBN 0-03-043081-X

Printed in the United States of America
2 3 038 9 8 7 6 5 4 3

To Beverly

PREFACE

The continued explosion of information in human genetics and in related scientific areas forces any textbook to be at best a small sample of the rich smorgasbord available. Many will not find their favorite dishes here. Some old preferences have had to yield to new choices that are more useful pedagogically. A number of interesting new findings were not included because they do not involve new principles. It is hoped that the instructor using this text will provide additional examples appropriate to the background and interests of the students.

As in the previous edition, certain chapters will be largely review for many students who have previously studied genetics or related fields. These may be omitted or assigned for review only. Chapters 2, 3, and 8 in particular will be familiar to many students.

Moreover, instructors may teach the chapters in a different sequence, since each chapter functions as an independent unit.

The order of chapters has been changed slightly in this edition. Mutation has been moved forward to a more molecular context, although some will prefer to delay it until after discussing population genetics. Analysis of pedigrees has been moved forward to Chapter 2, since virtually every student will have been exposed to Mendelian principles, including human pedigrees. One new chapter, on behavior genetics, has been added, pulling together material from several other chapters as well as adding new material. This area of human genetics is coming of age and deserves more systematic presentation than was possible earlier.

Any book is the work of a number of persons. Many suggestions and corrections have been made by users of the second edition. Among persons who have suggested specific improvements in the present edition are Drs. John Buettner-Janusch, New York University; William Daniel, University of Illinois at Urbana; Linda Dixon, University of Colorado at Denver; Jean MacCluer, Pennsylvania State University; O. J. Miller, Columbia Presbyterian Medical Center, New York; Dawson Mohler, University of Iowa; Muriel Nesbitt, University of California at San Diego; Henry Schaffer, North Carolina State University; Charles Shaw, M. D. Anderson Hospital, Houston, Texas; and Beverly Sutton, Austin State Hospital, Austin, Texas.

To all of these, my thanks, and my apologies if I did not always take their advice.

I am indebted to the Literary Executor of the late Sir Ronald Fisher,

F.R.S., Cambridge, and to Oliver and Boyd Ltd., Edinburgh, for their permission to reprint Table III from their book *Statistical Methods for Research Workers.*

H. E. S.

Austin, Texas
September, 1979

CONTENTS

AN INTRODUCTION TO HUMAN GENETICS

CHAPTER ONE

INTRODUCTION AND HISTORICAL BACKGROUND

THE ORIGINS OF HUMAN GENETICS

The science of genetics is generally said to have been founded by the publication of Mendel's paper in 1865 and to have become established with its rediscovery in 1900. These are the landmark years of discovery, but questions had been asked concerning the mechanisms of heredity for centuries. The brilliance of the contributions of Mendel and his twentieth-century followers has tended to obscure the more limited contributions of others prior to 1900. Nevertheless, man's interest in himself and his domestic animals and plants had led to appreciation of the importance of heredity in natural variation. In a few instances, rules were established empirically that accounted for inheritance of a rare trait, but these were never generalized into a valid theory of heredity.

Early Observations on Human Heredity

The first appreciation of heredity as a reliable expression of nature probably arose with the introduction of agriculture. The classical Greek authors knew that the best way to obtain domestic animals with desirable qualities is to start with parents with desirable qualities. In some instances, the suggestions on how to do this are quite specific. The maintenance of purebred strains and the production of hybrids by crossing strains became well-established practice prior to the modern scientific period, although the practice was often, and still is, embellished with rituals that are pure superstition. The application of these principles to man has often been suggested but never applied systematically. Sir W. Lawrence noted in 1823, "The hereditary transmission of physical and moral qualities, so well understood and familiarly acted on in the domestic animals, is usually true of man. A superior breed of human beings could only be produced by selections and exclusions similar to those so successfully employed in rearing our more valuable animals." (Quoted in Bateson, 1909, pp. 6–7.) Today we would argue with Lawrence about the definition of superior and the biological nature of morality, but the concept of transmission of traits still is valid.

There are many references in classical Greek literature to family resemblances, and the reasons for this were often the subject of speculation. Hippocrates and Aristotle, among others, discussed the determination of sex and the inheritance of disease. But their reasoning was distorted by belief in spontaneous generation of life—for example, of frogs and maggots—from nonliving materials and by complete ignorance of the anatomy and function of internal organs, such as testes and ovaries. It is small wonder then that some thought the sex of an offspring depended on whether the semen originated from the right or left testis and the semen carried messages from all parts of the body. Indeed, the latter idea was still accepted by Charles Darwin.

The earliest record of a correct analysis of the pattern of inheritance of a human trait appears to occur in the Talmud (cited in Stern, 1973). Hemophilia, an X-linked gene, causes failure of the clotting mechanism in blood. If two boys in a family died from bleeding following circumcision, the Talmud decreed that later born sons from the same mother or from her sisters need not be circumcised. Thus the Jews recognized that this trait, affecting only males, was transmitted through unaffected females.

The Scientific Era: The Pre-Mendelians

With the Age of Enlightenment came an appreciation for the importance of direct and careful observations of nature. Inevitably, human variation and the inheritance of peculiar traits came under scrutiny. Maupertuis, for example, recorded in 1752 a family in which polydactyly had occurred in four generations (see Glass, 1959). The occurrence in successive generations of rare dominant and X-linked traits was noted by many, and some of the patterns of transmission of these traits were recognized. The great chemist John Dalton read a paper before the Literary and Philosophical Society of Manchester in 1794 describing red–green colorblindness in himself, one of his two brothers, and in several other families. Dalton noted that only males are affected and that neither parents nor offspring are affected. He did not extend his observations to more distant relatives. Nasse, in 1820, described clearly X-linked recessive transmission of traits, and this subsequently became known as Nasse's law.

Of special interest is a book published in 1814 by Joseph Adams (1756–1818). The title is *A Treatise on the Supposed Hereditary Properties of Disease,* and the contents show remarkable perception by the author of many aspects of human genetics. He distinguished between dominant and recessive inheritance (referred to as *hereditary* and *familial,* respectively). His comments on genetic predisposition show special insight. Motulsky (1959), who rediscovered the work of Adams, presents an interesting analysis of Adams and the reasons for the small impact of his work on medicine and on biology in general.

Persons interested in human genetics prior to 1900 faced the same problems as persons working with experimental organisms, namely, that the traits of interest were all continuous traits, that is, measured on a continuous scale, rather than discontinuous traits or attributes. Without exception, these are complex in their heredity. To study them effectively requires an appreciation of probability and statistics, but the familiar statistical techniques of today had not been invented at the time of Mendel. Two brilliant English scientists, Francis Galton and then Karl Pearson, resolved the dilemma by inventing a series of statistics that could be used to compare observation with theory.

Francis Galton. The commanding figure in the study of human heredity prior to 1900 was Francis Galton (1822–1911), born in the same year as Gregor Mendel. Galton was a member of the British intelligentsia, a first cousin of Charles Darwin and grandson of Erasmus Darwin. He generally shared the attitudes of the upper class Englishmen of the nineteenth century, that is, that aptitude and morality are consequences of the class into which one is born and that nature rather than nurture is primarily responsible for the differences observed among persons.

Galton's penetrating analytical mind sought to explain human variation in terms of regular scientific laws. As had Mendel's predecessors, Galton chose socially interesting but exceedingly complex traits for his investigations. His first paper on eugenics was published in 1865, the year that Mendel's work was published. This was followed by a major work, the book *Hereditary Genius,* published in 1869. In this he assembled records of outstanding British families and concluded that heredity is the major factor in human achievement, a finding that, at the very best, fails to consider adequately the contributions of environmental variation. Among the series of publications from Galton are to be noted especially his 1876 paper on twins, in which the comparison of identical with nonidentical twins was recognized as a means of measuring the relative contributions of "nature vs. nurture" (a phrase that he took from Shakespeare), and his 1892 book *Finger Prints,* responsible in large part for initiating the widespread use of fingerprints in identification.

It is a tribute to Galton's genius that he recognized the need for careful observation and measurement. He further noted the element of chance in the transmission of genetic traits, and he saw the need to develop mathematical tools for describing the spectrum of possible outcomes from various types of human matings. He thus became the founder of statistical genetics. In his subsequent studies of resemblances among family members, Galton developed correlational analysis, a major technique of statistics used at present in nearly every field that requires statistics.

Galton was an "involved" scientist who believed strongly that human biological inheritance should be preserved and improved. He had a major role in founding the eugenics movement in England in the last part of the nineteenth century and was identified as its scientific leader during a period when it flourished at the beginning of this century. Galton differed from most persons associated with the eugenics movement in his recognition of the need for a sound scientific base, and the movement eventually foundered when it became part of political movements and was used to justify purely political ends.

Galton founded the Laboratory of National Eugenics at University College, London, in 1907, and, on his death in 1911, this became the Galton Laboratory of Eugenics supported by a large bequest from Galton. For several decades, this was virtually the only center where substantial research was carried out in human genetics, and it continues to be one of the leading centers.

It is an interesting fact of Galton's life that, when Mendelism was rediscovered in 1900, he refused to accept it. The traits that Galton studied were complex and did not occur as simple alternatives. The biometrics that he was instrumental in developing was of greater use in predicting stature and intelligence of offspring than was Mendelism, a fact that is still true today, and Galton, who above all was an empiricist, saw no reason to accept the findings of Mendel as the basis for a general theory of heredity. The decade following 1900 was characterized by sharp conflict in England between the "Biometric" school and the "Mendelian" school, the latter having equally vigorous proponents. An indication of the differences between these two groups is provided by a quotation from William Bateson, one of the champions of Mendelism, in his 1909 publication *Mendel's Principles of Heredity*:

> Of the so-called investigations of heredity pursued by extensions of Galton's non-analytical method and promoted by Professor Pearson and the English Biometrical school it is now scarcely necessary to speak. That such work may ultimately contribute to the development of statistical theory cannot be denied, but as applied to the problems of heredity the effort has resulted only in the concealment of that order which it was ostensibly undertaken to reveal.

Karl Pearson. Galton's successor as leader of the English biometrics group was Karl Pearson (1857–1936). Pearson, a statistician and mathematician, was director of the Galton Laboratory of Eugenics from 1911 until 1933. Pearson initially practiced law but soon received an appointment as a mathematician at University College, London. His 1892 book, *The Grammar of Science,* is a major treatment of the philosophy of science. His most significant contribution was development of the chi-square test of statistical significance (1900), and in 1904 he derived the intrafamily correlations that would be expected on the basis of Mendelian theory. Although much of Pearson's work was in the context of human heredity, his statistical contributions have application to all areas of statistics.

The Early Mendelians

A. E. Garrod. One person especially requires mention among the pioneers of human genetics—Sir Archibald E. Garrod (1858–1936). Garrod was an English physician who was interested in metabolic disease, although, at the turn of the century, the concept of enzyme catalysis was new, and very little was known of the nature of enzymes. Garrod was acquainted with the newly rediscovered laws of Mendel and was also familiar with certain human traits that occur as distinct attributes rather than as extremes of a continuous curve. These were some rare diseases—alkaptonuria, cystinuria, pentosuria, and albinism—that were present apparently from birth and that often affected two or more members of a sibship in spite of the rarity among the population as a whole. Parents ordinarily were not affected. Garrod reasoned that these might be recessive traits, and, at least with alkaptonuria, he proposed that an essential enzyme was missing. His studies on alkaptonuria, published in 1902, state clearly the hypothesis that alkaptonuria is inherited as a Mendelian recessive trait. Garrod further gave a clear explanation of the effects of consanguinity on the appearance of recessive traits. His contributions were summarized in the Croonian lectures of 1908, and then were subsequently expanded into a classic work, *Inborn Errors of Metabolism,* published in 1909.

As in the case of Mendel, Garrod's perception was well ahead of its time and as a consequence had little impact. His findings and ideas disappeared from the literature until they were rediscovered some four decades later. By that time, the field of genetics had caught up with Garrod and could appreciate him.

Although the early Mendelians were interested in human genetics, their real successes were with experimental organisms. Studies of human genetics were largely limited to identifying traits that followed Mendelian rules. However, a few brilliant persons—for example, J. B. S. Haldane (1892–1964) and R. A. Fisher (1890–1962)—continued to make highly original contributions to the methodology and concepts of human genetics. By the 1960s, when the mechanisms of gene action had become well understood, interest in human genetics became widespread. Recent years have seen a mating between basic genetics and medicine that has been highly productive for each.

THE LITERATURE OF HUMAN GENETICS

Original reports of research in human genetics are published in a great variety of journals. Most of the medical journals frequently have articles that are concerned with inherited disorders. In addition, the following journals and periodicals primarily publish articles on human genetics:

Acta Geneticae Medicae et Gemellologiae
Advances in Human Genetics
American Journal of Human Genetics
American Journal of Medical Genetics
Annales de Génétique
Annals of Human Genetics
Clinical Genetics
Cytogenetics and Cell Genetics
Human Genetics (Humangenetik)
Human Heredity
Japanese Journal of Human Genetics
Journal de Génétique Humaine
Journal of Medical Genetics
Progress in Medical Genetics

The following textbooks and monographs will prove useful to students and investigators:

BECKER, P. E. 1964–. *Humangenetik.* Ein kurzes Handbuch in 5 Bänden. George Thieme, Stuttgart.

BODMER, W. F., AND CAVALLI-SFORZA, L. L. 1976. *Genetics, Evolution, and Man.* San Francisco: W. H. Freeman, 782 pp.

FRASER, G. R., AND MAYO, O. (Eds). 1977. *Textbook of Human Genetics.* Oxford: Blackwell Scientific Publs., 524 pp.

GIBLETT, E. R. 1969. *Genetic Markers in Human Blood.* Philadelphia: F. A. Davis, Oxford; Blackwell Scientific Publs., 629 pp.

HARRIS, H. 1975. *The Principles of Human Biochemical Genetics,* 2nd Ed. Amsterdam, Oxford: North-Holland, 473 pp.

LEVITAN, M., AND MONTAGU, A. 1977. *Textbook of Human Genetics,* 2nd Ed. New York: Oxford Univ. Pr., 1012 pp.

LI, C. C. 1976. *First Course in Population Genetics.* Pacific Grove, Calif.: Boxwood Pr.

MC KUSICK, V. A. 1978. *Mendelian Inheritance in Man,* 5th Ed. Baltimore: Johns Hopkins Univ. Press, 976 pp.

NORA, J. J., AND FRASER, F. C. 1974. *Medical Genetics: Principles and Practices.* Philadelphia: Lea and Febiger, 399 pp.

NOVITSKI, E. 1977. *Human Genetics.* New York: Macmillan, 458 pp.

RACE, R. R., AND SANGER, R. 1975. *Blood Groups in Man,* 6th Ed. Philadelphia: J. P. Lippincott, 659 pp.
ROBERTS, J. A. F., AND PEMBREY, M. E. 1978. *An Introduction to Medical Genetics,* 7th Ed. New York: Oxford Univ. Pr., 336 pp.
STANBURY, J. B., WYNGAARDEN, J. B., AND FREDRICKSON, D. S. (Eds). 1978. *The Metabolic Basis of Inherited Disease,* 4th Ed. New York: McGraw-Hill. 1862 pp.
STERN, C. 1973. *Principles of Human Genetics,* 3rd Ed. San Francisco: W. H. Freeman, 891 pp.
THOMPSON, J. S., AND THOMPSON, M. W. 1973. *Genetics in Medicine,* 2nd Ed. Philadelphia: Saunders, 400 pp.

SUMMARY

1. Selective breeding has long been known as effective in improving domestic animals and plants. It was assumed to apply to human beings also.

2. Although earlier scientists were partially successful in working out patterns of heredity in human disease, Francis Galton (1822–1911) was the first to make a major systematic effort to understand the inheritance of complex traits in man. He recognized the importance of twins in genetic research and was largely responsible for the scientific study of fingerprints. His contributions include development of the biometrical approach to measurement of family resemblances.

3. Galton's colleague and successor as leader of the biometrical school of genetics was Karl Pearson (1857–1936), a mathematician whose contributions include development of the χ^2 test of significance.

4. Of the early Mendelians, A. E. Garrod (1858–1936) especially stands out for his understanding of the genetic control of metabolic diseases.

5. Some of the major reference sources in human genetics are listed.

REFERENCES AND SUGGESTED READING

BATESON, W. 1909. *Mendel's Principles of Heredity.* Cambridge: Cambridge Univ. Pr., 396 pp.
DALTON, J. 1798. Extraordinary facts relating to the vision of colours: With observations. *Mem. Lit. Philos. Soc. Manchester* 5 (part 1): 28-45.
GARROD, A. E. 1909. *Inborn Errors of Metabolism.* Reprinted 1963 with a supplement by H. Harris. London: Oxford Univ. Press.
GLASS, B. 1959. Maupertuis, pioneer of genetics and evolution. In *Forerunners of Darwin: 1745-1859* (B. Glass, O. Temkin, W. Straus, Jr, Eds.). Baltimore: Johns Hopkins Univ. Pr., pp. 51–83.

MOTULSKY, A. G. 1959. Joseph Adams (1756-1818). A forgotten founder of medical genetics. *Arch. Intern. Med.* 104: 490–496.

NEEL, J. V. 1976. Human Genetics. In *Advances in American Medicine: Essays at the Bicentennial*, Vol. I (J. Z. Bowers, E. F. Purcell, Eds.). New York: Josiah Macy, Jr. Foundation, pp. 39–99

STERN, C. 1973. *Principles of Human Genetics*, 3rd Ed. San Francisco: W. H. Freeman, p. 316

STUBBE, H. 1972. *History of Genetics.* Cambridge, Mass.: The MIT Press, 356 pp.

STURTEVANT, A. H. 1965. *A History of Genetics.* New York: Harper and Row, 165 pp.

CHAPTER TWO

MENDELIAN INHERITANCE

Throughout recorded history, the transmission of physical traits from parents to offspring has been recognized. The explanations for such family resemblances have been varied and need not be of concern here. Of more interest is a consideration of why investigations by competent scientists failed for so long to reveal the laws of heredity.

A major barrier to recognition of the laws of heredity was the assumption of a blending of hereditary traits, whether or not such assumptions were formally stated. The progeny of a cross between two different parental types frequently are a blend of parental characteristics, and it was thought that the inherited factors which determine these characteristics must blend together, losing their separate identities in the process. The progeny in turn would transmit to their offspring factors essentially different from those received from their own parents. This misconception had its origin in part in the attempts of experimenters to work with physical characteristics that have subsequently proved to be very complex in their inheritance, being influenced by many genes. Even today, when the basic laws of heredity are well-understood and sophisticated statistical methods are available, we still are largely unable to analyze complex traits in a manner that yields information regarding the action of individual genes.

The Austrian monk Gregor Mendel, working in a monastery in Brünn, succeeded where his predecessors had failed. A major factor in Mendel's success was his good judgment in selecting for study traits that occur as clearly alternate forms rather than on a continuous scale of differences. For his investigations he used the garden pea (*Pisum sativum*), in which it is possible to control pollination. Among the traits he observed were round versus wrinkled seeds, yellow versus green cotyledons, white versus brown seed coats, inflated versus constricted seed pods, green versus yellow pods, axial versus terminal flowers, and long versus short stems. For each of these traits, a single plant can be classified as belonging to one or the other category.

In 1865, after eight years of investigation, Mendel presented his conclusions before the Science Research Society at Brünn. Publication of his studies occurred the following year in the proceedings of that Society, but their significance was not immediately recognized and they lay forgotten until 1900. In that year, Karl Erich Correns (working with peas and maize), Hugo de Vries (working with *Oenothera*), and Erich von Tschermak (working with peas) each independently arrived at conclusions similar to those of Mendel. They also discovered his paper and recognized its great importance. The year 1900 thus marks the beginning of genetics as a science.

In his use of simple traits, Mendel was able to avoid the erroneous assumption that the factors of heredity transmitted by the parents to their offspring are blended in the offspring. Furthermore, he recognized that the transmission of traits is a statistical process that can be described only in terms of probability. As a result of his studies, he formulated some generalizations that have become the foundations of the science of genetics. These generalizations usually are expressed as the *Law of Genetic Segregation* and the *Law of Independent Assortment.*

Although Mendel continued his scientific studies after publication of his work on garden peas, he soon became so involved in administration of the monastery that there was little time for research. He died in 1884 without the recognition he so richly deserved.

MENDEL'S LAW OF GENETIC SEGREGATION

Mendel observed that if he crossed two true-breeding strains that were different for one of the traits under observation, the progeny (the F_1, or first filial generation) were all identical to each other. Furthermore, the F_1 generation was identical to, or very similar to, one of the parental stocks. Mendel allowed the F_1 generation to fertilize itself, producing an F_2 generation. In contrast to the F_1, the F_2 generation was composed of two types of plants. The majority (three-fourths) resembled the F_1 and, hence, one of the original parental types; the remainder (one-fourth) resembled the other parental stock (Figure 2-1). Further breeding of members of the latter group among themselves produced only offspring of the second parental type.

The recovery of traits in the F_2 generation that were not expressed in the F_1 clearly indicated that the hereditary factors contributed by each parent to the F_1 cross maintain their identity, even though they may not influence the appearance of the F_1 plant. This observation is the cornerstone of Mendelian theory. Mendel designated the expressed factor as the *dominant* form, the other factor as the *recessive* form. In the F_1 generation, each individual receives both dominant and recessive factors from its parents, but only the dominant trait is expressed. In a mating of two F_1 plants, some of the offspring receive only recessive factors and hence express the recessive trait.

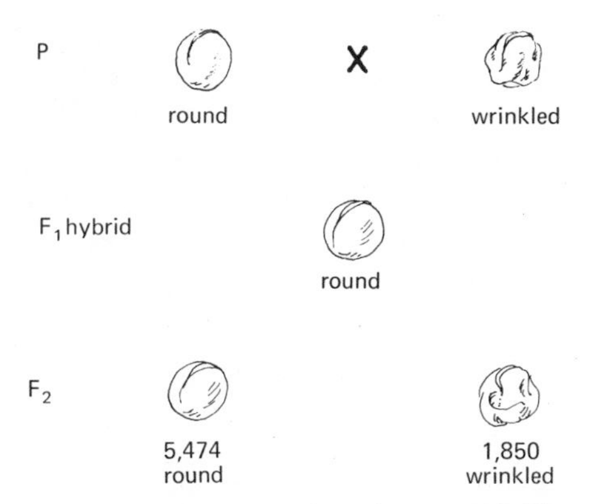

Figure 2-1 Diagram of one of Mendel's experiments with *Pisum sativum*, showing dominance of round seeds over wrinkled and recovery of wrinkled seeds in the F_2 generation.

It will be useful at this point to introduce the concepts of *genotype* and *phenotype*. In 1911, W. Johannsen used the word *gene* to denote the unit factor of heredity. The genotype of an organism refers therefore to the kinds of genes possessed, regardless of whether they are expressed. The haploid complement of genes is called a *genome*. The phenotype refers to the observable characteristics of an individual. This may be largely genetically controlled for a given trait, or it may be the result of genetic and environmental variation acting together. Occasionally, it is possible to infer part of the genotype of an individual from the phenotype, but genotype cannot be observed directly.* The alternate forms of a gene are designated *allelomorphs* or, more commonly, *alleles*. The position on a chromosome occupied by a particular series of homologous alleles is called the *locus* for those alleles.

As noted earlier, those members of the F_2 generation that expressed a phenotype corresponding to the recessive parent produced only recessive offspring when bred among themselves. The members of the F_2 exhibiting a dominant phenotype proved to be of two types: If allowed to self-fertilize, one-third of them produced only dominant offspring; the other two-thirds behaved as the F_1 generation, producing offspring three-fourths of which exhibited the dominant phenotype and one-fourth the recessive phenotype. This indicated to Mendel that there were three types of offspring in the F_2 generation: one-fourth with only recessive genes, one-fourth with only dominant genes, and one-half with both types. The seeds formed by the F_1 reflect a separation of the dominant and recessive genes into different gametes followed by chance recombination to form zygotes. The rule describing this separation of genes is known as the *Law of Genetic Segregation*.

* This statement, true for so many decades, is rapidly becoming incorrect as a result of recombinant DNA research.

The regular formation of dominant and recessive phenotypes in the F_2 and subsequent generations can be explained by assuming that each plant contains two alleles for each trait studied. If we represent the dominant form by A and the recessive by a, then a plant possessing the recessive phenotype would have an aa genotype. Plants with the dominant phenotype could be either AA or Aa. The original stocks used by Mendel were highly inbred and were therefore homozygous for most loci, either AA or aa. All plants in the F_1 generation would be heterozygous Aa, since they result from a cross of dominant $AA \times$ recessive aa. Each parent contributes equally to the offspring, whether it is the seed-bearing or pollinating parent.

The ratios in the F_2 can be explained by the assumption that each gene has an equal chance of appearing in a gamete, regardless of whether the gene is dominant or recessive. Thus an Aa plant would form equal numbers of A and a gametes (Figure 2-2). Random assortment of A and a gametes into new individuals would result in the ratio $1AA : 2Aa : 1aa$. The first two categories would be indistinguishable in phenotype, but two-thirds would be of type Aa and therefore capable of segregating into A and a gametes. These predictions correspond exactly to Mendel's results.

Mendel knew nothing about chromosomes. He proposed that gametes have one of each pair of genetic determiners and that zygotes therefore have two, because this hypothesis was adequate to predict the experimental results. It was later recognized that the sets of homolo-

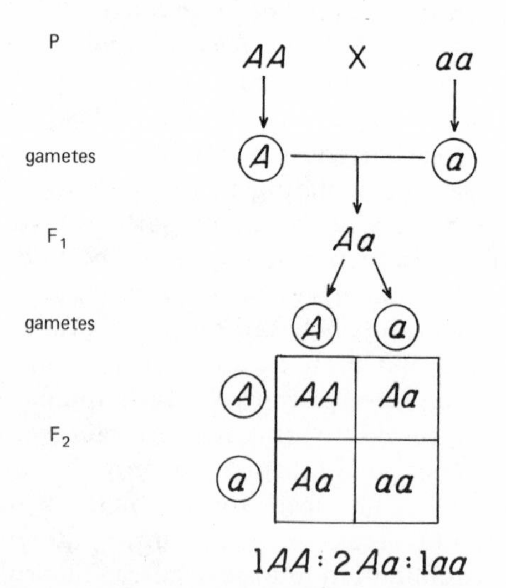

Figure 2-2 Diagram showing segregation of A and a in the gametes of the F_1 generation. All possible combinations of alleles occur in the F_2. The 2×2 chart in the F_2 generation is a convenient way of assuring that all possible combinations have been included and are in the correct proportions.

gous chromosomes in diploid cells would provide the physical basis for a double set of genes. At meiosis, each gamete receives only one of the sets of genes.

In Mendelian theory, the activity of a gene is independent of whether it was contributed to the zygote by sperm or ovum. Mendel found that for all the traits he studied it made no difference whether the trait was contributed by the pollen-bearing or the seed-bearing parent. When a new zygote is formed, the genes appear to act without any "memory" of previous cytoplasmic environments.

An important step in the analysis of an inherited trait is the formulation of a hypothesis of the mode of heredity. When a trait is designated dominant or recessive, this involves a definite prediction regarding the number and relationship of individuals with the trait. The validity of the hypothesis can be verified by an examination of the families of affected individuals. Should the results prove to be different from those predicted, then the hypothesis can be modified or discarded

The 3 : 1 Segretation Ratio. With many plants and animals, it is a relatively simple matter to establish genetic ratios through controlled matings. An example may be taken from Mendel's studies of peas. He crossed two parental (P) varieties, one having round seeds and the other wrinkled seeds. The F_1 hybrid seeds were all round, indicating round to be dominant over wrinkled. When the plants that developed from the F_1 seeds were allowed to self-fertilize, both round and wrinkled seeds were recovered. Of the 7324 F_2 seeds observed, 5474 were round and 1850 were wrinkled, a ratio of 2.96 to 1. A diagrammatic representation of the study is given in Figure 2-1. The observed 2.96 to 1 satisfies the 3:1 prediction, and it is not necessary to postulate additional factors operating in the inheritance of this trait.

The 3:1 ratio among offspring of two heterozygous parents is perhaps the most important means of recognizing simple recessive inheritance. A trait may be suspected of being recessive if it appears among the offspring of parents neither of whom has the trait. But it can be accepted as a simple recessive only if it occurs in one-fourth of the progeny of all appropriate matings. If a fraction other than one-fourth is affected, then consideration must be given to environmental influences as well as more complex genetic relationships. Differential survival of gametes or zygotes also may alter the ratio and cause a deviation from one-quarter.

The Backcross. A 3:1 ratio characterizes the dominant allele as well as the recessive in a mating of two heterozygotes. A more efficient means of demonstrating simple dominant inheritance is the *backcross* to the recessive parent. In experimental organisms, this consists in mating the F_1 generation to the parental stock that exhibits the apparently recessive phenotype. For example, in a cross between pigmented and albino mice, the F_1 is pigmented. If the F_1 is then crossed with the albino parental stock, and if pigment formation is a simple Mendelian trait with pigmentation dominant to albinism, both pigmented and albino offspring should occur in a ratio of 1:1. This can be symbolized as follows:

P	Pigmented (CC) × albino (cc)
F_1	Pigmented (Cc)
Backcross	Pigmented (Cc) × albino (cc)
Progeny	$\frac{1}{2}$ pigmented (Cc) + $\frac{1}{2}$ albino (cc)

MENDEL'S LAW OF INDEPENDENT ASSORTMENT

In the previous discussion, only one trait at a time was considered. When Mendel considered two or more traits together, he found that they were transmitted independently of each other. The inheritance of round instead of wrinkled seeds in the F_2 generation did not influence the probability of inheriting long versus short stems. In other words, the particular combinations of alleles transmitted by the parental gametes were not maintained in the F_1 generation.

Consider a mating of parental types that breed true for two traits: This may be written $AABB \times A'A'B'B'$, without specifying which alleles are dominant or recessive. Such a mating involving two loci is called a *dihybrid cross*. The F_1 generation will consist of $AA'BB'$ individuals (Figure 2-3). The gametes of the F_1 will consist half of A and half of A'; the same is true with respect to the B and B' alleles. If the two sets of alleles are independent, then half of the A gametes should be B and half should be B'. The same is true of the A' gametes. Thus there would be four types of gametes in equal numbers: AB, $A'B$, AB', and $A'B'$.

Formation of the F_2 zygotes would result in nine genetically different types in the ratio $1AABB:2AABB':2AA'BB:4AA'BB':1AAB'B':2AA'B'B':1A'A'BB:2A'A'BB:1A'A'B'B'$. All nine genotypes could be recognized only in the event both AA' and BB' are distinguishable from their respective homozygotes. If alternate forms are clearly dominant or recessive, as in the traits studied by Mendel, only four types of offspring can be recognized, in the ratio 9:3:3:1. For example, in a cross of round-seeded, yellow cotyledon × wrinkled, green cotyledon, 556 seeds were obtained in the F_2. These consisted of 315 round and yellow, 101 wrinkled and yellow, 108 round and green, and 32 wrinkled and green. Further breeding indicated, as predicted above, that nine genotypes were present among these seeds. Thus the assortment of genes from different loci into gametes appears to be entirely a random process.

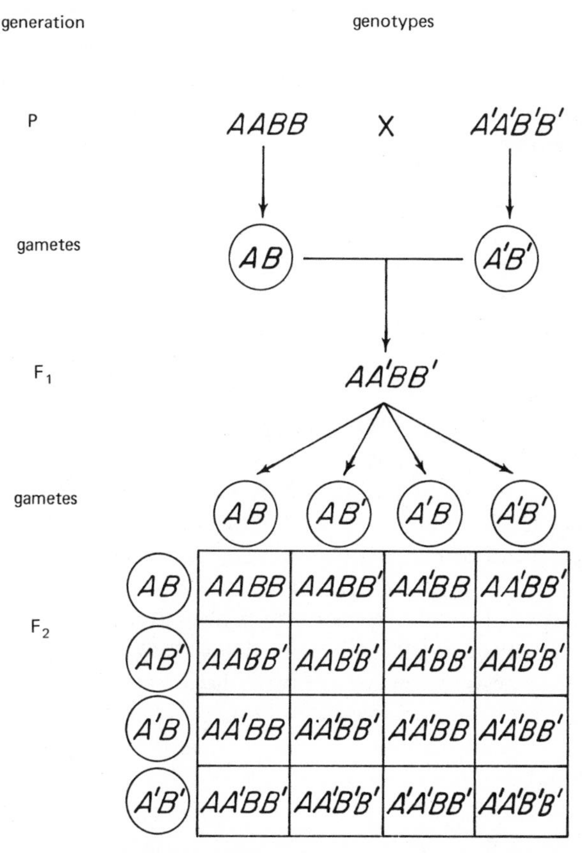

Figure 2-3 Diagram showing independent assortment of alleles at two loci in a dihybrid cross. Each of the F_1 gametes comprises $\frac{1}{4}$ of the total; therefore, each cell in the F_2 chart represents $\frac{1}{16}$ of the total F_2 generation.

Two loci may segregate independently and yet interact in the production of a single trait. For example, albinism in some plants may result from homozygosity for a recessive allele at any of several loci. In Figure 2-3, suppose that either $A'A'$ or $B'B'$ produces albinism and that any combination of alleles with at least one A and one B is normally pigmented. With independent segregation of A and B, there would be 9 pigmented and 7 albino offspring in the F_2 generation. Thus more complex situations may lead to ratios other than 3 : 1.

Although independent assortment is the rule for most loci, there are exceptions. If two genes A and B are located close together on the same chromosome, they ordinarily will go into the same gamete. Exchanges of segments of homologous chromosomes (*crossing over*) commonly occur at meiosis. If crossing over occurs between two loci, the allele combinations will be changed. If the loci are close together, the likelihood of a crossover between them is small. An individual who is heterozygous for two such loci (AB on one chromosome and $A'B'$ on the other) will produce the expected four types of gametes, but the AB and the $A'B'$ combinations will occur more frequently than AB' and $A'B$. None of the traits that Mendel studied happened to be *linked* on the same chromosome. The significance of linkage will be explored later (Chapter 19).

MENDELIAN INHERITANCE IN MAN

With the rediscovery of Mendelian heredity in 1900, a debate ensued as to whether the rather simple laws that applied to sharply contrasting traits of plants also applied to other organisms, including man. Many traits, such as human stature, do not occur as simple alternate forms—tall or short—and hence clearly do not follow Mendelian rules. But are any human traits inherited by Mendelian rules?

The first such trait was reported by Farabee in 1905. Brachydactyly is a condition in which the fingers are broad and short. In families in which brachydactyly occurs, persons with the trait can readily be distinguished from those without. The pattern of transmission over several generations was completely in accord with the expectations for a dominant trait (Figure 2-4). Thus Mendelian heredity was shown to occur for some traits at least. Many other examples were soon reported.

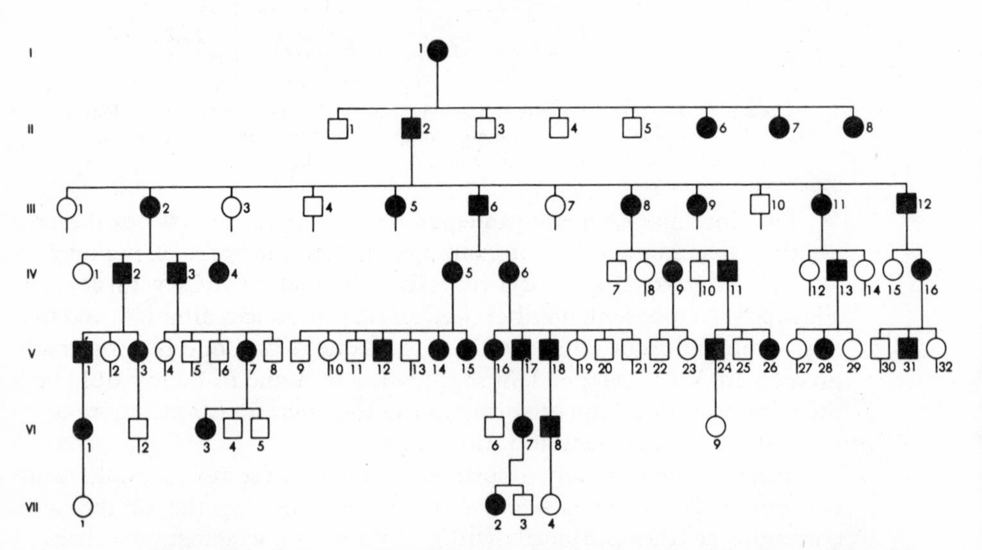

Figure 2-4 Pedigree of brachydactyly showing dominant inheritance. A portion of this pedigree was reported by Farabee in 1905 as the first example of Mendelian inheritance in man. The pedigree was revised and extended in 1962. (*From V. A. McKusick,* Human genetics, *2nd Ed. (1969). Reprinted by permission of Prentice-Hall, Inc., Englewood Cliffs, N.J.*)

Inheritance in man and other mammals is characterized as *autosomal* or *X-linked (sex-linked)*, depending on whether the gene is located on one of the autosomal chromosomes or on the X chromosome. In man there are 22 pairs of autosomes, alike in males and females, but females have two X chromosomes while males have one X and one Y chromosome. Many genes are known to be on the X chromosome, but only a few special genes at most appear to be on the Y chromosome. Females can be homozygous or heterozygous for X-linked genes, but males can only be *hemizygous*, since they have only one allele at each locus on the X chromosome.

It is often useful to characterize traits as dominant or recessive, autosomal or sex-linked, since this permits us to predict the risk that a particular trait will occur in an offspring or other relative. As later discussion will demonstrate, the terms *dominant* and *recessive* are descriptive of the phenotype and are not innate properties of the gene. For many so-called recessive traits, the heterozygote can be recognized with special tests. In some instances, it may be more useful to speak of the effect of having one copy of an allele as compared to two copies. Nevertheless, the terms dominant and recessive are useful in describing the transmission of phenotypes and the associated probabilities.

Autosomal Dominant Inheritance

A dominant trait can be recognized when its allele is in heterozygous combination with a different allele. Most "wild-type" alleles appear to be largely dominant. This implies that departures from wild type are likely to be recessive. More refined methods of observation show many recessive genes to have a partial dominant effect. The term dominant is sometimes interpreted to mean that the homozygous dominant and heterozygous phenotypes are similar—an idea that clearly is not true in many cases.

For uncommon or rare dominant traits, most of the matings are $Aa \times aa$, where A is the dominant allele. This type of mating is prevalent when the frequency of A is small compared with that of a. Most of the dominant genes of clinical significance are characterized by very low frequencies.

In a pedigree of dominant inheritance, it should be possible to construct a pedigree showing transmission of A through two or more generations. If the trait is due indeed to an autosomal dominant allele, several requirements must be met in such a pedigree: (1) Every affected person must have at least one affected parent. (2) Both males and females should be affected and should be capable of transmitting the trait, provided the trait does not cause sterility. (3) There is no skipping of generations or alternation of sexes. Father-to-son and mother-to-daughter transmission should be as frequent as father-to-daughter and mother-to-son. (4) An affected person should transmit the trait to half his offspring if the mating is $Aa \times aa$. A discussion of the difficulties of assessing ratios is given in Chapter 16.

In counseling, one is frequently asked to predict the genetic status of relatives of affected persons. For a dominant trait that is always expressed at an early age, there is no difficulty. Persons who do not have the trait will not develop it and will not transmit it to their children. It is necessary to modify such statements when dealing with traits of late onset or traits that may not always be expressed.

Autosomal Recessive Inheritance

As more sensitive techniques have been developed, it has become increasingly apparent that few genes are completely recessive. Many genes that must be in homozygous combination to show the "characteristic" expression can be detected nevertheless in heterozygous combination.

Galactosemia, a rare condition in which persons cannot convert galactose to glucose, behaves as a simple recessive. Affected persons lack the enzyme galactose-1-phosphate uridyl transferase. Persons heterozygous for the gene are able to metabolize galactose adequately and are therefore normal. Assays for the enzyme show them to possess only half normal activity. In terms of enzyme activity, the gene behaves as a codominant. There are still many genes for which no heterozygous effect is demonstrable.

The frequency of the gene in the population and the ability of homozygous persons to have offspring influence the kinds of pedigrees expected. Many recessives are deleterious, and affected persons may not survive. Or they may not be fertile if they do survive to adulthood. The elimination of recessive genes causes the frequency of homozygous affected persons to remain at very low levels. Under such circumstances, all homozygotes are produced by matings of the type $Aa \times Aa$. A typical pedigree therefore consists of a single sibship with affected persons, all other sibships being normal. An example of a pedigree showing inheritance of a relatively benign recessive trait is shown in Figure 2-5.

The genetic nature of recessive traits is sometimes difficult to prove. Some rare environmental factor might also lead to isolated sibships with more than one affected member. Evidence of the genetic basis is of two sorts. For more common traits, the most important is a 3 : 1 ratio of normal to affected (the necessity to correct for family size and method of ascertainment must be kept in mind). For very rare traits, there is an increase in consanguinity among parents of affected persons; that is, the parents are more likely to be related than other parents in the same population (Chapter 22). A few traits have been observed in only one

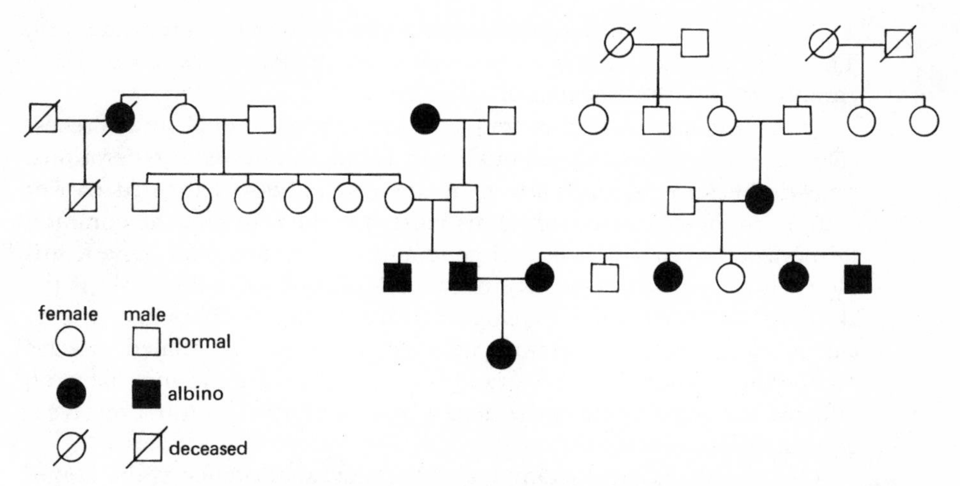

Figure 2-5 Pedigree of albinism showing recessive inheritance. This pedigree was obtained from the Hopi Indians, where the gene for albinism is unusually frequent. Therefore, pedigrees may show several generations to be affected. Superficially, such pedigrees may resemble dominant inheritance with reduced penetrance. (*From Woolf and Grant, 1962.*)

person. These have been attributed to recessive genes on the basis of highly specific enzyme defects. Such an interpretation is reasonable on the basis of what is known of gene action, but it should be remembered that direct evidence favoring genetic etiology is lacking.

Less deleterious genes may attain sufficient frequency so that $Aa \times aa$ matings sometimes occur. The presence of affected persons in two generations has been misinterpreted by some as evidence for a dominant form of the condition. However, a few pedigrees of this type will occur by chance, even if the gene is rare, and they may occur more often if the gene is common. There may indeed be both a dominant and a recessive form of a trait. Care should be exercised that the evidence for both forms is adequate.

X-Linked Recessive Inheritance

A special pattern characterizes genes on the X chromosome. Since females have two X chromosomes and males only one, genes that are recessive in females are expressed in males.

The transmission of X chromosomes provides the basis for the pattern of X-linked, or sex-linked, recessive inheritance. Affected males transmit the gene to all daughters and to no sons. Father to son transmission of a trait rules out X-linked genes, although the possibility should always be recognized that the son may by chance receive a similar gene from his mother. Unless the gene frequency is very high, most women carrying a gene a will be heterozygous Aa and will not express *a*. They will transmit *a* to half of their sons, who will express the trait. They will also transmit *a* to half their daughters, where it will be in combination with the paternal allele, usually *A*.

In addition to the higher frequency of expression of X-linked recessives in males, affected persons often occur in alternate generations. This is because of the pattern of affected father–unaffected carrier daughter–affected grandson. This pattern is the origin of the common misconception that skipping of generations is evidence for genetic etiology of a trait. It is evidence only for X-linked traits and only if the skipping follows a definite pattern. Since carrier mothers give rise to carrier daughters half the time, affected males may be related in ways other than grandfather and grandson. An affected male may have an affected maternal uncle if the gene was transmitted from his maternal grandmother.

One of the classical examples of X-linked inheritance is the hemophilia in many of the royal families of Europe. A pedigree is given in Figure 2-6. Affected males have a severe defect of the blood-clotting mechanism and are apt to bleed to death from minor cuts. Carrier females are normal. The first appearance of the hemophilia gene in the royal families was in a son of Queen Victoria. This is evidence that it probably arose as a mutation in Queen Victoria or in a segment of the germinal tissue of one of her parents.

X-Linked Dominant Inheritance

Although the catalogue of X-linked recessive genes is quite large, the first X-linked dominant genes, apart from the common "wild-type" alleles, were recognized very recently and only a few are known. The reason is probably because selection has favored survival on the X chromosome of alleles whose function can be fully satisfied by a single copy. In contrast, there are always two genes for each autosomal locus, and, if one is defective, the other frequently does not compensate 100 percent, even though it compensates adequately. With the discovery of X-chromosome inactivation, this argument will perhaps require modification, although inactivation may not affect the entire X chromosome.

The rules for X-linked dominant inheritance are similar to those for recessive, except that heterozygous females express the condition. Since females have twice as many X chromosomes as males, they have a higher frequency of affected persons. For rare traits and in the absence of interactions with other factors, two-thirds of affected persons should be female. An affected female, if she is heterozygous, will transmit the gene to half her offspring, regardless of sex. An affected male will not transmit to his sons, but all his daughters will be affected.

Probably the best example of an X-linked dominant trait is G6PD (glucose-6-phosphate dehydrogenase) deficiency. This enzyme occurs

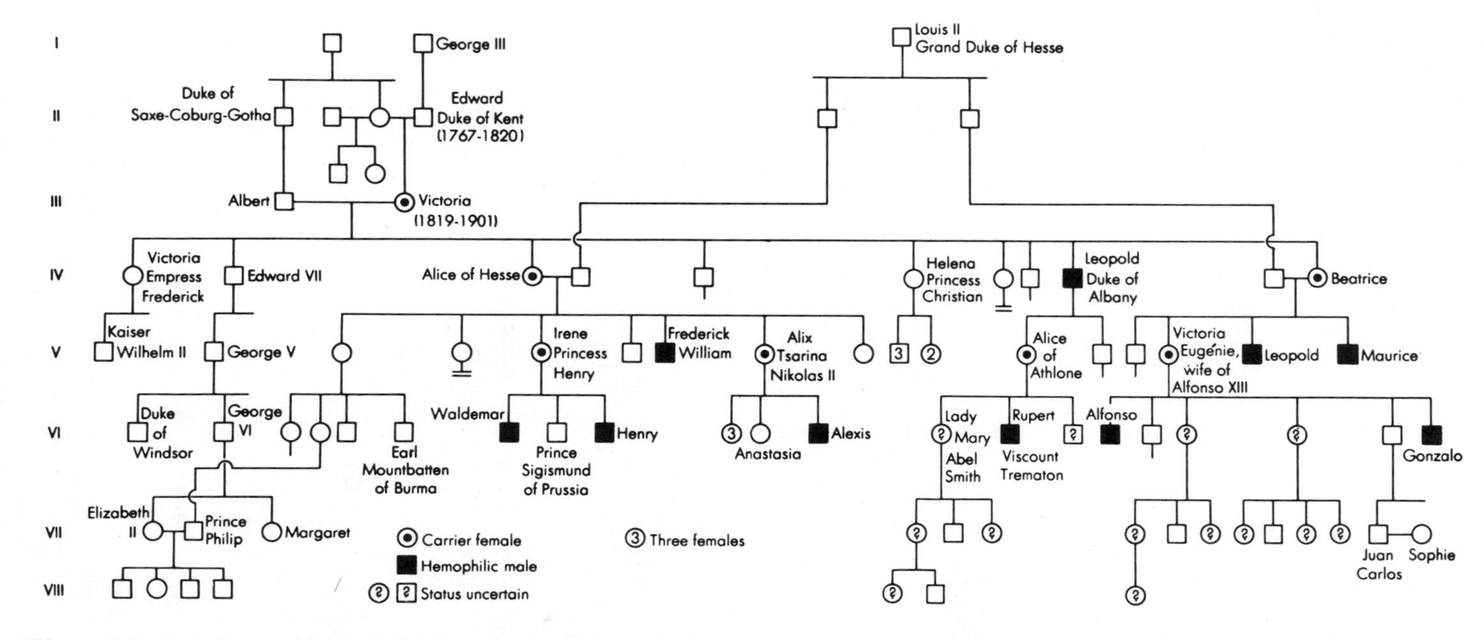

Figure 2-6 A pedigree of hemophilia, showing sex-linked transmission. The mutation presumably occurred in one of the parents of Queen Victoria, since she appears to have been heterozygous. (*From V. A. McKusick,* Human genetics, *2nd Ed.* (1969). *Reprinted by permission of Prentice-Hall, Inc., Englewood Cliffs, N.J.*)

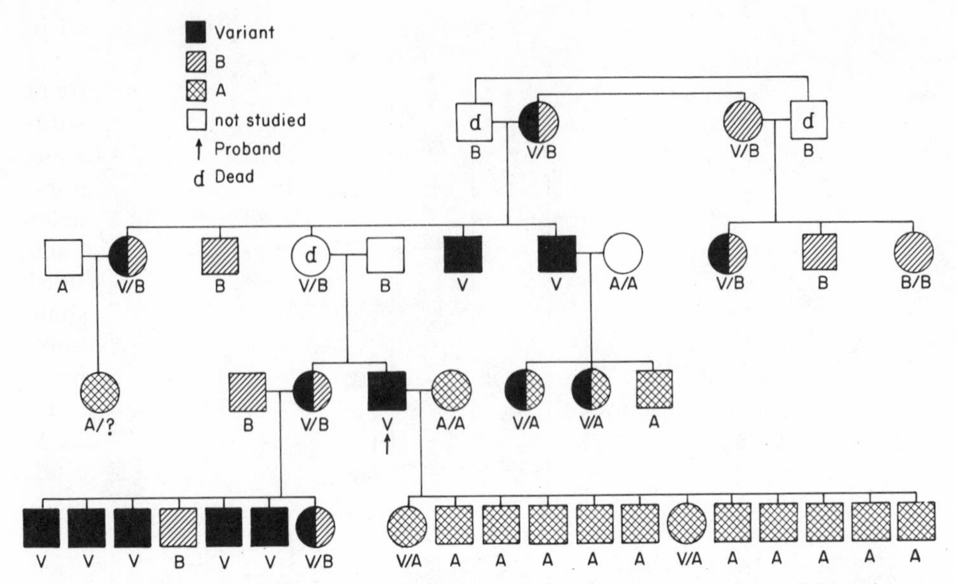

Figure 2-7 Pedigree of an inherited variation on the X chromosome. The various forms of the enzyme glucose-6-phosphate dehydrogenase are inherited as co-dominant traits, with great phenotypic variation in females because of X-chromosome inactivation (Chapter 3). Genotypes are indicated below the symbols. (*From Long et al., 1966.*)

in many tissues of the body, including red blood cells, where it is most readily assayed. Males with the deficiency allele show greatly reduced quantities of active enzyme. Homozygous deficient females are similar to males. Heterozygous females vary in phenotype from normal to fully defective, with most showing intermediate levels. The trait is dominant in that it can be detected in most although not all heterozygotes. The deficiency gene is sufficiently frequent in some populations to produce pedigrees with apparent father to son transmission. Additional study has shown that these families consist of heterozygous mothers with essentially normal phenotype married to affected fathers. The affected sons receive their gene from the apparently normal mother.

Several electrophoretic variants of G6PD are also known. These are structural alterations in the enzyme, detected by mobility differences in an electric field. They are controlled by genes on the X chromosome, presumably alleles of the G6PD deficiency gene. A pedigree of one of the uncommon variants is given in Figure 2-7.

Y-Linked (Holandric) Inheritance

Y-linked inheritance should be the easiest to recognize. Every son of an affected father should be affected, and no daughter should be affected or transmit the trait as a carrier. A number of examples of Y-linked inheritance have been suggested, but evidence in most cases is poor and in some is incompatible with true Y linkage.

The most famous example of Y linkage was the Lambert family of porcupine men. The condition consisted in changes in the skin sufficiently striking so that affected persons exhibited themselves in circuses. The Lamberts lived in England during the eighteenth century. The first pedigree prepared showed seven generations of affected males conforming to the exacting demands of Y linkage. In 1957, Stern and L. S. Penrose reexamined such records as were available and concluded that the traditional pedigree had been colored by the demands of show business and that the true pedigree was incompatible with Y linkage.

At present, there is no clear example of a genetic locus on the Y chromosome other than those thought to be associated with male sex determination. The locus that determines the H-Y antigen is on the Y chromosome (see Chapter 7), but this locus seems to be involved in sex determination. In any event, no alleles of the H-Y locus have been identified. The best evidence for a Y-linked gene other than sex-determining loci is for the trait hairy pinnae (hairy ears). This trait is illustrated in Figure 2-8. The trait is complex, however, since it does not appear in some affected males until late middle age. Many of the pedigrees can be interpreted in more than one way. Additional data will be necessary before the question can be completely resolved.

It should not be surprising that the Y chromosome is remarkably free of genes, since any function necessary for both men and women would have to be controlled from another chromosome. But, as will be discussed further in Chapter 7, the Y chromosome does determine maleness in human beings. Therefore, it is not entirely useless.

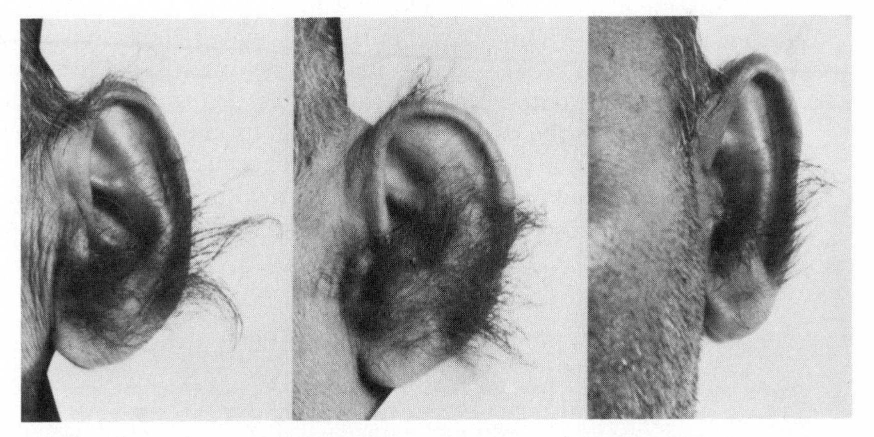

Figure 2-8 Photograph showing the trait "hairy ears," common in parts of India and in some other white groups. It has been suggested that this trait is controlled by a locus on the Y chromosome. (*From Stern et al., 1964.*)

Partial Sex Linkage

In some organisms with XY sex determination, portions of the X and Y chromosomes are homologous. Genes occurring within this region are said to be partially sex linked. They can recombine by crossing over.

Partial sex linkage in man has yet to be demonstrated unequivocally. At meiosis in males, the X and Y chromosomes pair end-to-end (Chapter 3). This suggests possibly a small region of homology. Further, there are several pedigrees of aberrant inheritance of the Xg blood group that could be explained if the *Xg* locus had been transferred to the Y chromosome. It is clear that any region of homology is very small and that partial sex linkage should be invoked as a mechanism only if other possibilities are rigorously excluded.

Genetic Diversity

Many inherited traits that cannot otherwise be distinguished have been shown by genetic analysis to be due to quite different gene loci and hence, presumably, to different primary defects. Muscular dystrophy was first clearly shown to be biologically heterogeneous by genetic analyses that indicated cases could be X-linked recessive, autosomal dominant, or autosomal recessive.

Even if the pattern of transmission is not different, traits can sometimes be differentiated by genetic means. Deafness in man may be inherited, usually as the result of homozygosity for any of several different loci. If two such deaf persons marry, the kinds of offspring reveal whether or not the parents are homozygous for deafness at the same or different loci. If, for one locus, we let *D* represent the dominant allele for normal hearing and *d* the recessive allele for deafness, then a parental mating of $dd \times dd$ can yield only *dd* offspring. On the other hand, if one parent is *ddEE* and the other is *DDee*, where *e* is the recessive allele for deafness at a second locus, then all offspring will be *DdEe* and none will be deaf. In both examples, there is no segregation of genes in the offspring, since the parents are homozygous at both loci and only one type of offspring is possible. If the heterozygous offspring were crossed with homozygous recessive mates or with other heterozygotes, it would be possible to observe segregation and, for two loci, independent assortment, but such matings are uncommon.

SUMMARY

1. By examining inheritance of alternative attributes rather than continuous metrical traits, Mendel was able to explain inheritance in

Pisum sativum on the basis of possession of two determinants for each trait. Each determinant exists in two forms (alleles).

2. If purebred plants of different type were crossed, the F_1 generation resembled the parent with the dominant determinant (gene). In subsequent breeding of the F_1, the original determinants could be recovered in the gametes in equal proportions. This is the basis of Mendel's First Law of Genetic Segregation.

3. If two traits were studied simultaneously, the assortment of their determinants in the F_1 gametes was independent. This is the basis for Mendel's Second Law of Independent Assortment.

4. The Mendelian ratios produced in the offspring in crosses of various parental types can be attributed to these two laws, taking into consideration the dominant/recessive relationships in the expression of these alleles.

5. Autosomal inheritance of an uncommon trait is the simplest pattern of genetic transmission to recognize in man. Persons need have only one copy of the variant allele to express the trait. Males and females should be equally often affected, and every affected person should have at least one affected parent unless a mutation has occurred.

6. Autosomal recessive inheritance is more difficult to establish. Males and females should be equally often affected. For rare traits, affected persons are typically found in a single sibship in a family, although inbred isolates may show more than one affected sibship.

7. X-linked recessive inheritance is known for many human traits. Males should be more often affected than females, and all affected males should be related through carrier females. Father-to-son transmission of X-linked traits cannot occur. Hence this relationship is especially useful in ruling out X-linked inheritance.

8. X-linked dominant inheritance is uncommon but is characteristic of loci producing variant enzymes that can be detected in heterozygous combination. All daughters of affected males would also be affected with an X-linked dominant trait.

9. Y-linked (holandric) inheritance is very rare but may occur for the trait hairy ears. It would not be possible to have a locus essential for individual survival on the Y chromosome, but some of the loci responsible for male development must be Y-linked. In Y linkage, a man transmits the trait to all sons but no daughters.

10. Partial sex linkage refers to inheritance by genes located on homologous regions of the X and Y chromosomes. This in not known to occur in human beings, although the human X and Y chromosomes pair end-to-end during meiosis, and in a few pedigrees aberrant transmission of the X-linked *Xg* locus could be explained by recombination with the Y chromosome.

REFERENCES AND SUGGESTED READING

The Birth of Genetics. English translation. Mendel, De Vries, Correns, Tschermak. *Genetics* 35 (1950): Supplement, 47 pp.

FARABEE, W. C. 1905. Inheritance of digital malformations in man. *Papers Peabody Museum Amer. Arch. Ethnol., Harvard Univ.* 3: 65–78.

GLASS, B., TEMKIN, O., AND STRAUS, W. L. Jr. (Eds.). 1959. *Forerunners of Darwin 1745–1859*. Baltimore: Johns Hopkins Univ. Pr., 471 pp.

GOODENOUGH, U. 1978. *Genetics*, 2nd Ed. New York: Holt, Rinehart and Winston, 840 pp.

KING, R. C. 1965. *Genetics*, 2nd Ed. New York: Oxford Univ. Pr., 450 pp.

LONG, W. K., KIRKMAN, H. N., AND SUTTON, H. E. 1965. Electrophoretically slow variants of glucose-6-phosphate dehydrogenase from red cells of Negroes. *J. Lab. Clin. Med.* 65: 81.

MC KUSICK, V. A. 1969. *Human Genetics*, 2nd Ed. Englewood Cliffs, N.J.: Prentice-Hall, 221 pp.

MENDEL, G. 1865. Experiments in plant hybridization. (Translation of the original which appeared in *Verhandlungen naturforschender Verein in Brünn*, Abhand. 4, 1865.(Reprinted in Peters, J. A. (Ed.). 1959. *Classic Papers in Genetics*. Englewood Cliffs, N.J.: Prentice-Hall, pp. 2–26; and in Sinnott, E. W., Dunn, L. C., Dobzhansky, T. 1958. *Principles of Genetics*, 5th Ed. New York: McGraw-Hill, pp. 419–443.

SLATIS, H. M., AND APELBAUM, A. 1963. Hairy pinna of the ear in Israeli populations. *Am. J. Hum. Genet.* 15: 74–85.

STERN, C. 1957. The problem of complete Y-linkage in man. *Am. J. Hum. Genet.* 9: 147–166.

STERN, C., CENTERWALL, W. R., AND SARKAR, S. S. 1964. New data on the problem of Y-linkage of hairy pinnae. *Am. J. Hum. Genet.* 16: 455–471.

STRICKBERGER, M. W. 1976. *Genetics*, 2nd Ed. New York: Macmillan, 914 pp.

WOOLF, C. M., AND GRANT, R. B. 1962. Albinism among the Hopi Indians in Arizona. *Am. J. Hum. Genet.* 14: 391–400.

REVIEW QUESTIONS

1. Define:

gamete	recombination	allele
zygote	genotype	backcross
dominant	phenotype	hybrid
recessive	genome	dihybrid cross
segregation	homozygous	heterozygous
segregation ratio	hemizygous	holandric
partial sex linkage		

2. Albinism is inherited as a recessive trait. If two normally pigmented parents produce an albino offspring, what ratio of albino to normal would be expected among subsequent offspring?

3. Albinism in man is now known to consist of at least two distinct genetic types—tyrosinase-positive and tyrosinase-negative—both recessive. If a normal offspring of a tyrosinase-negative albino marries a tyrosinase-negative albino, what would be the expected ratio of normal and albino phenotypes among their offspring? What would be the answer if the person married were a tyrosinase-positive albino?

4. If marriage occurs between two normally pigmented persons, both of whom had a tyrosinase-negative albino parent, what kinds of offspring would be expected and what proportions?

5. Draw a diagram similar to Figure 2-3 using the cross between round-seeded, yellow cotyledon peas × wrinkled, green cotyledon peas reported by Mendel (p. 14).

6. Two deaf persons marry. Both had normal parents. Write the possible genotypes of the persons in the pedigree if they have (a) only deaf children; (b) only normal children; (c) both normal and deaf children.

7. The ability to taste phenylthiocarbamide (PTC) is a dominant trait in man. In the Rh blood group system, Rh+ is dominant to Rh−. Two parents, both of whom are Rh+ and tasters, have an Rh− nontaster child. What are the genotypes of the persons in this family? What are the expected genotypes in subsequent children? In what ratio would these genotypes be expected?

8. Variations in height in human beings can be shown to be almost entirely inherited. Yet the exact mechanism of inheritance has yet to be established. Why should this be difficult in man when it is relatively simple in peas? In what instances in man would you expect the inheritance of height to be comparable to that in peas?

9. Indicate the genotypes for all persons in Figure 2-5. If more than one genotype is possible, indicate the probability of each.

10. Muscular dystrophy can be inherited as an autosomal dominant, an autosomal recessive, or a sex-linked recessive trait. For each of these modes of inheritance, which of the following would you expect to find: (a) father and son both affected, (b) males and females both affected, (c) onset in early childhood, (d) parental consanguinity, (e) excess of affected females?

11. A family with a small boy affected with muscular dystrophy seeks help at a genetics counseling center. They are especially concerned about the risk that subsequent children will be affected. What specific questions should be asked about other family members in order to establish the mode of inheritance in this family? What risk estimates should be given if the child turns out to be affected with the autosomal dominant form? The autosomal recessive form? The sex-linked recessive form?

12. What pattern(s) of inheritance can be ruled out for the following pedigrees? What is the most likely mode of inheritance in each case? Assume that the trait is very rare.

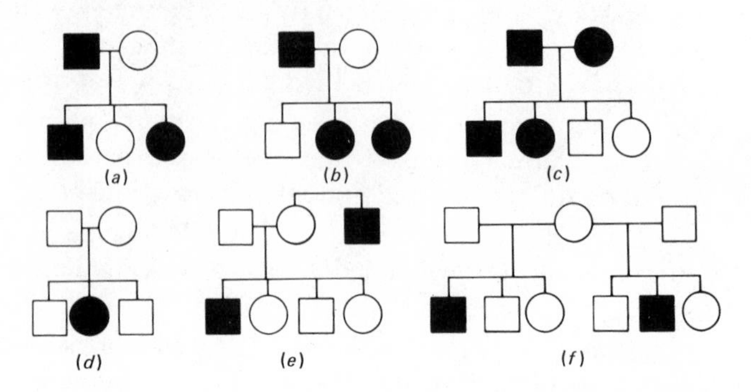

CHAPTER THREE

THE CHROMOSOMAL BASIS OF INHERITANCE

The cytologists of the late nineteenth century described in detail the events associated with cell division in many species of plants and animals. Among their observations was the elaborate process of nuclear division, a process that assures that each daughter cell receives an exact portion of the parental nuclear contents. When a cell is not dividing, the nucleus usually appears as a relatively homogeneous body. As cell division begins, the nucleus condenses into a series of elongated bodies, designated chromosomes because of their affinity for certain dyes. Both the number and types of chromosomes are characteristic of a given species.

The precision and pattern of chromosomal distribution in cell division is analogous to the manner in which inherited traits are transmitted, a fact that led W. S. Sutton and T. Boveri in 1902 to suggest that chromosomes are the physical structures which act as messengers of heredity. This hypothesis has been extensively verified.

THE STRUCTURE OF CHROMOSOMES

The classic work of Avery, MacLeod, and McCarty in 1944 established deoxyribonucleic acid (DNA) as the molecular material in which genetic information is stored in bacteria. They showed that characteristics of one pneumococcal strain could be transferred to another strain by means of DNA extracted from the donor strain. Additional experiments with viruses confirmed these results, although some viruses, the RNA viruses, use ribonucleic acid in place of DNA. With this one exception, DNA is the universal means for primary storage of genetic information. The information so stored is ultimately translated into proteins of vari-

Table 3-1 **DNA CONTENT PER HAPLOID GENOME OF VARIOUS SPECIES***

	Organism	Picograms	Number of nucleotide pairs
Viruses	ϕX174	2.6×10^{-6}	2,400
	Lambda	50×10^{-6}	45,000
	Fowlpox	382×10^{-6}	350,000
Bacteria	Mycoplasma	840×10^{-6}	760,000
	Escherichia coli	$4{,}000 \times 10^{-6}$	3.6×10^{6}
Fungi	Yeast	0.022	20×10^{6}
Molluscs	Snail (Tectarius)	0.7	0.64×10^{9}
Insects	Drosophila	0.2	0.2×10^{9}
Chordates	Amphioxus	0.6	0.55×10^{9}
Fishes	Lamprey (Petromyzon)	2.5	2.3×10^{9}
	Carp	1.7	1.5×10^{9}
Amphibians	Toad (*Bufo bufo*)	7.3	6.6×10^{9}
	Amphiuma	84.0	76×10^{9}
Birds	Domestic fowl	1.2	1.1×10^{9}
Mammals	Man	3.2	2.9×10^{9}
Plants	*Euglena gracilis*	3	2.7×10^{9}
	Lilium longiflorum	53	48×10^{9}
	Tradescantia	58	53×10^{9}

* Taken in part from Ris and Kubai (1970).

ous kinds and amounts, which in turn produce the great variety of biological forms and functions of living organisms.

In *prokaryotes* (viruses, bacteria, blue-green algae), DNA appears to exist in a very extended form, with little evidence of interaction with other molecules to form more complex structures. In *eukaryotes* (organisms with a nucleus), the DNA is found in combination with histones, and a small amount of RNA and nonhistone proteins. The amount of DNA per cell is the same in most cells of an organism and in most members of a species, although the variation among widely separated groups of organisms can be very large (Table 3-1). An exception, of course, is found in germ cells, which have half the DNA of somatic cells. Cells, such as liver, which may have a greater than diploid number of chromosomes, have a correspondingly increased amount of DNA.

An understanding of the storage of genetic information is dependent on some knowledge of the structure of nucleic acids. DNA and RNA are similar in many respects. They are both large polymers. The molecular weights of some types of RNA are in the range 30,000 to 50,000; others commonly have molecular weights of about 2 million. DNA generally is larger, with some types of approximately 2 million molecular weight but most 10 million or larger, perhaps much larger.

Nucleic Acid Composition

Both DNA and RNA are composed of similar types of chemical units. The differences are shown in Table 3-2. The primary (covalent) structure of each is a long chain of nucleotides. Each nucleotide consists of an organic base, a 5-carbon sugar, and phosphate (Figure 3-1). The bases are of two types—purines and pyrimidines. The purines are either adenine or guanine (Figure 3-2). The pyrimidines of DNA are cytosine and thymine; in RNA they are cytosine and uracil. DNA from some species, including man, has such bases as 5-methylcytosine, but these appear to be functionally equivalent to cytosine and occur at very low levels. Most RNA contains only the four bases given in Table 3-2, although transfer RNA contains other pyrimidine derivatives as well. DNA contains D-2-deoxyribose as the sugar moiety, whereas RNA contains D-ribose. Both contain phosphate.

In nucleic acids, nucleotides are joined together into long chains. This is accomplished by the phosphate of one nucleotide forming a bond with the 3′-hydroxyl of the adjacent nucleotide, resulting in a long chain with a "backbone" composed of alternating sugar and phosphate groups. A purine or pyrimidine base is attached to each sugar (Figure 3-3). There is no theoretical limit to the number of nucleotides that can enter into such a chain; in nucleic acids, the chains may consist of many thousands of nucleotides. At one end there will always be a phos-

Table 3-2 **THE CONSTITUENTS OF DNA AND RNA***

	DNA	RNA
Bases	Adenine	Adenine
	Guanine	Guanine
	Cytosine	Cytosine
	Thymine	*Uracil*
Sugars	D-2-deoxyribose	D-ribose
Acid	Phosphoric	Phosphoric

* In addition to the major constituents shown above, derivative pyrimidines occur in transfer RNA. Some viruses also have 5-methylcytosine or 5-hydroxymethylcytosine as a partial or complete replacement for cytosine, and it occurs in trace quantities in the DNA of many organisms. D-glucose may be conjugated to the hydroxyl group of 5-hydroxymethylcytosine.

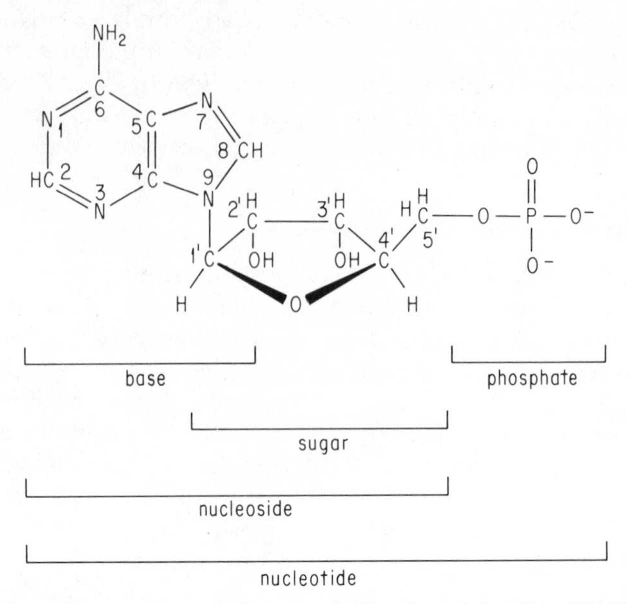

Figure 3-1 The structure of adenosine-5′-phosphate (adenylic acid). Other nucleotides have different bases attached to the ribose. Deoxyribonucleotides have deoxyribose as the sugar, that is, the hydroxyl attached at position 2′ is replaced by a hydrogen atom.

phoric acid group and at the other end an unbound 3′-hydroxyl on the sugar. The aspect of the structure in which every element is attached by covalent bonds is known as the *primary* structure.

DNA Structure

A single chain such as that pictured in Figure 3-3 would be quite flexible and resemble a tangled mass of string. The DNA of cells is normally a somewhat rigid rodlike structure. This property is conferred by interaction between DNA chains of forces other than covalent bonds. These noncovalent forces are responsible for the *secondary* structure of DNA.

Native DNA consists of two nucleotide chains coiled about each other to form a double helix. In the structure proposed by Watson and Crick (1953a), the two strands of the double helix are held together by hydrogen bonds between pairs of bases. Hydrogen bonds are weak bonds formed when a hydrogen atom covalently bound to one atom, such as nitrogen or oxygen, also has an affinity for another oxygen or nitrogen atom having unshared electrons. The groups must be precisely oriented spatially in order for hydrogen bond formation to occur. The purines and pyrimidines found in nucleic acids can combine in several ways to form hydrogen-bonded structures. The distance between the

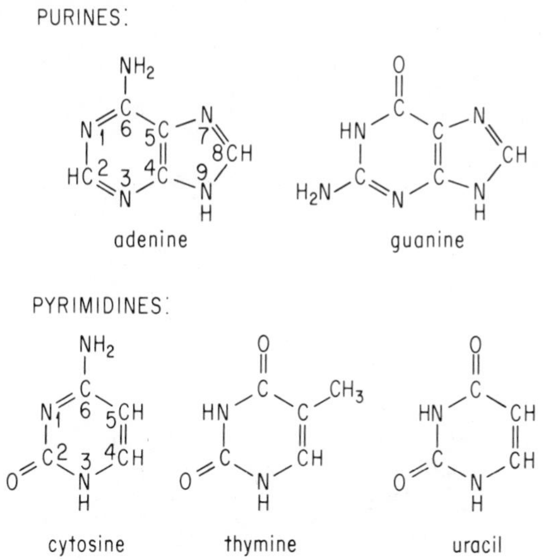

PURINES:

adenine guanine

PYRIMIDINES:

cytosine thymine uracil

Figure 3-2 The purine and pyrimidine bases that occur commonly in nucleic acids.

sugar-phosphate backbones, however, is such that only a purine-pyrimidine combination can fit properly. Two purines would be too large and too pyrimidines too small. Only two combinations meet all the physical and chemical requirements of the DNA structure—adenine bonded to thymine and cytosine bonded to guanine (Figure 3-4). In these two paired structures, the distances between pyrimidine N^3 and purine N^9 are the same and correspond to the distance between the sugar-phosphate backbones.

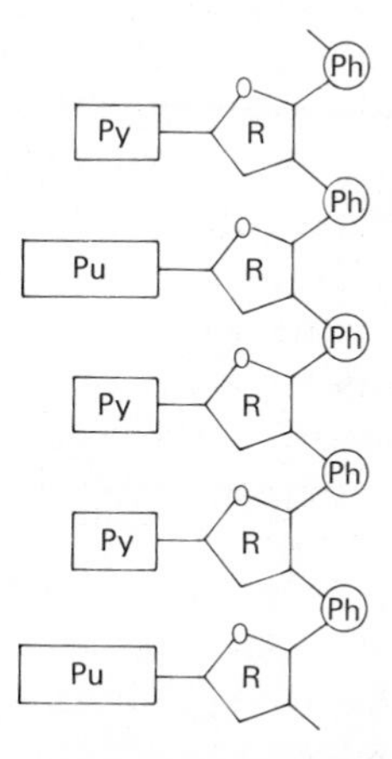

Figure 3-3 Diagram of the structure of a polynucleotide chain. A free 5′ phosphate is indicated at top and a free 3′-hydroxyl at the bottom. Note that the chain has polarity because of the asymmetry of the sugar units.

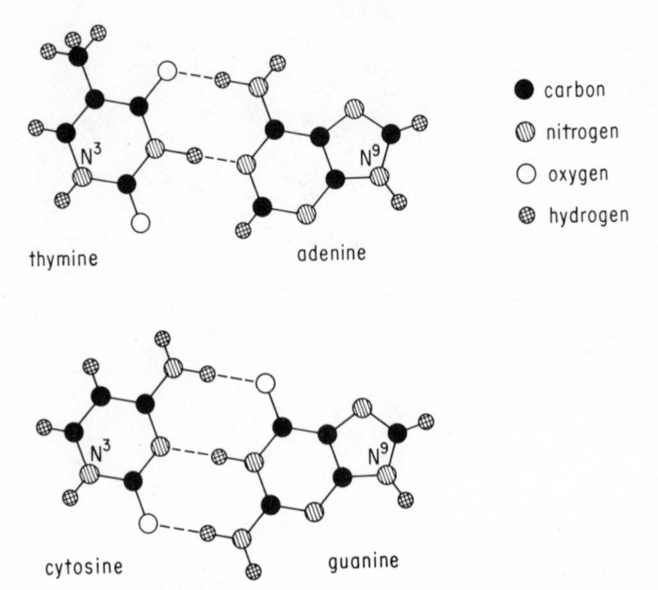

Figure 3-4 The pairing of bases observed in DNA. Hydrogen bonds are indicated by dotted lines. The distance between pyrimidine N^3 and purine N^9 is the same in these two structures.

One consequence of this specific pairing is that the amount of thymine in a particular source of DNA should always be equal to the amount of adenine, and the amount of cytosine should equal the amount of guanine. These relationships have been widely verified. A segment of a DNA molecule is illustrated in Figure 3-5.

The sequence of bases is an important feature of the molecule. At any one position, any of the four bases may be present. Thus, for any one chain, any sequence of bases is possible. However, the sequence in one chain requires a specific reciprocal sequence in the other. If adenine is present at a given position in one chain, thymine must be present at the corresponding position in the other. The sequence AATCGGG (using initial letters as symbols for the bases) in one chain must be matched by the sequence TTAGCCC in the other. The two chains are of opposite polarity, however, i.e. the 3′-hydroxyl at the end of one chain corresponds to the 5′-hydroxyl (or phosphate) on the other.

Since any one of the four bases may be present at any one of the nucleotide positions, an enormous number of DNA forms is possible. The number is 4^n, where n is the number of nucleotide positions. For only two positions—a dinucleotide—16 structures could exist: AA, AT, AC, AG, TA, TT, and so on. For three positions, the number is 64, and for $n = 10$, the number is 1,048,576. Therefore, DNA can exist in the many forms necessary to store the information of a vast number of different genes. Stated otherwise, the sequence of nucleotides may be viewed as the means by which genetic information is coded and stored.

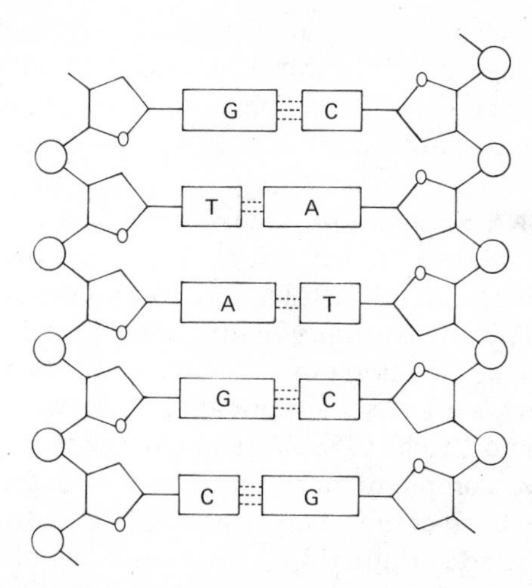

Figure 3-5 Diagram of a segment of a double-stranded DNA molecule. Dotted lines represent hydrogen bonds. In the crystalline form and under many biological conditions, the two strands are twisted around each other to form a helix.

RNA Structure

The primary structure of RNA is similar to that of DNA (Table 3-2). RNA normally is found to be a single-strand structure, in contrast to the double helix of DNA. Certain types of RNA show hydrogen bonding between bases, resulting from twisting of a single RNA strand—in the same way that a suspended rope that is twisted will coil to form a double helix.

Four types of RNA have been observed. The largest type has a molecular weight of approximately 2 million and is a component of *ribosome* particles (rRNA). Ribosome particles are complexes of RNA and protein found primarily in cytoplasm but also to some extent in nuclei. A second type is *messenger RNA* (mRNA), which is synthesized in nuclei and migrates to the cytoplasm. This RNA carries genetic information from the DNA of chromosomes to ribosomes, where protein is synthesized. The third type is *transfer RNA* (tRNA), of molecular weight 30,000 to 50,000. It functions in the transfer of amino acids in protein synthesis. The functions of these three types of RNA will be discussed in more detail in the next chapter. *Heterogeneous nuclear RNA* (HnRNA) is a high molecular weight form that occurs in eukaryote nuclei and appears to be a precursor to messenger RNA, but the function is not well understood.

Replication of Nucleic Acid

The chemical replication of DNA and the visible replication of chromosomes are distinct processes. DNA completes replication several hours prior to the beginning of cell division. The replication can be viewed in terms of replication of genetic information and replication of chemical structures. In this discussion, we will be concerned only with the former and will omit reference to the many chemical reactions in nucleic acid synthesis.

DNA Replication. DNA has two important functions as the carrier of genetic information: a *heterocatalytic* function concerned with the direction of cellular activities of other kinds of chemical components, and an *autocatalytic* function concerned with the replication of itself. The double helix of the Watson–Crick structure suggested to its discoverers a means of replication that would serve both functions (Watson and Crick, 1953b). As noted, the two strands of a DNA double helix have complementary sequences of bases. If the two strands are separated, each could serve as a template to direct a copy of its complementary strand (Figure 3-6). The result would be two complete DNA molecules, each consisting of a strand of the original molecule and a strand of newly synthesized material.

In order for such reactions to occur, specific enzymes and building blocks must be present. A primer—preformed DNA—is also necessary to act as a template. Various DNA polymerases have been detected in

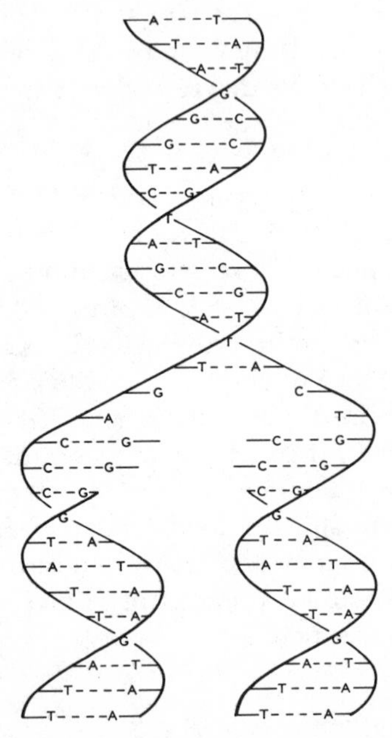

Figure 3-6 Proposed replication of DNA. As the double-stranded parent DNA unwinds, the separated strands serve as templates for the alignment of nucleotides, which combine to form new strands complementary to the parental strands.

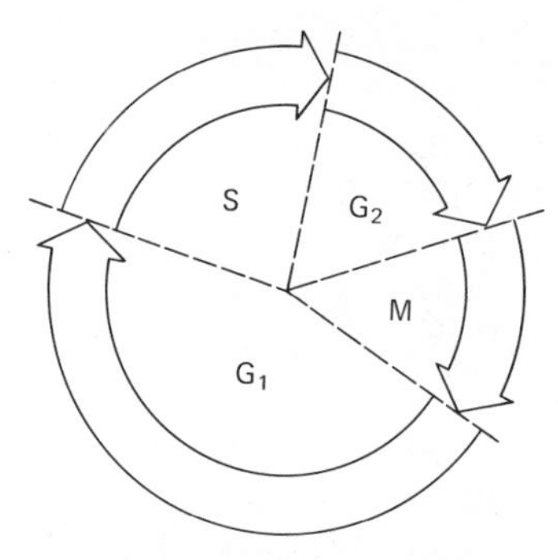

Figure 3-7 Diagrammatic representation of cell growth. M is the period of mitosis. G_1 is the period following mitosis and before DNA synthesis. S is the period of DNA synthesis. G_2 is the period after DNA synthesis and before mitosis. In cultured human cells, S lasts 6 to 12 hours and G_2 lasts 3 to 6 hours.

biological materials, some of which function in regular gene replication and some in gene repair (Chapter 9).

The time at which DNA replicates is quite separate from chromosome division (Figure 3-7). This period of DNA synthesis, the S period, can be identified readily because of the unique occurrence of thymine in DNA. If one adds the nucleoside thymidine to a cell culture, it is incorporated only into those cells that are replicating DNA.

This event can be followed cytologically rather simply. If some of the hydrogen atoms of thymine are replaced with the radioactive isotope tritium (^3H), the fixing of tritium reflects DNA synthesis. Due to the very short track of decay particles from tritium, the location of the newly incorporated thymidine within a cell can be pinpointed accurately. In practice, tritium-labeled thymidine is added to a growing culture of cells, either as a pulse or for longer periods. After several hours, depending on the objective of the experiment, the cells are harvested and transferred to a slide for staining and examination of chromosomes. A photographic emulsion is placed on the slide and is exposed by the radioactivity of the tritium. After development, the location of thymine is recognized by the appearance of silver granules in the "autoradiograph."

If ^3H-thymidine is introduced during the G_1 period and allowed to remain through the S period, then all newly synthesized DNA will be labeled and all chromosomes will have a dense cluster of granules in the autoradiograph. On the other hand, addition of ^3H-thymidine in the G_2 period results in absence of granules associated with chromosomes. The beginning and ending of the S period can be defined rather exactly in this manner. If ^3H-thymidine is added during the S period, that DNA yet to be replicated will be labeled, but DNA that has replicated before addition of isotope will not be labeled.

At any one time, cells in a culture will be in all stages of the growth cycle. If ^{3}H-labeled thymidine is added shortly before harvesting, most metaphase chromosomes will show no incorporation of label while others will show various degrees of labeling, depending on their status at the time of addition of the label. For cells in the S period, the distribution of label among the chromosomes is nonuniform, indicating that DNA synthesis occurs earlier in some—or in segments of some—and later in others. If the cell is near the end of the S period when the label is introduced, only those chromosomes or portions of chromosomes in which DNA is synthesized last will incorporate the label.

Figure 3-8 shows the results observed in a cell labeled late in the S period. The most heavily labeled chromosome is always one of the X chromosomes in females. The other X chromosome in females and the single X chromosome in males are not *late-labeling.* As will be discussed in Chapter 8, only one X chromosome is fully active in any cell. If there are two or more X chromosomes present, the inactive X chromosomes are the last of the complement to replicate and so become heavily labeled. The active X chromosomes replicate early. Autosomes that show late-labeling are a region in the long arm of 1, all of 4, the short arm of 5, part of the long arm of 13, the centromeric region of 14, and 18. The long arm of the Y chromosome is also late-labeling. Autoradiography was especially useful in distinguishing between 4 and 5 and among the D group before the discovery of banding techniques.

It has also been possible to study replication of individual DNA molecules in mammalian cell cultures using highly radioactive precur-

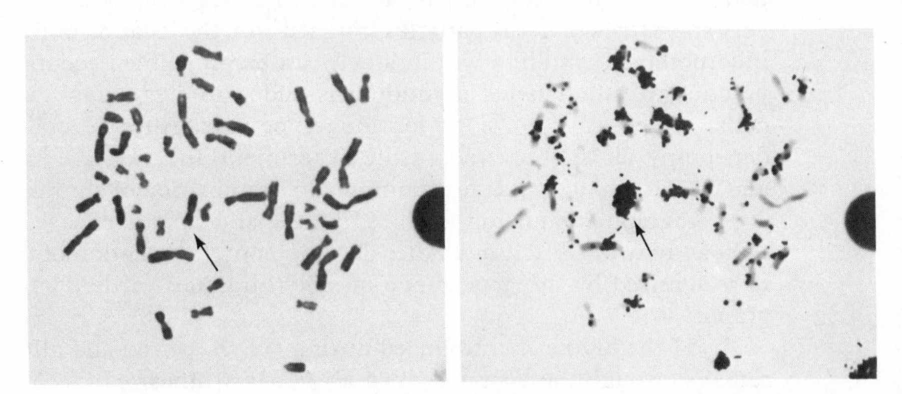

Figure 3-8 Photomicrograph of female chromosome spread with one X chromosome labeled with ^{3}H-thymidine. The ^{3}H-thymidine was introduced into the culture late in the cycle of DNA replication. Although several autosomes are labeled (right), only one of the X chromosomes is very heavily labeled. Since the number of late-labeling X chromosomes is equal to the number of sex chromatin bodies, inactive X chromosomes probably form the sex chromatin bodies and also replicate later than the rest of the chromosome complement. (This photograph was prepared by the duplicate photography technique described in J. German, *J. Cell Biol.* 20: 37–55, 1964, from which this figure was taken. A conventional stain of the same chromosome spread is on the left.)

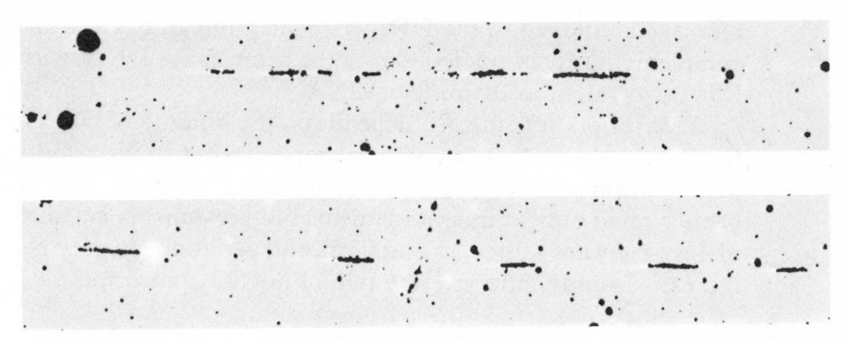

Figure 3-9 Autoradiographs of DNA in Chinese hamster fibroblasts. Highly radioactive ^{3}H-thymidine was added to the cultures shortly before harvesting. The segments of DNA synthesized subsequent to that time incorporated the tritium. The cells were then disrupted and the contents were dispersed on a filter and covered with film. The decay of the tritium exposed the film. A single molecule of DNA, represented by the interrupted alignment of tritium, is replicating at several different points. Regions that did not incorporate tritium either had already replicated or had yet to start replication. (*From Huberman and Riggs, 1968.*)

sors. Figure 3-9 shows a molecule of DNA after it has replicated for a short period in radioactive medium. It can be seen that synthesis starts at several points along the molecule. An interpretation of the replication is shown in the diagram in Figure 3-10.

RNA Replication. RNA replicates by complementary pairing in much the same way as DNA, but DNA serves as the primer for RNA as well. The enzymes catalyzing RNA synthesis are designated DNA-dependent RNA polymerases and provide a means of transcribing the

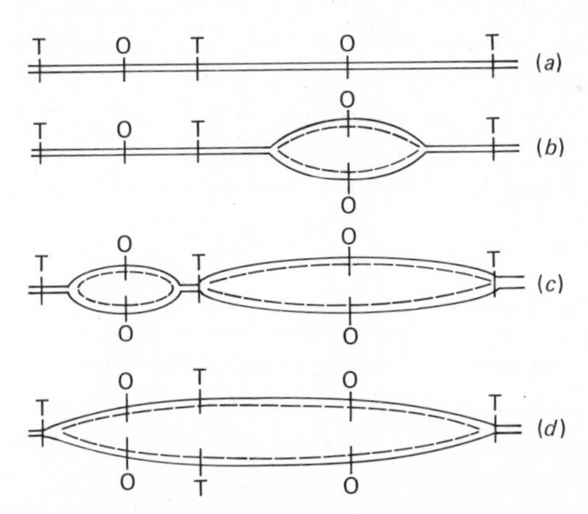

Figure 3-10 Bidirectional model for DNA replication. Each pair of double lines represents the two polynucleotide strands of a DNA molecule. Replication origins are indicated by O and termini by T. (*a*) Prior to replication; (*b*) replication started in right-hand unit; (*c*) replication started in left-hand unit and completed at termini of right-hand unit: (*d*) replication completed in both units with separation at the common terminus. (*From Huberman and Riggs, 1968.*)

message contained in the DNA sequence into an RNA sequence. The complementarity of nucleotides is the same as for DNA, with uracil in RNA equivalent to thymine of DNA.

Transcription of mRNA depends on the binding of DNA-dependent RNA polymerase to a *promoter* region adjacent to the region that ultimately is translated into protein sequences. The promoter regions have been studied only in microorganisms but presumably occur in all organisms. They are subject to mutation, and variations in polymerase binding can strongly influence the rate of mRNA transcription.

RNA-Dependent DNA Synthesis. There is an additional mechanism of DNA synthesis used by RNA viruses. This is with the enzyme RNA-dependent DNA polymerase, more popularly known as reverse transcriptase. This permits RNA viruses to make a DNA copy of their genetic information, which then can be used by the enzymes of the host cell for making additional RNA copies and ultimately the proteins that are coded in the viral genome. The presence of reverse transcriptase in mammalian cells has been used as evidence of viral infection, since mammals appear not to have this enzyme ordinarily.

The ability to make DNA copies from RNA raised the possibility of making copies of genes from messenger RNA. The genes of choice are those responsible for the globin chains of hemoglobin, since the mRNA for globin can be isolated in relatively pure form from reticulocytes. When added to a system with reverse transcriptase and deoxyribonucleoside triphosphates, a complete DNA strand complementary to the mRNA is formed (Marks and co-workers, 1974). Interestingly, the mRNA appears to have a sequence of polyadenylate before the beginning of the globin message, and it is necessary to add a comparable segment of polydeoxythymidylate. The poly-dT apparently forms a double strand with the poly-A and provides a starting point for the reverse transcriptase. This technique should be applicable to any purified mRNA and has obvious implications for gene therapy.

DNA in Chromosomes

DNA and histone are present throughout the length of a chromosome. Furthermore, the DNA in a chromosome appears to be a single continuous molecule, unbroken by non-DNA connections. This amount of DNA is too great to be compatible with a straight, unfolded structure. For example, the DNA in the smallest human chromosomes (21 and 22) has a total molecular weight of 2×10^{10} and a length of 1 cm. The lengths of the chromosomes are only 1.5 μ during metaphase, and, although they may be much longer during interphase, they could not ac-

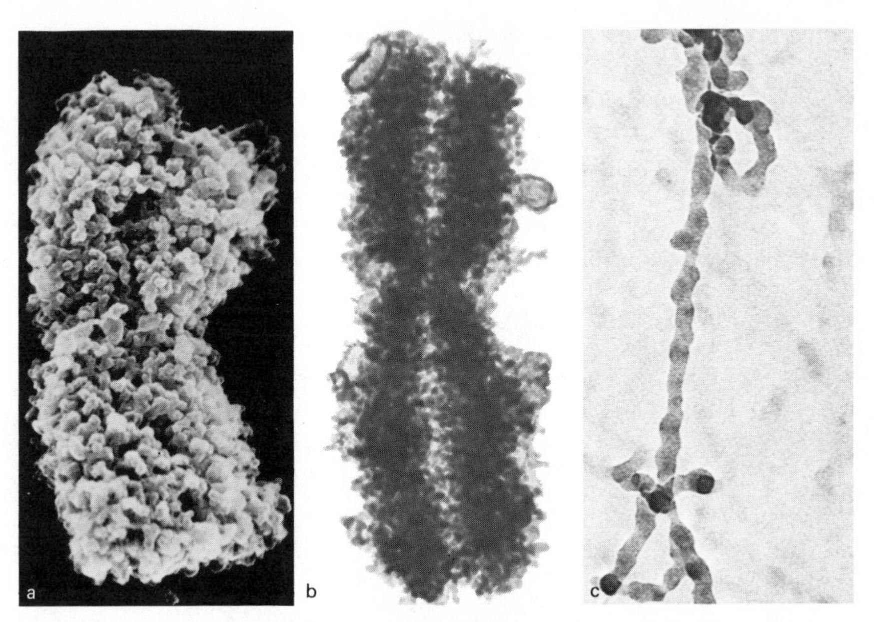

Figure 3-11 (*a*) Scanning electron miscroscope image of a metaphase human chromosome. (*From M. L. Mace, Jr, Y. Dascal, H. Busch, V. P. Wray, and W. Wray,* Cytobios *19: 27, 1978.*) (*b*) A human metaphase chromosome shown in a transmission electron micrograph. (*Supplied by E. Stubblefield and W. Wray.*) (*c*) Chromatin fiber from a human interphase nucleus. The average diameter of the fiber is 200 Å. (*From J. J. Yunis and G. F. Bahr,* Exp. Cell Res. *122: 63, 1979.*)

commodate 1 cm of DNA unless it were coiled or made more compact. An electron micrograph of a human chromosome is shown in Figure 3-11.

A number of stains show chromosomes to vary along their length. The principal difference has been the relative affinity for certain dyes. *Chromatin,* the material that stains, is divided into two types: *heterochromatin,* stainable in very early prophase and in some cells during interphase, and *euchromatin,* which stains less readily. Most of the chromosomal material in mammals is euchromatin. Certain portions—the Y chromosome, a segment of the X chromosome, and centromere regions of autosomes—are characteristically heterochromatic, but there is much variation among species. These will be recognized as corresponding to the C-bands (Chapter 4). In *Drosophila* and seemingly in man as well, virtually all the known genes are located in euchromatic regions.

DNA from eukaryotes can be divided also into repetitive and unique sequence (nonrepetitive) DNA. From 5 to 15% of the genome is repetitive, consisting of clusters of nucleotide sequences some 300 nucleotides long. Within a cluster the sequences are highly repetitious, although this conservatism is not noted among species. How the uniformity within a species is maintained in the face of evolutionary diversity is unknown. The highly repetitive DNA occurs primarily in the heterochromatic regions. Separating the clusters of repetitive DNA are sequences that appear to occur a single time per haploid genome. The structural genes are thought to be located in the unique sequences.

DNA within a chromosome always occurs in association with histones, proteins that are very basic because of the high content of positively-charged lysine and arginine residues. There are five major classes of histones, H1, H2A, H2B, H3, and H4, that appear to be the same in all tissues of an organism, suggesting that histones do not play a major role in tissue differentiation. By careful extraction of chromatin from chromosomes, it is possible to obtain fibers that consist of strings of small bodies (*nu* bodies or *nucleosomes*) connected by strands of DNA. These nucleosomes are histone-DNA complexes, held together as shown in Figure 3-12. Their exact function is unknown, but their ubiquitous occurrence in eukaryotes is evidence of an essential role, possibly in maintaining the structural integrity of the chromosome.

Each nucleosome contains eight histone molecules—two each of H2A, H2B, H3, and H4—around which there are two loops of DNA equivalent to 140 base pairs total. A single molecule of histone H1 is attached to the DNA on one side of the nucleosome. The DNA between nucleosomes is approximately 60 base pairs long but varies in different species.

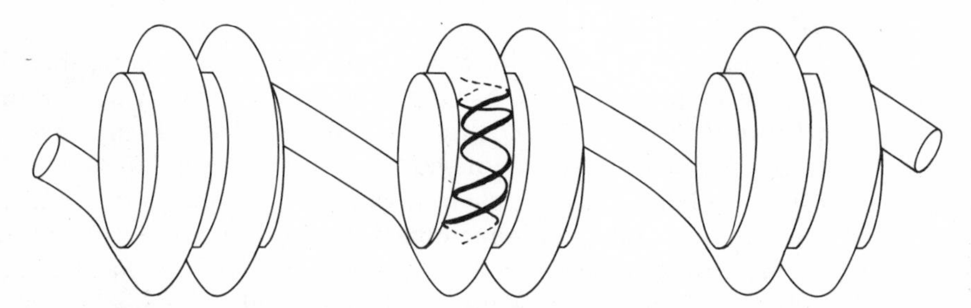

Figure 3-12 Diagram of a chain of nucleosomes, showing a portion of a molecule of DNA wrapped around histone units. The diameter of the histone units is 70 Å, and each is an octomer composed of two molecules each of histones H2A, H2B, H3, and H4. The segment of DNA associated with each histone unit is approximately 140 base-pairs long and is wrapped around the histone $1\frac{3}{4}$ turns. Another histone molecule, H1, is attached to the DNA in the region between the histone octomers, but H1 is not shown in the diagram. This drawing is not meant to represent the precise geometric details of nucleosomes, which are not known, but shows the general relationship of histones to DNA. The nucleosomes probably are stacked close together in vivo. (*The model shown is based on results of Finch et al., 1977.*)

CHROMOSOMES IN CELL DIVISION

The chromosomes of most organisms, both plants and animals and including man, occur in pairs. That is, the genetic information found in any one chromosome is duplicated in another chromosome ordinarily identical in size and shape. There appears to be considerable advantage in having genetic information present in duplicate, although in many cases a single copy is adequate for the direction of specific cellular activities.

A complement of unpaired chromosomes is called a *haploid* complement. In man, in whom there are 46 chromosomes, the haploid complement is 23 chromosomes. The double set of chromosomes is a *diploid* complement. In females, there are 23 pairs of homologous chromosomes, including one pair of sex chromosomes (the X chromosomes). In males, the sex chromosomes are nonhomologous, consisting of an X and a Y chromosome. Chromosomes other than the X and Y are called *autosomes*. Occasionally, there are aberrant *triploid* cells containing three chromosomes of each type. Higher degrees of *polyploidy* are observed in special situations.

The chromosomes of nondividing cells are not usually individually visible. It is thought that they are uncoiled into very long, narrow structures whose diameters are below the limits of resolution of ordinary light microscopes. This period during which no division occurs is designated *interphase*. It is probably during interphase that chromosomes carry out most of their chemical activities. It is also during interphase that the chemical replication of chromosomes occurs.

Although minor variations occur, nuclear division in most higher plants and animals is similar. Most organisms have two kinds of cell division. *Mitosis* occurs in somatic cells, yielding two daughter cells identical to the mother cell in their chromosomal complement. *Meiosis* occurs in germ cells, producing *gametes* (eggs and sperm) with half the full complement of chromosomes, that is, one of each chromosome pair. Furthermore, each gamete has a unique assortment of parental chromosomes. A consideration of these two processes will provide a basis for understanding transmission of inherited variation.

Mitosis

The earliest stage of cell division that is recognized is designated *prophase*. As shown in Figure 3-13, this is characterized by the appearance of the chromosomes first as long, thin threads. These gradually contract, becoming shorter and thicker and more readily visible. As prophase continues, each chromosome appears double along its longitudinal axis, and each half is termed a *chromatid*. Initially, the chromatids

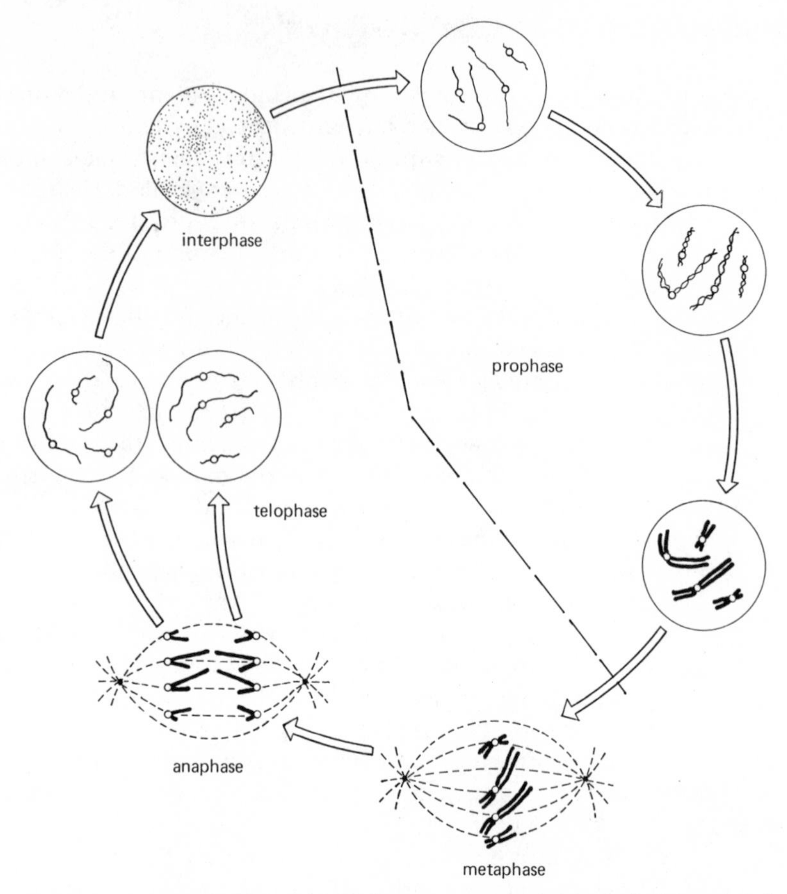

Figure 3-13 Mitosis. Two pairs of chromosomes are shown. Only the nucleus is represented. Following anaphase, new nuclear membranes are formed and the cell cleaves to give two daughter cells identical in genetic content to the parent cell. These diagrams are intended to show the processes rather than the actual appearance of the chromosomes. For example, centromeres are not ordinarily seen in the microscope, although their presence and location can be deduced readily from the chromosome movement during anaphase.

are coiled around each other, but as contraction of the chromosomes continues, the chromatids separate, remaining attached to each other only at the *centromere* (also called *kinetochore*), which has not yet divided. The centromere is a region present in every chromosome, apparently free of genes but functioning in the distribution of chromosomes into daughter cells.

At the time of chromatid separation, other structures also undergo changes. The nuclear membrane disappears. The *centrioles,* two small bodies lying outside the nucleus, migrate to opposite poles of the cell, and serve as centers for the formation of protein fibers. These fibers, issuing radially from the centrioles, join together in the center of the cell, giving rise to a continuous structure, the *spindle*. As the spindle is formed, the chromosomes align between the two poles of the spindle in a plane that is perpendicular to the spindle axis. This part of the cycle is designated *metaphase*. At this time, the chromosomes are maximally contracted and, therefore, most readily observable microscopically. The plane in which the chromosomes lie is known as the *equatorial* plane, and the array of chromosomes at this time is referred to as a *metaphase plate*.

At the conclusion of metaphase, the centromeres divide, completing the formation of two new chromosomes from each original chromosome. It also becomes apparent at this time that the centromeres are attached to spindle fibers. These fibers contract, pulling the chromosomes toward the two poles of the spindle. The process is designated *anaphase*. During anaphase, each pair of newly formed chromosomes separates, the two migrating to opposite poles. The result is two identical and complete chromosome complements.

Following separation of the two sets of chromosomes, a new nuclear membrane is formed around each and the chromosomes become more extended, gradually losing their individual visibility and forming a new interphase nucleus. This period of reorganization is known as *telophase*. With the formation of two nuclei, the cytoplasm divides, forming two complete cells.

Meiosis

Most organisms, both plant and animal, are characterized by alternating cycles of haploid and diploid existence. In some, such as ferns, both haploid and diploid phases persist for appreciable periods of growth. In other organisms, one of the phases is transient. For example, in the mold *Neurospora*, diploid nuclei are present during one cycle of cell division only. In mammals, it is the haploid phase that persists only for one cell cycle. These haploid cells, the gametes, are incapable of mitotic division and must either fuse to form a diploid zygote or perish.

Meiosis is the process of cell division that yields haploid daughter cells from a diploid parent cell. It is characterized by two divisions of the nucleus but only one of the chromosomes. In many ways, it is similar to mitosis, but there are very important differences. An outline of meiosis is given in Figure 3-14.

First meiotic prophase is characterized by several stages. In *leptotene* the chromosomes become visible as long extended threads that begin to coil, becoming shorter and thicker. Differential coiling, apparently in a pattern consistent for a particular chromosome, produces structures called *chromomeres* along the length of the chromosome. The next stage, *zygotene,* is signaled by the beginning of lengthwise pairing of homologous chromosomes. This highly precise, point-by-point pairing,

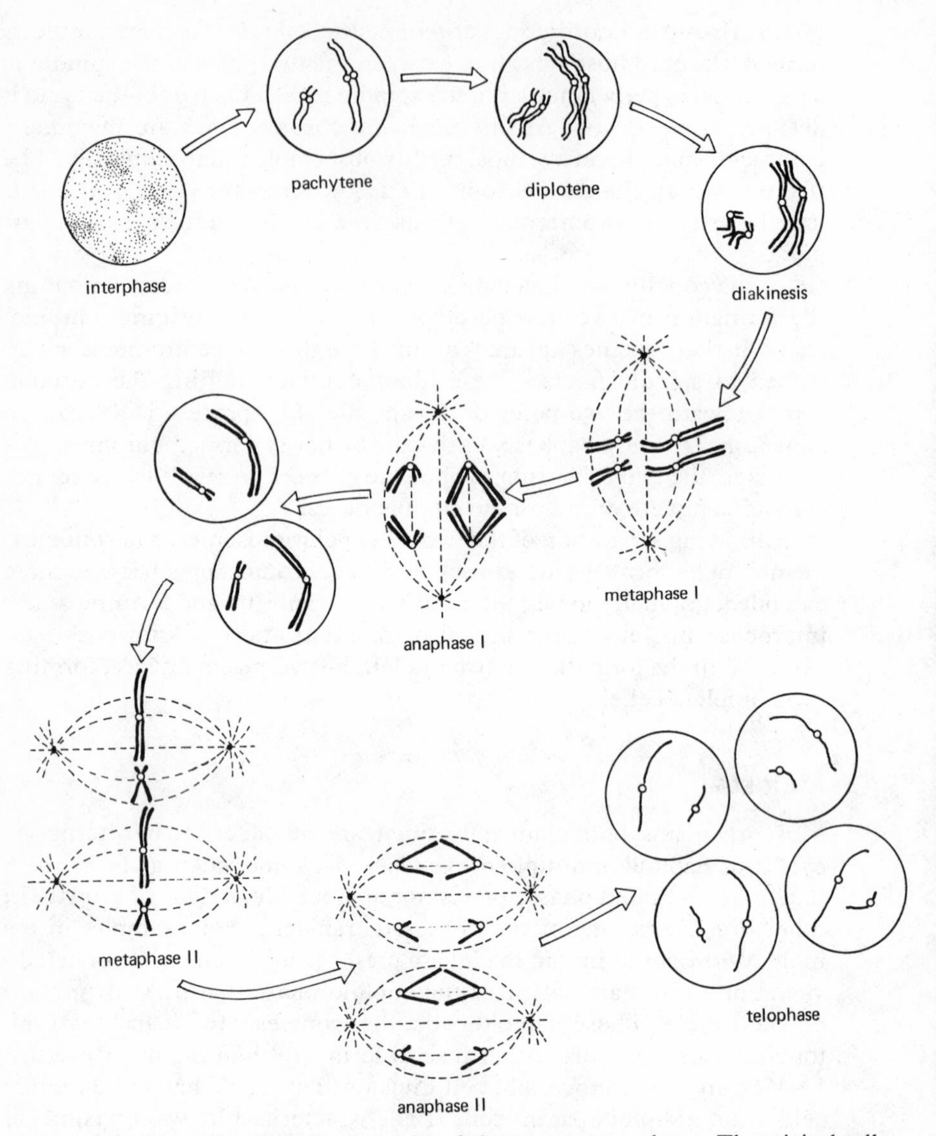

Figure 3-14 Meiosis. Two pairs of chromosomes are shown. The original cell gives rise to four haploid cells (gametes). Some of the steps recognized by cytologists have been omitted.

known as *synapsis,* may begin at one or several points and proceeds in zipper fashion until completed. An exception occurs in males with XY sex chromosomes, since these are nonhomologous. They often show end-to-end pairing involving the short arms of each, suggesting some limited homology.

On the completion of synapsis, the chromosomes enter *pachytene,* a relatively stable period during which they greatly condense (Figure 3-

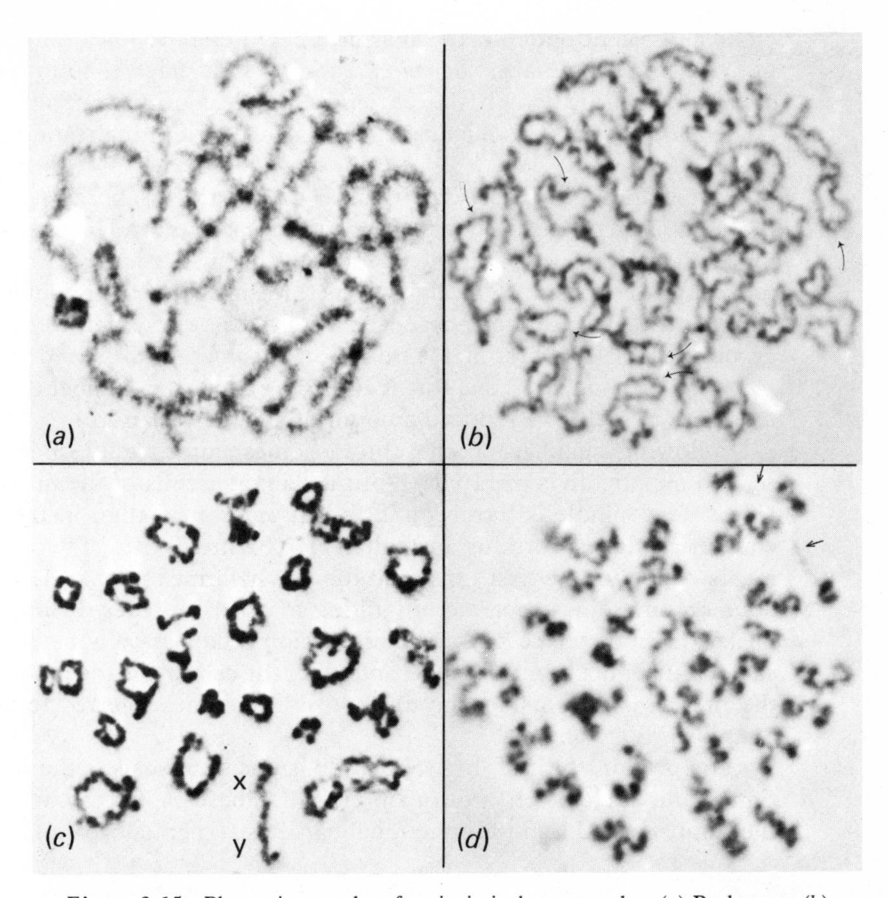

Figure 3-15 Photomicrographs of meiosis in human males. (*a*) Pachytene. (*b*) Diplotene. Arrows indicate several chiasmata in terminal positions. (*c*) Diakinesis. There are 23 bivalents, including the XY bivalent in which there is end-to-end pairing. In many preparations, the X and Y chromosomes are not paired. (*d*) Metaphase II. The arrows indicate an unusually extended secondary constriction on chromosome 9. (*From Hultén and Lindsten, 1973.*)

15*a*). Each chromosome pair is called a *bivalent,* and of course the number of bivalents equals the haploid number. During the period of contraction the chromomeres become very pronounced, and the chromomere patterns have been used to characterize individual chromosomes. The beginning of *diplotene* (Figure 3-15*b*) occurs when the pairing forces cease and the homologous chromosomes are held together only by *chiasmata* (s. *chiasma*). These bridges may form between any two nonsister chromatids, and several chiasmata may be observed in any one *tetrad* involving any combination of pairs. They are the points at which *crossing over* occurs. In this process, homologous segments of chromosomes may be exchanged among the four chromatids. Near the end of diplotene, the homologous chromosomes repel each other, especially at the centromere. The chiasmata begin to move toward the ends of the chromosome arms (*terminalization*). This stage is followed by *diakinesis,* during which the chromosomes become maximally contracted and the number of chiasmata is reduced by terminalization (Figure 3-15*c*).

At the end of prophase, the nuclear membrane disappears and the first of two metaphases occurs. In metaphase I, the bivalents align on the equatorial plane. During anaphase I, homologous centromeres move toward opposite poles, pulling apart the bivalents. This yields two complements of chromosomes, each with half the original number of chromosomes. The first meiotic division is sometimes referred to as a *reduction* division, each daughter cell having a reduced number of chromosomes as well as a reduced amount of genetic variation.

Following anaphase I, the chromosomes immediately undergo a second meiotic division. In each of the daughter cells of the first division a new spindle is formed, and the chromosomes align in the new equatorial planes, forming metaphase II (Figure 3-15*d*). The centromeres divide for the first time, and the newly formed pairs of homologous chromosomes separate, migrating to opposite poles during anaphase II. Since the second meiotic division produces cells having the same number of chromosomes (haploid) as the cell that is the product of the first division, this second division is sometimes referred to as an *equational* division.

The overall result of the two cell divisions of meiosis is the formation of four haploid cells from a diploid cell. These meiotic products are then transformed into the functional gametes—spermatozoa and ova.

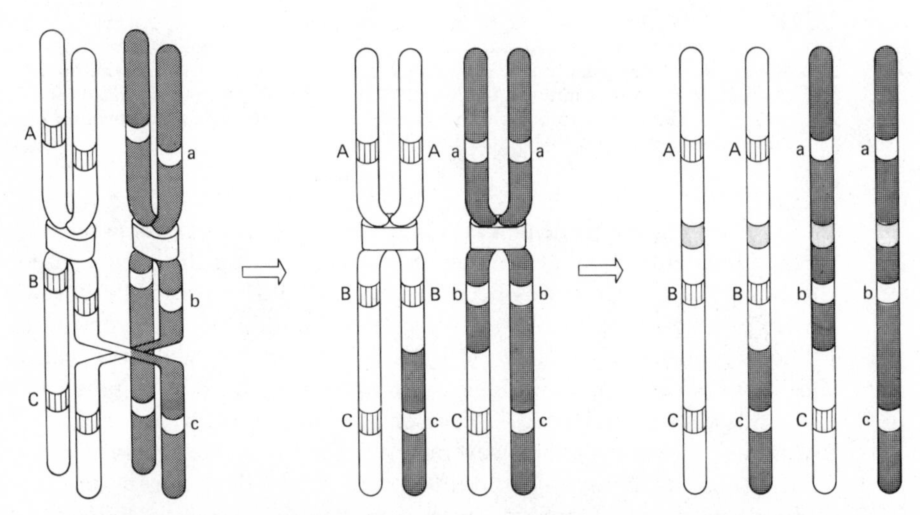

Figure 3-16 A single crossover involving two chromatid strands showing reassortment to give four different combinations in the haploid products.

Crossing Over

One of the regular features of meiosis is crossing over. As noted earlier, this involves the exchange of chromatid segments preceding metaphase I. If two homologous chromosomes differ for a series of genes, as in Figure 3-16, then the effect of crossing over is to produce four different chromosome combinations. Figure 3-16 illustrates the results of a single crossover involving two of the four chromatids of a chromosome pair. The existence of a crossover in one position does not prevent other crossovers occurring elsewhere among the chromatids, and multiple crossovers commonly are observed. Figure 3-17 shows some of the types that are possible.

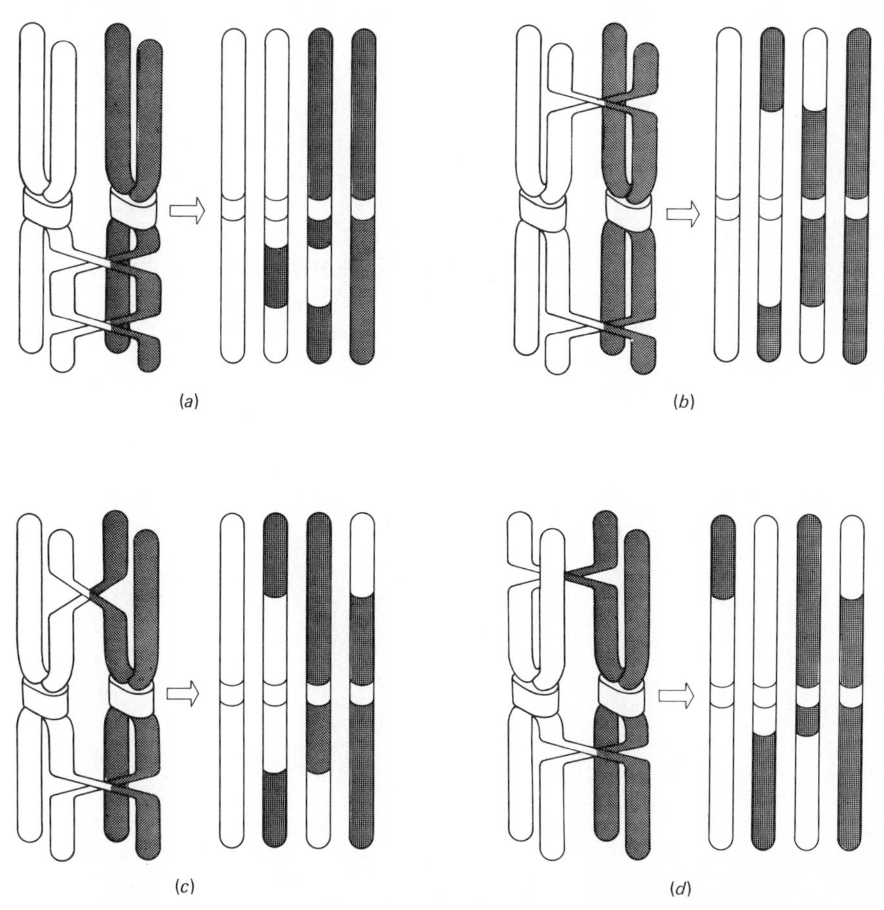

Figure 3-17 Some varieties of crossing over that have been observed. Other combinations, including triple or higher crossovers, may also occur. (*a*) Two-strand double crossover, both on same side of centromere. (*b*) Two-strand double crossover, involving different arms of chromosome. (*c*) Three-strand double crossover. (*d*) Four-strand double crossover.

Crossing over plays an important role in reassorting alleles into new combinations. Were it not for crossing over, the combination of genes on a particular chromosome would remain together indefinitely except for an occasional mutation. Since most mutations are deleterious, chromosomes gradually would accumulate mutations, eventually becoming incompatible with normal life. Crossing over permits a continual reshuffling of genes, so that new and sometimes advantageous combinations can occur. The genetic consequences of crossing over are considered in more detail in Chapter 19.

Crossing over between homologs may occur in somatic as well as germ cells, although the frequency is much less, since homologs do not ordinarily pair during mitosis. The principal genetic significance of somatic crossing over is the frequent production of daughter cells homozygous for loci for which the parent cell was heterozygous. For example, if the parent cell is heterozygous *Aa*, both sister chromatids will be *A* in one chromosome and *a* in the homologous chromosome (Figure 3-18). If somatic crossing over occurs between the centromere and the *A* locus, both homologs will then have *A* and *a* sister chromatids. At anaphase, both *A* chromosomes may go into one daughter cell and both *a* chromosomes into the other, or both daughter cells may receive an *A* and an *a*. In the former case, the daughter cells will be homozygous for all loci beyond the point of the crossover, with one clone of cells homozygous for the *A* segment, the other for the *a* segment. This is the origin of the so-called twin spots observed when appropriate markers are used to detect somatic crossing over.

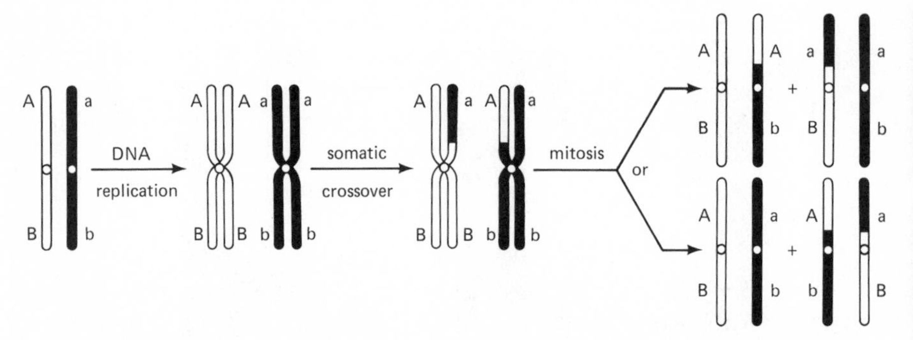

Figure 3-18 Diagram showing effects of somatic crossing over, assumed to occur between the centromere and the *A* locus. Two distributions of chromosomes may occur. That shown in the upper path produces two daughter cells homozygous for genes distal to the point of crossing over. The lower path would produce daughter cells identical in gene content but with alleles combined differently (i.e., one daughter cell now has *A* on the same chromosome with *b* and *a* with *B*). If somatic crossing over occurred prior to DNA replication, only one type of daughter cell could result.

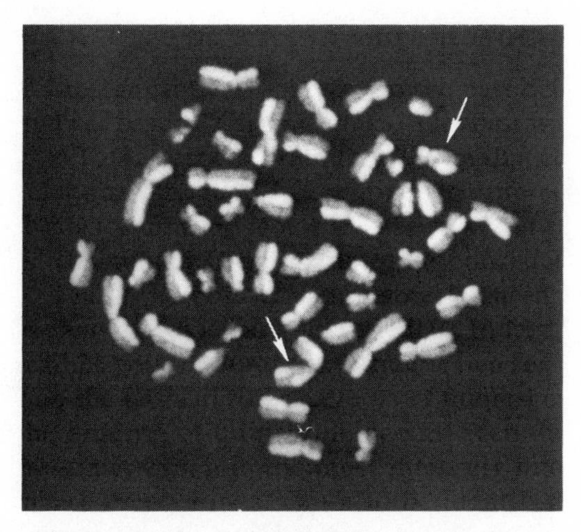

Figure 3-19 A human cell in metaphase stained to show sister chromatid exchange. The cells were exposed to bromodeoxyuridine (BUDR) for one or two mitotic cycles, followed by growth in the absence of BUDR through one or two additional cycles. Under these circumstances, BUDR, which is incorporated into DNA in place of thymine, segregates into only one of the two chromatids. The fluorescent dye Hoechst 33258 is partially quenched by BUDR, so that the chromatid with BUDR does not fluoresce as brightly as its sister chromatid. It is clear that exchange has occurred in several chromosomes, two of which are indicated by arrows. (*Furnished by Ann P. Craig-Holmes.*)

Crossing over, or, more properly, chromatid exchange, may also occur between sister chromatids, although in this instance there should be no genetic consequences since the exchanged segments are identical (Wolff, 1977). The phenomenon has been useful in attaining greater understanding of the structure of chromosomes and the replication of DNA. With the recent development of efficient methods of observing sister chromatid exchange, as illustrated in Figure 3-19, this technique has become important in the assessment of chromosome stability and the detection of environmental agents that break chromosomes (Chapter 9).

FORMATION OF GAMETES AND ZYGOTES

Spermatogenesis

In most male animals, meiosis occurs in the testes. In vertebrates, testicular tissue consists of two types—seminiferous tubules and interstitial tissue. These are illustrated in Figure 3-20. Interstitial tissue

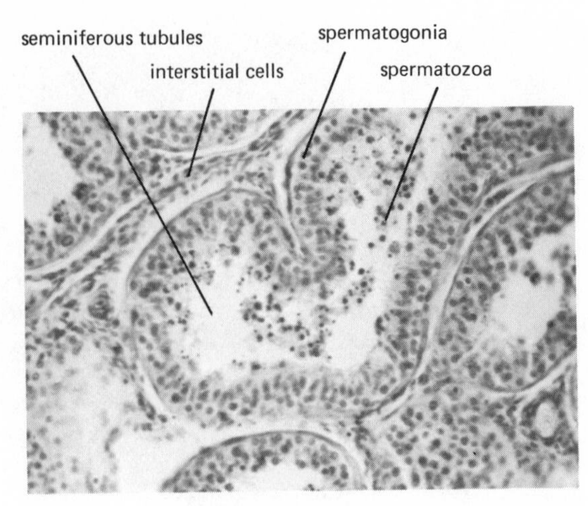

Figure 3-20 Section of testicular tissue, showing seminiferous tubules and interstitial tissue. (*Furnished by C. de Chenar.*)

functions in the synthesis of male hormones, which influence many activities of the body. Sperm are produced in the seminiferous tubules.

The cells lining the tubules are known as *spermatogonia* and are the stem cells of the germinal tissue. They replicate mitotically, maintaining a supply of cells for the production of gametes. Half of the spermatogonia in each cell cycle undergo meiosis. Once on the pathway of meiosis, they are known as *spermatocytes*. At the completion of meiosis, the haploid products resemble other cells in possessing a nucleus surrounded by appreciable quantities of cytoplasm. At this stage they are known as *spermatids*. Transformation into *spermatozoa* involves loss of virtually all of the cytoplasm and formation of a long whiplike structure, the tail, that confers motility on the mature sperm. In man, the time required to go from spermatogonia to mature spermatozoa is approximately 64 days.

The near absence of cytoplasm in a mature sperm has implications for the control of inherited variation. The nuclear contribution to an offspring is essentially equal for each parent. The cytoplasmic contribution from the father is very much less than that from the mother. For those traits that show equal contributions of the father and mother to the variation, it may be assumed that the control is transmitted by the nucleus.

Spermatogenesis is a quantitatively important process. An average human ejaculate may contain 200,000,000 sperm, and ejaculation may occur several times a week. The rate of production is therefore very high. As the sperm mature, they pass from the seminiferous tubules into the epididymis, where they are stored until released into the semen by ejaculation. Sperm that are not ejaculated are eventually resorbed.

Oögenesis

Gametogenesis in females occurs in the ovaries. During embryonic development of a human female, several million *ovarian follicles* are formed in the ovaries. Each follicle consists of a cell, or *oögonium,* which is destined to become the ovum and which is surrounded by a layer of follicular cells. By the time of birth, only two million follicles remain, many obviously degenerating. When a female reaches sexual maturity, one of these follicles matures approximately every month. The oögonium enlarges into a primary oöcyte, and the surrounding cells form a fluid-filled vesicle. The oöcyte is embedded in a corona of cells attached to the inner wall of the follicle. When the much enlarged follicle is fully mature, it bursts, releasing the ovum, which then enters one of the Fallopian tubes. Thus only about 400 follicles actually mature.

Meiosis is initiated in all oögonia during the fetal period. It proceeds to late prophase, then becomes dormant until maturation of the follicles. Thus all oöcytes remain in prophase for at least 12 years, and many are dormant much longer. With maturation, each oöcyte completes the first meiotic division to become a secondary oöcyte. These secondary oöcytes plus the corona of cells constitute the ova released from the follicles. The second meiotic division occurs immediately after penetration of the ovum by a spermatozoon. Without fertilization, it may not occur at all.

The nuclear aspects of meiosis are similar in males and females. There is a marked difference in the cytoplasmic aspects, however, as indicated in Figure 3-21. In the female, following anaphase I, virtually all the cytoplasm stays with one of the two daughter nuclei, giving rise to one very large and one very small cell, the latter being designated a *polar body.* Both cells may then undergo the second meiotic division, and again the cytoplasm of the large cell remains intact. The end products of oögenesis consist of a haploid cell with virtually all the original cytoplasm (the ovum) and two (or three) polar bodies, also haploid but with very little cytoplasm. Apparently, the polar bodies are by-products,

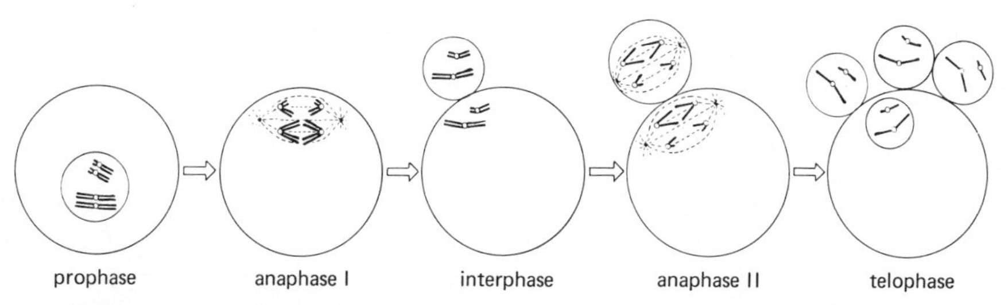

Figure 3-21 Oögenesis, showing unequal division of cytoplasm to give one ovum and three polar bodies.

not contributing further to oögenesis. The cytoplasm appears to function as a source of nourishment for the zygote prior to attachment to the uterine wall, somewhat analogous to the yolk of a hen's egg.

Fertilization

If sperm are present during the few days that an ovum is in the Fallopian tube, then one of the sperm may fuse with the ovum to form a zygote, a process called fertilization or *syngamy.* Penetration of the sperm through the corona of cells surrounding the ovum is accomplished in part by the enzyme hyaluronidase that is present in the sperm. When the head of the sperm reaches the cell membrane of the ovum, it enters the ovum, forming a pronucleus. The ovum is then impenetrable to other sperm.

As the pronucleus of the sperm enters the cytoplasm, the second meiotic division of the ovum occurs, giving rise to a pronucleus and a second polar body. The two haploid pronuclei are drawn to each other, and, during the first mitotic division, fuse into a single diploid structure.

Several cases have been reported in which a person was a mosaic of cells derived from two different sperm. In one case, about half the cells had two X chromosomes and the remainder had one X and one Y. A number of other genetic markers also differed in the two cells, giving results that could only be explained by two separate haploid contributions from the patient's father. The patient was a hermaphrodite, having both an ovary and an ovotestis. The authors interpreted this patient as resulting from mitosis of a haploid egg followed by fertilization of each daughter cell by separate sperm. Other possibilities were not ruled out.

In a second case, only a small portion of cells were XX, the majority being XY. The patient was a male of normal phenotype. It was possible to show by blood group antigens that both father and mother each contributed two haploid complements to their offspring. These authors suggested that one sperm fertilized the regular egg nucleus, and the other fertilized a polar body. The resulting person was, in a genetic sense, the product of two zygotes joined together.

The extreme rarity of these exceptional cases is indicative of the strict regularity of fertilization.

Parthenogenesis

The question sometimes arises as to whether an unfertilized ovum can develop into a functional organism. In some cases, this is known to happen. For example, if frog eggs are stimulated by various mechanical or chemical means, development is initiated and live tadpoles may hatch. The resulting individuals are haploid. In various species of *Hymenop-*

tera, males normally are haploid and are produced by development of unfertilized eggs, whereas fertilized eggs give rise to females.

The development of an unfertilized egg, called *parthenogenesis,* has been verified in only a few animal groups. Conceivably, individuals that develop parthenogenetically could be either haploid, as in the above examples, or diploid. In the latter case, the originally haploid chromosome complement of the egg might duplicate without cell division, or there might be fusion between an egg and a polar body. Whatever the mechanisms visualized, all lead to one conclusion: Since all of the genetic material arises from the mother, it would be impossible for the offspring to possess genes not present in the mother. Thus, if the offspring has red cell antigens not present in the mother, this would be sufficient evidence that genetic material has been introduced from an individual other than the mother. In mammals, in which maleness depends on the presence of a Y chromosome, only females could arise by parthenogenesis, since ova do not contain a Y chromosome.

In the case of humans, the development of benign ovarian teratomas (dermoid cysts) has been attributed to parthenogenesis (Linder et al., 1975). These ovarian tumors consist of tissues of a variety of histologic types: connective tissue, neural tissue, adipose tissue, bones, teeth, and hair. The chromosomes are the normal female diploid complement. Examination of a number of loci indicate the tumors always to be homozygous for chromosome markers close to the centromere when the patient is heterozygous, but they may be heterozygous for more distal markers. It was concluded that the tumors arise from single germ cells after the first meiotic division, with equational division of the chromosomes to form a diploid complement.

CYTOPLASMIC INHERITANCE

In addition to chromosomes, various other structures are transmitted through the egg and sperm to offspring. This is particularly true of the egg, which is a large cell containing a large amount of cytoplasm with mitochondria, ribosomes, Golgi apparatus, and other organelles. Is it possible that the characteristics of some of these are transmitted independent of nuclear control?

The usual procedure for testing for non-Mendelian (extranuclear, maternal, or cytoplasmic) inheritance is to compare the progeny from reciprocal crosses of two different strains. A prediction of chromosomal inheritance is that the offspring phenotype is independent of which parent serves as the maternal or paternal source of genes. Thus offspring from the cross ♂A × ♀B should be the same as offspring from ♂B × ♀A.

The exceptions to this prediction are very rare. Maternal inheritance was first demonstrated in certain crosses of plants where the plant size is dependent on the strain serving as female parent. Additional examples are known in other organisms. In *Neurospora*, some mitochondrial characteristics are transmitted by the maternal cytoplasm. *Chlamydomonas*, a genus of green algae, has also been shown to exhibit cytoplasmic inheritance.

Earlier, it was thought that deoxyribonucleic acid (DNA) did not occur outside the nucleus. In 1965, mitochondria were shown to have DNA distinct from that of the nucleus. With this discovery, the observed independent inheritance of mitochondria became understandable in terms of conventional models of gene action. Indeed, the mitochondria have been viewed by some as "infecting" particles. They have established a symbiotic relationship with the host cell, the mitochondria providing the machinery for oxidizing cellular substrates to yield energy, the host cell providing those substrates as well as other components required by the mitochondria. Whatever the origin of mitochondria, they appear to have some genes of their own and are thus not entirely dependent on nuclear heredity. On the other hand, many of the "mitochondrial" enzymes, such as malate dehydrogenase in man, show genetic variation that is Mendelian, indicating that at least some of the proteins associated with mitochondria are coded in the nucleus.

The importance of cytoplasmic inheritance for man has not been fully ascertained. An important feature of cytoplasmic inheritance is that all offspring in a sibship resemble the mother. Thus, if segregation occurs, cytoplasmic inheritance can be ruled out. On the basis of this rule, it seems likely that little if any of the known human variation can be attributed to inheritance via cytoplasm.

SUMMARY

1. The similarity of transmission of genes and of chromosomes led W. S. Sutton and T. Boveri to suggest that the genetic determinants are located on chromosomes.

2. Deoxyribonucleic acid (DNA) serves for the primary storage of genetic information in all species except the RNA viruses. The basic structure of nucleic acids is a long-chain polymer of nucleotides. Each nucleotide consists of a purine or pyrimidine base, a sugar (ribose or 2-deoxyribose), and phosphate.

3. In DNA the bases are adenine, guanine, cytosine, and thymine. DNA ordinarily exists as a double helix of two such polymers, the bases pointing toward the interior of the helix. Opposing bases form hydrogen bonds, but only the base pairs adenine–thymine and guanine–cytosine can fit into a double helix. Thus the two polymeric strands of a double helix have complementary structures.

4. There are several kinds of ribonucleic acid (RNA). In most, the bases are adenine, guanine, cytosine, and uracil. RNA is a single polymeric strand, although in some instances, such as transfer RNA, the strand is folded to form a double helix with itself.

5. DNA replicates by unwinding, each single strand then serving as a template on which a new complementary strand is formed. This produces two identical daughter molecules, each consisting of one parental strand and one newly synthesized strand. RNA is replicated as a strand complementary to a single DNA strand. In cells infected with RNA viruses, an enzyme, called reverse transcriptase, is produced that can catalyze the synthesis of DNA using RNA as a template.

6. DNA replication occurs during interphase when chromosomes are highly dispersed. The exact time of replication of each chromosome or portion thereof can be established with tritium-labeled thymidine. The most pronounced deviation in time of synthesis occurs when there are two or more X chromosomes in a cell. In these instances, all but one of the X's is late-labeling, that is, the DNA replicates at the very end of the period of synthesis.

7. Chromosomes can be differentiated into euchromatin and heterochromatin based on staining properties. Most genes appear to be in the euchromatin. DNA from chromosomes is found in close association with histones, forming complexes called nucleosomes.

8. The replication and distribution of chromosomes is very precise. In man there are 23 pairs of chromosomes. The sex chromosome pair is homologous in females (XX) but different in males (XY). The remaining 22 pairs are homologous and called autosomes.

9. Mitosis occurs in somatic cells and involves duplication and distribution of chromosomes during cell division to yield daughter cells identical to the parental type.

10. Meiosis occurs in germ cells and involves a single duplication of chromosomes combined with two cell divisions to give four daughter cells with only a single copy of each chromosome. During meiosis, homologous chromosomes pair and exchange segments (crossing over), permitting reassortment of genes located on the same chromosome.

11. In spermatogenesis, all four meiotic products form spermatozoa. In oögenesis, only one meiotic product is associated with the cytoplasmic mass of the ovum. The remaining three become polar bodies and are not fertilized. While spermatogenesis is a continuous process through most of the life of the male, oögenesis proceeds to late prophase I during embryogenesis and remains at that stage until sexual maturity of the woman. Then once a month one of the oöcytes is released and continues meiosis.

12. In fertilization, one sperm enters the ovum, and the chromosomal material of the sperm fuses with one of the products of the second meiotic division of the ovum, giving a diploid nucleus. Rarely, in nonhuman organisms, unfertilized ova may be stimulated to undergo embryogenesis (parthenogenesis).

13. Although nearly all examples of genetic variation are attributed to chromosomal genes, cytoplasmic organelles such as mitochondria have traces of DNA and appear to transmit certain traits.

REFERENCES AND SUGGESTED READING

AVERY, O. T., MAC LEOD, C. M., AND MC CARTY, M. 1944. Studies on the chemical nature of the substance inducing transformation of pneumococcal types. *J. Exp. Med.* 79: 137–156.

BEATTY, R. A. 1957. *Parthenogenesis and Polyploidy in Mammalian Development.* New York: Cambridge Univ. Pr. 132 pp.

BROWN, S. W. 1966. Heterochromatin. *Science* 151: 417–425.

CLERMONT, Y. 1966. Renewal of spermatogonia in man. *Am. J. Anat.* 118: 509–524.

DAVERN, C. I., AND MESELSON, M. 1960. The molecular conservation of ribonucleic acid during bacterial growth. *J. Molec. Biol.* 2: 153–160.

DE LANGE, R. J., AND SMITH, E. L. 1971. Histones: structure and function. *Annu. Rev. Biochem.* 40: 279–314.

DU PRAW, E. J. 1970. *DNA and Chromosomes.* New York: Holt, Rinehart and Winston, 340 pp.

EVANS, H. J. 1977. Some facts and fancies relating to chromosome structure in man. *Adv. Hum. Genet.* 8: 347–438.

FINCH, J. T., LUTTER, L. C., RHODES, D., BROWN, R. S., RUSHTON, B., LEVITT, M., AND KLUG, A. 1977. Structure of nucleosome core particles of chromatin. *Nature* (Lond.) 269: 29–36.

FOWLER, R. E., AND EDWARDS, R. G. 1973. The genetics of early human development. *Prog. Med. Genet.* 9: 49–112.

GARTLER, S. M., WAXMAN, S. H., AND GIBLETT, E. R. 1962. An XX/XY human hermaphrodite resulting from double fertilization. *Proc. Nat. Acad. Sci.* (U.S.A.) 48: 332–335.

GERMAN, J. 1964. The pattern of DNA synthesis in the chromosomes of human blood cells. *J. Cell Biol.* 20: 37–55.

GOODENOUGH, U. 1978. *Genetics,* 2nd Ed. New York: Holt, Rinehart and Winston, 840 pp.

GRIFFITH, F. 1928. The significance of pneumococcal types. *J. Hyg.* (*Camb.*) 27: 113–159.

HAMERTON, J. L. 1971. *Human Cytogenetics,* Vol. I. *General Cytogenetics.* New York: Academic Press, 412 pp.

HARRISON, R. J. 1967. *Reproduction and Man.* New York: W. W. Norton, 134 pp.

HERSHEY, A. D., AND CHASE, M. 1952. Independent functions of viral pro-

tein and nucleic acid in growth of bacteriophage. *J. Gen. Physiol.* 36: 39–56.
HUBERMAN, J. A., AND RIGGS, A. D. 1968. On the mechanism of DNA replication in mammalian chromosomes. *J. Molec. Biol.* 32: 327–341.
HULTÉN, M., AND LINDSTEN, J. 1973. Cytogenetic aspects of human male meiosis. *Adv. Hum. Genet.* 4: 327–387.
KORNBERG, R. 1977. Structure of chromatin. *Annu. Rev. Biochem.* 46: 931–954.
LINDER, D., MC CAW, B. K., AND HECHT, F. 1975. Parthenogenic origin of benign ovarian teratomas. *N Engl. J. Med.* 292: 63–66.
MARKS, P. A., RIFKIND, R. A., SPIEGELMAN, S., KACIAN, D. L., AND BANK, A. 1974. Isolation and synthesis of human genes. In *Birth Defects:* (A. Motulsky, W. Lenz, Eds.). Amsterdam: Excerpta Medica, pp. 73–80.
MILLER, O. J. 1970. Autoradiography in human cytogenetics. *Adv. Hum. Genet.* 1: 35–130.
MOORE, J. A. 1972. *Heredity and Development,* 2nd Ed. London, New York: Oxford Univ. Pr., 292 pp.
OHNO, S., KLINGER, H. P., AND ATKIN, N. G. 1962. Human oögenesis. *Cytogenetics* 1: 42–51.
PAULSON, J. R., AND LAEMMLI, U. K. 1977. The structure of histone-depleted metaphase chromosomes. *Cell* 12: 817–828.
PRESCOTT, D. M. 1970. The structure and replication of eukaryotic chromosomes. *Adv. Cell Biol.* 1: 57–117.
RIS, H., AND KUBAI, D. F. 1970. Chromosome structure. *Annu. Rev. Genet.* 4: 263–294.
SWANSON, C. P. 1957. *Cytology and Cytogenetics.* Englewood Cliffs, N.J.: Prentice-Hall, 596 pp.
TAYLOR, J. H. 1968. Rates of chain growth and units of replication in DNA of mammalian chromosomes. *J. Molec. Biol.* 31: 579–594.
THOMAS, C. A. Jr. 1971. The genetic organization of chromosomes. *Annu. Rev. Genet.* 5: 237–256.
WAGNER, R. P. 1969. Genetics and phenogenetics of mitochondria. *Science* 163: 1026–1031.
WATSON, J. D. 1976. *Molecular Biology of the Gene,* 3rd Ed. Menlo Park, Calif.: W. A. Benjamin, 739 pp.
WATSON, J. D., AND CRICK, F. H. C. 1953a. A structure for deoxyribose nucleic acid. *Nature* (Lond.) 171: 737–738.
WATSON, J. D., AND CRICK, F. H. C. 1953b. Genetical implications of the structure of deoxyribonucleic acid. *Nature* (Lond.) 171: 964–967.
WHITE, M. J. D. 1973. *The Chromosomes,* 6th Ed. London: Chapman and Hall, 214 pp.
WILKINS, M. H. F. 1963. Molecular configuration of nucleic acids. *Science* 140: 941–950.
WOLFF, S. 1977. Sister chromatid exchange. *Annu. Rev. Genet.* 11: 183–201.
YUNIS, J. J. (Ed.). 1977. *Molecular Structure of Human Chromosomes.* New York: Academic Press, 336 pp.
ZUBAY, G. L., AND MARMUR, J. (Eds.) 1973. *Papers in Biochemical Genetics,* 2nd Ed. New York: Holt, Rinehart and Winston. 622 pp.

ZUELZER, W. W., BEATTIE, K. M., AND REISMAN, L. E. 1964. Generalized unbalanced mosaicism attributable to dispermy and probably fertilization of a polar body. *Am. J. Hum. Genet.* 16: 38–51.

REVIEW QUESTIONS

1. Define:

purine
pyrimidine
ribose
deoxyribose
nucleoside
nucleotide
hydrogen bond
ribonucleic acid
deoxyribonucleic acid
histone
euchromatin
heterochromatin
messenger RNA
transfer RNA
ribosome
heterocatalysis
autocatalysis
template
polymerase
reverse transcriptase
chromosome
chromatid
centromere
centriole
haploid
diploid
mitosis
meiosis
interphase
prophase
metaphase
anaphase
telophase
leptotene
zygotene
pachytene
diplotene
synapsis
bivalent
tetrad
chiasma
crossing over
spermatogonium
spermatocyte
spermatid
ovarian follicle
oögonium
oöcyte
polar body
gamete
zygote
syngamy
parthenogenesis
mitochondria
diakinesis

2. **a.** The average weight of a DNA nucleotide is approximately 330 daltons. Using this figure, how many nucleotides are in one human haploid genome? (1 dalton = 1.66×10^{-24} grams.)

b. If a typical structural gene requires 1000 nucleotide pairs, what is the number of such genes that could be coded in one human genome?

3. Human DNA consists of 30% adenylate residues on a molar basis. What are the percentages of thymidylate, guanylate, and cytidylate?

4. There is evidence that the distribution of types of nucleotides is not uniform in human chromosomes. Which regions would you expect to be more resistant to thermal denaturation? Why?

5. The pairing of homologous regions of chromosomes during meiosis is highly specific. Can you design a scheme based on the structure of DNA that accounts for pairing, including the specificity? Does your scheme also account for terminalization of chiasmata?

6. There is evidence that crossing over can occur within genes as well as between. What mechanisms might be imagined that would insure crossing over at exactly homologous points along the DNA?

7. A diploid organism is heterozygous at three loci on the same chromosome (*ABC*/*abc*). With respect to these loci, what kinds of gametes could be formed by this organism? Indicate the number of crossovers necessary to produce each gamete. When does crossing over occur?

8. Why is the first meiotic division called a reduction division?

9. What is the relative duration of the first meiotic prophase in females and males?

10. Why would crossing over between sister chromatids ordinarily not be detectable by genetic means?

11. What are the genetic disadvantages of parthenogenesis?

12. From an evolutionary point of view, what would be the advantage of the great rate of meiosis in males compared with the very low rate in females?

CHAPTER FOUR

THE NORMAL HUMAN CHROMOSOME COMPLEMENT

For many years it has been recognized that the numbers and kinds of chromosomes are highly characteristic for a given species. In some organisms (of which *Drosophila* is the outstanding example), it has been possible to map chromosomes both genetically by breeding experiments and cytologically by microscopic observation of chromosomes so that identification of minute structural deviations is possible. For most mammals—including man—direct examination of chromosomes has been a technically difficult procedure that for a long time gave results of limited utility.

One difficulty to be conquered was the procurement of satisfactory preparations of dividing cells at metaphase, for, in most organisms, chromosomes can be most readily examined during this limited part of the growth cycle when they are maximally contracted. In most tissues, only a very small portion of cells are in metaphase at any one time. The traditional solution to this problem was to examine testicular tissue, since the spermatocytes of mature males show a high frequency of cells in meiosis, and bivalents at metaphase I can be more readily counted than mitotic chromosomes. However, such material is difficult to work with and would fail to provide information on many clinically important problems.

A second difficulty was that, even in the best preparations of metaphase chromosomes, the large number of chromosomes of man and many other species made accurate observation almost impossible. The clumping and overlapping of chromosomes always lent uncertainty to the results.

In spite of such technical difficulties, the estimates of several investigators came close to the correct number of chromosomes for man. The best preparations seemed to indicate the number 48, and for many years this was accepted.

THE IDENTIFICATION OF HUMAN CHROMOSOMES

Earlier Methods

In the 1950s, several important technical advances were made that allowed the question of chromosome number and morphology to be reopened. The first of these was the development of techniques for culturing tissues and cells *in vitro*. Although a few strains of cancer cells had been grown *in vitro* for some years, it was only after a systematic investigation of the fastidious nutritional requirements of mammalian cells that it became possible to culture a variety of normal tissues.

The rate of cell growth and division is generally higher in cultured cells than in most tissues of a normal animal, thus providing more opportunities for formation of metaphase figures.

A second technical development was the application to tissue cultures of the plant alkaloid colchicine. For over 2000 years, this substance has been used in the treatment of gout. Although its action in gout is unknown, it does have the property of arresting cell division at the metaphase stage through interference with the spindle function. Thus, if colchicine is present in a cell culture, division will proceed normally until formation of a metaphase plate. The arrested metaphase figures are sufficiently stable so that it is possible to accumulate many more of them than would exist at any one time without colchicine. This greatly increases the likelihood of finding cells suitable for examination.

A third development was the use of hypotonic treatment of cells prior to fixing on a slide. This treatment has the effect of swelling the cells and spreading the chromosomes out over a much larger area, reducing the amount of clumping and overlapping. Treatment consists of suspending the cells briefly in a salt solution of concentration lower than that of the intracellular fluid. Water moves into the cell, causing expansion and separation of various structures, including chromosomes. These structures may then be examined with much greater reliability.

A later discovery in the case of lymphocyte cultures was the use of *phytohemagglutinin* (PHA). One of the problems of blood culture is the great excess of red cells, which are not of interest since they are devoid of nuclei, compared with white cells, which still have the potential for growth. Extracts of certain plant seeds have the ability to agglutinate red cells, and these phytohemagglutinins were used as an inexpensive means of removing red cells. Eventually it was recognized that PHA also stimulates growth of lymphocytes, which seems to be initiated in response to antigenic stimulation. PHA is now routinely added to lymphocyte cultures primarily because of its *mitogenic* activity.

Karyotype Analysis

Adapting these techniques to cultured fibroblasts of human embryonic lung tissue, Tjio and Levan in 1956 found that earlier reports had erred in reporting the number of chromosomes of man; they found 46 rather than 48. The chromosomal constitution (*karyotype*) of females is 46,XX (that is, a total of 46 chromosomes, of which two are X's); that of males is 46,XY. The X and Y chromosomes are the *sex chromosomes*; the remaining 44 chromosomes consist of 22 pairs of *autosomes*. These unexpected findings were quickly confirmed by other investigators.

Simple enumeration of the chromosomes is an inadequate characterization of the karyotype because it does not reveal abnormalities in chromosome structure. If a person is to be normal, the 46 chromosomes must be the correct 46. The techniques available between 1958 and 1971 did not reveal the internal structure of human chromosomes. Rather, they were more comparable to shadowgrams. An example is given in Figure 4-1. With this technique it is possible to classify chromosomes into seven distinct groups and to recognize some individual chromosomes. Figure 4-1 illustrates the manner in which this is accomplished. Each chromosome is characterized by two major variables: the length and the position of the centromere. The centromere may divide the chromosome into two arms of equal length, in which case it is *metacentric*. Or the centromere may be nearly terminal (*acrocentric*). Truly terminal centromeres (*telocentric*) do not occur in man but are found in other species. Chromosomes with centromeres located between the median and terminal positions are designated *subterminal* or *submedian*. In some preparations, secondary constrictions (the primary constriction being the location of the centromere) are observed at specific positions on the chromosomes and are useful in identifying those chromosomes.

Analysis of a karyotype is based on a photograph, as in Figure 4-1*a*. Ordinarily, peripheral white blood cells are used for karyotyping, but fibroblasts and bone marrow cells are often used, and any cells that undergo mitosis should be usable. Typically, the cells are grown for several days in artificial medium, then treated with colchicine for two hours, and finally immersed in hypotonic salt solution. The cells are then spread on a glass microscope slide and allowed to air dry in order to flatten the cells and spread the chromosomes. The slide is fixed and stained with a nuclear stain. If a karyotype is to be prepared, suitable spreads are photographed, and the individual chromosomes are cut out from the photograph and arranged in the standard sequence shown in Figure 4-lb.

In 1960, a number of the foremost investigators in human cytogenetics met in Denver, Colorado, to adopt a standard system of chromosome nomenclature. It was agreed that the longest chromosomes are to

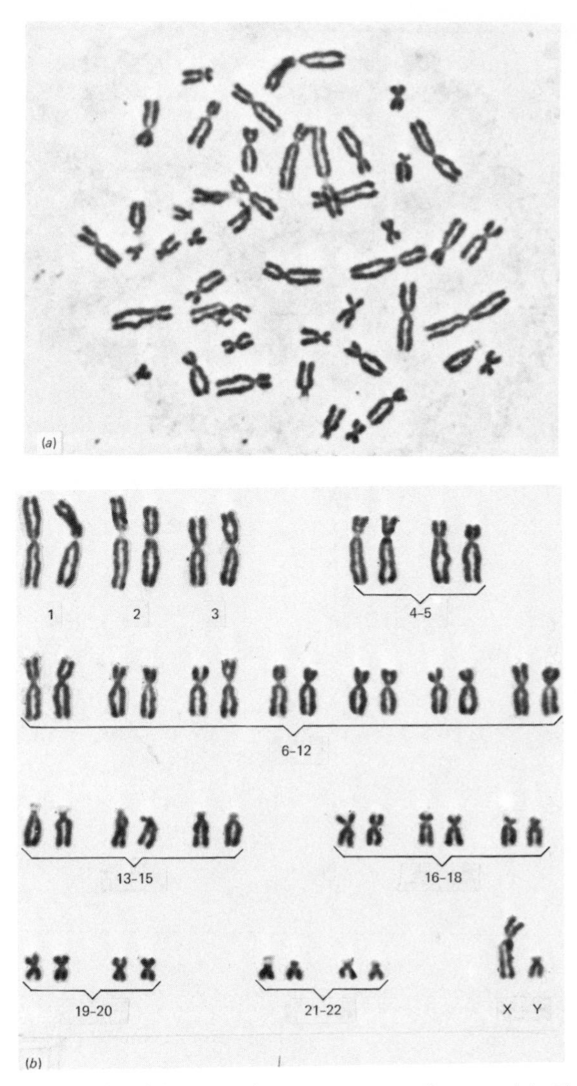

Figure 4-1 Analysis of human chromosome complement. (*a*) Chromosome spread from lymphocyte culture. (*b*) Arrangement of chromosomes into standard karyotype. (*Furnished by J. J. Biesele.*)

be placed at the beginning of the sequence, the shortest at the end. If two pairs of chromosomes are the same length, the pair with the more centrally located centromere is placed first. This system, known as the "Denver system," is given in Table 4-1. Included also are the divisions into the seven major morphological groups.

With staining techniques available at the time of the Denver Conference, most of the chromosomes could not be uniquely identified. It was

Table 4-1 **NOMENCLATURE AND MEASUREMENTS OF HUMAN MITOTIC CHROMOSOMES***

Group	Chromosome number	Relative length†	Centromeric index‡
A	1	8.50	48.4
	2	8.08	39.2
	3	6.84	47.0
B	4	6.32	28.9
	5	6.09	29.0
C	6	5.92	38.9
	7	5.37	38.9
	X	5.12	39.9
	8	4.93	34.0
	9	4.80	35.2
	10	4.61	33.8
	11	4.63	40.2
	12	4.64	29.9
D	13	3.75	17.0
	14	3.58	18.7
	15	3.44	20.0
E	16	3.34	41.4
	17	3.22	33.7
	18	2.92	30.5
F	19	2.65	46.4
	20	2.53	45.5
G	21	1.88	30.7
	22	2.02	30.3
	Y	2.16	26.8

* Measurements are taken from the *Paris Conference (1971)*, p. 42. The values shown here are weighted means of columns C and D of Table 5 from that publication.
† Expressed as percent of the haploid autosomal complement.
‡ Ratio of short arm to total length of chromosome, expressed as percent.

not possible to recognize structural changes in chromosomes unless the change altered the overall length or the position of the centromere. The chromosomes in group A are each identifiable, but the rest ordinarily could only be assigned to a group. Groups A, B, C, E, and F are submetacentric, with a few, such as chromosome 1, being almost metacentric. The members of groups D and G are acrocentric and, with the exception of the Y chromosome, have a darkly staining region, called a *satellite,* attached to the end of the short arm. The satellite often appears to be attached by no more than a thin strand, but newer studies show that the region between the satellite and the centromere is more substantial but does not stain readily.

The Banding Patterns of Chromosomes

Several technical developments in the past decade have revolutionized the study of human chromosomes. The first was a report from the laboratory of T. O. Caspersson in Stockholm that certain fluorescent derivatives of quinacrine bind differentially to different parts of chromosomes. The basis for binding is not well known, but the pattern of fluorescent bands, designated Q-bands, is quite consistent for any one chromosome. These patterns can therefore be used to distinguish chromosomes that otherwise are very similar. Indeed, it is possible to distinguish all 22 pairs of autosomes and the two sex chromosomes from each other on the basis of their patterns of fluorescent Q-bands (Figure 4-2).

The binding of quinacrine occurs even during interphase. The Y chromosome, which is the most intensely fluorescent chromosome in metaphase preparations, can also be identified as a bright fluorescent spot in interphase nuclei (Figure 4-3). Other chromosomes are too diffuse to be recognized at interphase.

Not long after the fluorescent technique became available, several other laboratories developed alternate techniques for detecting the same or related banding patterns. Arrighi and Hsu showed that chromosome preparations treated with denaturing agents such as heat, acid, or alkali

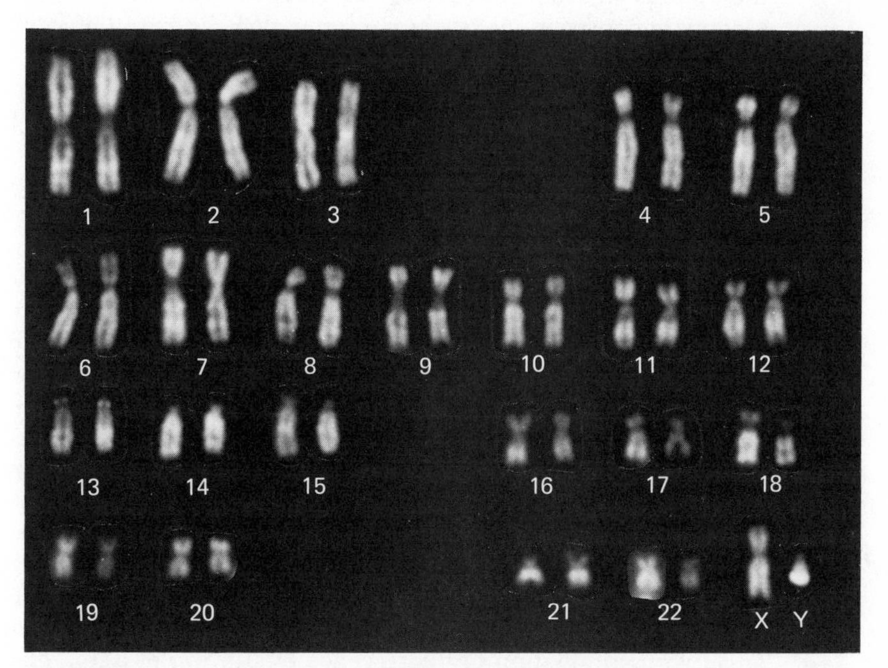

Figure 4-2 Karyotype of a 46,XY male showing Q-bands (quinacrine stain). The long arm of the Y chromosome is very bright. The C-band regions of chromosomes 1, 9, and 16 are dull. (*Furnished by T. R. Chen.*)

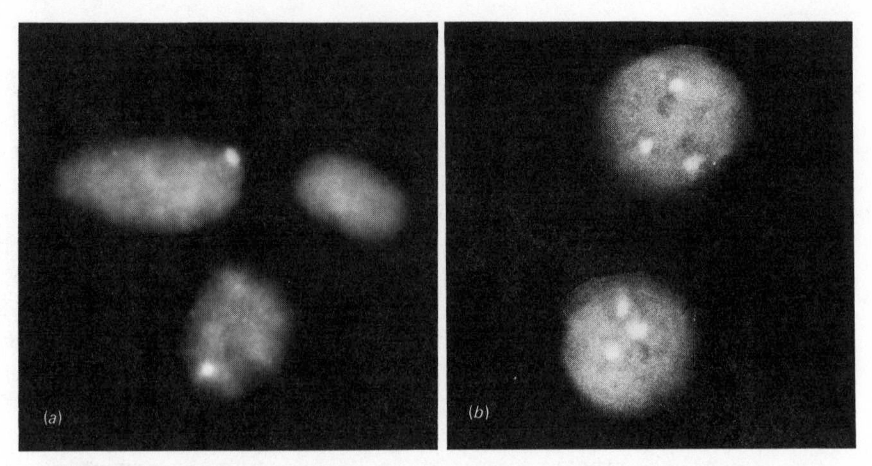

Figure 4-3 Human interphase nuclei showing (*a*) one fluorescent Y body in fibroblasts from a normal XY male, and (*b*) three Y bodies in lymphocytes of an XYYY male. (*Figure 4-3a furnished by T. R. Chen; Figure 4-3b from Schoepflin and Centerwall, 1972.*)

lose their ability to be stained by the usual nuclear stains. However, if the preparations are then maintained at 37°and neutral *p*H for a period, the regions adjacent to centromeres become stainable again, as do the secondary constrictions of chromosomes 1, 9, and 16, the satellites of the acrocentrics, and the distal long arm of the Y chromosome. These regions of high affinity for stain have been referred to as block chromatin or C-bands (C for constitutive heterochromatin). A typical karyotype showing C-bands is shown in Figure 4-4.

Further variation in treatment of slide preparations prior to staining produced a series of additional bands that are fainter but much more numerous than C-bands and are distributed in all parts of the chromosome arms. This discovery, made independently in five different laboratories in 1971, has completely changed the study of human chromosomes. Although the bands shown by this technique—usually called the modified Giemsa stain—are the same as those revealed by the fluorescent technique, no special equipment is required, and the procedures can be carried out in any laboratory equipped for routine karyotyping. These bands are designated G-bands or R-bands, depending on the techniques used to demonstrate them. (G-bands are so named because of the Giemsa stain. R-bands refer to the "reverse banding" which occurs with heat denaturation in which the dark R-bands correspond to light G-bands and vice versa.) The R-bands show up especially well with a technique that depends on binding of the fluorescent dye acridine orange, with stimulation of fluorescence by ultraviolet light. The bands appear as various shades of orange, yellow, and green.

Figure 4-5 shows the chromosomes as they appear using one of the

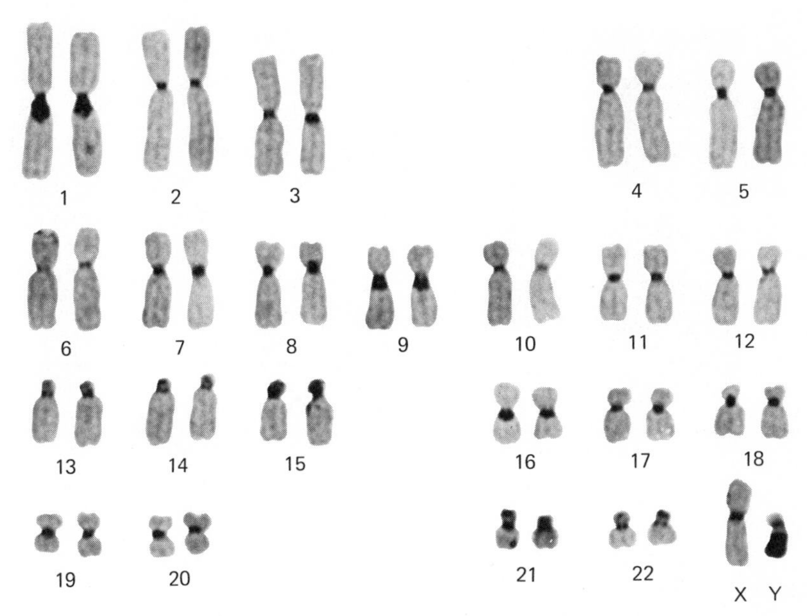

Figure 4-4 Karyotype of a 46,XY male stained to show C-bands. There are four regions with prominent C-bands: (1) the centromere regions of all chromosomes; (2) the proximal long arms (secondary constrictions) of autosomal pairs 1, 9, and 16; (3) the satellite regions of the acrocentric chromosomes (13, 14, 15, 21, and 22); and (4) the distal part of the long arm of Y. (*Furnished by Ann P. Craig-Holmes.*)

techniques to produce G-bands. Each chromosome has a unique pattern of bands, making it possible to identify every chromosome and, in many cases, segments of chromosomes. This is very useful in identifying abnormal chromosome complements, especially structural rearrangements.

The mechanism for band formation is not understood. Initially, it was thought that mild denaturation caused the DNA strands to separate, and that during subsequent incubation certain regions tended to form a stainable double helix more rapidly than others. Doubt has been cast on this idea with the discovery that proteolytic enzymes such as trypsin cause the same banding pattern. Several lines of evidence now suggest that the bands are regions rich in G-C base pairs (R-bands) or A-T base pairs (Q- and G-bands). Fluorescent antibodies prepared against poly-GC bind to the R band sites, as do antibiotics such as olivomycin that bind to G-C pairs. Conversely, fluorescent labels that bind to A-T pairs produce Q-bands. The bands therefore do appear to reflect differences in the DNA sequences of the chromosomes.

While metaphase chromosomes are the easiest to observe, more detail is visible at stages when the chromosomes are less condensed. Yunis (1976) studied prophase chromosomes and found that many of the metaphase bands originate by coalescence of smaller bands visible

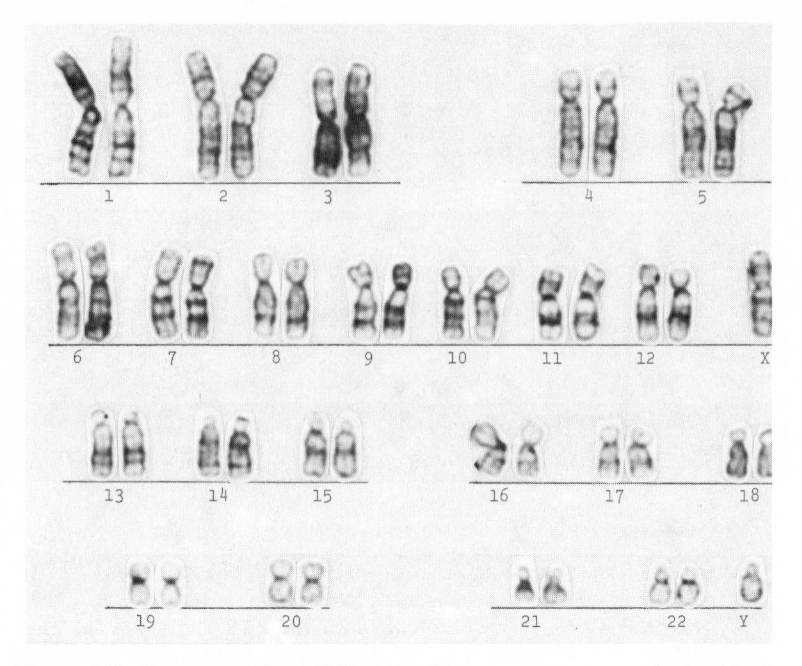

Figure 4-5 Karyotype of a 46,XY male. Chromosomes were stained with the modified Giemsa technique to show the presence of G-bands. Each chromosome pair has a unique banding pattern which allows positive identification of all elements. (*Furnished by Patricia N. Howard-Peebles.*)

at earlier stages. In early prophase, he was able to identify some 3000 bands. A comparison of prophase and metaphase chromosomes is shown in Figure 4-6.

If cultured cells are grown one cycle in BUDR before staining with Hoechst 33258 (see Figure 3-19), *lateral asymmetry* can be seen in some of the regions of constitutive heterochromatin. Presumably this is due to AT-rich segments, in which there is preponderance of A in one strand and a corresponding preponderance of T in the other. Replication of such a segment would yield one daughter chromatid rich in BUDR:

$$A - T \rightarrow A - B^{\star} + A^{\star} - T$$

where the B* and A* represent newly synthesized strands. Fluorescence of the dye would be quenched in the B strand. The pattern of lateral asymmetry is variable and can be used as an inherited chromosome marker. For example, Angell and Jacobs (1978) found that 20 of 44 persons were heterozygous for lateral symmetry involving chromosome 1.

It should be remembered that the banding seen in insect polytene chromosomes is not the same as the banding of human chromosomes, since polytene chromosomes represent a many-fold replication of chro-

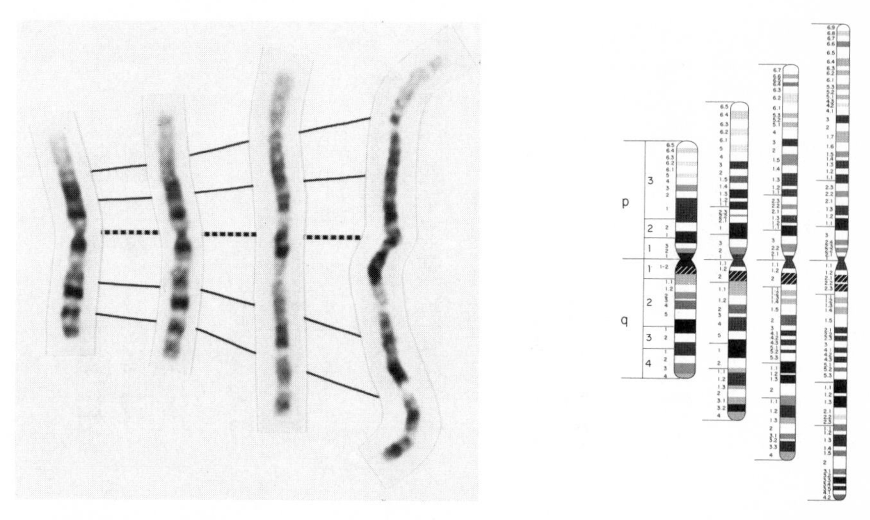

Figure 4-6 G-band preparation and diagram of chromosome 1 at different stages of contraction. From right to left: late prophase, prometaphase, early metaphase, mid-metaphase. The bands of late prophase appear to coalesce to form the larger bands of metaphase. (*From J. J. Yunis, J. R. Sawyer, and D. W. Ball,* Chromosoma *67: 293, 1978.*)

mosomes in cells that have lost the ability to divide further. The banded human chromosomes are in cells that are in the process of division, and each chromatid is a single copy of the genetic information.

With the advent of banding techniques, it has become convenient to designate the bands with a standard nomenclature. Cytogeneticists have agreed on the pattern of bands and on their designation as shown in Figure 4-7. These agreements, including some earlier conventions, were published in 1972 as the Paris Conference (1971): Standardization in Human Cytogenetics, to which reference should be made for a fuller description. Briefly, the short arm of a chromosome is symbolized by p, the long arm by q. The chromosome arms are divided into *regions* by *landmarks.* A landmark is a consistent feature useful in identifying chromosomes, such as the ends of chromosome arms, the centromere, and certain major bands. A region is the space between two adjacent landmarks. Regions are numbered consecutively from the centromere, and bands are numbered consecutively within each region. Both the light and dark bands are given separate numbers. For example, band 2q23 would refer to chromosome 2, the long arm, region 2, band 3.

As the study of karyotypes has progressed, particularly karyotypes with numerical or structural abnormalities, a number of standard symbols have been developed as part of the nomenclature. A list of symbols is given in Table 4-2. These include symbols for abnormal complements and rearrangements to be discussed in later chapters.

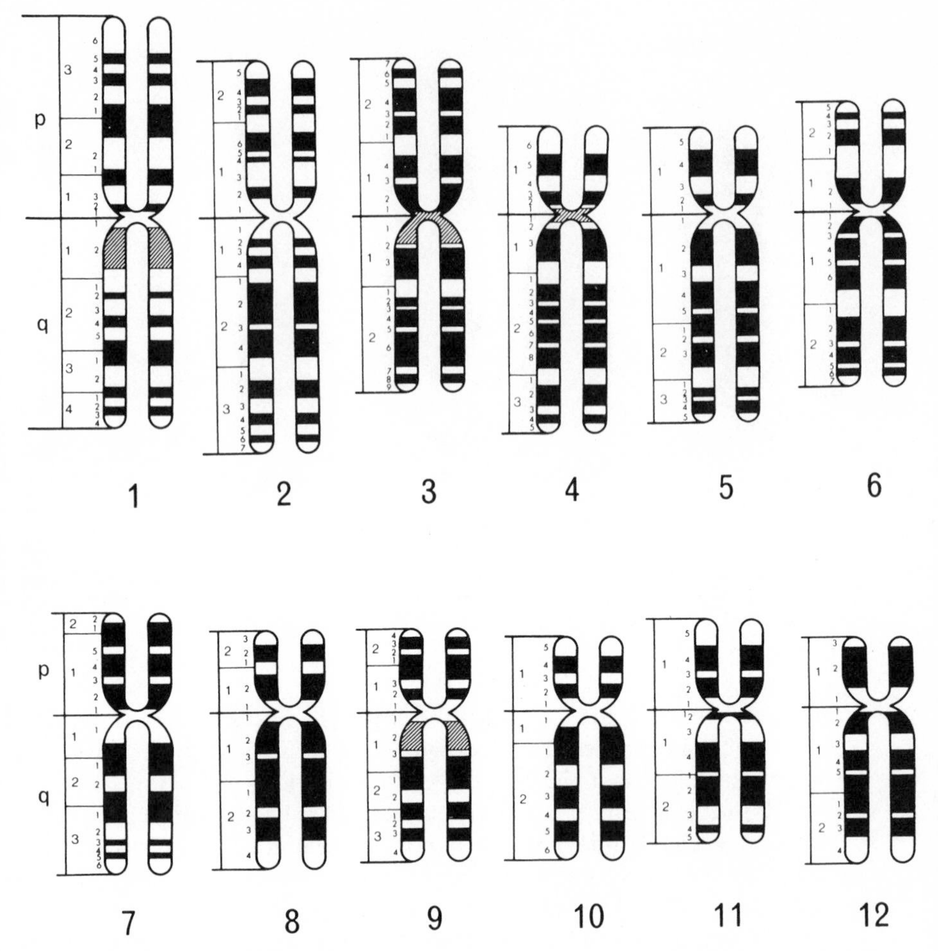

Figure 4-7 Diagrammatic representation of chromosome bands as observed with the Q-, G-, and R-staining methods; centromere representative of Q-staining only. *(From Paris Conference, 1971; reproduced by permission of the National Foundation—March of Dimes.)*

Normal Variations in Chromosome Morphology

The karyotypes of normal persons are generally indistinguishable from karyotypes of other normal persons of the same sex. Occasional normal variants are observed, however (Selles and co-workers, 1974; Jacobs, 1977). The most variable of the chromosomes is the Y. Many examples of Y chromosomes with long arms that are unusually short or unusually long are known (Figure 4-8). These include pedigrees in which the distinctive chromosome can be identified in all males related as first degree relatives. In one case in France, a marked difference in appearance of

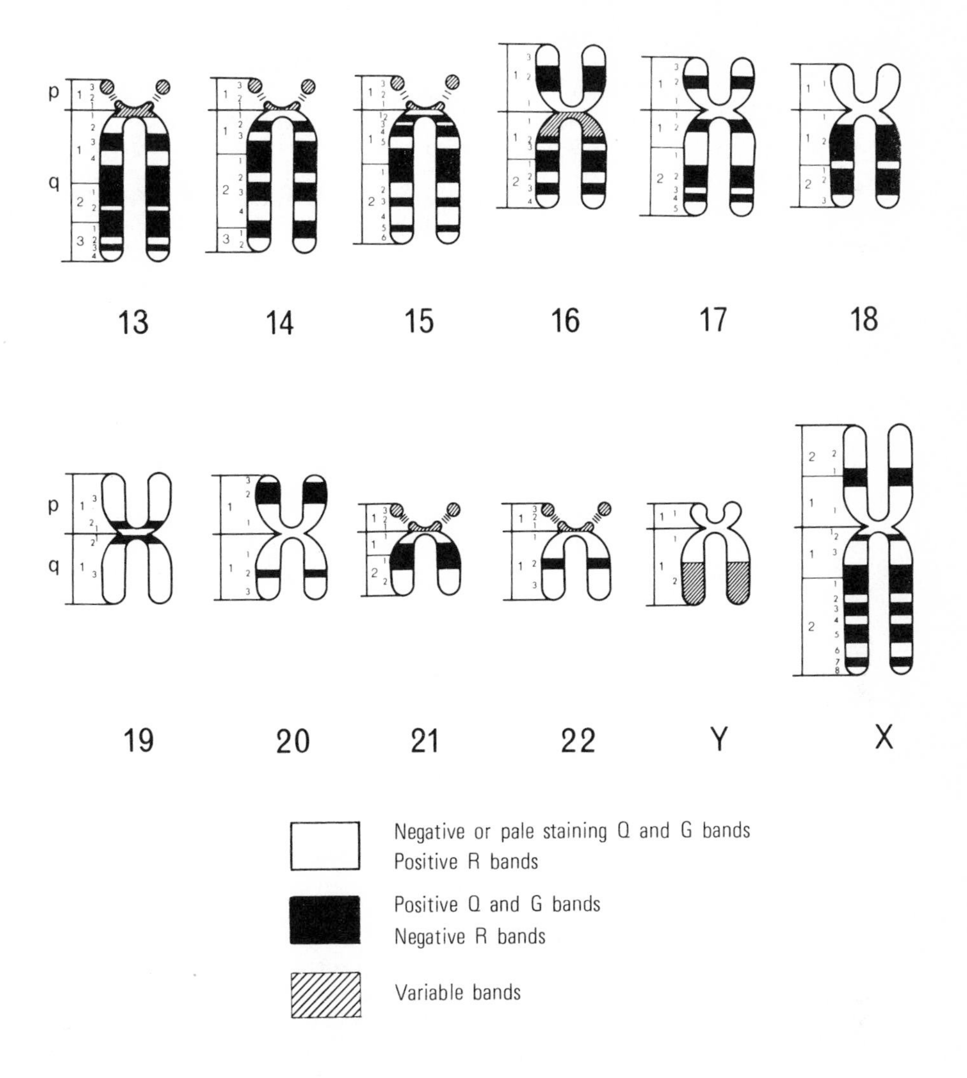

the Y chromosome in a child and his putative father was accepted by the courts as evidence of nonpaternity (Grouchy, 1977). The C-bands also have proved to be remarkable variable. In a study of four persons, seven variant C-bands were observed (Craig-Holmes and Shaw, 1971). In the case of the Yq chromosome arm and the C-bands, the variation appears to be in the heterochromatin, thought to be relatively devoid of genes and thus perhaps more tolerant of variation. Variation in the C-bands is undoubtedly an expression of the same variation that is observed in lateral asymmetry, alluded to above. Secondary constrictions on chromosomes 1, 9, and 16 occur as normal inherited variations and can be treated as dominant traits. Indeed, they have been used successfully in

Table 4-2 **SYMBOLS USED IN DESIGNATING KARYOTYPES***

Symbol	Meaning
A–G	The chromosome groups
1–22	The autosome numbers (Denver system)
X, Y	The sex chromosomes
Diagonal (/)	Separates cell lines in describing mosaicism
Plus sign (+) or minus sign (−)	When placed immediately before the autosome number or group letter designation indicates that the particular chromosome is extra or missing; when placed immediately after the arm or structural designation indicates that the particular arm or structure is larger or smaller than normal
Question mark (?)	Indicates questionable identification of chromosome or chromosome structure
ace	Acentric
cen	Centromere
dic	Dicentric
end	Endoreduplication
h	Secondary constriction or negatively staining region
i	Isochromosome
inv	Inversion
inv(p + q−) or inv(p − q+)	Pericentric inversion
mar	Marker chromosome
mat	Maternal origin
pat	Paternal origin
p	Short arm of chromosome
q	Long arm of chromosome
r	Ring chromosome
s	Satellite
t	Translocation
tri	Tricentric
del	Deletion
dup	Duplication
der	Derivative chromosome
ins	Insertion
inv ins	Inverted insertion
rec	Recombinant chromosome
rcp	Reciprocal translocation (when desired to be more precise than t)
rob	Robertsonian translocation (when desired to be more precise than t)
tan	Tandem translocation (when desired to be more precise than t)
ter	Terminal or end (pter = end of short arm; qter = end of long arm)
ter rea	Terminal rearrangement
dir dup	Direct duplication
inv dup	Inverted duplication
:	Break (no reunion, as in a terminal deletion)
: :	Break and join
→	From–to
;	Separates 2 or more chromosomes that have been altered
mos	Mosaic
chi	Chimera
repeated symbols	Duplicated chromosome structure

* From Chicago Conference (1966) and Paris Conference (1971).

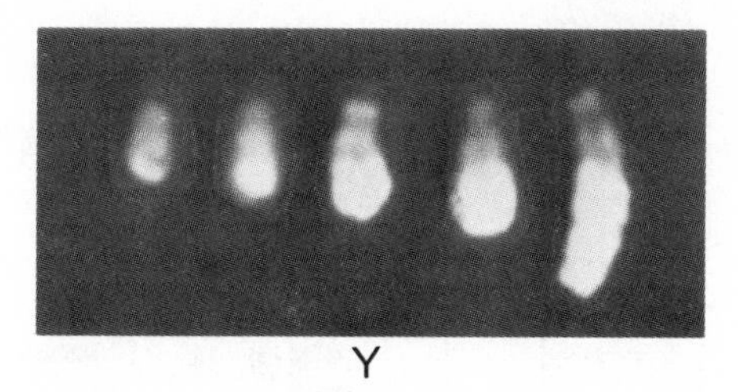

Figure 4-8 Variation in morphology of normal human Y chromosomes. (*From B. Dutrillaux, 1974,* Birth Defects. *Amsterdam: Excerpta Medica.*)

linkage studies. Finally, individual persons and pedigrees are occasionally observed to have a chromosome with distinctly variant morphology unassociated with any phenotypic expression.

EVOLUTION OF CHROMOSOMES

One use of chromosome structure is to study the relationship among species by means of the degree of similarity of their chromosomes. This has been done most effectively in Drosophila, where polytene chromosomes possess very detailed banding patterns. In higher mammals, such studies have received increased attention since the development of the modified Giemsa stain.

The diploid amount of DNA in all mammals is approximately 6 picograms (10^{-12} g) per cell. One might conclude therefore that all mammals have approximately the same number of genes. There is no direct evidence on this point, but it is thought that only a small portion of the DNA actually codes for genes. Therefore, the number of genes in different species could vary without noticeably changing the DNA content.

With the advent of the banding techniques, it has become possible to make gross comparisons of karyotypes from different species. Most interesting for man is the comparison with his closest relatives, the apes. This has been done for the chimpanzee (*Pan troglodytes*), the gorilla (*Gorilla gorilla*), and the orangutan (*Pongo pygmaeus*). In contrast to man, the three apes have 48 chromosomes. However, the similarities in banding patterns are striking.

Figure 4-9 compares the karyotypes of man with the great apes for human chromosomes 1 and 2. The overall homologies are apparent, but some details are consistently different. Most striking in these instances is the homology between the large submetacentric human chromosome 2 and two acrocentric chromosomes in the apes. Apparently the evolution of the human line involved, among other changes, centric fusion of

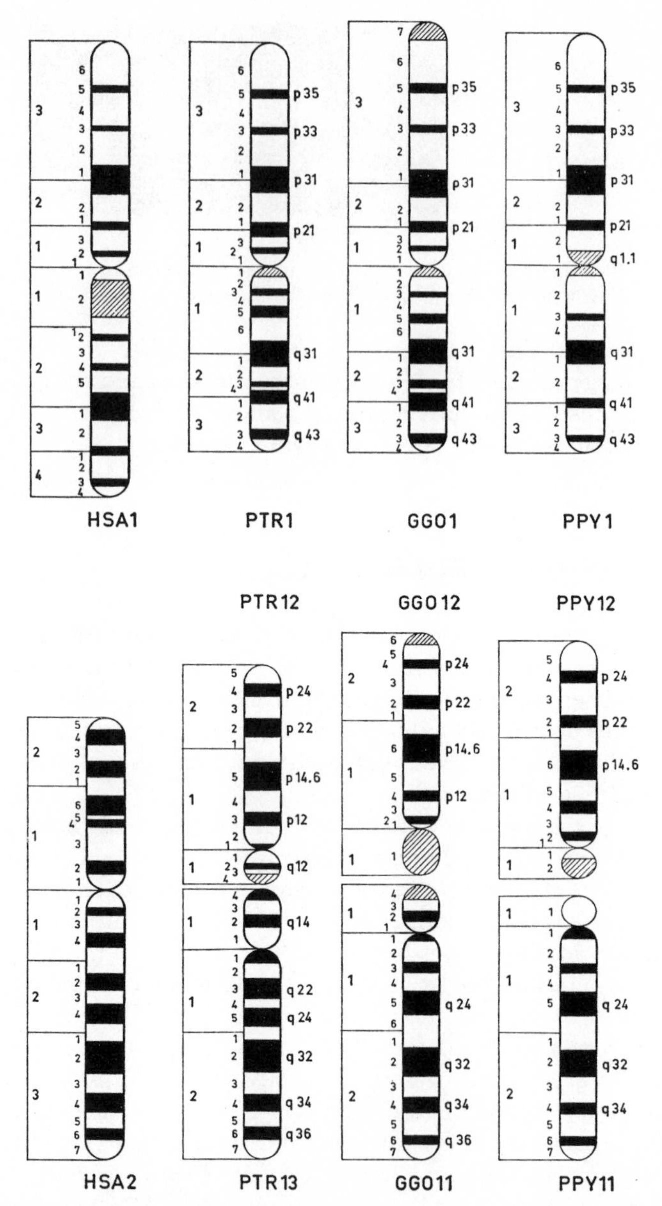

Figure 4-9 Comparison of banding patterns of human chromosomes 1 and 2 (HSA1, HSA2) with homologous chromosomes of *Pan troglodytes* (PTR), *Gorilla gorilla* (GGO), and *Pongo pygmaeus* (PPY). The light bands are Q- and G-negative and R-positive. The dark bands are Q- and G-positive and R-negative. Cross-hatched areas are variable. (*From Paris Conference, 1971, Supplement, 1975; p. 22.*)

two acrocentrics to give a single large chromosome containing the combined genetic content of the long arms of the two acrocentrics. This would account for the reduction in number of chromosomes from 48 in some common human/ape progenitor to 46 in man. Such a process is well known in the evolution of other organisms and is often referred to as a *Robertsonian fusion.* Pericentric inversions also appear to have played an important part in primate divergence. Humans, chimpanzees, and gorillas differ from each other by six to eight pericentric inversions, while the orangutan differs from this group by nine to ten.

This and other cytogenetic differences point out the early separation of the orangutan line from the line that subsequently gave rise to man and the other great apes. The order of separation of man, gorilla, and chimpanzee has not been established. The similarity in magnitude of the differences among these three species suggests that the separation into three stocks occurred within a relatively short period on the evolutionary time scale. The gibbon (*Hylobates lar*), also an ape, has 44 chromosomes that are quite distinct from those of the great apes, indicating a much earlier evolutionary divergence. One proposal for evolution of hominoid chromosomes is given in Figure 4-10.

Formation of new combinations of chromosomes serves as an isolating mechanism in evolution. A zygote rarely is viable unless it possesses a balanced complement of genes. In diploid organisms, this usually means two for each genetic locus. A hybrid zygote from gametes of two different species may develop into a functional adult if the two species are very similar in gene complement. The mule, a hybrid between horse and donkey, is a familiar example. Horses (*Equus ca-*

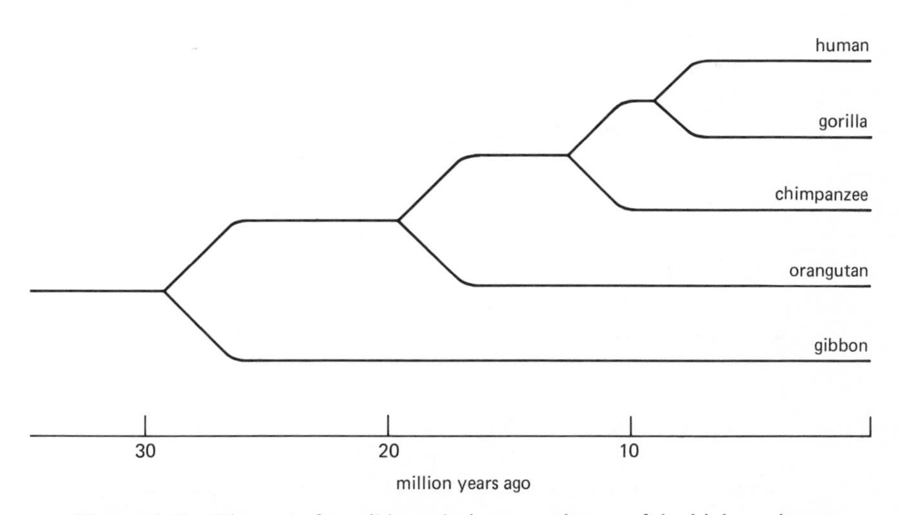

Figure 4-10 Diagram of possible evolutionary pathways of the higher primates, based on chromosome homologies. (*Based on Miller, 1977.*)

ballus) have 64 chromosomes, and donkeys (*E. asinus*) have 62. The gametes have 32 and 31, respectively. A zygote formed from these gametes has 63 chromosomes. But the chromosomes of such hybrids may not be sufficiently homologous to pair successfully at meiosis. As a consequence, separation of chromosomes at anaphase will only rarely and by chance lead to gametes with a normal complement of genes. Interspecific hybrids are thus usually sterile. Considering their karyotypic similarity, one might expect viable but sterile offspring from interspecific matings of the higher primates.

SUMMARY

1. Man has 46 chromosomes, of which 44 are autosomes and 2 are sex chromosomes. The karyotype formula for males is 46,XY and for females is 46,XX.

2. Human metaphase chromosomes are distinguished on the basis of size, location of the centromere, and banding patterns. They are classified into seven major groups (A–G) on the basis of size and location of the centromere. Techniques available before 1970 could not distinguish most of the individual chromosomes.

3. Several procedures have been developed that produce differentially stained bands in metaphase chromosomes. These are mostly variations on use of quinacrine fluorescence (Q-bands), a modified Giemsa stain (G-bands), and heat denaturation (R-bands). In general, Q-bands and G-bands are similar and are the reverse of R-bands. With the banding techniques, all human chromosomes can be recognized. The fluorescence of the Y chromosomes with quinacrine is especially strong and can be seen in interphase nuclei (Y bodies).

4. C-bands (for constitutive heterochromatin) are primarily located in the centromeric regions. C-bands show much stable inherited variations in size and can be used as chromosome markers.

5. The Y chromosome shows the greatest variability in size among normal persons as compared to other chromosomes. The size of the Y can occasionally be used as a dominant marker.

6. Banding techniques applied to great apes and other primates show close homology among man, the gorilla, and the chimpanzee. The human chromosome 2 is represented in the great apes by two acrocentrics that underwent centric fusion in the human line. Pericentric inversions appear to have been important in the evolution of the apes. The orangutan shows many homologies to man and the other great apes but appears to have separated earlier from the primate stock that gave rise to these species. The gibbon has a karyotype very distinct from the great apes, with few obvious karyotypic homologies.

REFERENCES AND SUGGESTED READING

ANGELL, R. R., AND JACOBS, P. A. 1978. Lateral asymmetry in human constitutive heterochromatin: frequency and inheritance. *Am. J. Hum. Genet.* 30: 144–152.

ARRIGHI, F. E., AND HSU, T. C. 1971. Localization of heterochromatin in human chromosomes. *Cytogenetics* 10: 81–86.

CASPERSSON, T., LOMAKKA, G., AND ZECH, L. 1971. 24 fluorescent patterns of human metaphase chromosomes—distinguishing characters and variability. *Hereditas* 67: 89–102.

Chicago Conference: Standardization in Human Cytogenetics. 1966. *Birth Defects: Original Article Series* 2 (No. 2) The National Foundation—March of Dimes, New York. 21 pp.

CRAIG-HOLMES, A., AND SHAW, M. W. 1971. Polymorphism of human constitutive heterochromatin. *Science* 174: 702–704.

Denver Conference. 1960. A proposed standard system of nomenclature of human mitotic chromosomes. *Am. J. Hum. Genet.* 12: 384–388.

DRETS, M. E., AND SHAW, M. W. 1971. Specific banding patterns of human chromosomes. *Proc. Natl. Acad. Sci. (U.S.A.)* 68: 2073–2077.

DUTRILLAUX, B., AND LEJEUNE, J. 1971. Sur une nouvelle technique d'analyse du caryotype humain. *C. R. Acad. Sci.* (Paris) 272: 2638–2640.

DUTRILLAUX, B., AND LEJEUNE, J. 1975. New techniques in the study of human chromosomes: methods and applications. *Adv. Hum. Genet.* 5: 119–156.

EVANS, H. J. 1977. Some facts and fancies relating to chromosome structure in man. *Adv. Hum. Genet.* 8: 347–438.

GROUCHY, J. DE. 1977. Annulation de reconnaissance de paternité fondée sur la non identité des chromosomes Y. *Ann. Génét.* 20: 133–135.

GROUCHY, J. DE, TURLEAU, C., AND FINAZ, C. 1978. Chromosomal phylogeny of the primates. *Annu. Rev. Genet.* 12: 289–328.

GROUCHY, J. DE, TURLEAU, C., ROUBIN, M., AND KLEIN, M. 1972. Evolutions caryotypiques de l'homme et du chimpanzé. Etude comparative des topographies de bandes après dénaturation ménagée. *Ann. Génét.* 15: 79.

JACOBS, P. A. 1977. Human chromosome heteromorphisms (variants). *Prog. Med. Genet.* n.s. 2: 251–274.

HSU, T. C. 1973. Longitudinal differentiation of chromosomes. *Annu. Rev. Genet.* 7: 153–176.

London Conference on The Normal Human Karyotype. 1964. *Am. J. Hum. Genet.* 16: 156–158.

MILLER, D. A. 1977. Evolution of primate chromosomes. *Science* 198: 1116–1124.

MAKINO, S. 1975. *Human Chromosomes.* Tokyo: Igaku Shoin; Amsterdam: North-Holland, New York: American Elsevier, 600 pp.

PAINTER, T. S. 1923. Studies in mammalian spermatogenesis. II. The spermatogenesis of man. *J. Exp. Zool.* 37: 291–335.

Paris Conference (1971): Standardization in Human Cytogenetics. 1972. *Birth Defects: Original Article Series* 8 (no. 7). The National Foundation—March of Dimes, New York. 46 pp.

Paris Conference (1971), Supplement (1975): Standardization in Human Cytogenetics. 1975. *Birth Defects: Original Article Series* 11 (no. 9). The National Foundation, New York. 36 pp.

PATIL, S. R., MERRICK, S., AND LUBS, H. A. 1971. Identification of each human chromosome with a modified Giemsa stain. *Science* 173: 821–833.

PEARSON, P. 1972. The use of new staining techniques for human chromosome identification. *J. Med. Genet.* 9: 264–275.

PRIEST, J. L. 1977. *Medical Cytogenetics and Cell Culture,* 2nd Ed. Philadelphia: Lea & Febiger, 344 pp.

SCHNEDL, W. 1971. Analysis of the human karyotype using a reassociation technique. *Chromosoma* 34: 448–454.

SCHOEPFLIN, G. S., AND CENTERWALL, W. R. 1972. 48,XYYY: a new syndrome? *J. Med. Genet.* 9: 356–360.

SCHWARZACHER, H. G., AND WOLF, U. (Eds.). 1974. *Methods in Human Cytogenetics.* Berlin: Springer-Verlag, 295 pp.

SELLES, W. D., MARIMUTHU, K. M., AND NEURATH, P. W. 1974. Variations in normal human chromosomes. *Humangenetik* 22: 1–5.

SUMNER, A. T., EVANS, H. J., AND BUCKLAND, R. A. 1971. A new technique for distinguishing between human chromosomes. *Nature New Biol.* 232: 31–32.

TJIO, J. H., AND LEVAN, A. 1956. The chromosome number of man. *Hereditas* 42: 1–6.

TRUJILLO, J. M. 1962. Chromosomes of the horse, the donkey, and the mule. *Chromosoma* 13: 243–248.

TURLEAU, C., AND GROUCHY, J. DE. 1973. New observations on the human and chimpanzee karyotypes. *Humangenetik* 20: 151–157.

YUNIS, J. J. (Ed.). 1974. *Human Chromosome Methodology,* 2nd Ed. New York: Academic Press, 377 pp.

YUNIS, J. J. 1976. High resolution of human chromosomes. *Science* 191: 1268–1270.

YUNIS, J. J. (Ed.). 1977. *Molecular Structure of Human Chromosomes.* New York: Academic Press, 336 pp.

REVIEW QUESTIONS

1. Define:

karyotype
metacentric
acrocentric
telocentric
centromeric index
satellite
primary constriction
secondary constriction
Q-bands
C-bands
R-bands
G-bands
heterochromatin
chromosome landmarks
chromosome regions
Y body
mitogen
lateral asymmetry
Robertsonian translocation

2. How can one distinguish between a primary and a secondary constriction?

3. Why might one expect the Y chromosome to consist mostly of heterochromatin?

4. Arrange the following in order of increasing centromeric indices: acrocentric chromosomes, metacentric chromosomes, submetacentric chromosomes, telocentric chromosomes.

5. Would you expect R-bands or Q-bands to be more resistant to heat denaturation?

6. Why are chromosome rearrangements superior to gene mutations as isolating mechanisms in the formation of new species?

CHAPTER FIVE

CHROMOSOMAL ABNORMALITIES: NUMERICAL

Deviations from normal chromosome complements can be grouped into those involving an abnormal number of normal chromosomes and those involving a structural change in one or more chromosomes. Either of these two types of deviation may on occasion lead to the same phenotypic result. However, the mechanisms by which they arise are different and will be considered separately.

During normal cell division, each daughter cell receives an exact complement of chromosomes, either a copy of each chromosome (in mitosis) or one of each pair of chromosomes (in meiosis). Should an accident prevent normal functioning of the spindle apparatus, one or more chromosomes may fail to migrate properly during anaphase. As a result of this *nondisjunction,* one daughter cell receives both chromosomes of a pair and the other daughter cell none. *Chromosome lag* during cell division also causes a chromosome not to be incorporated into the daughter nucleus on occasion. This leads to deficiency of the chromosome in one daughter cell.

These phenomena have been observed in many species, including man. The effects of nondisjunction depend on the chromosomes involved and on the time during development when nondisjunction occurs. If it occurs late in the somatic development of an organism, then only the cells that are direct descendants of the defective mitosis are *aneuploid.* The individual is a *mosaic,* some tissues being *euploid* (with a normal diploid complement of chromosomes), some being *monosomic* (deficient in a chromosome), and some being *trisomic* (with an extra chromosome). If only a small portion of the total somatic tissue is aneuploid, there may be very little or no effect on the development of the individual, and the aneuploidy may go unrecognized. If nondisjunction occurs early in development, the individual may show marked effects of the aneuploidy.

Although both monosomic and trisomic cells are produced by somatic nondisjunction, it should not be assumed that they are equally

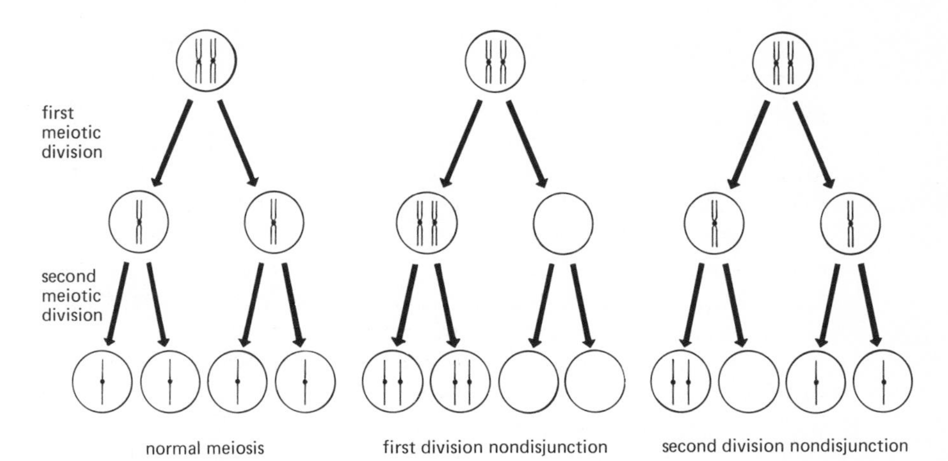

Figure 5-1 Nondisjunction at first and second meiotic divisions, producing gametes with excess chromosomes or deficient in chromosomes. A single homologous pair is shown. Usually the remainder of the chromosomes segregate normally.

represented in the derived tissues. Generally, the two types of cells are not equally viable. Cells seem to tolerate presence of an extra chromosome more readily that absence of one. As a result, monosomic cells often do not survive.

If nondisjunction occurs during meiosis, the gametes are abnormal. If nondisjunction occurs at the first meiotic division, all four products are abnormal, two having an extra chromosome and two being deficient. With nondisjunction occurring at the second meiotic division, only two are abnormal; the other two have a normal haploid complement of chromosomes (Figure 5-1). A zygote derived from an abnormal gamete is aneuploid, and all the cells of the organism derived from that zygote are aneuploid.

Examples of aneuploidy in organisms other than man have long been recognized. In *Datura stramonium,* the Jimson weed, there are 12 pairs of chromosomes. Plants trisomic for each of these chromosomes have been observed, each leading to characteristic morphological changes. In some organisms, such as *Drosophila,* trisomics are known for some chromosomes but are nonviable in the case of the other chromosomes.

ABNORMALITIES OF AUTOSOMES

Down's Syndrome

The first example of trisomy in man was discovered in 1959 by Lejeune and his associates. Their paper stands as one of the classics in human genetics, in that it opened the area of human cytogenetics to intensive investigation.

Clinical Expression. The condition known as Down's syndrome (mongolian idiocy; trisomy 21 syndrome) had long been an enigma. Approximately one in every 600 to 700 births is a child with Down's syndrome, and these comprise some 5 to 15% of institutionalized mental defectives. A variety of stigmata in addition to mental deficiency characterize the disease, including such features as short stature; short, broad fingers and toes; a round face with epicanthal folds and a long protruding tongue; and unusual finger and palm prints, including a single transverse palmar crease (simian crease). Most patients do not have all these features. Furthermore, the features occur occasionally in individuals who do not have Down's syndrome. Nevertheless, an experienced observer can usually recognize such a patient rather easily because of the typical appearance (Figure 5-2).

É. Séguin in 1844 was apparently the first to give a detailed description of the disease. Langdon Down in 1866 considered these patients to resemble members of the Mongolian race, and he interpreted the condition as an atavism toward a more primitive Mongoloid type of individual. The terms mongolism and mongolian idiocy were applied to the disease and are still widely used.

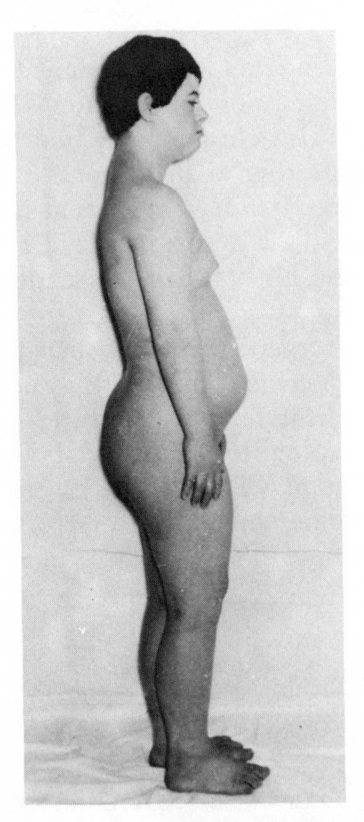
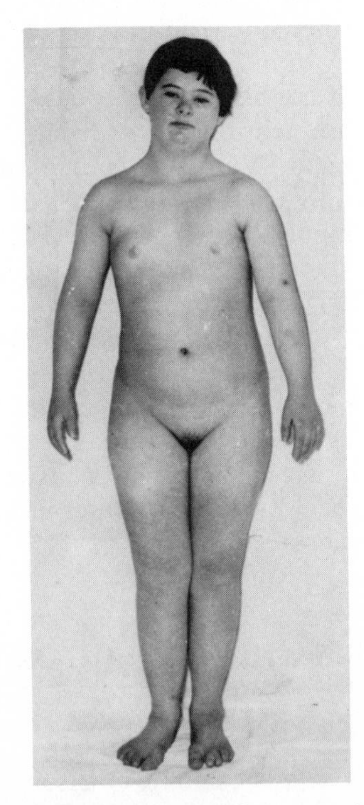

Figure 5-2a A patient with Down's syndrome showing many features typical of this condition. Note particularly the face. (*Courtesy J. Lejeune.*)

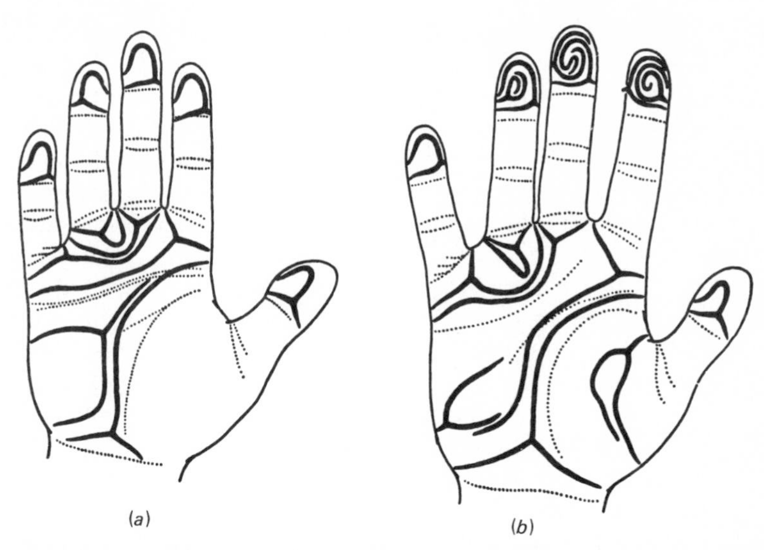

Figure 5-2b (*a*) Palm of Down's syndrome patient showing (1) triradius near center, (2) absence of pattern in thenar area, (3) digital loop between digits III and IV, (4) radial loop on tip of digits IV, (5) ulnar loops on all other digits. (*b*) Palm of normal person showing features rare in Down's syndrome: (1) triradius at ulnar edge of hypothenar region, (2) pattern in thenar area, (3) numerous whorls on tips of digits. (*From Penrose, 1961.*)

Etiology of Down's Syndrome. Many investigators attempted to discover the cause of Down's syndrome but with no success. Because it is congenital, a hereditary basis was suggested. Supporting a genetic etiology was the observation that identical twins (who have identical genetic makeup) are nearly always alike in being either both affected or both nonaffected. On the other hand, nonidentical twins are rarely both affected, and more than one affected person in a sibship is uncommon. This would rule out any simple genetic mechanism as a cause.

The risk of having a child with Down's syndrome increases with age of the mother. The incidence of affected offspring is plotted against maternal age in Figure 5-3. This led to the hypothesis, now ruled out, that older women are less able to provide a prenatal environment suitable for normal development of embryos.

As early as 1932, Waardenberg suggested that Down's syndrome might be the consequence of a chromosomal aberration, but the techniques then available were not sufficiently sensitive to reveal a departure from normal. When Lejeune examined the chromosome complement with newer techniques, he found that persons with Down's syndrome consistently had 47 chromosomes. The extra chromosome was one of the small acrocentrics in group G (21 or 22). This observation was quickly confirmed in many laboratories. Until the development of the fluorescent technique in 1971, it was not possible to distinguish

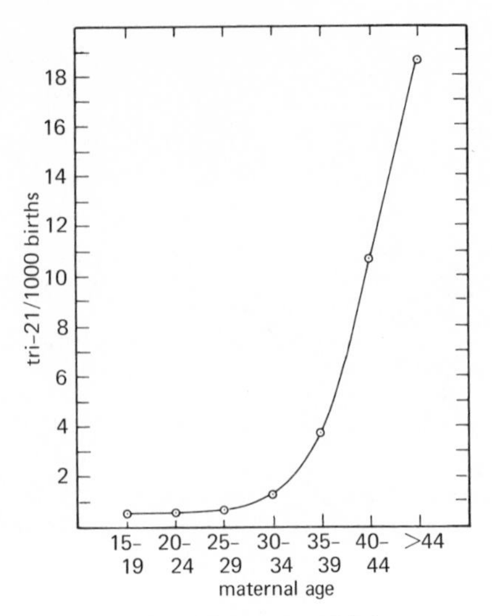

Figure 5-3 The incidence of Down's syndrome births versus maternal age, showing the much greater risk of nondisjunction among older mothers. (*Data from C. O. Carter and K. A. Evans,* Lancet *2:785–797.*)

21 from 22 with certainty, although some of the early evidence was thought to favor 21 as the trisomic chromosome in Down's syndrome. By general agreement, the chromosome trisomic in Down's syndrome was assigned number 21, and the condition is therefore also known as trisomy 21 syndrome. Chromosomes 21 and 22 are readily distinguished by the modified Giemsa stain now in use. Figure 5-4 is a karyotype of such a patient. In current nomenclature, the karyotype formula is 47, + 21 (see Table 4-2).

The recognition that Down's syndrome results from an autosomal trisomy has led to the understanding of many of the features of the disease. For example, identical twins would both be affected if the trisomy were present in the original zygote before it separated into two individuals. But since fraternal twins arise from entirely different zygotes, the likelihood of trisomy occurring independently in both zygotes is very small.

Reproduction in Down's Syndrome. A person who is trisomic for an autosome should form two types of gametes in equal numbers—those that are normal and those with an extra chromosome. Their offspring should therefore consist of both normal and trisomic individuals. In the case of Down's syndrome, fertility is greatly reduced and males ordinarily are infertile. A few females have produced offspring, however. The offspring consist equally of normal children and children affected with Down's syndrome. Such a 1 : 1 ratio would be expected only if the abnormal gametes and zygotes were of normal viability. The transmission of an extra chromosome from a trisomic parent is referred to as *secondary nondisjunction.*

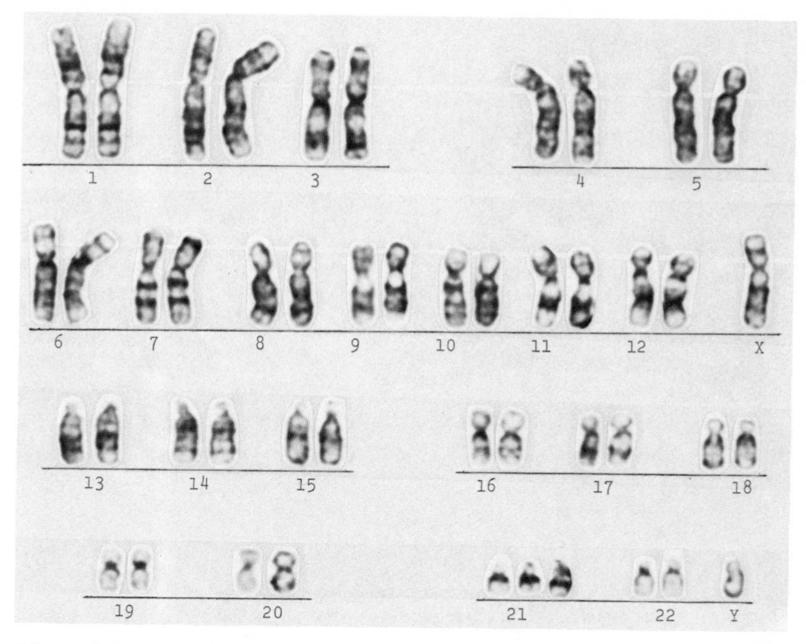

Figure 5-4 Karyotype of a boy with Down's syndrome showing 47,XY,+21. The arrow points to the extra chromosome. (*Furnished by Patricia N. Howard-Peebles.*)

Meiotic studies in Down's syndrome have not often been possible since males generally are sterile and meiotic studies in females are done only if an ovary is being removed. Occasionally a male is identified in whom spermatogenesis does occur, and testicular biopsy has permitted study of the behavior of the three homologous chromosomes in meiosis. In some cells, the three number 21 chromosomes join to form a trivalent. In a somewhat larger fraction, the three chromosomes form a bivalent plus a univalent. Figure 5-5 shows some of the meiotic figures seen in these persons.

Phenotypic Effects of Trisomy. Trisomic conditions are characterized by multiple defects involving many organs. In a normal person, the complement of genes is balanced. An extra chromosome means that a large number of genes are present three times rather than the balanced two. Since the genes on a particular chromosome may be quite unrelated in their primary action, trisomy would disturb many apparently unrelated functions. But any affected person may deviate from the group characteristics because of his particular combination of genes. It is probably safe to state that all trisomic conditions show multiple, seemingly unrelated effects, but it cannot be said that all conditions with multiple effects result from chromosomal abnormalities. This is because a single primary defect may secondarily influence many other physiological reactions. Unless the primary defect has been recog-

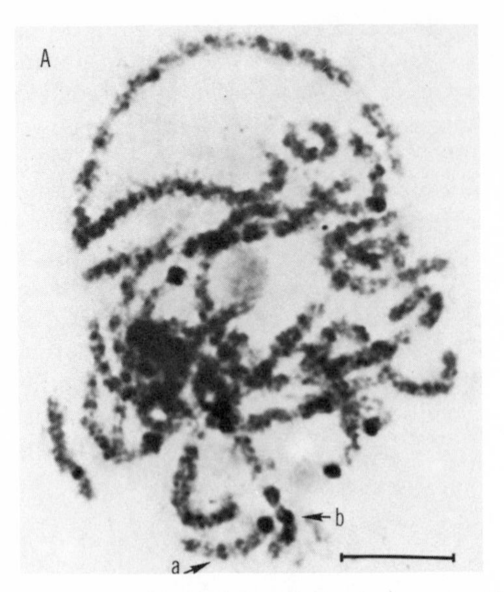

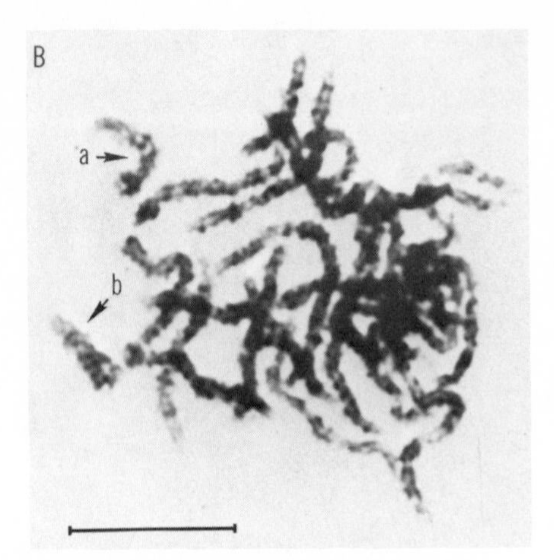

Figure 5-5 Primary spermatocytes from a male mosaic for trisomy 21. The chromosomes are in the pachytene stage. The longer group G chromosome pair (number 22) is indicated by arrow *a* and the shorter (number 21) by arrow *b*, In *A*, number 21 is present as a bivalent and in *B* as a trivalent. In other studies, trisomic cells have also been observed with a bivalent plus a univalent. The scale indicates 10 μ. (*From Hungerford and co-workers, 1970; the subject was a patient of W. J. Mellman.*)

nized, it may not be possible to relate the secondary defects in a meaningful way. They might then appear to be unrelated when in fact it is our ignorance that makes them seem so.

It is probably not by chance that the most prevalent autosomal trisomic condition involves one of the two smallest chromosomes. If it is assumed that the size of a chromosome is related to the number of genes located on it, then fewer genes would be unbalanced in trisomy of a small chromosome compared with trisomy of a large chromosome. To be sure, a single unbalanced locus might have such devastating effects as to be incompatible with life, regardless of the chromosome in which it is located. But the imbalance of many loci is apt to have more profound effects than imbalance in a few. In fact, only a few trisomic conditions have been found in liveborn children, in spite of an intensive search. None involves the larger chromosomes.

A single example has been found of Down's syndrome in a chimpanzee. As in human cases, the animal was trisomic for the small chromosome homologous to human chromosome 21, and the phenotypic effects were also similar, confirming the observation of homologous gene arrangement in hominoid lines.

Mosaicism in Down's Syndrome. Postzygotic nondisjunction should lead to two cell lines, or *clones*, differing in chromosomal constitution in the same individual. Many examples of such chromosomal *mosaicism* are known. Numerous persons who are mosaic for normal and tri-21 cells have been observed, with a karyotype formula of 46/47,+21. These persons characteristically show only mild expression of the stigmata of Down's syndrome. Without karyotype analysis, diagnosis would be very uncertain. It is possible that many persons are mosaic for trisomy 21 without expressing any significant clinical signs. Such persons might have only a small portion of their cells trisomic. If this small portion included gonadal tissue, the risk of producing a trisomic offspring could be high. A few families have been observed in which several trisomy 21 offspring were produced by seemingly normal parents. Whether these families have a high tendency to nondisjunction or whether one of the parents is mosaic for the trisomic condition has not been ascertained.

Other Autosomal Trisomies

Studies of spontaneous abortuses indicate that all chromosomes may be involved in trisomy. Most are incompatible with development through the fetal period and are not seen in live births. The following are occasionally compatible with survival to birth and beyond.

Trisomy 8. Several dozen persons have been identified as trisomy 8, although the majority are mosaic normal/trisomy 8. Indeed, since the exclusion of mosaicism requires examination of several different tissues, a procedure not carried out on many of the "nonmosaic" patients, it is possible that all are in fact mosaic. Children with trisomy 8 have a large skull with prominent forehead and frequently hypertelorism. The body is slender with occasional skeletal defects. There are flexion deformities on the fingers, with characteristic deep furrows in the palms and soles of infants (Figure 5-6). Most are mentally retarded. The lifespan is in the normal range. Parental age is slightly increased, the significance of which is not clear if most cases arise postzygotically to produce mosaics. It is possible, of course, that the zygote was trisomic 8, with loss of a chromosomes 8 postzygotically to produce a more balanced euploid clone.

Trisomy 9. Only five liveborn cases have been reported, of which four (and perhaps all) were mosaic. Only one survived beyond early infancy, dying at 9 years of age. He was mentally retarded. Patients have had microcephaly, hypertelorism, and multiple, severe deformities.

Trisomy 13 (Patau's Syndrome; Trisomy D_1). The first infant with this disorder was described in 1960 by Patau and co-workers. With the availability of karyotyping, a large number of additional cases have

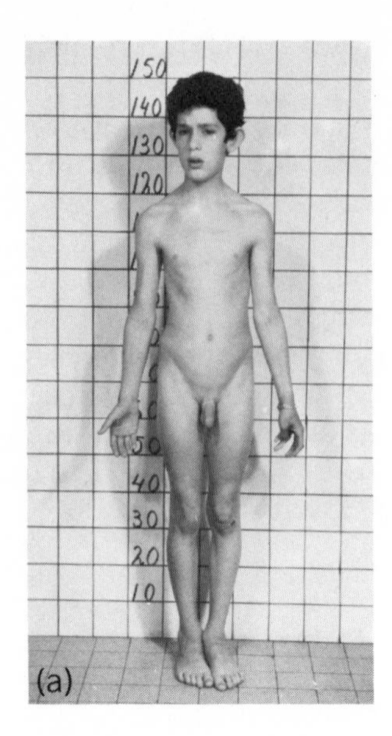

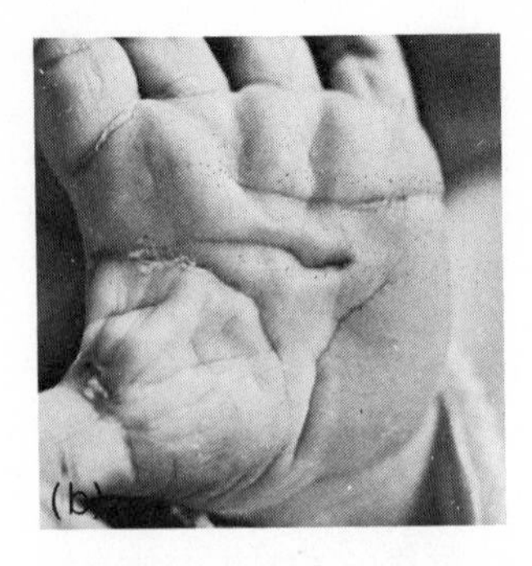

Figure 5-6 Examples of partial trisomy 8. (*a*) A young boy with trisomy 8. (*b*) The palm of another patient with trisomy 8, showing the deep creases often observed in this syndrome. (*Fig. 5-6a furnished by J. de Grouchy; Fig. 5-6b furnished by M. Rethoré.*)

been identified. All patients have markedly retarded mental and physical development. Other features found in most patients are hypertelorism, various eye defects (including anophthalmia), cleft palate and cleft lip, polydactyly, vascular nevi, micrognathia, and low-set, malformed ears. A variety of defects of internal organs is found. Dermatoglyphic features include an increased frequency of digital arches, a distal axial triradius, and frequently simian flexion creases of the palm. A more complete listing of abnormal findings is given by Levine (1971). Survival is very limited, approximately half the cases dying by one month of age. Rare instances of survival beyond three years have been reported, however, with one patient still alive at ten years.

The incidence of trisomy 13 is not accurately known and depends in part on the age structure of the parental population. Most studies give estimates of 1 in 4000 to 1 in 10,000, with the higher frequencies more common. There appears to be a slight excess of females detected, possibly due to higher prenatal mortality of affected males. It should be remembered that those detected as liveborn infants represent only an unknown portion of the zygotes that start out with trisomy 13. As with trisomy 21, the risk is higher for older mothers.

Trisomy 18 (Edwards' Syndrome). The first example of this condition was described in 1960 by Edwards and co-workers. Many others have since been described. Features include severe mental retardation; hypertonicity; low-set, malformed ears; micrognathia; congenital heart defects; and rocker-bottom feet. Most have arches on six or more fingers, and there is often a single flexion crease on one or more fingers. So high a frequency of arches is not known for any other condition. Levine (1971) provides a comprehensive list of defects that have been observed in trisomy 18. A photograph of the head and hand of a patient is shown in Figure 5-7. Compared with male births, more than twice as many trisomy 18 females are born. This is thought to reflect a greater male mortality in the fetal period. Fifty percent of the patients die by two months of age, and only a few have been known to survive beyond several years of age. Estimates of the frequency of trisomy 18 have varied from 0.2 to 2 per thousand live births. The lower figure probably is closer to the true value. Again it should be noted that many of the affected fetuses may be aborted and that liveborn frequencies represent only a portion of trisomic 18 zygotes. Older mothers have an increased risk of trisomy 18 offspring.

Trisomy 22. Several reports of patients with trisomy 22 were made prior to the discovery of banding techniques, but most of these now appear not to have had trisomy 22. The identification was made on the basis of an extra G group chromosome without stigmata of Down's syndrome. With the advent of banding techniques, true trisomy 22 could be positively identified, and several dozen examples are now known. Frequent features are mental and growth retardation, microcephaly, low-set malformed ears, congenital heart disease, and cleft palate. Survival appears to be good. Parental age is increased somewhat. Mosaicism has not been reported in affected children, but one pheno-

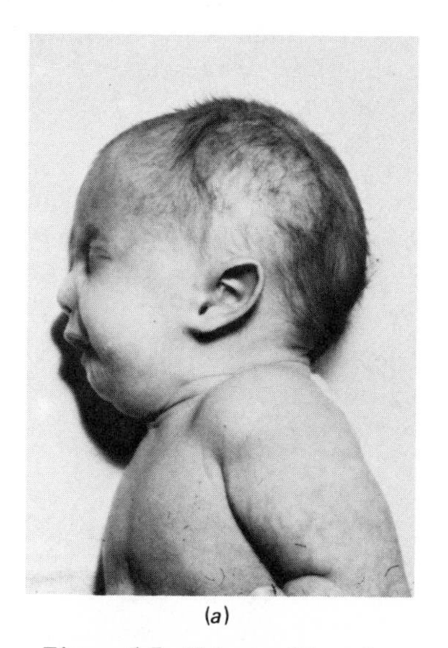

(a)

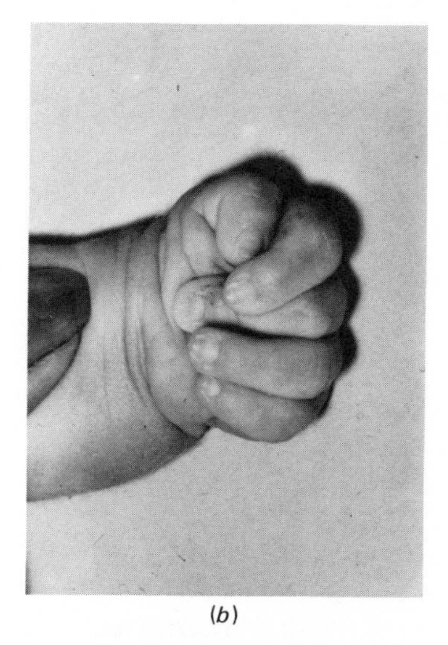

(b)

Figure 5-7 Trisomy 18 syndrome. (*a*) A patient, showing characteristic facial structure and low-set ears. (*b*) Unusual hand position found in trisomy 18 syndrome (*Courtesy J. Lejeune.*)

typically normal mother who produced two trisomy 22 offspring was mosaic 46,XX/47,XX,+22 (Uchida et al., 1968).

Lethal Trisomies. Only the above trisomies appear to be viable at birth. There is ample evidence from abortuses that nondisjunction may involve any chromosome, but most are lethal in the embryonic or fetal stage. Even the above viable trisomies have a greatly increased risk of abortion, this apparently being one reason for the rarity of trisomy 22 among newborns as compared to trisomy 21, which is more likely to survive the fetal period.

Monosomy of Autosomes. Nondisjunction should give rise to two types of abnormal zygotes—trisomic and monosomic. It appears, though, that an extra chromosome can be tolerated much more easily than a missing chromosome. This would be due not only to the dosage effect of normal alleles but also to the fact that all recessive genes on the one chromosome present would be expressed. Many of the recessive genes are likely to be lethal when hemizygous (present only a single time in the gene complement). No clearly valid example of autosomal monosomy is known in man. Two cases of apparent monosomy for a group G chromosome have been reported: One involved multiple severe defects, but the other was only mildly abnormal.

The difficulties in accepting these as monosomies are (1) the possibility of undetected mosaicism involving euploid cells and (2) the possibility that most of the missing G group chromosome is translocated to a larger autosome. In the latter case, the translocation might go undetected with techniques in use at the time of the reports. The newer banding techniques should make the possibility of undetected translocation less likely in future studies.

ABNORMALITIES OF SEX CHROMOSOMES

Among the numerous congenital defects in which abnormal karyotypes might be expected are those involving aberrant sexual development. Of the types of sexual abnormalities, two were outstanding as candidates for sex chromosome studies. These were Klinefelter's syndrome and Turner's syndrome. Both of these conditions are characterized by a *sex chromatin* status opposite to that of the external sex.

Sex Chromatin

In 1949, Barr and Bertram reported a variation in the appearance of the interphase nuclei, a difference that was found to be associated with sex.

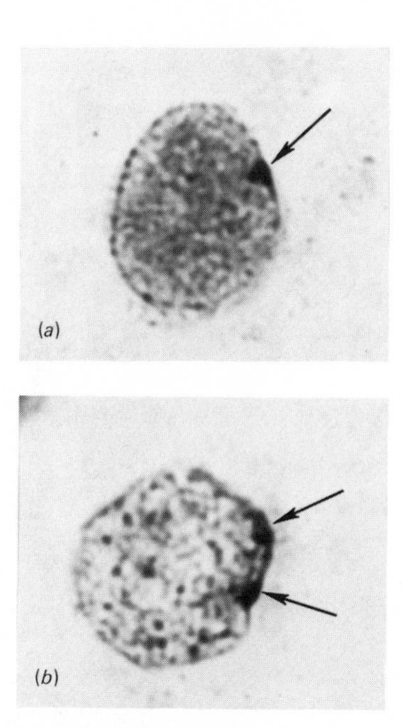

Figure 5-8 Interphase nuclei showing sex chromatin bodies. (*a*) A single sex chromatin body, typical of cells with two X chromosomes. (*b*) Two sex chromatin bodies, characteristic of cells with three X chromosomes. (*Courtesy J. de Grouchy.*)

Most of the nuclei of females have a deeply staining body attached to the inside of the nuclear membrane (Figure 5-8). The nuclei of males lack this body, known as the *sex chromatin* (*Barr body or X-body*). Persons are said to be sex chromatin positive if some 30 percent or more of their cells in a buccal smear possess the body; they are sex chromatin negative if only an occasional cell appears to possess the structure. Normally, females are positive and males are negative. The sex chromatin body is now known to be the inactive X chromosome (Chapter 3).

Syndromes Involving Sex Chromosome Aneuploidy

Klinefelter's Syndrome. Persons with this condition are phenotypic males, being somewhat eunuchoid, however (Figure 5-9). Their limbs are longer than average, genitalia are underdeveloped, and body hair is sparse. The majority of patients are of diminished intelligence, and many are placed in institutions for the mentally defective. The condition occurs approximately once in every 600 male births and accounts for some 1% of male patients in such institutions. Many patients with apparently normal physical development and normal intelligence have been described. Such patients are nearly always sterile, however.

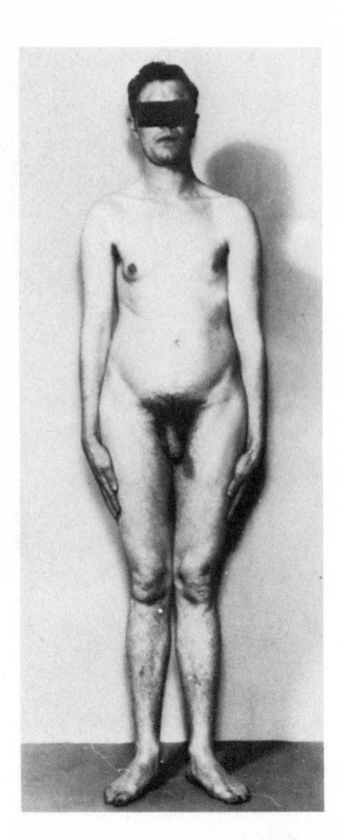

Figure 5-9 A patient with Klinefelter's syndrome (47,XXY). The patient has typical eunuchoid features. (*From M. A. Ferguson-Smith, 1973. Intersexual states and allied conditions. In A. Sorsby, Ed.,* Clinical Genetics, *2d ed. London: Butterworth.*)

Most Klinefelter patients are sex chromatin positive, suggesting an XX constitution. In 1959, shortly after the first report on trisomy 21 syndrome, Jacobs and Strong published the observations that these patients have 47 chromosomes, including a normal Y chromosome. Many additional studies have confirmed that the Klinefelter's karyotype is 47,XXY. The condition therefore is another example of nondisjunction. Like the autosomal trisomies, the risk of having a child with Klinefelter's syndrome is higher in older mothers.

Turner's Syndrome. Concurrent with the studies of Klinefelter patients were studies of patients with Turner's syndrome (Ford and co-workers, 1959). These persons occur with a frequency of approximately 1 in 2000 female births. They are females phenotypically, but ovaries are not properly developed: hence, the frequently used term *ovarian* or *gonadal dysgenesis.* Individuals with this condition are characterized by short stature, pronounced webbing of the neck, infantile genitalia, and lack of development of secondary sexual characteristics (Figure 5-10). Although their mental development is usually within the normal range, they often appear to have difficulty in certain types of

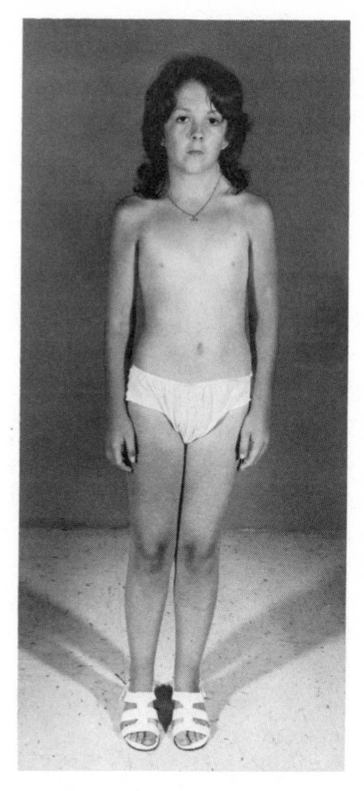

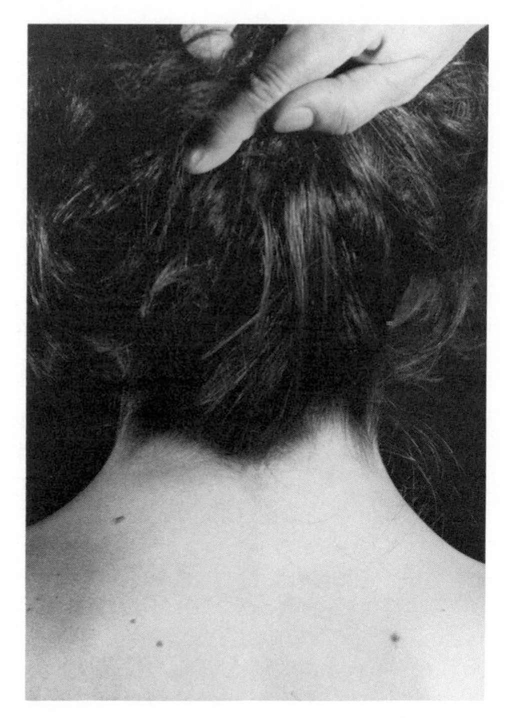

Figure 5-10 A patient with Turner's syndrome (45,XO). (*Furnished by J. L. Simpson.*)

abstractions as, for example, the perception of space (Money and Alexander, 1966).

Patients with Turner's syndrome are sex chromatin negative, suggesting an abnormal sex chromosome complement. It was earlier thought they might result from XY zygotes that appeared as females through some developmental error. Consistent with this was the observation that the frequencies of X-linked recessive traits, such as colorblindness, were the same as in the male population rather than in the female population. However, karyotype analysis revealed a complement of 45 chromosomes, with only one X and no Y. Therefore, the karyotype formula is 45,XO, where O indicates absence of the other member of the chromosome pair. With only one X chromosome, Turner's patients are hemizygous for X-linked genes, as are males. This accounts for the high frequencies of colorblindness and other X-linked traits among them as compared with 46,XX females.

Although XO zygotes can arise from meiotic nondisjunction, most appear to result from postzygotic errors in chromosome distribution. As would be expected from a postzygotic event, mosaicism involving XX and XO tissue or other combinations is very frequent in these patients. In contrast to trisomies, there is no effect of maternal age on the incidence of Turner's syndrome.

The Triplo-X Syndrome. The second trisomy involving sex chromosomes was discovered in a female of decreased intelligence who was also sterile (Jacobs et al., 1959). She had 47 chromosomes, with only four small acrocentrics but with an extra chromosome of morphology similar to an X. That it was indeed an X was suggested by the presence of *two* sex chromatin bodies in the nuclei of her cells. Her karyotype was 47,XXX. Figure 5-11 shows a cell from a triplo-X female labeled late in the S-period with ^{3}H-thymidine.

The frequency of triplo-X females is approximately 1 per 1600 female births. The frequency of triplo-X females among mental defectives and among psychotic patients is slightly greater, being three to four per thousand. However, many triplo-X females are of normal intelligence and fertility. The mean age of parents of triplo-X females is greater than for the general population, as in Down's syndrome.

An unexpected finding in the case of fertile triplo-X females is the absence of aneuploid offspring. One would expect that, as in the case of trisomy 21, the three X chromosomes would segregate to produce equal numbers of secondary oöcytes with one and two X chromosomes.

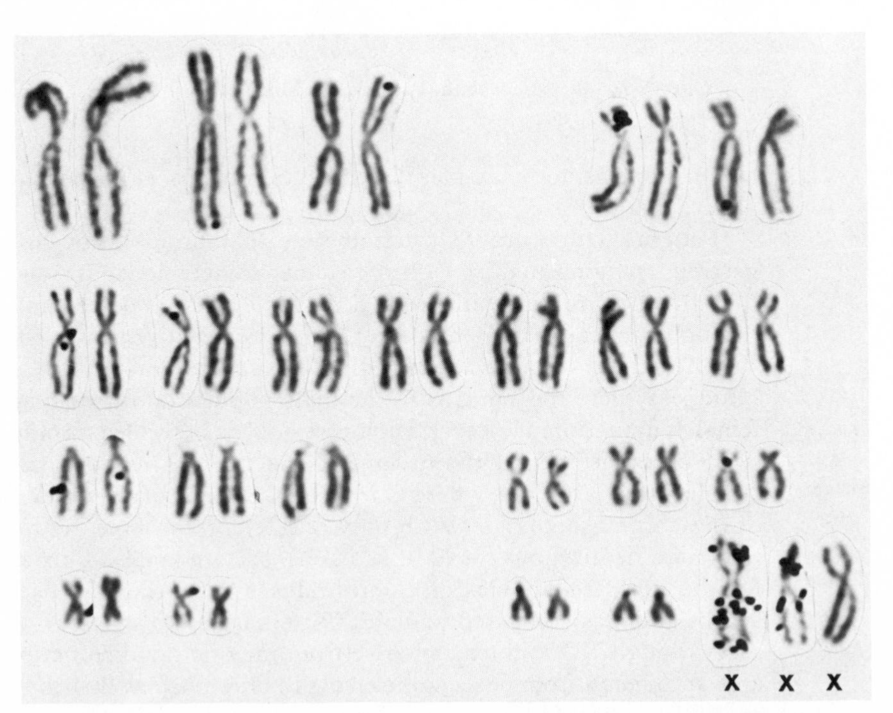

Figure 5-11 Karyotype of a 47,XXX female labeled with ^{3}H-thymidine late in the S period. Only two of the three X chromosomes are appreciably labeled, indicating that they replicated late and are "inactive." The chromosomes are stained with the standard Giemsa stain, which does not show bands. (*Furnished by Margery W. Shaw.*)

Those with two, when fertilized, should give rise to offspring with either 47,XXX or 47,XXY karyotypes. However, this is not the case. There is at most a small increased risk of such offspring from triplo-X mothers. It is possible that in XXX oöcytes, segregation products with two X chromosomes regularly go into the polar body, leaving the ovum to develop from the normal haploid set.

The XYY Syndrome. Persons with 47,XYY constitution have been observed on many occasions in apparently normal males. Attention was focused on this genotype in 1965 when Jacobs and co-workers reported that the frequency of 47,XYY persons was much higher among inmates of a maximum security prison than in the general population. This was especially true among tall inmates.

Karyotype studies in randomly chosen newborns gives a frequency of 47,XYY of approximately one per thousand males. The frequencies in surveys of adult males in the general population are approximately the same. Hook (1973, 1978) has reviewed the findings of a number of penal, mental-penal, and mental institutions and obtained overall frequencies among inmates not selected for increased height of 2.9 per thousand in mental institutions, 4.4 per thousand in penal institutions, and 20 per thousand in mental-penal institutions. This is greater than the newborn frequency or the normal adult frequency. Whatever the chain of events leading to imprisonment or other restraint, 47,XYY males clearly suffer a higher risk.

The psychological basis of the increased risk of antisocial behavior is obscure (reviewed in Borgaonkar and Shah, 1974). Many have lower intelligence than the general population. Since 47,XYY males are taller than their 46,XY counterparts, it has been suggested that increased size alone tends to induce aggressive behavior in response to the expectations and treatment from society. On the other hand, this does not appear to happen in 46,XY males who are equally tall. A comparison of the sibs of institutionalized XYY and XY males gives lower frequencies of antisocial behavior among sibs of XYY as compared with those of XY, suggesting that the environment may contribute to the problems of XYY males less than does their biology. Attempts to characterize XYY males psychologically have not been very helpful. The aggressive behavior that they display has been described as heavily impulsive. XYY males who get into trouble tend to do so much earlier than XY males, the average age in one study being 13 years as compared with 18 years.

Meiotic nondisjunction as a cause of 47,XYY necessarily occurs in the second division of the fathers. Mitotic nondisjunction in the early postzygotic stages of development could give rise to mosaic 45,XO/47,XYY along with 46,XY cells. The 47,XYY cells, if they became the predominant type, might be responsible for a few cases classified simply as 47,XYY. The paternal age does not appear to be significantly increased.

XYY males are fertile, and one might expect segregation to produce a high frequency of abnormal offspring. With rare exceptions, all offspring have had normal chromosome constitutions, either XX or XY, indicating that a mechanism operates against the incorporation of an extra Y into functional sperm. Meiotic studies in XYY males indicate that ordinarily the two Y chromosomes form a bivalent in prophase I. In diakinesis, association occurs between the short arms, but chiasma formation between the long arms is rare (Hultén and Pearson, 1971).

Several patients have been described who were 48,XYYY. In addition to having average or above normal height, they exhibited the aggressiveness of the XYY syndrome and had clearly diminished intelligence.

Other Numerical Abnormalities of Sex Chromosomes. Simple nondisjunction in females, either at first or second meiotic division, can give rise to gametes with two X chromosomes or with none. In males, the potential gametes are XX, XY, YY, or nullisomic for sex chromosomes. Combinations of these abnormal gametes with normal account for the conditions already described. However, persons with additional combinations of sex chromosomes are occasionally noted, and these can be accounted for by two nondisjunctional events, either in tandem in one parent or one in each parent. Nondisjunction at *both* first and second divisions can produce the additional combinations XXY, XYY, and XXYY in sperm and XXX and XXXX in ova. The additional zygotic combinations produced by uniting these abnormal gametes with normal have all been observed, although they are rare.

Several tetra-X females (48,XXXX) have been observed. The principal characteristic shared by these was mental deficiency. Sexual development was essentially normal. Rare patients have also been described who were penta-X (49,XXXXX). They were mentally retarded.

A number of patients of karyotype 48,XXXY have been observed. These closely resemble Klinefelter's syndrome patients, but they have a more consistent and greater degree of mental deficiency. Testicular development closely resembles 47,XXY Klinefelter's syndrome. A substantial number of patients with 49,XXXXY karyotypes have been reported. These have severe mental retardation, marked underdevelopment of testes and other sexual characteristics, and skeletal defects.

Many patients have been described with 48,XXYY karyotypes. In general, these resemble 47,XXY Klinefelter's syndrome patients, but they tend to be taller, as one might expect with an extra Y chromosome. They share some of the features of XYY males, being more often mentally retarded, and are unduly represented in institutions for persons with criminally agressive behavior. Only one patient has been observed with a 49,XXXYY karyotype. He was mentally defective, tall, and very underdeveloped sexually.

In theory, additional combinations of sex chromosomes might be observed if nondisjunction occurred in both parents, including both divisions of meiosis in at least one parent. Such combinations must be exceedingly rare, and the zygote, if viable, would develop abnormally.

ORIGINS OF NONDISJUNCTION

As exemplified by the sex chromosome anomalies, nondisjunction may occur either at first or second meiotic division or at both divisions and in either parent. It is sometimes possible to locate the point at which nondisjunction occurred by examining the consequences. For example, a zygote with two Y chromosomes indicates nondisjunction in the second meiotic division of the father, this being the only way for two Y chromosomes to get into the same zygote. One would not expect a maternal age effect in the occurrence of 47,XYY offspring, and such is the case. Other sex chromosome trisomies are ambiguous in their origins generally, but the maternal age effect suggests oögenesis as a common site of nondisjunction.

In a few favorable cases, it is possible to use genetic markers to trace the origin of chromosomes. In the example shown in Figure 5-12, the nondisjunction must have occurred in the mother. It occurred at the second meiotic division if crossing over between the colorblind locus and the centromere did not occur, or it could have occurred at the first division if such crossing-over did occur.

The X-linked dominant blood group Xg also has been useful in tracing nondisjunction of the X chromosomes. Race and Sanger (1969) reported three families in which the father was Xg(a+), the mother Xg(a−), and the 47,XXY offspring Xg(a+), which could only happen if at least one X and the Y chromosome were paternal. Further, the nondisjunction must have happened in the first division. Analysis of their entire data by these authors indicated that 40% of the Klinefelter's cases are $X^{M}X^{P}Y$ (where X^{M} is a maternally derived and X^{P} a paternally derived X chromosome), 50% are $X^{M1}X^{M2}Y$ (X^{M1} and X^{M2} are different maternal X chromosomes), and 10% are $X^{M1}X^{M1}Y$ or $X^{M2}X^{M2}Y$. Similar studies of three informative 49,XXXXY patients revealed that all four X chromosomes were maternal, requiring consecutive nondisjunction in both first and second meiotic divisions. Similarly, in two informative 48,XXYY patients, one X chromosome was paternal, indicating consecutive nondisjunction in the father.

The existence of families with a trisomy 21 child who also are segregating a morphological variant of chromosome 21 has enabled investigators to pinpoint the occurrence of the nondisjunction in many such cases (Figure 5-13). Of 48 informative families studied through 1976,

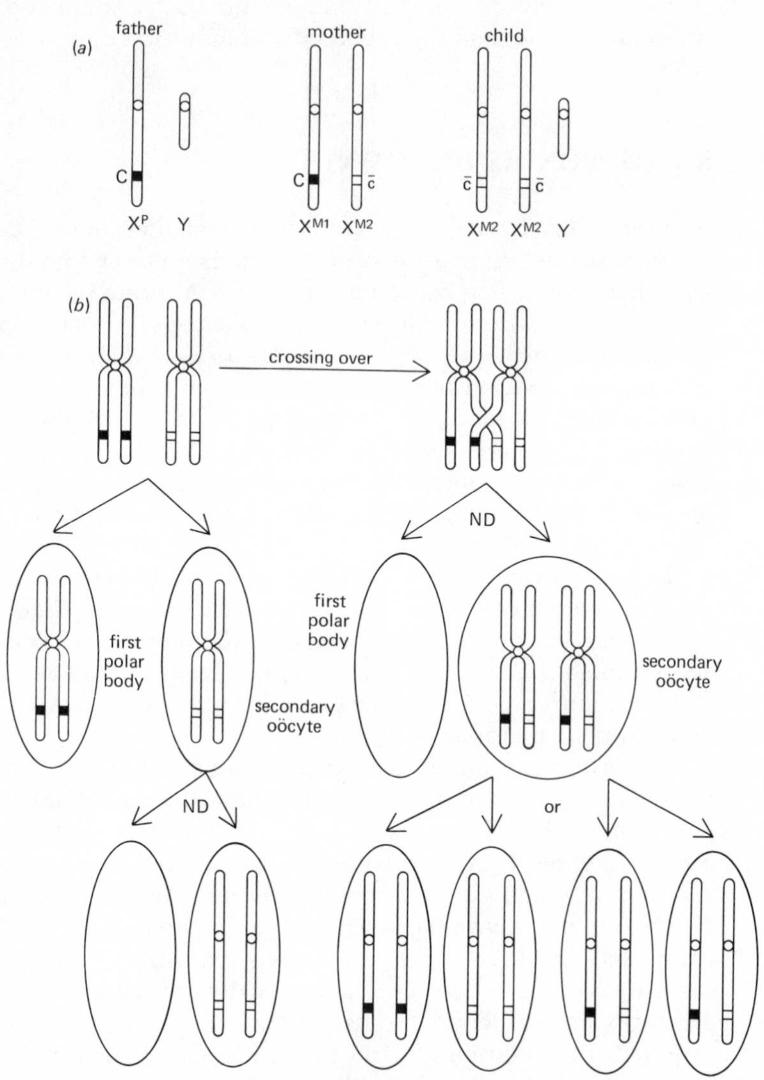

Figure 5-12 Diagram showing origin of 47,XXY colorblind male from parents both with normal color vision. Colorblindness is due to a recessive gene on the X chromosome. The mother must therefore have been heterozygous for colorblindness. Both X chromosomes in the child carry the allele for colorblindness, indicating maternal origin of both. (*a*) Sex chromosomes of family members. (*b*) Production of ovum containing two alleles for colorblindness either by nondisjunction at the second division or, if crossing over between the centromere and colorblind locus occurs, by nondisjunction at the first division. ND = nondisjunction. (*Based on pedigree published by Nowakowski, Lenz, and Parada, 1958.*)

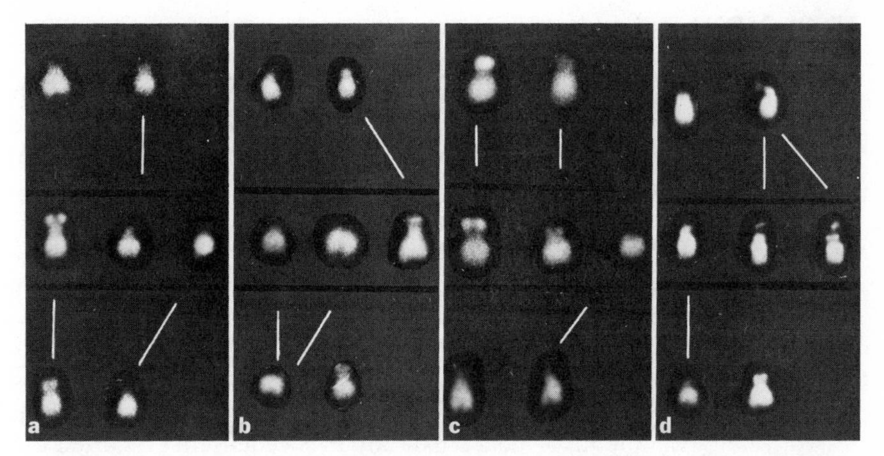

Figure 5-13 Chromosome "pedigrees" in four trisomy 21 families. The No. 21 chromosomes are shown for the mother (top row), father (bottom row), and patient (center row). In each case, it is possible to tell in which parent nondisjunction occurred and whether in meiosis I or II. (*a*) Paternal M I; (*b*) paternal M II; (*c*) maternal M I; (*d*) maternal M II. (*Supplied by Professor W. Schnedl.*)

29 resulted from maternal nondisjunction and 19 from paternal (summarized in Erickson, 1978). These involved nondisjunction at both first and second meiotic divisions. In instances in which nondisjunction occurred in the father, parental age is not substantially greater than for the general population. However, it is increased for maternal nondisjunction.

An observation made in several populations is the clustering of trisomy 21 and of sex chromosome trisomies both geographically and in time. Various explanations have been advanced for such "epidemics," including viral disease, but none has received strong support. The extent of the clustering in fact is very low. There also appears to be clustering of simple trisomies in families, suggesting a tendency for nondisjunction, possibly inherited. In some of these families, the trisomies are of different type, that is, 47,+21 and 47,XXY. The frequency of such families is sufficiently low to make it uncertain whether a substantially increased risk exists or whether the multiple occurrence in a family is coincidence.

There are conflicting reports on the increase of human nondisjunction as a result of ionizing radiation (Uchida, 1977). It has long been known that x-rays increase nondisjunction in aged *Drosophila* females. It remains to be seen whether this is generally true in human females or males. *In vitro* studies show that human lymphocytes exposed to low levels of radiation have an increased frequency of mitotic nondisjunction, which may or may not be relevant to meiotic nondisjunction (Uchida et al., 1975).

TRIPLOIDY

Three full sets of chromosomes (*triploidy*) does not appear to be compatible with normal embryonic and fetal development. Triploidy does occur with substantial frequency, however. Jacobs and co-workers (1978) found 26 of 340 spontaneously aborted fetuses to be triploid. Of these, two-thirds arose from dispermy (fertilization of a single egg by two sperm). The remainder arose from failure of chromosomes to separate during meiosis I, either in spermatogenesis or oögenesis, producing diploid sperm or eggs. Other studies have reported even greater frequency of triploidy (Carr and Gedeon, 1977). Thus more than 1% of recognized pregnancies involve triploid zygotes.

Rare examples of mosaicism for triploidy have been reported. One such example involved a small boy in Sweden, first reported to be simple triploidy. Subsequently it was established that he was mosaic for diploid and triploid cell lines. Although most tissues of the body consist of both cell lines, the white cells were entirely diploid. This is thought to be because there is a high rate of replication in the blood cell-forming tissue, and the triploid line would be unable to replicate as fast as diploid cells. It would therefore fail to keep pace and would gradually be replaced by the diploid line. In slower replicating tissues, there would be less diploid advantage.

CHROMOSOMES AND ABORTION

Reference has been made to particular chromosome complements being lethal or leading to fetal loss, and chromosome abnormalities are clearly a major factor in the cause of abortion. In a recent summary of studies of spontaneous abortions, it was concluded that approximately 50% of all abortuses have a recognizable cytogenetic abnormality (Carr and Gedeon, 1977). About half of these are trisomies, with all chromosomes except numbers 1 and 5 having been involved. It may be presumed that trisomy 1 and trisomy 5 also occur, leading to very early embryonic death, too early to be recognized as an abortion. Trisomy 16 is the most frequent of the trisomies found. Of particular note is the high frequency of 45,XO abortuses, nearly 20%, in view of the fact that 45,XO liveborns have excellent survival. Unbalanced translocations (Chapter 6) are found in 2 to 4% of abortuses.

Most of the chromosomally abnormal fetuses are aborted during the first trimester. Therefore the frequency of cytogenetic abnormalities among later abortions is much lower. Even so, a fourth of the trisomy 21's that survive the first trimester are aborted before birth. Among stillborn infants, the frequency of cytogenetic abnormalities is about 5%.

Since approximately 15% of recognized conceptions result in spontaneous abortion, it follows that 7 to 8% of such conceptions are cytogenetically abnormal.

PRENATAL DIAGNOSIS

Chromosomal abnormalities are among the easiest to diagnose prenatally. The general procedure for prenatal diagnosis of any genetic defect begins with *amniocentesis* at the twelfth to fourteenth week of gestation. After the position of the fetus is established, preferably with the aid of ultrasound, a transabdominal tap is made for amniotic fluid. This fluid contains live cells from the fetus as well as many dissolved substances that reflect the metabolism of the fetus. The cells are centrifuged down and placed in culture medium to stimulate growth. In approximately two weeks, there should be sufficient cell growth to subculture, treat with colchicine, and prepare slides for karyotype analysis, using generally the same techniques as for lymphocyte cultures (Chapter 4). The cultured cells may also be used for assay of various enzymes, and the supernatant from the amniotic fluid can be assayed directly for substances that indicate the presence of genetic or other defects in the fetus.

Prenatal diagnosis has proved to be a very safe procedure, but because of the limited number of medical centers with the required technical skills, it is ordinarily carried out only if the fetus has an increased risk of abnormality. For simple trisomies, the usual reason is advanced age of the mother. In Figure 5-3, the risk is seen to increase rather sharply starting about age 35. Pregnant women above this age are considered high risk for a child with Down's syndrome. For most, prenatal diagnosis provides the assurance of a chromosomally normal child. If a chromosome abnormality is present, the parents may choose to terminate the pregnancy. Other high risk pregnancies include earlier birth of a child with a chromosomal abnormality or knowledge that one of the parents carries a structurally abnormal chromosome (Chapter 6). A number of inherited metabolic diseases can also be diagnosed prenatally as will be discussed in later chapters.

SUMMARY

1. Errors in distribution of chromosomes may lead to cells with loss or gain of chromosomes. In Down's syndrome, there are three copies of chromosome 21 (trisomy 21). The nondisjunction responsible for

Down's syndrome may arise during meiosis in either parent. It is much more likely to happen in older mothers.

2. Other autosomal trisomies observed in newborns involve chromosomes 8, 9, 13, 18, and 22. The malformations in these syndromes are severe. Most are probably spontaneously aborted, and survival is greatly reduced among liveborns trisomic for these chromosomes. Monosomy for autosomes is exceedingly rare at best among liveborn offspring and may not occur.

3. Nondisjunction of sex chromosomes leads to a variety of aneuploid states. The first discovered was Klinefelter's syndrome, a hypogonadal condition in males in which the karyotype formula is 47,XXY. As in trisomy 21, the risk increases with maternal age. However, using X-linked genes as markers, some cases have been shown to arise by paternal nondisjunction.

4. Females with Turner's syndrome have the karyotype formula 45,XO. Most appear to arise postzygotically through loss of an X or Y chromosome, and many are mosaic for sex chromosome complements.

5. Other relatively common aneuploid states involving sex chromosomes include the triplo-X females (47,XXX), with essentially normal phenotype and a frequency of 1 per 1600 female births, and XYY males, with a frequency of 1 per 1000 male births. Some XYY males appear to have a higher than usual risk of aggressive behavior.

6. The causes of nondisjunction are unknown. The transmission of genetic traits or morphological variants of chromosomes has demonstrated that nondisjunction may occur during meiosis in either parent. The risk seems to increase greatly with age of the mother but not the father. Clustering of cases of trisomy in space and time have suggested the influence of environmental factors on nondisjunction. Some, but not all, studies implicate ionizing radiation as a factor.

7. Some 50% of spontaneous abortions are associated with abnormal karyotypes, primarily numerical. These include trisomies for most chromosomes and triploidy.

8. By means of amniocentesis at the beginning of the second trimester of pregnancy and culture of the amniotic cells, it is possible to prepare karyotypes of fetuses and to identify those with abnormal chromosome complements. Such prenatal diagnosis is carried out in high risk pregnancies, as, for example, when the mother is older than 35 years of age or when there has been an earlier birth of a chromosomally abnormal child.

REFERENCES AND SUGGESTED READING

APGAR, V. (Ed.). 1970. Down's syndrome (mongolism). *Ann. N.Y. Acad. Sci.* 171: 303–688.

BARR, M. L., AND BERTRAM, E. G. 1949. A morphological distinction between neurones of the male and female, and the behaviour of the nucleolar satellite during accelerated nucleoprotein synthesis. *Nature* (Lond.) 163: 676.

BEATTY, R. A. 1978. The origin of human triploidy: An integration of qualitative and quantitative evidence. *Ann. Hum. Genet.* 41: 299–314.

BÖÖK, J. A., MASTERSON, J. G., AND SANTESSON, B. 1962. Malformation syndrome associated with triploidy—further chromosome studies of the patient and his family. *Acta Genet.* 12: 193–201.

BORGAONKAR, D. S., AND SHAH, S. A. 1974. The XYY chromosome male—or syndrome? *Prog. Med. Genet.* 10: 135–222.

CAROTHERS, A. D., COLLYER, S., DE MEY, R., AND FRACKIEWICZ, A. 1978. Parental age and birth order in the aetiology of some sex chromosome aneuploidies. *Ann. Hum. Genet.* 41: 277–287.

CARR, D. H., AND GEDEON, M. 1977. Population cytogenetics of human abortuses. In *Population Cytogenetics* (E. B. Hook, I. H. Porter, Eds.). New York: Academic Press. pp. 1–9.

COLLMANN, R. D., AND STOLLER, A. 1963. Comparison of age distributions for mothers of mongols born in high and in low birth incidence areas and years in Victoria, 1942–57. *J. Ment. Defic. Res.* 7: 79–83.

COURT BROWN, W. M. 1967. *Human Population Cytogenetics.* Amsterdam: North Holland; New York: John Wiley, 107 pp.

EDWARDS, J. H., HARNDEN, D. G., CAMERON, A. H., CROSSE V. M., AND WOLFF, O. H. 1960. A new trisomic syndrome. *Lancet* 1: 787–790.

ERICKSON, J. D. 1978. Down syndrome, paternal age and birth order. *Ann. Hum. Genet.* 41: 289–298.

FORD, C. E., JONES, K. W., POLANI, P. E., DE ALMEIDA, J. C., AND BRIGGS, J. H. 1959. A sex-chromosome anomaly in a case of gonadal dysgenesis (Turner's syndrome). *Lancet* 1: 711.

Geneva Conference. 1966. Standardization of procedures for chromosome studies in abortion. *Cytogenetics* (Basel) 5: 361–393.

GROUCHY, J. DE, AND TURLEAU, C. 1977. *Clinical Atlas of Human Chromosomes.* New York: John Wiley, 319 pp.

GOAD, W. B., ROBINSON, A., AND PUCK, T. T. 1976. Incidence of aneuploidy in a human population. *Am. J. Hum. Genet.* 28: 62–68.

HAMERTON, J. L. 1971. *Human Cytogenetics. Vol. 1, General Cytogenetics,* 412 pp; *Vol. 2, Clinical Cytogenetics,* 545 pp. New York: Academic Press.

HAMERTON, J. L. 1976. Human population cytogenetics: dilemmas and problems. *Am. J. Hum. Genet.* 28: 107–122.

HOOK, E. B. 1973. Behavioral implications of the human XYY genotype. *Science* 179: 139–150.

HOOK, E. B. 1978. Extra sex chromosomes and human behavior: The nature of the evidence regarding XYY, XXY, XXYY, and XXX genotypes. In

Genetic Mechanisms of Sexual Development (H. L. Vallet, I. H. Porter, Eds.). New York: Academic Press.

HOOK, E. B., AND LINDSJÖ, A. 1978. Down syndrome in live births by single year maternal age interval in a Swedish study: Comparison with results from a New York State study. *Am. J. Hum. Genet.* 30: 19–27.

HOOK, E. B., AND PORTER, I. H. (Eds.). 1977. *Population Cytogenetics*. New York: Academic Press, 374 pp.

HULTÉN, M., AND PEARSON, P. L. 1971. Fluorescent evidence for spermatocytes with two Y chromosomes in an XYY male. *Ann. Hum. Genet.* 34: 273–276.

HUNGERFORD, D. A., MELLMAN, W. J., BALABAN, G. B., LA BADIE, G. U., MESSATZZIA, L. R., AND HALLER, G. 1970. Chromosome structure and function in man, III. Pachytene analysis and identification of the supernumerary chromosome in a case of Down's syndrome (mongolism). *Proc. Natl. Acad. Sci.* (U.S.A) 67: 221–224.

HUNTER, H., AND QUAIFE, R. 1973. A 48,XYYY male: a somatic and psychiatric description. *J. Med. Genet.* 10: 80–96.

JACOBS, P. A., ANGELL, R. R., BUCHANAN, I. M., HASSOLD, T. J., MATSUYAMA, A. M., AND MANUEL, B. 1978. The origin of human triploids. *Ann. Hum. Genet.* 42: 49–57.

JACOBS, P. A., BAIKIE, A. G., COURT BROWN, W. M., MAC GREGOR, T. N., MACLEAN, N., AND HARNDEN, D. G. 1959. Evidence for the existence of the human "super female." *Lancet* 2: 423.

JACOBS, P. A., BRUNTON, M., MELVILLE, M. M., BRITTAIN, R. P., AND MC CLEMONT, W. F. 1965. Aggressive behaviour, mental subnormality and the XYY male. *Nature* 208: 1351–1352.

JACOBS, P. A., AND STRONG, J. A. 1959. A case of human intersexuality having a possible XXY sex-determining mechanism. *Nature* (Lond.) 183: 302.

LEJEUNE, J. 1964. The 21 trisomy—current stage of chromosomal research. *Prog. Med. Genet.* 3: 144–177.

LEJEUNE, J., GAUTIER, M., AND TURPIN, R. 1959. Etude des chromosomes somatiques de neuf enfants mongoliens. *C. R. Acad. Sci.* (Paris) 248: 1721–1722.

LEVINE, H. 1971. *Clinical Cytogenetics*. Boston: Little, Brown, 572 pp.

LICZNERSKI, G., AND LINDSTEN, J. 1972. Trisomy 21 in man due to maternal nondisjunction during the first meiotic division. *Hereditas* 70: 153–154.

LILIENFELD, A. M., AND BENESCH, C. H. 1969. *Epidemiology of mongolism*. Baltimore: Johns Hopkins University Press, 145 pp.

MC CLURE, H. M., BELDEN, K. H., PIEPER, W. A., AND JACOBSON, C. B. 1969. Autosomal trisomy in a champanzee: resemblance to Down's syndrome. *Science* 165: 1010–1012.

MIKKELSEN, M., HALLBERG, A., AND POULSEN, H. 1976. Maternal and paternal origin of extra chromosome in trisomy 21. *Hum. Genet.* 32: 17–21.

MONEY, J., AND ALEXANDER, D. 1966. Turner's syndrome: further demonstration of the presence of specific cognitional deficienceis. *J. Med. Genet.* 3: 47–48.

NOWAKOWSKI, H., LENZ, W., AND PARADA, J. 1958. Diskrepanz

zwischen Chromatinbefund und chromosomalen Geschlecht beim Klinefelter Syndrom. *Klin. Wochenschr.* 36: 683–684.
PATAU, K., SMITH, D. W., THERMAN, E., INHORN, S. L., AND WAGNER, H. P. 1960. Multiple cogenital anomaly caused by an extra autosome. *Lancet* 1: 790.
PENROSE, L. S. 1961. Mongolism. *Br. Med. Bull.* 17: 184–189.
PENROSE, L. S., AND SMITH, G. F. 1966. *Down's Anomaly.* London: Churchill, 218 pp.
RACE, R. R., AND SANGER, R. 1969. Xg and sex-chromosome abnormalities. *Br. Med. Bull.* 25: 99–103.
TAYLOR, A. I. 1968. Autosomal trisomy syndromes: a detailed study of 27 cases of Edwards' syndrome and 27 cases of Patau's syndrome. *J. Med. Genet.* 5: 227–252.
TENNES, K., PUCK, M., BRYANT, K., FRANKENBURG, W., AND ROBINSON, A. 1975. A developmental study of girls with trisomy X. *Am. J. Hum. Genet.* 27: 71–80.
TURPIN, R., AND LEJEUNE, J. 1965. *Les Chromosomes Humains.* Paris: Gauthier-Villars, 535 pp.
UCHIDA, I. A. 1977. Maternal radiation and trisomy 21. In *Population Cytogenetics* (E. B. Hook, I. H. Porter, Eds.). New York: Academic Press, pp. 285–299.
UCHIDA, I. A., RAY, M., MC RAE, K. N., AND BESANT, D. F. 1968. Familial occurrence of trisomy 22. *Am. J. Hum. Genet.* 20: 107–118.
UCHIDA, I. A., LEE, C. P. V., AND BYRNES, E. M. 1975. Chromosome aberrations induced in vitro by low doses of radiation: Nondisjunction in lymphocytes of young adults. *Am. J. Hum. Genet.* 27: 419–429.
WAGENBICHLER, P., KILLIAN, W., RETT, A., AND SCHNEDL, W. 1976. Origin of the extra chromosome No. 21 in Down's syndrome. *Hum. Genet.* 32: 13–16.
YUNIS, J. J. (Ed.). 1977. *New Chromosomal Syndromes.* New York: Academic Press, 404 pp.

REVIEW QUESTIONS

1. Define:

euploid
aneuploid
nullisomy
monosomy
trisomy
triploidy
nondisjunction
secondary nondisjunction
chromosome lag
clone
simian crease
sex chromatin
Barr body
mosaic
dispermy

2. What would be the gametic products of first meiotic nondisjunction of sex chromosomes in males? Of second division nondisjunction? Of consecutive first and second division nondisjunction? Answer the above for females also.

3. What sex chromosome aneuploidies in addition to those reported in the text might be observed, assuming nondisjunction in both parents and in both divisions of meiosis?

4. In the following families identify, if possible, the parent in whom the nondisjunction occurred.

a. 46,XY Xg(a+) father × 46,XX Xg(a−) mother → 47,XXY Xg(a+) child

b. 46,XY Xg(a+) father × 46,XX Xg(a+) mother → 47,XXY Xg(a−) child

c. 46,XY Xg(a−) father × 46,XX Xg(a+) mother → 47,XXY Xg(a−) child

d. 46,XY Xg(a−) father × 47,XXX Xg(a+) mother → 47,XYY Xg(a+) child

e. 47,XYY Xg(a+) father × 46,XX Xg(a−) mother → 47,XXY Xg(a−) child

5. A 47,XXX woman gives birth to a 47,XXY child. What Xg types in this family would exclude secondary nondisjunction as the cause of the aneuploidy in the child?

6. How would you account for the existence of 48,XYYY persons?

7. What would you expect to be the effect of family limitation on incidence of Down's syndrome as compared to a situation in which the size of families is not limited?

8. It has been suggested that the incidence of Down's syndrome among children of young unwed mothers is slightly higher than among children of married women of the same age. What hypotheses might be offered to explain this difference? How would you test your hypotheses?

9. What explanation might be offered for the age effect in maternal nondisjunction in contrast to the small or absent age effect in paternal nondisjunction?

CHAPTER SIX

CHROMOSOMAL ABNORMALITIES: STRUCTURAL

Chromosomes are considered to have a precise amount of genetic information arranged linearly in a precise manner. Although this is usually true, variant forms have long been known in experimental organisms, and a number of variant chromosomes have been described in human beings. Most of these do not function as well as normal chromosomes and hence are designated aberrant. Occasional chromosomes are found that are distinctive morphologically but that do appear to function normally.

Structurally altered chromosomes arise through errors in breakage and reunion. This may happen as part of crossing over if mispairing of homologs occurs in meiotic prophase. In this case, the exchange of chromatid segments would be nonreciprocal, the usual term applied being *unequal crossing over*. Or the chromosome may break during some other part of the cell cycle. Correct rejoining of the broken ends may occur, but often it does not. Nonrejoined chromosomal segments are unstable and usually are lost. A summary of some of the more common terminology and symbols used to describe structural rearrangements is given in Table 6-1. A more complete list is given in Table 4-2.

Alterations of structure may involve only a few base pairs, or they may involve entire chromosome arms. Many so-called gene mutations in experimental organisms have proved on more detailed analysis to involve deletions of small segments of chromosomes containing more than one structural gene but still too small to affect chromosome morphology as viewed microscopically. Accordingly, the distinction between chromosome mutations and gene mutations is somewhat arbitrary, the former referring to gross changes in chromosome structure that visibly alter the chromosome, the latter to changes that are not visible. The difference often is one of size only and not quality.

Table 6-1 TERMINOLOGY USED WITH STRUCTURAL REARRANGEMENTS OF CHROMOSOMES*

Deletion (*del*)

A *terminal* deletion is indicated by the chromosome number in parenthesis followed by designation of the breakpoint; e.g. 46,XY,del(18) (q22) would be a male with 46 chromosomes, of which one of the 18 chromosomes would have a deletion on the long arm of band 22 and all material distal to that point. A more detailed system would be 46,XY,del(18) (pter → q22:), which indicates that the deleted chromosome 18 *consists* of the material beginning with the terminus of the short arm (pter) and going to band 22 on the long arm (q22). A single colon indicates a break.

An *interstitial* deletion is indicated by the two boundaries of the deletion in parenthesis; e.g. 46,XY,del(1) (q21q31) would involve deletion of material between bands 21 and 31 on the long arm of chromosome 1. The more detailed system would be 46,XY,del(1)(pter → q21::q31 → qter), where the material in the last parenthesis indicates the composition of the deleted chromosome 1. The double colon indicates breakage and reunion.

The symbolism for deletions, as recommended by the Chicago Conference (p. 74), consisted of a minus sign placed after the chromosome arm. Thus del(18q) would be 18q−. Since the development of banding techniques, the use of *del* before the chromosome number is preferred. The minus sign is still frequently used, especially if the break points are not specified.

Ring chromosomes (*r*)

The points of the breaks on the two arms are indicated in parenthesis; e.g. 46,XY,r(2) (p21q21) would mean loss of the material beyond band 21 of the short arm and band 31 of the long arm of chromosome 2, with joining of the proximal segments to form a ring. The more detailed version would be 46,XY,r(2)(p21 → q31), with no double colon to indicate breakage and reunion.

Inversions (*inv*)

An inversion is indicated by the chromosome number in parenthesis followed by the two break points in parenthesis. Thus, 46,XX,inv(5) (p12q23) would be a pericentric inversion of chromosome 5 with breakage and reunion at band 5p12 and 5q23. A more detailed symbol would be 46,XX,inv(5) (pter → p12::q23 → p12::q23 → qter).

Translocations (*t*)

The two chromosome involved are listed in parentheses after t, followed by the break points separated by a semicolon, also in parentheses. 46,XY,t(7;9) (q31;q22) would be a male with a balanced translocation involving exchange of the segments distal to 7q31 and 9q22. The more detailed symbolism would be 46,XY,t(7;9)(7pter → 7q31::9q22 → 9qter;9pter → 9q22::7q31 → 7qter). *rcp* can be used in place of *t* to be more specific.

A Robertsonian translocation involves loss of the short arms of two chromosomes and centromere of one. This results in a reduction in the total number of chromosomes. 45,XX,t(14q21q) would be a female with a Robertsonian translocation, with the derivative chromosome having the long arms of 14 and 21. The origin of the centromere is not indicated. A more detailed symbolism would be 45,XX,t(14;21) (14qter → cen → 21qter). *rob* can be used in place of *t* to be more specific.

Isochromosomes (*i*)

The arm that is the origin of the isochromosome is listed in parentheses after i. Thus 46,X,i(Xq) would involve an isochromosome consisting of the long arms of X.

* Additional rules and examples are found in the *Paris Conference (1971)* and the *Paris Conference (1971), Supplement (1975)*.

TYPES OF STRUCTURAL CHANGES

Deletions and Duplications

Loss of chromosomal segments commonly happens by two mechanisms. The simplest is chromosome breakage with failure to rejoin the broken ends. In that case, one fragment will be acentric and will become lost in subsequent cell divisions. The centric fragment has the potential for proper distribution if the broken end is "healed" and will persist as a chromosome having a *terminal deletion.* Alternatively, two breaks may occur within an arm, with rejoining of the distal fragment to the centric fragment and loss of the intermediate region (*interstitial deletion*).

The second mechanism involves mispairing of chromatids and unequal crossing over in meiosis (Figure 6-1). The genetic consequences of unequal crossing over are a deletion in one chromatid and a reciprocal *duplication* in the other. The consequences of duplications depend on the dosage sensitivity of the genes involved. Small duplications involving only one or a few genes are an important mechanism for the evolution of new genes, since the duplicate genes are released from the need to perform some essential function. Several examples of partial or complete duplication of genes in man will be discussed later. No clear

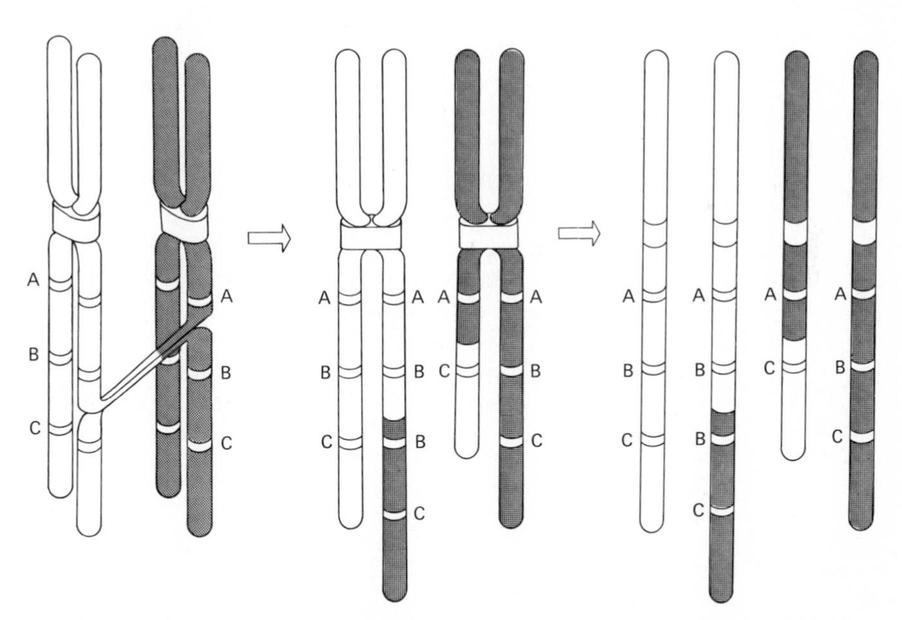

Figure 6-1 Unequal crossing over in meiosis, resulting in duplication of a chromosome segment in one product and deletion of a segment in another.

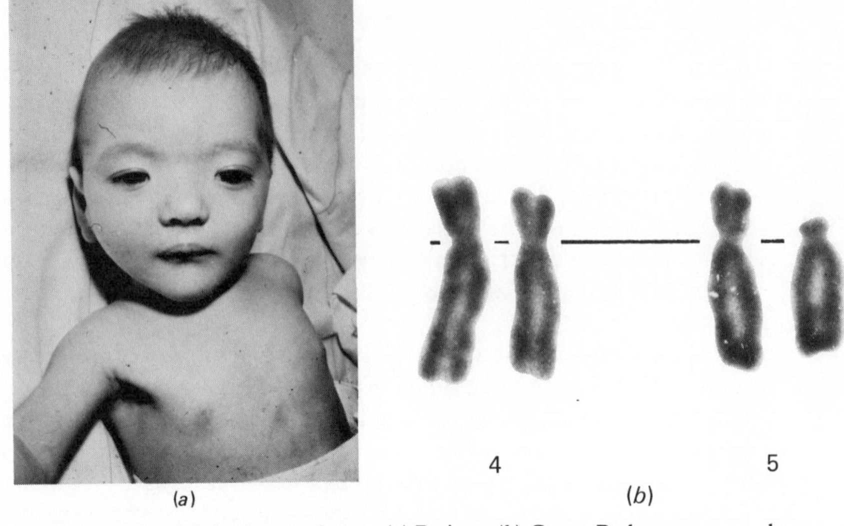

Figure 6-2 *Cri-du-chat* syndrome. (*a*) Patient. (*b*) Group B chromosomes, showing partial deletion of the short arm of number 5. (*Courtesy J. Lejeune.*)

example of duplication of an arm segment in man is known, although the Bar Eye mutation in *Drosophila* is a classical example of a small duplication that can be observed readily in the giant polytene salivary chromosomes. Unequal crossing over produces *tandem* duplications. Through translocation (see the following), these may be separated and become parts of different chromosomes.

Deficiencies are likely to have more serious consequences than duplications. The zygote will be hemizygous for all the genes covered by the deletion, and this may create a problem of dosage imbalance. Also, the recessive genes in the corresponding segment of the normal homolog will be expressed. This leads to a situation where members of a family with the same deleted chromosome may have quite different phenotypes because in each case the abnormal chromosome is paired with a different normal chromosome.

Most deficiencies probably are nonviable, but many have been observed in man, and certain ones are sufficiently often viable so as to give rise to recognized syndromes. Among the more frequent and one of the earliest deletion syndromes to be described is the *cri-du-chat* syndrome (Figure 6-2) first reported in 1963 (Lejeune et al.). The many cases of *cri-du-chat* syndrome all have a partial deletion of the short arm of chromosome 5. Other deletion syndromes are listed in Table 6-2. Additional rare deletions have been reported but too infrequently to provide a reliable description of a syndrome.

Table 6-2 **A PARTIAL LIST OF DELETIONS OBSERVED IN HUMAN LIVE BIRTHS. IN MOST CASES, THE DELETIONS DO NOT INCLUDE THE ENTIRE ARM***

Chromosomal deletion	No. of cases found	Phenotype
4p	several dozen	microcephaly, severe mental retardation, seizures, characteristic wide nose bridge, frequent cleft lip and palate, early death but may survive several decades
5p	100	*cri-du-chat* syndrome: "moon face," microcephaly, hypertelorism, small mandible, severe mental retardation, distinctive cry like kitten in infants, good survival
9p	10	trigonocephaly, frequent cardiac malformations, IQ 30–60, good survival
11q	6	trigonocephaly, mental retardation
12p	6	microcephaly, moderate mental retardation
13q	several dozen	microcephaly, moderate to severe mental retardation, bridge of nose broad and high, frequent hypertelorism, ocular anomalies, bone anomalies, retinoblastoma
18p	100	small head and body size, moderate to severe mental retardation, some have severe malformations of head, good survival
18q	100	hypotonia, moderate microcephaly, depressed midface, ocular malformations, mild to severe mental retardation, high childhood death rate with some survival to adulthood
Xp†	several dozen	Turner's syndrome
Xq†	several dozen	streak gonads

* More complete descriptions, as well as additional examples of partial monosomy, can be found in Grouchy and Turleau (1977) and in Yunis (1977).
† Viable only in combination with a normal X chromosome. These conditions are discussed further in Chapter 7.

Ring Chromosomes

If a chromosome breaks on both sides of the centromere, occasionally the two proximal ends join to form a ring and the two distal fragments, being acentric, are lost (Figure 6-3). The extent of the phenotype defect depends on the specific genes lost in the deleted fragments. Many patients have been found to have ring chromosomes, confirming that some small deletions are consistent with life. The presence of a ring is

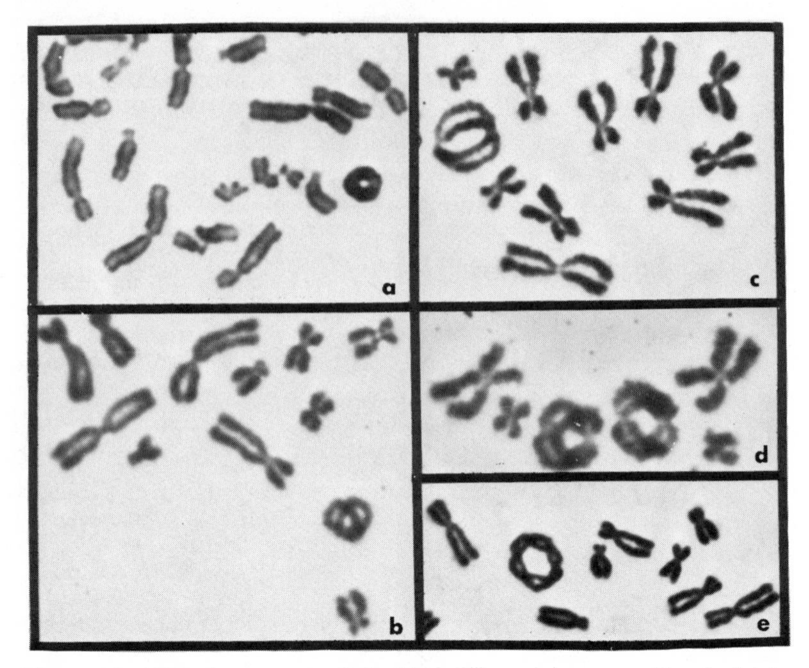

Figure 6-3 Ring chromosomes derived from a human autosome. Terminal deletions with healing of the proximal ends produces a ring. Rings are unstable during mitosis, resulting in additional rings of variable size and structure as shown here. (*a*) A ring chromosome. (*b*) An interlocking ring. (*c*) A double-size ring. (*d*) Two dicentric rings. (*e*) A ring with multiple chromatid twists. (*Furnished by Margery W. Shaw.*)

qualitative proof that material has been deleted from both arms, whereas comparable deletions not involved in ring formation might be too small to detect by their effects on chromosome morphology. Chromosome 18 is best known for its involvement in ring chromosomes, as might be expected from the existence of viable deletion syndromes for the distal regions of both arms. The 46,18r patients have many of the features of both 18p− and 18q− patients. Rings originating from several other chromosomes have also been reported, especially chromosome 13.

The existence of ring chromosomes raises interesting questions regarding chromosome replication, most of which cannot be answered at present. The ring chromosome does not appear to have marked disadvantage with respect to replication. Ring chromosomes have been known in lower animals and plants for many years. If a "twist" occurs during replication, the two sister chromatids of the ring are interlocked at anaphase, resulting in an anaphase bridge followed by breakage at the point of stress and subsequent fusion. Ring chromosomes are often unstable, generating larger and smaller rings and polycentric rings (Figure 6-3).

Inversions

If two breaks occur in a chromosome, the intermediate segment may become inverted prior to repair of the breaks. Such an inversion is *paracentric* if the centromere is not included in the inverted segment; it is *pericentric* if the centromere is included. A paracentric inversion does not change the arm ratio of a chromosome and is difficult to detect in unpaired chromosomes unless it is relatively large and an alteration in banding pattern can be observed. Pericentric inversions likely will change the position of the centromere, unless the break points are equidistant from the centromere.

A chromosome containing an inversion still has a full complement of genes and should function normally. Problems arise only during meiosis, when normal pairing cannot occur. In order for homologous regions to pair in a cell heterozygous for an inversion, the chromosomes must form an inversion loop, as shown in Figure 6-4. If crossing over does not occur within the loop, the anaphase chromosomes have a balanced complement of genes. Crossing over within the loops leads to unbalanced complements. Many examples have now been identified in which normal parents, one of whom carries a balanced inversion, produce children with an unbalanced duplication/deficiency through crossing over in the inversion loop. The specific phenotypic abnormalities depend on the chromosomal segments involved in the duplication/deficiency.

(*a*) paracentric inversion

(*b*) pericentric inversion

inversion heterozygote — zygotene — pachytene — anaphase I — potential gamete

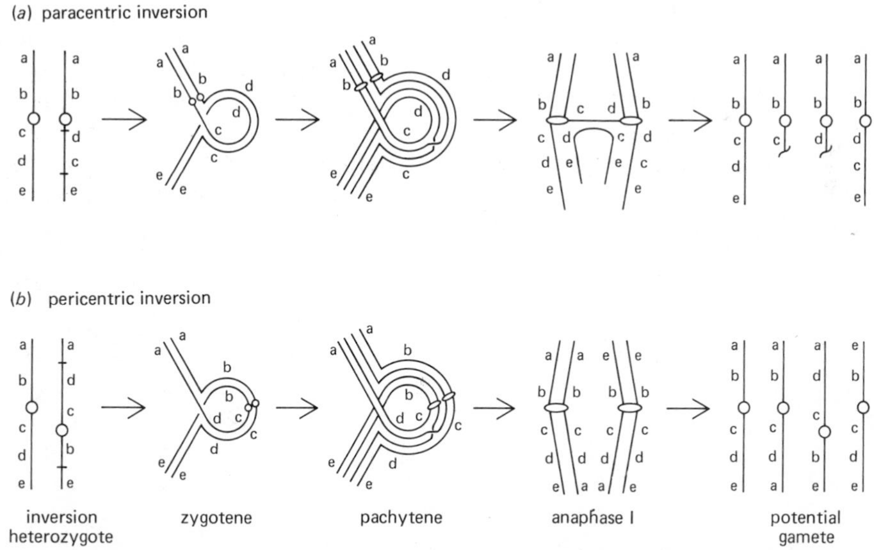

Figure 6-4 Diagram showing effects of crossing over within inversion loops. Both types of inversions give rise to some unbalanced gametes, leading in some organisms to reduced fertility.

In view of the problems generated in meiosis by inversions, it is somewhat surprising that inversion chromosomes occur with substantial frequency in some populations. In a study carried out in Finland in 1973, over 1% of the population was estimated to be heterozygous for small pericentric inversions of chromosomes 9 and 10 (de la Chapelle et al., 1974). In one marriage between distantly related persons, both were heterozygous for an inversion of chromosome 9 with break points in region 9p1 and band 9q13, producing a chromosome that can be represented as inv(9) (p1q13). One offspring of this marriage was homozygous for the inversion. Most of the persons heterozygous for these inversions were phenotypically normal, as was the homozygote. There was no obvious increase in fetal loss from heterozygous parents.

Inversions are thought to play a significant role as isolating mechanisms in the formation of species. It is therefore of special interest that the most frequent difference between human and ape karyotypes appears to be pericentric inversions.

Translocation

Translocation occurs when two chromosomes break and then rejoin in the wrong combinations (see Figure 6-5). If no genetic information is lost during the breakage and reunion, the translocation is *reciprocal* or *balanced*. There may be no cellular malfunction, and balanced function will be maintained through subsequent mitosis. But, at meiosis, pairing is possible only if the two pairs of chromosomes synapse into a single quadriradial figure, as shown in Figure 6-5. At anaphase the chromosomes may segregate three ways. In adjacent-1 segregation, daughter cells receive nonhomologous centromeres from adjoining chromosomes, as in the combinations I, III and II, IV of Figure 6-5. Adjacent-2 segregation pairs homologous centromeres, as in I, II, and III, IV. Alternate segregation pairs opposite chromosomes, which also have nonhomologous centromeres. The four types of gametes from adjacent segregation are unbalanced. Only alternate segregation gives a balanced genome, either with normal chromosomes or with a balanced translocation like the original.

Crossing over may occur between chromosomes in a *quadrivalent*. No new combinations will be generated unless the crossover occurs in the *interstitial* region (between the point of translocation and the centromeres). Crossing over followed by either alternate or adjacent-1 segregation causes these two to become equivalent. However, if adjacent-2 segregation occurs, four gametic combinations are possible in addition to the two that occur in the absence of crossing over.

The best example of translocation in man is found in Down's syndrome. Although the great majority of such patients have 47 chromosomes and are trisomic for chromosome 21, approximately 5%

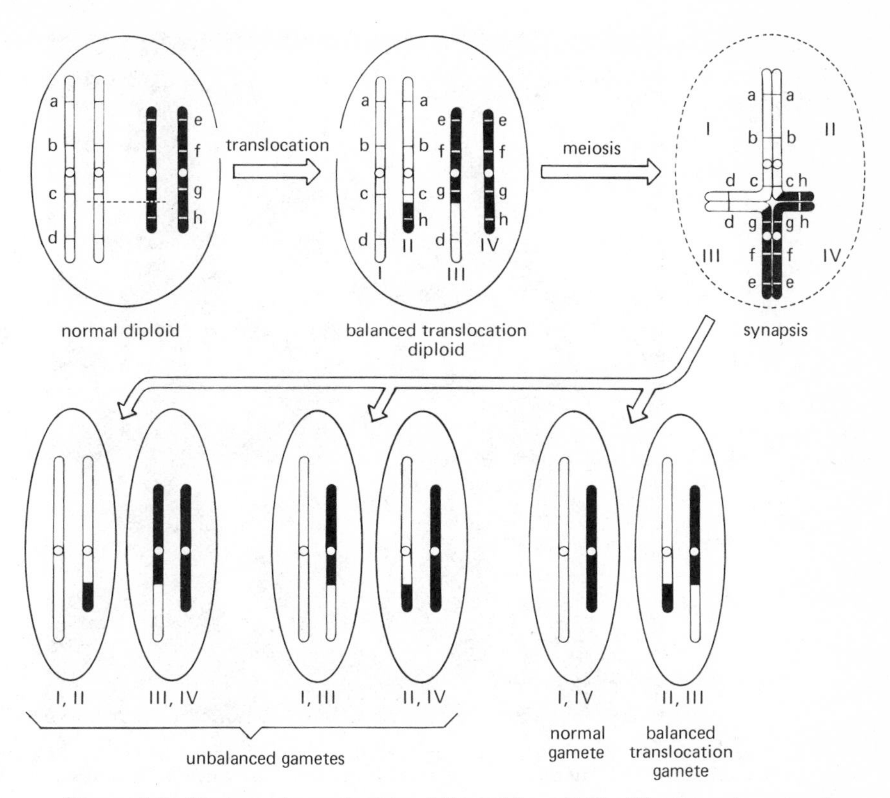

Figure 6-5 Reciprocal translocation followed by meiosis. There are three possibilities for segregation, giving six possible gametes. Only two types of gametes would carry a normal complement of genes, one with normal chromosomes and the other with a balanced translocation.

possess 46 chromosomes. Analysis of their karyotypes gives the pattern shown in Figure 6-6. The number of chromosomes is normal, but the morphology is not. In many cases, a chromosome of the D group has been replaced by one with a long arm typical of this group but with a much larger short arm. Analysis of the banding pattern indicates this chromosome to be derivative of the long arm of the D chromosome and the long arm of chromosome 21. The chromosome may thus be designated t(Dq21q) and the full karyotype formula 46,−D, + t(Dq21q). There would be two complements of the D group chromosome—one normal and one additional copy of the long arm in the translocation chromosome. There would be three complements of chromosome 21—two normal copies plus the translocation chromosome. This type of translocation involving fusion of the long arms of two acrocentric chromosomes is known as a Robertsonian fusion. Figure 6-7 illustrates meiosis in a person with a balanced translocation involving a D chromosome and 21.

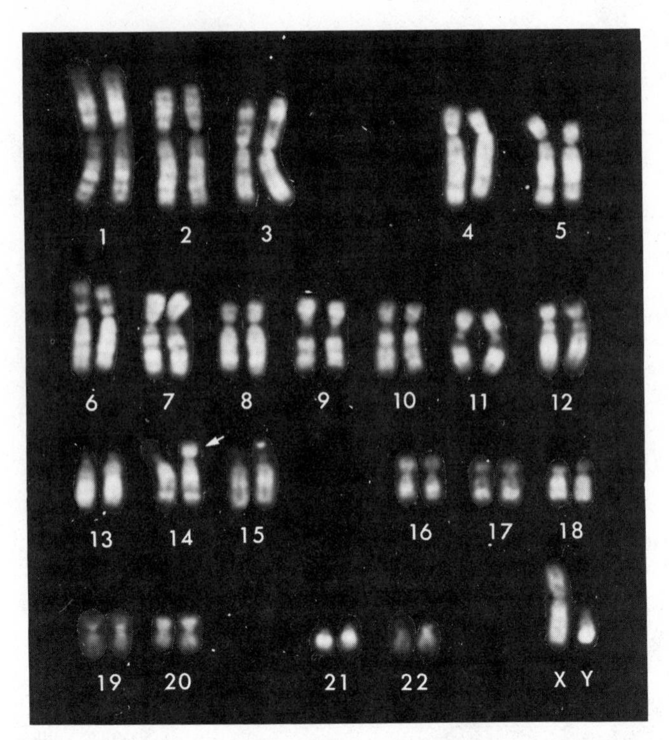

Figure 6-6 Karyotype of a translocation trisomy 21 syndrome. There are only 46 chromosomes, but there is an extra chromosome (indicated by the arrow) and only one chromosome 14. The karyotype formula is 46,− 14,+ t(14q21q). The chromosomes are stained to show Q-bands. (*Furnished by Irene Uchida.*)

Approximately 55% of the translocations in Down's syndrome involve a translocation of 21q to members of the D group (13-15). Forty-one percent involve a translocation with another G chromosome. About 60% of the D translocations involve chromosome 14, the remaining 40% equally divided between 13 and 15. Most of the G translocations (83%) are with another 21 chromosome, the remainder with chromosome 22.

A most interesting feature of the translocation-type Down's syndrome is its transmission by apparently normal persons. The meiotic products of a cell carrying the original translocation (or derived from the original cell by mitosis) would be six types of gametes, as shown in Figure 6-7. When combined with a normal gamete, these would produce six types of zygotes. Three of the zygotes appear to be lethal. Individuals derived from the second type of zygote would be identical to the parent in chromosomal constitution. The total genetic information is normal, so they would develop normally but would possess only 45 chromosomes. They would produce the same gametes as the parent with the original translocation and, hence, could produce children who are normal with 46 chromosomes, who are affected with 46 chromo-

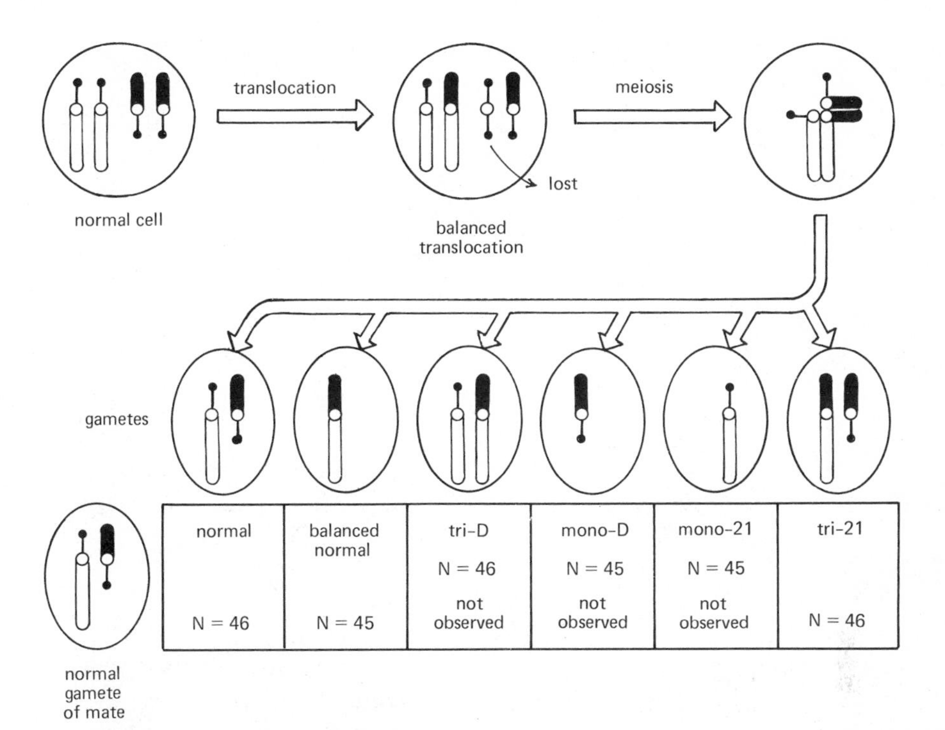

Figure 6-7 Offspring from person who is a carrier of a balanced Robertsonian translocation involving chromosome 21 and a D chromosome. Six types of offspring should be possible, but only three types are observed. The remainder are presumably lethal at a very early stage, since they involve major chromosome imbalance. The balanced translocation offspring is similar to the parent and therefore has a high risk of producing offspring with Down's syndrome.

somes, and who are carriers with 45 chromosomes. A pedigree of translocation-type Down's syndrome is given in Figure 6-8. Figure 6-9 illustrates meiotic pairing involving a translocation.

There are two important differences in the occurrence of translocation Down's syndrome and nondisjunction trisomy 21. In translocation families, there is likely to be more than one affected person. If the six types of gametes produced by a 45-chromosome parent were equally likely to be produced, one-third of the viable offspring would show Down's syndrome. As yet, it has not been established that the gametes in fact occur with equal frequency. Studies of chromosomal anomalies in other organisms indicate that various gametic combinations are not equally probable. Analysis of the families available suggests that the risk of an affected offspring is 10 to 15% if the mother is the translocation heterozygote but less than 5% if the father carries the translocation. This contrasts with nondisjunction trisomy 21, where risk of giving birth to an affected child is about 1/700 and increases only slightly for later children born to the same sibship.

A second difference between the two causes for Down's syndrome is

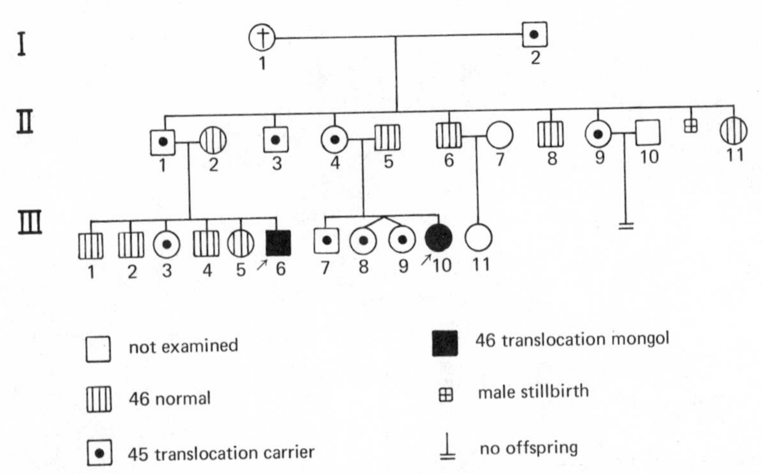

Figure 6-8 Pedigree showing transmission of Down's syndrome through persons who have balanced translocations of chromosomes 21 and 22. *(Redrawn from M. Shaw, 1962, with permission of the author and publisher, S. Karger, Basel/New York.)*

the absence of a maternal age effect in translocation Down's syndrome. Each trisomic patient may be considered the consequence of a new nondisjunction, an event that occurs more frequently in older females. But an individual who has 45 chromosomes, including the translocation, may produce affected offspring in high frequency as soon as sexual maturity is attained.

No clinical distinction has been made between the two types of Down's syndrome, and therefore patients are treated alike regardless of their chromosomal status. However, it is useful to parents to know which of the types is present in a child. If it is the more common trisomy, the risk of another affected child is low, about 2%. Parents of a translocation patient, on the other hand, face a much higher risk of recurrence and would be candidates for prenatal diagnosis for any subsequent pregnancy.

A variety of translocations has been observed in man. Typically, these are identified through birth of a severely defective child or through a tendency for spontaneous abortion of defective fetuses. Screening of normal populations has indicated a frequency of translocation heterozygosity of 1.6 per thousand live births (Hamerton, 1971, Vol. 1, p. 386).

Dicentric Chromosomes

When translocation between two chromosomes occurs, it is possible for the two fragments with centromeres to fuse, leaving the two acentric fragments to fuse into a large acentric fragment. The acentric fragment would become lost, and the dicentric chromosome would often form

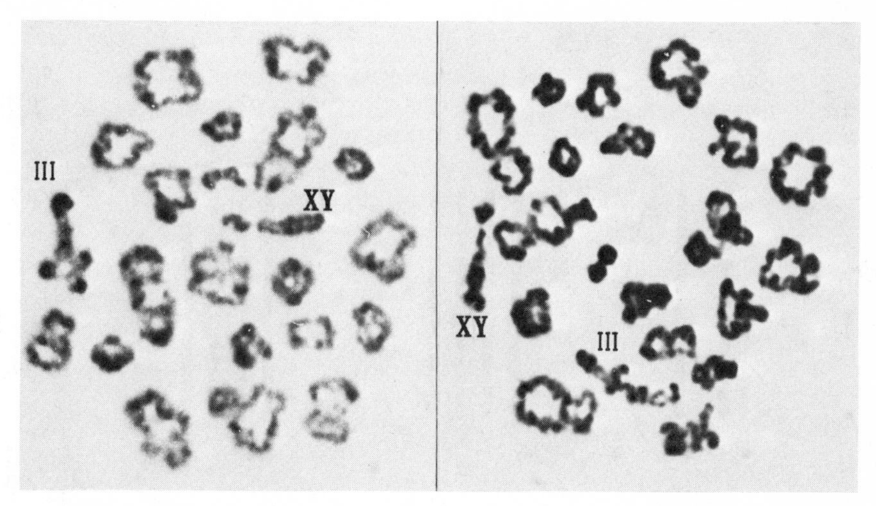

Figure 6-9 Primary spermatocytes from a carrier of a 14/21 translocation. There are 21 autosomal structures, including a trivalent (III). (*From Hultén and Lindsten, 1973.*)

bridges at anaphase leading to breakage. Dicentric chromosomes are therefore rare, although they have been observed in several species and are able to function reasonably well if the two centromeres are very close to each other. In man there have been examples of dicentric chromosomes in which one centromere has been inactivated, resulting in a secondary constriction that stains as an interstitial C-band.

Dicentrics may also be formed from a single chromosome. The most common dicentric chromosome observed in man involves only the Y chromosome. A number of cases have been reported with the karyotype formula 45,X/46,X,dic(Y), either Yp or Yq. It is noteworthy that only one patient with a dicentric Y has not been shown to be mosaic. The origin of the dic(Yq) is uncertain, but it may arise by breakage of the short arm followed by fusion of the proximal ends of the two chromatids. In such a case, the 45,X mosaic line probably arises by loss of the dicentric Y in the early postzygotic stage. Alternatively, the dicentric Y may arise postzygotically, although it is difficult in this case to account for the almost uniform mosaicism involving a 45,X line.

Persons with 45,X/46,X,dic(Y) may express Turner's syndrome, or they may have a predominantly male phenotype. The variation undoubtedly arises from differences in the origin of the mosaicism, both in terms of time of development and tissues involved.

Isochromosomes

Several examples of *isochromosomes* have been found in human beings. An isochromosome is a chromosome in which the two arms are identi-

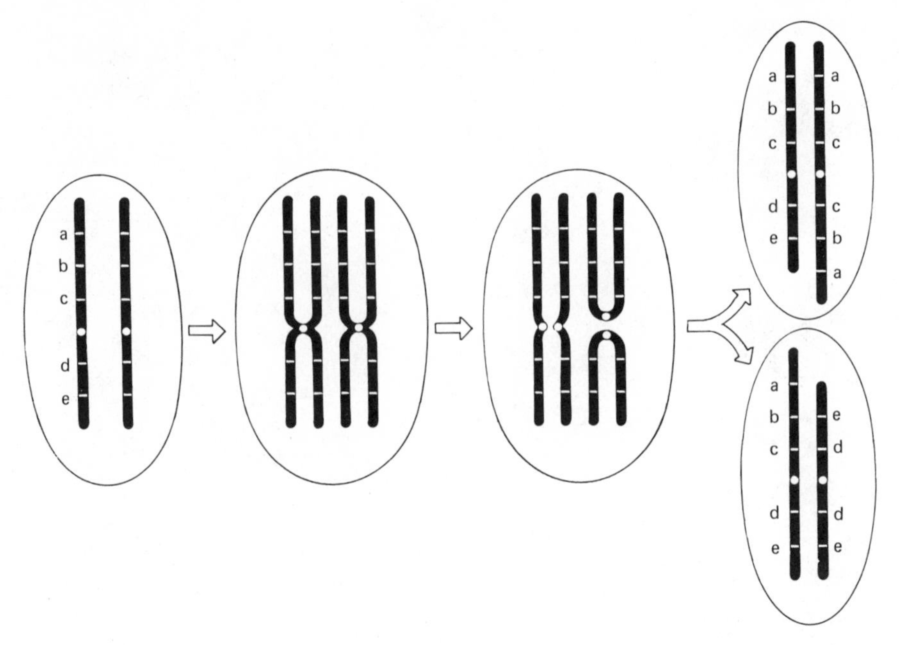

Figure 6-10 Formation of an isochromosome by division of the centromere in the wrong plane. The daughter cells will each be partial trisomic and partial monosomic. Since the two arms of an isochromosome are identical, the chromosome always is metacentric.

cal. It is thought to arise when a centromere divides in the wrong plane, yielding two daughter chromosomes, each of which carries the information of one arm only but present twice. This process is illustrated in Figure 6-10. Formation of isochromosomes presumably can occur either during mitosis or meiosis. Fertilization of an isochromosome-bearing gamete by a normal gamete would lead to an unbalanced karyotype. The zygote would have the information of one arm present three times and that of the other arm only once. Thus, the individual developing from such a zygote would be partially trisomic and partially monosomic.

Isochromosome formation by the long arm of the X chromosome is a common abnormality. Such 46,X,i(Xq) patients have defects that are typical of Turner's syndrome and usually cannot be distinguished clinically. They are sex chromatin positive, the i(Xq) chromosome regularly forming the Barr body. This is expected, since the i(Xq) lacks the genes of the short arm. If, by chance, the normal X became inactive in a 46,X,i(Xq) embryo, those cells in which this happened would not survive, leaving the cells with the normal X active to form the embryo.

Isochromosomes formed from the long arms of the Y chromosome and the acrocentric chromosomes have been reported, but these are rarer than i(Xq) isochromosomes. A 46,XY,i(18q) has also been reported.

CHROMOSOMES AND CANCER

Among the mammalian cells most readily cultivated *in vitro* are those derived from malignant tumors. Long before it was possible to cultivate normal tissues routinely, a number of viable cell lines had been established starting with tumor cells. Several of these cell lines have been maintained in the laboratory for a period of years and have proved invaluable in investigations of mammalian metabolism.

When techniques for chromosomal analysis were introduced it was discovered that the established cell lines had grossly abnormal numbers of chromosomes. An obvious question was whether the abnormal chromosome complements were characteristic of malignant tissues and, if so, whether the abnormal chromosomes were related to the origin of the cancer.

Contrary to the idea of a direct relationship was the observation that different cell lines originally derived from a single source had acquired complements of chromosomes different from each other. Although this probably happened subsequent to laboratory culture, it seemed possible that an aberration originally might have been responsible for the malignancy. There was also the possibility that cancerous cells might be more prone to develop abnormal chromosomes.

With the application of the newer banding techniques it is apparent that many more chromosomes in cancer cells are rearranged than previously suspected. Furthermore, these abnormal chromosomes have a tendency to undergo further rearrangements in culture in contrast to the normal chromosomes which remain quite stable. For the most part, specific abnormal karyotypes are not associated with a certain type of malignancy. Some degree of specificity is occasionally observed, as, for example, in acute nonlymphocytic leukemia, in which 50% of the patients have some type of karyotypic abnormality in the leukemic cells (Golomb et al., 1978), the most frequent being a 15;17 translocation. The remaining patients have normal karyotypes.

Relevant to this issue are inherited diseases that lead to chromosome instability and subsequently to cancer. Bloom's syndrome, Fanconi's anemia, ataxia-telangiectasia, and xeroderma pigmentosum are autosomal recessive disorders with different primary defects. These patients have a very high risk of malignancy. Karyotypes show a high frequency of metaphases with broken chromosomes and rearrangements. In addition, patients with Bloom's syndrome show increased sister chromatid exchanges (Figure 6-11). Clearly, the chromosome instability precedes the malignancy, and most of the aberrant cells are not malignant. It is nevertheless tempting to suppose that occasionally a cell with an abnormal complement of chromosomes becomes malignant *because* of the chromosome complement.

The first specific association between a malignancy and aberrant

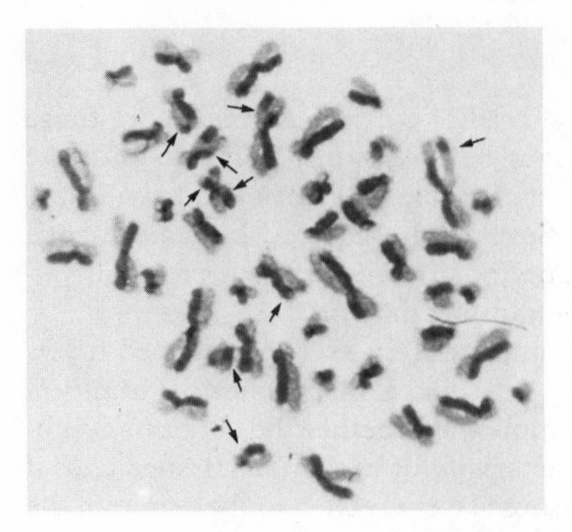
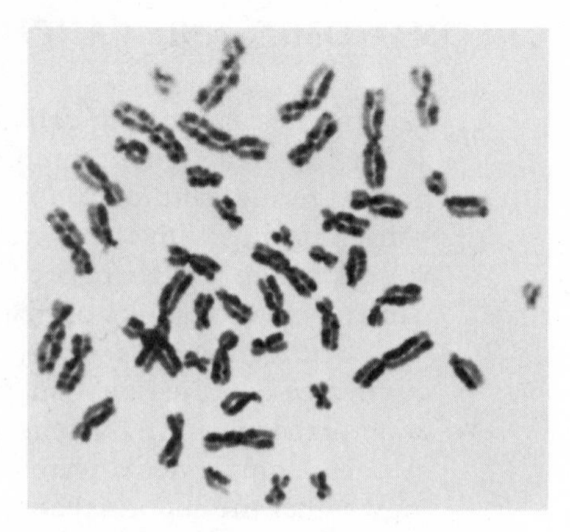

Figure 6-11 Metaphase chromosomes from a normal person (left) and a patient with Bloom's syndrome (right). Lymphocytes were treated as in Figure 3-19. The frequency of sister chromatid exchanges is greatly elevated in Bloom's syndrome. (*From R. S. K. Chaganti, S. Schonberg, and J. German, III, 1974.* Proc. Natl. Acad. Sci., *U.S., 71: 4508–4512.*)

chromosomes involved chronic myelocytic leukemia. Dividing bone marrow cells from approximately 90% of patients with this uncommon type of leukemia are characterized by two cell lines, one with normal and one with 46,del(22q) karyotypes. The resulting abnormal chromosome is known as a Philadelphia or Ph^1 chromosome (Figure 6-12). Individuals with Ph^1 chromosomes also have many cells with a normal complement of chromosomes, suggesting that the Ph^1 chromosome arose from originally normal cells. Although rigorous proof is lacking, the abnormal chromosome is thought to play a role in the development of the leukemia in those cases where it is found. The proportion of cells with abnormal karyotype is associated with the progression of the disease. Recent studies of patients using the banding technique suggest that the missing portion of the long arm of chromosome 22 is translocated to one of the larger chromosomes, most often to the long arm of number 9 (Rowley, 1973), but also to other chromosomes. It has also been demonstrated recently that the cells containing the Ph^1 chromosome constitute a clone, having descended from a single cell in which the chromosomal rearrangement originally occurred.

Another specific association involves chromosome 13 and retinoblastoma. Ordinarily, the predisposition to retinoblastoma is inherited as a simple dominant trait with no associated abnormalities. Retinoblastoma also occurs in association with a complex of congenital defects due to deletion of segments of the long arm of 13. Comparison of the segments deleted in different patients indicates that the risk of retinoblastoma is associated with loss of material in or near 13q14 (Knudson et al., 1976). The retinoblastoma mutation presumably involves loss of function of some specific locus, either by structural alteration of the gene or deletion.

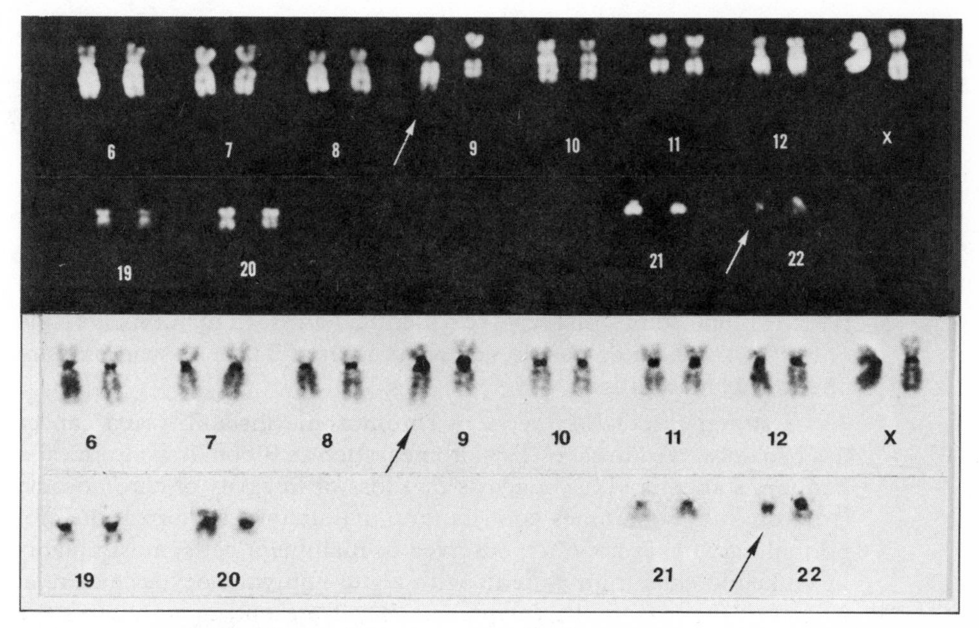

Figure 6-12 Partial karyotype of leukemic cells from a person with chronic myelocytic leukemia. The Philadelphia (Ph^1) chromosome has shorter long arms than normal, the distal segment of 22 being translocated to a number 9. (*Furnished by Janet Rowley.*)

SUMMARY

1. Chromosome breakage followed by reunion leads to a variety of structural abnormalities: deletions, ring chromosomes, inversions, translocations, dicentric chromosomes, and isochromosomes.

2. Deletions (including ring chromosomes) usually lead to phenotypic defect if the deletion involves a large segment.

3. Inversions may be associated with normal phenotype if no genetic material was lost during the rearrangement. At meiosis, a chromosome with an inversion can pair with a normal chromosome by forming a "loop." If crossing over occurs within the loop, some of the meiotic products will be genetically unbalanced.

4. Translocations may be balanced if the exchange of segments between nonhomologous chromosomes involves no loss of genes. Many such balanced translocation carriers are known in man. At meiosis, pairing requires formation of a quadriradial figure consisting of four chromosomes. Segregation from such a figure may lead to grossly unbalanced gene complements. Some familial cases of Down's syndrome are due to unbalanced segregation from balanced translocation carriers, who also may produce phenotypically normal children who are balanced carriers.

5. Translocation sometimes produces chromosomes with two centromeres, that is, they are dicentric. Such chromosomes do not function well if both centromeres are active, since they will often go to opposite poles in anaphase. Instances are known in humans in which one of the centromeres is inactive, with a resulting chromosome that can segregate normally in anaphase.

6. Isochromosomes result from misdivision of the centromere, giving rise to chromosomes that have two identical arms. In man this is viable only with the X chromosome, and some cases of Turner's syndrome are due to this mechanism.

7. A strong association between chromosome instability and cancer has been observed in several syndromes, such as Bloom's syndrome and Fanconi's anemia. This suggests that loss of integrity of chromosome structure may sometimes contribute to malignant transformation. Abnormal karyotypes are often observed in malignant cells, most notably in leukemic cells from patients with acute nonlymphocytic leukemia. However, the specific karyotypic change, if present, is not always the same. An exception is chronic myelocytic leukemia, in which the leukemic cells typically have the distal region of 22q translocated to another chromosome. Many cases of retinoblastoma also are due to deletion of a segment of chromosome 13 in or near 13q14.

REFERENCES AND SUGGESTED READING

BARR, R. D., AND FIALKOW, P. J. 1973. Clonal origin of chronic myelocytic leukemia. *N. Engl. J. Med.* 289: 307–309.

BORGAONKAR, D. S. 1977. *Chromosomal Variation in Man,* 2nd Ed. New York: Alan R. Liss, 403 pp.

COHEN, M. M., MAC GILLIVRAY, M. H., CAPRARO, V. J., AND ACETO, J. A. 1973. Human dicentric Y chromosomes. *J. Med. Genet.* 10: 74–79.

DE LA CHAPELLE, A., SCHRÖDER, J., STENSTRAND, K., FELLMAN, J., HERVA, R., SAARNI, M., ANTTOLAINEN, I., TALLILA, I.,TERVILÄ, L., HUSA, L., TALLQVIST, G., ROBSON, E. B., COOK, P. J. L., AND SANGER, R. 1974. Pericentric inversions of human chromosomes 9 and 10. *Am. J. Hum. Genet.* 26: 746–766.

FERGUSON-SMITH, M. A. 1961. Chromosomes and human disease. *Prog. Med. Genet.* 1: 292–334.

FORD, C. E., AND CLEGG, H. M. 1969. Reciprocal translocations. *Br. Med. Bull.* 25: 110–114.

GERMAN, J. 1972. Genes which increase chromosomal instability in somatic cells and predispose to cancer. *Prog. Med. Genet.* 8: 61–101.

GOLOMB, H. M., VARDIMAN, J. W., ROWLEY, J. D., TESTA, J. R., AND MINTZ, U. 1978. Correlation of clinical findings with quinacrine-banded chromosomes in 90 adults with acute nonlymphocytic leukemia. *N. Engl. J. Med.* 299: 613–619.

GROUCHY, J. DE, LAMY, M., THIEFFRY, S., ARTHUIS, M., AND SALMON, C. 1963. Dysmorphie complexe avec oligophrénie: délétion des bras courts d'un chromosome 17-18. *C. R. Acad. Sci.* (Paris) 256: 1028–1029.

GROUCHY, J. DE, AND TURLEAU, C. 1977. *Clinical Atlas of Human Chromosomes.* New York: John Wiley, 319 pp.

HAMERTON, J. L. 1971. *Human Cytogenetics. Vol.* 1, *General Cytogenetics,* 412 pp; *Vol.* 2, *Clinical Cytogenetics,* New York: Academic Press, 545 pp.

HULTÉN, M., AND LINDSTEN, J. 1973. Cytogenetic aspects of human male meiosis. *Adv. Hum. Genet.* 4: 327–410.

KNUDSON, A. G., Jr, MEADOWS, A. T., NICHOLS, W. W., AND HILL, R. 1976. Chromosomal deletion and retinoblastoma. *N. Engl. J. Med.* 295: 1120–1123.

LEJEUNE, J., LAFOURCADE, J., BERGER, R., VIALATTE, J., BOESWEILLWALD, M., SERINGE, P., AND TURPIN, R. 1963. Trois cas de délétion partielle du bras court du chromosome 5. *C. R. Acad. Sci.* (Paris) 257: 3098–3102.

LEVINE, H. 1971. *Clinical Cytogenetics.* Boston: Little, Brown, 572 pp.

NIEBUHR, E. 1978. The Cri du Chat syndrome. *Hum. Genet.* 44: 227–275.

Paris Conference (1971): Standardization in Human Cytogenetics. *Birth Defects: Original Article Series* 8 (No. 7) 1972. The National Foundation, New York. 46 pp.

Paris Conference (1971), Supplement (1975): Standardization in Human Cytogenetics. *Birth Defects: Original Article Series* 11 (No. 9) 1975. The National Foundation, New York. 36 pp.

PENROSE, L. S., ELLIS, J. R., AND DELHANTY, J. D. A. 1960. Chromosomal translocation in mongolism and in normal relatives. *Lancet* 2: 409–410.

POLANI, P. E., BRIGGS, J. H., FORD, C. E., CLARKE, C. M., AND BERG, J. M. 1960. A mongol girl with 46 chromosomes. *Lancet* 1: 721–724.

ROWLEY, J. D. 1973. A new consistent chromosomal abnormality in chronic myelogenous leukaemia identified by quinacrine fluorescence and Giemsa staining. *Nature* (Lond.) 243: 290–293.

RUSSELL, L. B. 1962. Chromosome aberrations in experimental mammals. *Prog. Med. Genet.* 2: 230–294.

SHAW, M W. 1962. Familial mongolism. *Cytogenetics* 1: 141–179.

THERMAN, E., SARTO, G. E., AND PATAU, K. 1974. Apparently isodicentric but functionally monocentric X chromosome in man. *Am. J. Hum. Genet.* 26: 83–92.

WANG, H-C., MELNYK, J., MC DONALD, L. T., UCHIDA, I. A., CARR, D. H., AND GOLDBERG, B. 1962. Ring chromosomes in human beings. *Nature* (Lond.) 195: 733–734.

YUNIS, J. J. (Ed.). 1977. *New Chromosomal Syndromes.* Academic Press, New York. 404 pp.

REVIEW QUESTIONS

1. Define:

acentric	isochromosome	ring chromosome
deletion	paracentric	translocation
dicentric	pericentric	unequal crossing over
duplication	Ph^1 chromosome	interstitial

2. Give the karyotype formulas of the following:

a. Down's syndrome involving Robertsonian translocation between chromosomes 21 and 22.

b. Turner's syndrome involving a ring-X chromosome.

c. *Cri-cu-chat* syndrome

d. A male with a reciprocal translocation between the distal segments of the short arms of chromosomes 1 and 7.

e. A female with a terminal deletion of the short arm of chromosome 8, starting with band 21.

3. A woman is mosaic for normal cells and for cells with a balanced translocation in which the long arm of the X chromosome is attached to the short arm of number 13. Assuming her to be fertile but with gonadal mosaicism, what would be the potential gametes? Which would be balanced? If she were mated with a normal male, what kinds of offspring might be produced? How many sex chromatin bodies would each have?

4. For a balanced reciprocal translocation:

a. Show that crossing over in the interstitial region leads to no new gametic combinations if followed by alternate or adjacent-1 segregation.

b. Give the six gametes that may result from crossing over in the interstitial region followed by adjacent-2 segregation.

5. Prior to the development of banding techniques, it was impossible to distinguish cytologically between 45,−21,−21,+t(21q21q) and 45,−21,−22,+t(21q22q). A woman with karyotype 45,−G,−G,+t(GqGq) had one child with Down's syndrome and one phenotypically normal child. Which type of translocation does she have?

CHAPTER SEVEN

DETERMINATION OF SEX

Reproductive mechanisms are divided into two major categories—sexual and asexual. In asexual reproduction, an organism gives rise to one or more offspring identical in genetic content to the parent. In sexual reproduction, offspring receive a genetic contribution from more than one parent. Many organisms can reproduce either sexually or asexually. Most organisms that rely primarily on asexual mechanisms in reproduction undergo sexual reproduction on occasion.

The universal occurrence of sexual reproduction among higher plants and animals suggests that the system is biologically beneficial. This benefit would seem to be the greatly increased opportunity to experiment with new combinations of genes. In asexual reproduction, new genetic combinations arise only by mutation of genes. Evolution by the accumulation of mutations is at best a slow process, since favorable new complexes of mutant genes must arise by stepwise accumulation of individual mutations. Unless these mutations are individually favorable, they are not likely to persist.

By contrast, in sexual reproduction, mutations arising in different organisms have a chance to reassort and come together into the same genome. Such assortment permits evolution to proceed at a much faster rate than is possible by sequential mutation, thus providing the population with greater flexibility to respond to environmental changes.

SEXUAL REPRODUCTION IN PROKARYOTES

The mechanisms of sexual reproduction are quite varied. Probably the simplest occurs among viruses. In bacteriophage, for example, if two related but genetically distinguishable virus particles both infect the same bacterial cell, many of the new virus particles produced will carry genes from both parents. There is no basic difference in the sexual and asexual processes in viruses, the only distinction being whether the progeny

DNA has more than one parental template from which to copy. Furthermore, there are no "sexes," since there are no restrictions involving combinations of parental particles and there are no functional differences among various particles.

A slightly more complex system is found in the bacteria in which sexual processes can be observed. In certain strains of *Escherichia coli*, conjugation and transfer of genetic material depend on the presence in one of the cells of tiny DNA particles that may be in the cytoplasm or attached to the chromosome. These F (fertility) particles promote conjugation and lead to the transfer of genes from F-infected (F+) cells to noninfected cells. The recipient cell is then comparable to a zygote, although it may have received only a portion of the genome of the F+ cell. This partially diploid state may be maintained through several divisions, but eventually a haploid state is restored. The new haploid cell may exhibit genetic traits characteristic of both the original parental cells.

Two other sexual processes are observed in bacteria in the laboratory and may also occur in nature. These are *transformation* and *transduction*. In transformation, the genetic material (DNA) from one bacterial cell is present in solution in the environment of another related cell. This intact cell absorbs the DNA, and after cell division the progeny have acquired some of their genes from the absorbed DNA. Transduction is a similar process except that viruses serve as vectors in transporting bacterial DNA from one cell to another. The virus must be temperate, that is, it must be able to infect without destroying the host cell.

SEXUAL REPRODUCTION IN EUKARYOTES

In prokaryotes, sexual reproduction is the exception rather than the rule. As evolution led to more complex organisms, selection favored those groups in which sexual reproduction became an integral part of the life cycle. This assured that each generation was a new complex of genes and hence a new experiment in survival advantage. Along with the acquisition of sexual reproduction, most species developed special processes for transmitting nuclear and cytoplasmic material to offspring. These processes generally can be classified as "male" or "female," the distinction being that in males, only nuclear material is contributed, whereas in females, both nuclear and cytoplasmic material are contributed to the progeny. In many organisms, particularly in plants, both male and female organs are found in the same individual, and self-fertilization may occur. In others, an individual is either male or female. In some plants, special genes prevent fertilization of the ova by pollen of the same plant, thus assuring greater variety of genes in the offspring.

Sex Determination in Nonmammals

Although sexual dimorphism is characteristic of a very large segment of the biological world, the mechanism by which sexual distinction occurs is remarkably variable in different species. An example has already been given of primitive sex determination of some bacteria owing to presence or absence of viruslike particles. A "female" cell lacking particles can be transformed into a "male" cell by becoming infected with the particles.

Environmental control of sex is found in the marine worm *Bonellia viridis*. Larvae that develop on the proboscis of a female become males, whereas those that develop away from females become females. Presumably some chemical secretion of the females influences the direction of development.

Genetic control of sex is common, although the precise mechanisms vary. In *Hymenoptera*, fertilized eggs have a diploid complement of chromosomes and become females. Unfertilized eggs can develop, however, and always yield haploid males. Thus diploidy versus haploidy appears to be the feature responsible for sex determination.

A simpler genetic mechansim is found in organisms such as the mold *Neurospora*. Although this mold commonly reproduces asexually, it can undergo meiosis and sexual reproduction. There are two sexes, exactly equivalent and designated *A* and *a*. Mating occurs only between gametes of unlike sex. The mating type, *A* or *a*, is determined by a pair of autosomal alleles and follows simple Mendelian inheritance.

A chromosomal basis for sex determination is found in mammals, birds, Drosophila, and many other animals. In mammals and *Drosophila,* sex determination is of the XY type. Females have two X chromosomes and are said to be *homogametic*; that is, all of the gametes produced by females have an X chromosome and are therefore equivalent with respect to sex determination. Males are *heterogametic* in that they have only one X chromosome plus one Y chromosome. Gametes may carry either an X or a Y chromosome. As indicated in Chapter 4, human X chromosomes are fairly large, whereas Y chromosomes are in the smallest group.

Birds have a similar mechanism of sex determination, except that females are the heterogametic sex. To avoid confusion with the XY mechanism, the symbols WZ are used, males being ZZ and females ZW. A third type of chromosomal mechanism is the XO type found in some fishes. Absence of a chromosome is designated O, and in these species the females are XX, the males XO. Thus, the males have one chromosome fewer than the females.

The statement that human or *Drosophila* females are XX and males are XY leaves unanswered an interesting question regarding sex determination. What attribute of the chromosomal constitution actually is responsible for maleness or femaleness? This question is legitimate be-

Table 7-1 **SEX DETERMINATION IN *DROSOPHILA****

Phenotypic sex	X chromosomes	Sets of autosomes	Ratio X/autosomes
Superfemale (sterile)	3	2	1.5
Normal female (tetraploid)	4	4	1.0
Normal female (triploid)	3	3	1.0
Normal female (diploid)	2	2	1.0
Intersex (sterile)	2	3	0.67
Normal male	1	2	0.5
Supermale (sterile)	1	3	0.33

* From Bridges, 1925.

cause at least five parameters are different in the XX and XY karyotypes: (1) the number of X chromosomes, (2) the number of Y chromosomes, (3) the ratio of X chromosomes to autosomes, (4) the ratio of Y chromosomes to autosomes, and (5) the ratio of X chromosomes to Y chromosomes. It is difficult to separate these variables in normal persons, but consideration of chromosomally abnormal persons can answer these questions, as suggested in the previous chapter.

Sex Determination in Drosophila. A variety of aneuploid states involving the sex chromosomes is known in *Drosophila.* In 1922, Bridges analyzed some of these to arrive at a theory of sex determination. Flies that were polyploid were normal females, provided the number of X chromosomes and the number of sets of autosomes were equal. Flies with three X chromosomes but the normal number of autosomes ($2n$) were sterile "superfemales." In the presence of three sets of autosomes, two X chromosomes produced a sterile intersex and one X chromosome produced a sterile "supermale." These results are tabulated in Table 7-1.

The number of Y chromosomes does not seem to play an important role in sex determination in *Drosophila.* Bridges concluded that the critical factor seemed to be the ratio of X chromosomes to sets of autosomes. The autosomes appear to have genes for maleness, whereas the X chromosome has genes for femaleness. In an XX diploid fly, the balance is correct to produce a functional female; in an XY diploid, the balance is right for a functional male. Any deviation from this ratio produces a sexually nonfunctional fly.

Sex Determination in Mammals

Since *Drosophila* and mammals both have an XX-XY type of sex determination, it might be supposed that the mechanism in each case is the same. However, the evolutionary distance between them is so large as to necessitate independent verification. The analysis used in mammals is the same as in *Drosophila*.

A discussion of human aneuploidy involving sex chromosomes was given in Chapter 5. A variety of aneuploid conditions is known, all of which follow the rule that presence of a Y chromosome results in a male phenotype. Absence of a Y chromosome results in a female. The number of X chromosomes is important, but the effects are small compared with the effects of Y chromosomes. For example, persons with Klinefelter's syndrome, XXY, are clearly males, and even XXXXY persons are males. But XXY males have greatly reduced fertility, and whether they can be fertile at all is yet to be ascertained.

The contrast between sex determination in man and *Drosophila* is shown in Table 7-2. Sex determination in house mice resembles that in human beings, although the abnormal complements produce somewhat different phenotypes (Cattanach, 1974).

The H-Y Antigen and Sex

Individuals that are genetically identical, whether mice or human beings, can exchange skin or other organ grafts without formation of antibodies to and rejection of the graft. In man, this occurs only if the persons are identical twins. In experimental organisms, especially mice, there exist strains that have been inbred so many generations that the entire strain is *isogenic,* that is, all individuals have the same genotype. This is almost true, at least, but not quite, since females and males differ in their sex chromosomes. Ignoring the number of X chromosomes, males should have all the genes that females have and they would be identical. But in addition males would have genes on the Y chromosome not shared by females.

Using this fact, Eichwald and Silmser (1955) demonstrated that, in inbred mouse strains, female skin can be transplanted to male mice

Table 7-2 **SEX AND CHROMOSOME STATUS IN THREE ORGANISMS WITH XX–XY SEX DETERMINATION**

Sex chromosome status	Human phenotype	Mouse phenotype	*Drosophila* phenotype
XO	Sterile female*	Fertile female	Sterile male
XX	Normal female	Normal female	Normal female
XXX	Fertile female	Unknown	Sterile female
XY	Normal male	Normal male	Normal male
XXY	Sterile male	Sterile male	Fertile female
XYY	Fertile male	Semisterile male	Fertile male

* A few cases of fertile XO females have been reported, although these may have been mosaic XO/XX.

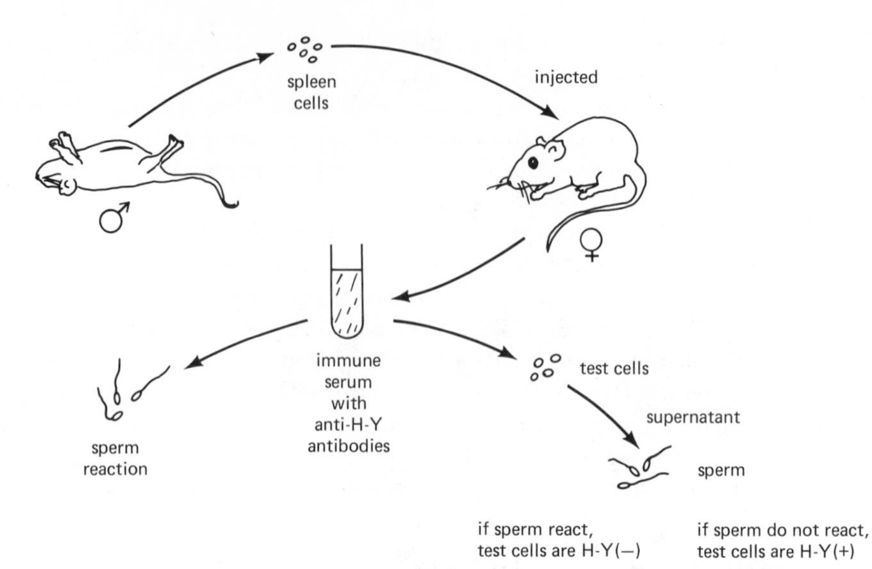

Figure 7-1 Diagram of H-Y test procedure. Cells from a male mouse are injected into a female from the same isogenic strain. The immune serum produced will react with male cells, indicating the presence of antigens in males that are not present in females. One of the common ways of detecting anti–H-Y antibodies is with sperm, which are rich in H-Y antigen. A positive reaction may be detected by sperm killing, mixed agglutination with red cells, etc. Other cells can be tested indirectly by their ability to remove the antibodies from the immune test serum, which can then no longer react with sperm.

without rejection, but male skin transplanted to female mice is eventually rejected. It was concluded that there must be antigenic determiners in males dependant on Y-linked genes. In keeping with other *histocompatibility* loci, the Y-linked locus was named H-Y.

Evidence for Y-linked histocompatibility loci was obtained in several other species (reviewed in Silvers and Wachtel, 1977). However, direct testing of histocompatibility can only be done within species, preventing cross-species comparisons. This problem was solved with the development of an *in vitro* assay for cell-bound H-Y antigens. Briefly, serum from female mice immunized against mouse H-Y (usually spleen cells from males) is exposed to a test tissue, such as white cells, sperm, or fibroblasts. The "absorbed" serum is then tested for its ability to react with mouse sperm. If the test tissue is H-Y(+), the mouse serum can no longer react with sperm. If it is H-Y(−), the antibodies against H-Y can still react (Figure 7-1). Using the test, it has been shown that H-Y is present on cells of human males, and it is immunologically identical in all vertebrates tested. It is noteworthy that in chickens and in the South African clawed frog, *Xenopus laevis,* the females rather than the males are heterogametic. In these instances, females are H-Y(+) and males are H-Y(−).

As will be noted further in Chapter 13, cellular antigens are complex structures, only a few of which have been studied chemically. Many show variation that is inherited as a simple Mendelian trait. One may presume that such is also the case here, i.e. a locus on the Y chromosome specifies the structure of an antigen on cells of mammalian males. One may use the presence of H-Y antigen as evidence for Y-derived chromosomal material in persons who lack an identifiable Y chromosome. This is useful in resolving questions about some of the apparent discrepancies between sex and karyotype discussed later. Still to be resolved is whether the gene that specifies structure of the H-Y antigen is itself on the Y chromosome or whether that gene is on an autosome or X chromosome, with a regulator on the Y that turns it on.

SEX DIFFERENTIATION IN MAMMALS

The events that lead to differentiation into males and females are incompletely understood. A scheme is presented in Figure 7-2 that gives some of the principal relationships established. Initially the gonad appears to have the potential of developing either into a testis or an ovary. If a Y chromosome is present, the undifferentiated gonad becomes a testis. There is presumably a male inducer substance, but its nature is unknown. There appears to be no counterpart female inducer substance, development of an ovary being the consequence of absence of male induction.

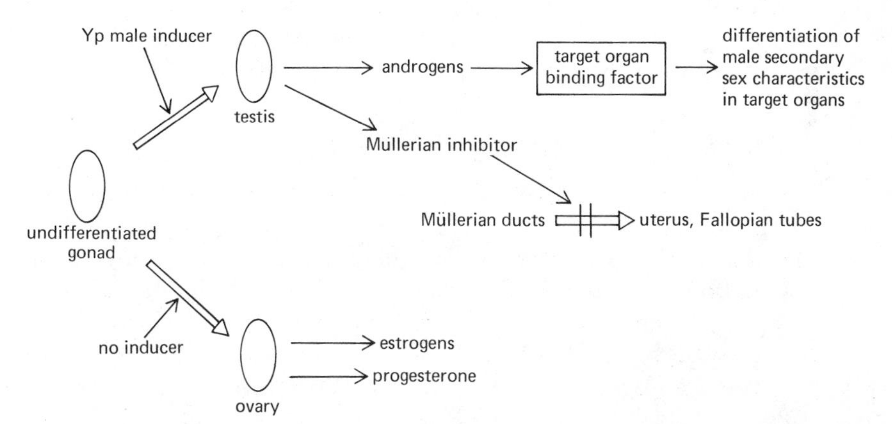

Figure 7-2 Diagram showing differentiation of undifferentiated gonad under the influence of male inducer to produce a testis, with subsequent effects on development of secondary sex characteristics and inhibition of Müllerian ducts. The male inducer appears to result from action of the short arm of the Y chromosome (Yp). Pituitary and other endocrine interactions not directly involved in the testis/ovary differentiation are omitted.

Testes elaborate two substances—androgen from the interstitial tissue and Müllerian inhibitor from the seminiferous tubules. Androgens, principally testosterone, are secreted into the circulation and cause the target organs to develop into male structures or to have male characteristics. The ability of tissues to respond to androgen is inherited, as evidenced by the condition testicular feminization, in which the various tissues fail to respond to normal levels of androgen. The result is an external female phenotype.

Most male characteristics can be induced in castrated fetuses by administration of androgen. One process cannot. This is the inhibition of development of Müllerian ducts into uterus and Fallopian tubes. The seminiferous tubules of the fetal testis secrete a factor, the function of which is to cause regression of the Müllerian ducts. Little is known of the nature of this substance. In its absence, the Müllerian duct of the embryo develops without additional stimulus into uterus, Fallopian tubes, and upper vagina.

One may summarize the above observations by saying that an embryo develops into a male if male inducer directs the undifferentiated gonad to become a testis early in development; otherwise the gonad becomes an ovary and the embryo a female.

GENETIC VARIATION IN HUMAN SEXUAL DEVELOPMENT

Various genetic defects in human beings interrupt the normal sequence of sex development. Some of these have been very instructive in separating the steps of sex development. In addition to abnormal development due to chromosomal aneuploidy, many simply inherited disorders have been identified (Goldstein and Wilson, 1974; Simpson, 1977).

Turner's Syndrome and Its Variants

As noted in Chapter 5, persons of XO chromosome constitution are phenotypic females. The main features of Turner's syndrome are streak gonads, short stature, sexual infantilism, and a variety of congenital malformations, including webbing of the neck, shield chest, and multiple pigmented nevi. However, many variations of Turner's syndrome are observed, and some of these can be associated with karyotypes other than standard monosomy for X (Ferguson-Smith, 1965).

It is clear that either a Y chromosome or a second X chromosome is essential for gonadal development. Without these, the gonad remains an undifferentiated streak gonad and the Müllerian ducts consequently differentiate only into infantile structures. Accordingly, neither the second X nor the Y chromosome is genetically completely inactive, and at least two copies of a locus on the X chromosome or two copies on the X and Y combined are necessary for normal gonadal development. It may be that the regions on the short arms of the X and Y chromosomes that are involved in pachytene pairing are genetically homologous and that genes in this region are concerned with gonadal development. Direct proof is lacking, but several kinds of chromosomal aberrations are consistent with this idea. For example, deletion of Xp or of the distal tip of Yp both result in short stature, but this is not necessarily the result of homologous genes (German and co-workers, 1973).

XXp− Females. A number of females with partial monosomy of the short arm of the X chromosome because of deletion have been reported. All have typical Turner's syndrome (Simpson, 1977), as do persons with 46,X,i(Xq), who also are monosomic for Xp.

XXq− Females. These patients have streak gonads but usually lack other features of Turner's syndrome. In particular, they are of normal height, as are persons with 46,X,i(Xp) karyotypes (for example, see de la Chapelle and co-workers, 1972). There must be factors on the Xq arm concerned with gonadal development, but these appear to be distinct from the loci whose imbalance leads to short stature and other stigmata of Turner's syndrome. It should be noted though that several XXq− females have been reported with some of the stigmata of Turner's syndrome (Hecht and co-workers, 1970).

XX Gonadal Dysgenesis. A number of females have been observed to have streak gonads associated with apparently nonmosaic 46,XX karyotypes (Simpson, 1977). Many of these persons are short, but often height is normal, and other stigmata of Turner's syndrome are generally lacking. Although affected persons are rare, there sometimes are two or more in a single sibship. This, coupled with the fact that the parents of affected persons often are related, suggests inheritance as an autosomal recessive trait.

Males with XX Sex Chromosome Constitution

A number of males have been observed to have karyotypes typical of normal of 46,XX females. The condition is rare, occurring on the order of 1 in 10,000 male births. Phenotypically, they somewhat resemble persons with Klinefelter's syndrome, having small testes that lack spermatogenesis. They are taller on the average than 46,XX females but shorter than 46,XY males or 47,XXY males. De la Chapelle (1972) has summarized the clinical findings on 45 cases of 46,XX males.

There have been several suggestions on the origin of XX males. One possibility is the presence of a mutant gene that causes masculinization. Such autosomal genes are known in goats, pigs, and mice. If a dominant mutation were responsible for cases in man, it should lead only occasionally to familial concentration of 46,XX males, since the mutant gene could not be transmitted to offspring because of associated sterility. For the most part, the cases found are isolated, that is, nonfamilial. This would rule out autosomal recessive inheritance as a common explanation. The rarity of XX males is also consistent with a mutational event.

One instructive pedigree has been reported, however (Figure 7-3). This pedigree is complicated by the presence of a small marker chromosome of unknown origin in addition to the expected 46 chromosomes, but the marker chromosome appears to be independent of the sex reversal, since II-I, a 46,XX male, lacks it. If the marker chromosome had male-determining genes for a Y chromosome, it could explain the pedigree, assuming that the marker chromosome was originally present in II-I but became lost after having set the course of gonadal development. The time in embryonic development at which this would have to occur is unknown. Translocation of male-determining loci from a Y chromosome to the X chromosome also seems ruled out by the normal female phenotype of II-2. The most likely explanation is that an autosomal dominant gene exists in this family that, in combination with an XX karyotype, causes male development but has no obvious effect when combined with an XY karyotype. Such a gene, an allele at the *Sxr* (*sex-reversed*) locus, is known in mice.

In another family containing three XX male sibs, the mother appeared to be weakly H-Y(+). It was suggested that in this family, the parents had an autosomal gene that could not produce males in heterozygous combination (as in the mother) but could when homozygous (de

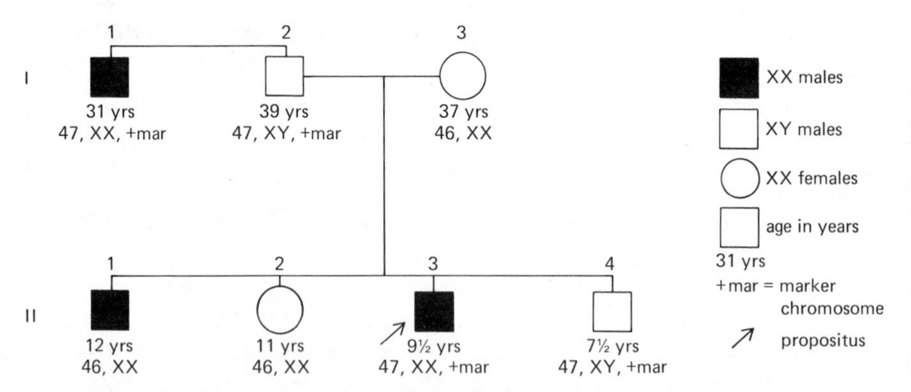

Figure 7-3 Pedigree showing transmission through an XY male of the tendency for XX male phenotype. Most probably an autosomal locus is involved. (*Redrawn by permission from Kasdan and co-workers, 1973.* N. Engl. J. Med. *288: 539.*)

la Chapelle et al., 1978). Maleness in these patients would therefore be an autosomal recessive trait in XX zygotes.

Since most XX males seem not to be familial, another mechanism must be more often involved. One possibility is that a reciprocal translocation might have occurred between the presumably homologous regions of the X and Y chromosomes, in the process of which the male-determining locus of the Y became transferred to the X chromosome. This would always give independent origin for markers on the Xq arm, with resulting genotype frequencies identical with those of XX females. Actually, the frequency of Xg(a+) phenotypes among XX males is the same as among XXY males, although the observations are too few to rule out chance deviation from XX female frequencies. This suggests, though, that the origin of the two X chromosomes may result from parental nondisjunction, and that the fathers contribute a Y chromosome. This would suppose that XX males ordinarily start out as XXY zygotes, and that the Y chromosome became lost in the major line of cells after having initiated testicular development. These persons would then be 46,XX/47,XXY mosaics, with 46,XX tissue as the predominant type. Contrary to this hypothesis is the failure to detect increased maternal age in 46,XX males, in contrast to the well-documented increased maternal age in Klinefelter's syndrome. Also, XX males are now known to be H-Y(+). This suggests that a critical part of the Y chromosome has been translocated to another chromosome, even though it is not visible.

There are many things yet to be learned about this interesting group of patients, and it is possible that some new mechanisms will need to be considered in order to explain fully the characteristics of the group.

Females with XY Chromosome Constitution

Testicular Feminization. The readiness with which phenotypic sex can be genetically influenced is illustrated by the rare condition known as testicular feminization. Persons with this condition have an apparently normal XY karyotype. Yet, externally they appear to be normal females. They are reared as females and have typical feminine attitudes and identification. In most cases, they have come to medical attention because of primary amenorrhea. Internally, they lack ovaries and associated Müllerian-derived structures typical of normal females. The vagina ends blindly, and the gonads are testes.

Testicular feminization is inherited as a simple monogenic trait. A pedigree is shown in Figure 7-4. Affected persons often occur in more than one generation. For many years, it was not known whether the gene is on the X chromosome or an autosome. This distinction, usually simple to make, is difficult here because affected persons of XY karyo-

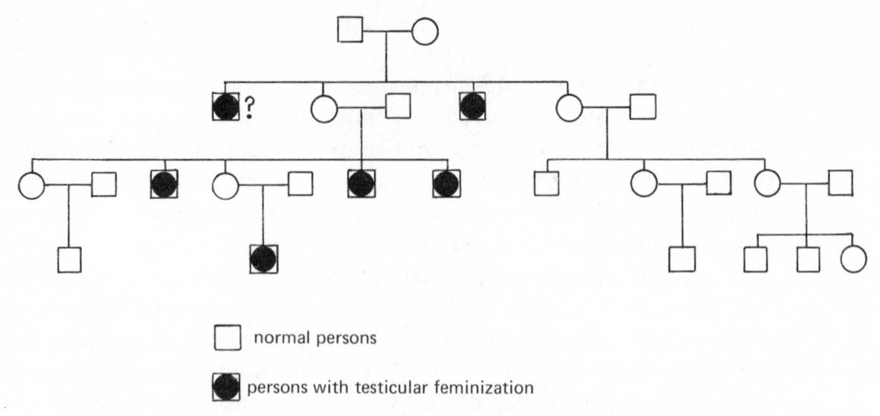

Figure 7-4 Pedigree of testicular feminization. Affected persons are females externally but have testes and an XY karyotype. They are related to each other through normal female carriers. This pattern of heredity is compatible with either a sex-linked recessive gene or an autosomal dominant gene with expression limited to males. (*Redrawn from Pettersson and Bonnier, 1937.*)

type are sterile. Normal females, of XX constitution, do not show effects of the gene, although they may transmit it to their offspring. Pedigrees are consistent with X chromosome inheritance, but the condition itself prohibits fertile matings that would provide the evidence of father to son transmission necessary to eliminate the X chromosome as the site of the gene. A similar gene also occurs in mice as well as in several other species. In mice it is clearly X-linked, and, in view of the fact that structural evolution of the X chromosome in mammals has been very conservative (Ohno, 1967), it was proposed that the human gene also is X-linked. Comparable studies of linkage to other loci on the X chromosome have not been informative in man.

The primary defect, in both human and mouse testicular feminization, is in the response of tissues to normal levels of androgen. Gonadal development is essentially normal, with normal androgen secretion, although sperm are not produced. However, the various end organs are completely insensitive to androgen. The problem lies in lack of activity of a receptor protein specific for dihydrotestosterone. Studies of fibroblast clones of cells from females heterozygous for testicular feminization indicate that some cells are normally responsive to androgen while others are insensitive, supporting the idea that the locus is on the X chromosome (Meyer et al., 1975). White blood cells from these androgen-insensitive patients are H-Y(+), consistent with the presence of the Y chromosome and indicating that androgen binding is not a necessary step in the production of the H-Y antigen (Koo et al., 1977).

Several milder forms of defective androgen action have been distinguished clinically (summarized in Goldstein and Wilson, 1974). These include such disorders as the Gilbert-Dreyfus syndrome, the Lubs syndrome, and the Rosewater syndrome. They are characterized by various degrees of sexual ambiguity, ranging from an essentially female phenotype with slight virilization to an essentially male phenotype with hypospadias and gynecomastia. In each case the gonads are testes and blood testosterone is normal. The traits are familial and are consistent with X-linkage. It is tempting to speculate that they represent different alleles at the locus for testicular feminization, but proof is lacking.

XY Gonadal Dysgenesis (The Swyer Syndrome). Persons with this disorder are females with apparently normal 46,XY karyotypes (Cohen and Shaw, 1965; Simpson, 1977). Swyer syndrome patients differ from persons with testicular feminization in having streak gonads rather than testes, and they have a complete vagina and Fallopian tubes. Breast development is minimal or does not occur. Height is normal. The one case tested for H-Y was positive (Dorus et al., 1977). A number of families have been observed in which there are multiple cases, sometimes in two generations. At least some of the cases appear to be the result of an X-linked recessive gene. An example of a pedigree is given in Figure 7-5. As has been noted by German and co-workers (1978), these cases point out the necessity of both a functional gene on the X chromosome as well as the Y-linked factor in the induction of the male phenotype in humans.

XYP− Gonadal Dysgenesis. Several females with streak gonads have been reported to be effectively 46,XYp−. The two patients of Jacobs and Ross (1966) had isochromosomes for the long arm of Y (Robinson and Buckton, 1971) as did the patient of Böök and co-workers (1973). This suggests that the male-determining factor is located on

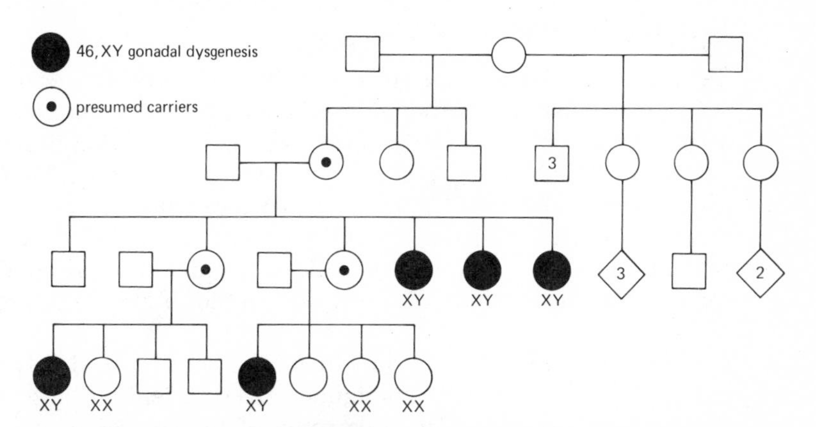

Figure 7-5 Pedigree of 46,XY gonadal dysgenesis. The transmission is consistent with an X-linked recessive trait but is formally consistent also with an autosomal dominant trait expressed only in males. (*Redrawn by permission from Espiner and co-workers, 1970.* N. Engl. J. Med. *283: 6.*)

the short arm of the Y chromosome. This is consistent with the observation that much of the long arm of the Y is heterochromatic and presumably has few active genes. A deletion of most of the Yq arm is consistent with normal male development (Meisner and Inhorn, 1972).

The XY Gonadal Agenesis Syndrome. A small number of persons have been reported with absent vagina and uterus and rudimentary Fallopian tubes (Sarto and Opitz, 1973). No gonadal tissue could be identified. The karyotypes are 46,XY. Developmentally, these persons are consistent with functioning of a testis for a short period, enough for Müllerian duct inhibition but not for Wolffian duct development. The disorder is possibly inherited, but too few patients have been observed to be sure.

Defects in Müllerian Inhibition

Since the testis has two distinct functions in development—secretion of androgen and secretion of of Müllerian inhibitor—it is not surprising that defects limited to the Müllerian inhibitor should have been observed. A number of sibships with two or more affected males have been reported (Goldstein and Wilson, 1974). They are characterized by bilateral cryptorchidism and inguinal hernias which, on repair, show a uterus and Fallopian tubes in the inguinal canal. Wolffian duct development is normal; therefore, these patients appear to be normal males. It is not known whether the defect is in synthesis of the Müllerian inhibitor or in response of tissues to normal levels of the inhibitor.

Hermaphroditism

True hermaphrodites—persons with both ovarian and testicular tissue—are uncommon, although a number have been observed (Polani, 1970). While some true hermaphrodites are XX/XY mosaics, most appear to be 46,XX. The reason for the presence of testicular tissue in these latter is unknown. It may be that these also are cases of undetected mosaicism and that often one gonad contains Y-bearing cells while the other does not. This is difficult to prove or disprove. True hermaphrodites are generally H-Y(+), even in the absence of demonstrable Y mosaicism.

Pseudohermaphroditism, in which both gonads are clearly testes or ovaries but associated with ambiguous external genitalia or genitalia of the opposite sex, would include some of the conditions already discussed. *Male pseudohermaphrodites* are persons with testes but some external female characteristics. These include the testicular feminization and related syndromes as well as several disorders of androgen synthesis. In these latter, the person has an XY karyotype with testes, but specific enzyme defects in the biosynthesis of testosterone result in defective differentiation of sex organs and secondary sex characteristics. These defects are inherited as autosomal recessive traits and are comparable to the metabolic blocks discussed in Chapter 11.

Female pseudohermaphrodites are genetic females with uterus, Fallopian tubes, and ovaries, but with various degrees of masculinization of external genitalia. The inherited forms are usually the result of enzyme blocks in the fetal adrenal gland leading to overproduction of adrenal androgens. In males, the defect is expressed as precocious puberty, but in females the effects on the external genitalia are more striking. The defects are inherited as autosomal recessive traits.

In many additional cases, the basis for the abnormal development of sexual characteristics is unknown. Some may well be genetic. Others may be due to nongenetic disturbances in endocrine function during a crucial period of development.

Correlation between Phenotype, Karyotype, and H-Y Status

Any effort to derive general rules about sex determination must contend with the fact that there are many examples of aberrant sex development that are poorly understood, some of which appear to contradict the general rules. Some of the "exceptions" in the past have disappeared with development of new test techniques. Some also proved very informative with more detailed study. In spite of the unexplained exceptions, it is useful to search for general rules against which to test new observations as they become available.

1. *Simple Mendelian genes can cause persons with normal male or female karyotype to develop as the opposite sex.* Examples are testicular feminization in humans, mice, and rats; sex-reversed in mice (*Sxr*) and in humans (some XX males); a number of defects in hormone synthesis.

2. *In the case of abnormal karyotypes, presence of the proximal region of the short arm of the Y chromosome produces a male.* Males occasionally are produced when the Yp segment cannot be detected, but no example is known of female development if the karyotype contains Yp. Yq, or most of it at least, is not important in sex determination.

3. *If the proximal segment of Yp is present, the person is H-Y*(+), *even though other genes may have caused female phenotype.*

4. *All males, even those who lack a detectable Yp, are H-Y*(+). This suggests the presence of Yp through translocation to some other chromosome.

These generalizations underscore the importance of male-determining factors on the short arms of normal Y chromosomes. These factors may be translocated to other chromosomes, perhaps most often to a homologous region of the X chromosome. It has been proposed that the H-Y antigen itself is the male factor, but it is awkward to reconcile both the conservative nature of H-Y among species and the presence of H-Y in females when they are the heterogametic sex.

THE SEX RATIO

The ratio of males to females has been extensively studied in man and experimental animals. It is useful to distinguish between the primary sex ratio, that is, the ratio at conception, and the secondary sex ratio, that is, the ratio at birth. The tertiary sex ratio is sometimes used for specified later ages.

Considering only the process of meiosis, there should be equal numbers of X- and Y-bearing sperm, giving rise to equal numbers of males and females (a sex ratio of 1.0). In fact, the secondary sex ratio of whites is approximately 1.06, and it is greater than 1.0 in all populations. The reasons for such an excess of males is unknown. It was thought at one time that the primary sex ratio was still higher, based on the fact that there appeared to be a larger number of male abortuses than female. This observation, however, seems to have been attributable to the difficulty in determining the sex of early fetuses, females being often misclassified as males. More recent studies using karyotype analysis to determine the sex have given ratios much closer to 1.0. Although the primary sex ratio cannot be determined because of the impossibility of obtaining zygotes in large numbers, there is no convincing evidence at present to suggest that the sex ratio changes substantially from early embryo to birth. Nor are the data adequate to rule out a moderate change (Hamerton, 1971).

Although the cause of sex ratio deviation is unknown, several associations have been noted. The variation over time is especially well documented (Figure 7-6). An increase in male births in Europe following World Wars I and II has intrigued investigators for many years. It has been suggested that in time of war, with disruption of family life and reduced opportunities for impregnation, those males with more "vigorous" sperm are likely to impregnate more readily and that the outcome of such pregnancies is more likely to be a male offspring. It is difficult to justify such an explanation in genetic terms, however. No properties of Y-bearing sperm or XY zygotes are known that would be likely to lead to these consequences. Indeed, one would expect the survival of an XY zygote to be more influenced by the genetic composition of its X chromosome than its Y chromosome.

Other associations of the sex ratio that have been noted are with

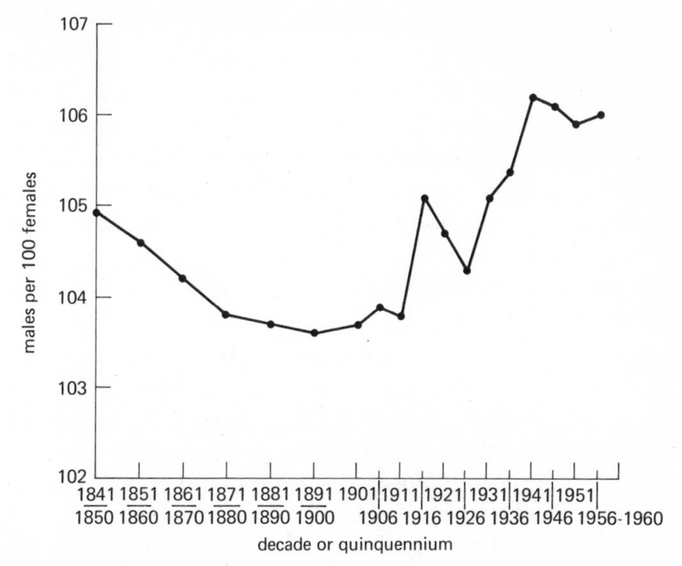

Figure 7-6 Sex ratio for live births in England and Wales, 1841–1960. (*Redrawn from Parkes, 1963.*)

race, socioeconomic status, parity, and interval between insemination and ovulation. Alterations in the sex ratio have been especially suggested as a means of detecting lethal mutations on the X chromosome and are so used in some experimental organisms. This is discussed further in Chapter 9.

SUMMARY

1. Sex is determined by a variety of mechanisms. In some bacteria it depends large on "infective" particles. In eukaryotes it may depend on environmental factors, single genes, or chromosome complements. In mammals, the crucial factor is presence of a Y chromosome (in males) or absence of a Y chromosome (in females).

2. Cells from XY males have an antigen, H-Y, not found on XX cells. The locus for this antigen appears to be on Yp, near to and perhaps identical with the locus for male gonadal determination. Exceptions are birds and *Xenopus laevis,* in which the heterogametic females are H-Y(+).

3. Females who lack a second X chromosome or who lack either the short arm only (XXp−; X,iXq) or the long arm only (XXq−; X,iXp) have streak gonads rather than ovaries. Apparently genes on both Xp and Xq are necessary for gonadal development. Deletion of Xp is associated with the stigmata of Turner's syndrome, especially short stature. Persons with deletion of Xq usually lack stigmata of Turner's syndrome. Gonadal dysgenesis without Turner's stigmata also appears to be inherited on rare occasions as a recessive trait.

4. Males with XX constitution occur, although infrequently. The mechanism is unknown, although in a few families it is clearly genetic, perhaps due to an autosomal dominant gene expressed only in XX persons.

5. In several conditions, a female phenotype is associated with an XY karyotype. In testicular feminization, inherited as an X-linked recessive trait, somatic tissues are unable to respond to the normal levels of androgen secreted by the testes. Such persons have normal female external genitalia. They lack a uterus and Fallopian tubes, indicating normal response to Müllerian inhibitor. In XY gonadal dysgenesis, patients have streak gonads rather than testes, and they have a complete vagina and Fallopian tubules, reflecting lack of Müllerian inhibition. Some of the cases appear due to an X-linked recessive gene. Both conditions are H-Y(+). A phenotype similar to XY gonadal dysgenesis can occur in XYp− patients, but XYq− persons develop as normal males. This is consistent with the location of the male sex-determining gene(s) on Yp.

6. Abnormal sexual development may also occur as a result of enzyme deficiencies in the biosynthesis of testosterone in males, giving rise to genetic males with testes but with female genitalia. Similarly, a block in the biosynthesis of certain adrenal hormones may lead to excessive secretion of adrenal androgens, resulting in virilization of external genitalia in females. Both groups of defects are inherited as autosomal recessive traits.

7. The primary sex ratio (the ratio of males to females at conception) is approximately 1.0, although it is not accurately known. The secondary sex ratio (the ratio at birth) is slightly greater than 1.0, being 1.06 in whites. However, it fluctuates, having reached peaks after World Wars I and II. The reasons for the fluctuations are unknown.

REFERENCES AND SUGGESTED READING

BÖÖK, J. A., EILON, B., HALBRECHT, I., KOMLOS, L., AND SHABTAY, F. 1973. Isochromosome Y [46,X,i(Yq)] and female phenotype. *Clin. Genet.* 4: 410–414.

BRIDGES, C. B. 1925. Sex in relation to chromosomes and genes. *Am. Naturalist* 59: 127–137.

CATTANACH, B. M. 1974. Genetic disorders of sex determination in mice and other mammals. In *Birth Defects* (A. G. Motulsky, W. Lenz, Eds.). Amsterdam: Excerpta Medica, pp. 129–141.

COHEN, M.M., AND SHAW, M. W. 1965. Two XY siblings with gonadal dysgenesis and a female phenotype. *N. Engl. J. Med.* 272: 1083–1088.

DE LA CHAPELLE, A. 1972. Nature and origin of males with XX sex chromosomes. *Am. J. Hum. Genet.* 24: 71–105.

DE LA CHAPELLE, A., KOO, G. C., AND WACHTEL, S. S. 1978. Recessive sex-determining genes in human XX male syndrome. *Cell* 15: 837–842.

DE LA CHAPELLE, A., SCHRODER, J., AND PERNU, M. 1972. Isochromosome for the short arm of X, a human 46,*XXpi* syndrome. *Ann. Hum. Genet.* 36: 79–87.

DORUS, E., AMAROSE, A. P., KOO, G. C., AND WACHTEL, S. S. 1977. Clinical, pathologic, and genetic findings in a case of pure 46,XY gonadal dysgenesis (Swyer's syndrome). *Am. J. Obstet. Gynecol.* 127: 829–831.

EDWARDS, A. W. F. 1962. Genetics and the human sex ratio. *Adv. Genet.* 11: 239–272.

ESPINER, E. A., VEALE, M. O., SANDS, V. E., AND FITZGERALD, P. H. 1970. Familial syndrome of streak gonads and normal male karyotype in five phenotypic females. *N. Engl. J. Med.* 283: 6–11.

FEDERMAN, D. D. 1973. Genetic control of sexual difference. *Prog. Med. Genet.* 9: 215–235.

FERGUSON-SMITH, M. A. 1965. Karyotype-phenotype correlations in gonadal dysgenesis and their bearing in the pathogenesis of malformations. *J. Med. Genet.* 2: 142.

GERMAN, J., SIMPSON, J. L., CHAGANTI, R. S. K., SUMMITT, R. L., REID, L. B., AND MERKATZ, I. R. 1978. Genetically determined sex-reversal in 46,XY humans. *Science* 202: 53–56.

GERMAN, J., SIMPSON, J. L., AND MCLEMORE, G. A. JR. 1973. Abnormalities of human sex chromosomes. I. A. ring Y without mosaicism. *Ann. Génét.* 16: 225–231.

GOLDSTEIN, J. L., AND WILSON, J. D. 1974. Hereditary disorders of sexual development in man. In *Birth Defects* (A. G. Motulsky, W. Lenz, Eds.). Amsterdam: Excerpta Medica, pp. 165–173.

HAMERTON, J. L. 1971. *Human Cytogenetics.* Vol. II. *Clinical Cytogenetics.* New York: Academic Press, p. 393.

HECHT, F. H., JONES, O. L., DELAY, M., AND KLEVIT, H. 1970. Xq− Turner's syndrome: reconsideration of hypothesis that Xp− causes somatic features in Turner's syndrome. *J. Med. Genet.* 7: 1–4.

JACOBS, P. A., AND ROSS, A. 1966. Structural abnormalities of the Y chromosome in man. *Nature* (Lond.) 210: 352–354.

JOSSO, N. 1973. *In vitro* synthesis of Müllerian-inhibiting hormone by seminiferous tubules isolated from the calf fetal testis. *Endocrinology* 93: 829–834.

JOST, A. 1972. A new look at the mechanisms controlling sex differentiation in mammals. *Johns Hopkins Med. J.* 130: 38–53.

JOST, A. 1974. Development and endocrinological aspects of sex differentiation. In *Birth Defects* (A. G. Motulsky, W. Lenz, Eds.). Amsterdam: Excerpta Medica, pp. 142–147.

KASDAN, R., NANKIN, H. R., TROEN, P., WALD, N., PAN, S., AND YANAIHARA, T. 1973. Paternal transmission of maleness in XX human beings. *N. Engl. J. Med.* 288: 539–545.

KOO, G. C., WACHTEL, S. S., KRUPEN-BROWN, K., MITTL, L. R., BREG, W. R., GENEL, M., ROSENTHAL, I. M., BORGAONKAR, D. S., MILLER, D. A., TANTRAVAHI, R., SCHRECK, R. R., ERLANGER, B. F., AND MILLER, O. J. 1977. Mapping the locus of the *H-Y* gene on the human Y chromosome. *Science* 198: 940–942.

KOO, G. C., WACHTEL, S. S., SAENGER, P., NEW, M. I., DOSIK, H., AMAROSE, A. P., DORUS, E., AND VENTRUTO, V. 1977. H-Y antigen: Expression in human subjects with the testicular feminization syndrome. *Science* 196: 655–656.

MAYNARD SMITH, J. 1978. *The Evolution of Sex.* Cambridge: Cambridge Univ. Pr., 222 pp.

MEISNER, L. F., AND INHORN, S. L. 1972. Normal male development with Y chromosome long arm deletion (Yq−). *J. Med. Genet.* 9: 373–377.

MEYER, W. J. III, MIGEON, B. R., MIGEON, C. J. 1975. Locus on human X chromosome for dihydrotestosterone receptor and androgen insensitivity. *Proc. Natl. Acad. Sci.* (U.S.A.) 72: 1469–1472.

MITTWOCH, U. 1973. *Genetics of Sex Differentiation.* New York and London: Academic Press, 253 pp.

OHNO, S. 1967. *Sex Chromosomes and Sex-linked Genes.* Berlin, Heidelberg, New York: Springer-Verlag, 192 pp.

OHNO, S. 1971. An argument for the simplicity of mammalian regulatory systems: single gene determination of male and female phenotypes. *Nature* (Lond.) 234: 134–137.

PARKES, A. S. 1963. The sex-ratio in human populations. In *Man and His Future* (G. Wolstenholme, Ed.). A Ciba Foundation Volume. Boston: Little, Brown, pp. 91–99.

PETTERSSON, G., AND BONNIER, G. 1937. Inherited sex-mosaic in man. *Hereditas* 23: 49–69.

POLANI, P. E. 1970. Hormonal and clinical aspects of hermaphroditism and the testicular feminization syndrome. *Phil. Trans. Roy. Soc. London B* 259: 187–206.

ROBINSON, J. A., AND BUCKTON, K. E. 1971. Quinacrine fluorescence of variant and abnormal human Y chromosomes. *Chromosoma* 35: 342–352.

RUSSELL, L. B. 1961. Genetics of mammalian sex chromosomes. *Science* 133: 1795–1803.

SARTO, G. E., AND OPITZ, J. M. 1973. The XY gonadal agenesis syndrome. *J. Med. Genet.* 10: 288–293.

SILVERS, W. K., AND WACHTEL, S. S. 1977. H-Y antigen: Behavior and function. *Science* 195: 956–960.

SIMPSON, J. L. 1977. *Disorders of Sexual Differentiation.* New York: Academic Press, 466 pp.

SIMPSON, J. L. 1978. Male pseudohermaphroditism: genetics and clinical delineation. *Hum. Genet.* 44: 1–49.

TEITELBAUM, M. S. 1972. Factors associated with the sex ratio in human populations. In *The Structure of Human Populations* (G. A. Harrison, A. J. Boyce, Eds.). Oxford: Clarendon Press, pp. 90–109.
WACHTEL, S. S. 1977. H-Y antigen and the genetics of sex determination. *Science* 198: 797–799.
WELSHONS, W. J., AND RUSSELL, L. B. 1959. The Y-chromosome as the bearer of male determining factors in the mouse. *Proc. Natl. Acad. Sci.* (U.S.A.) 45: 560–566.

REVIEW QUESTIONS

1. Define:

male
female
hermaphrodite
pseudohermaphrodite
homogametic
heterogametic
streak gonad
Müllerian ducts
gonadal dysgenesis
gonadal agenesis
seminiferous tubules
interstitial cells
androgen
estrogen
testicular feminization
primary sex ratio
secondary sex ratio
histocompatibility

2. Draw the pedigree in Figure 7-4, indicating the genotypes of all persons on the assumption that (a) testicular feminization is an autosomal dominant gene expressed only in males, and (b) it is an X-linked recessive gene.

3. In the case of 46,XX males, why would one expect a different frequency of persons with Xg(a+) phenotypes depending on whether (a) all cases originate by translocation of a male-determining locus from the Y to the X chromosome or (b) all cases originate as XXY zygotes with subsequent loss of the Y chromosome?

4. What would be the predicted sexual characteristics of persons with the following karyotypes?

a. 46,X,r(X)
b. 46,XXp−
c. 46,XXq−
d. 46,XYp−
e. 46,XYq−
f. 46,X,+t(XqYp)
g. 46,X,+t(XpYq)
h. 47,XX,+i(Yq)

5. Suppose that two parents with normal color vision and normal karyotypes have a 46,XX male offspring who is colorblind. Which theories of origin of these patients are consistent with this family? If false paternity is proved, would that alter your answer?

6. In some cultures, having at least one son is highly desired by parents. If parents continued to have children until at least one son were born in every family, what effect would this have on the secondary sex ratio? (Assume that the likelihoods of having a boy or a girl are equal and that it is the same in all families.)

CHAPTER EIGHT

GENE ACTIVITY AND PROTEIN SYNTHESIS

A. E. Garrod was seemingly the first to propose that homozygosity for a mutant gene may lead to loss of activity of a specific enzyme (see Chapter 1). From his studies of alkaptonuria, he concluded that persons homozygous for this rare gene lack an enzyme necessary for the complete metabolism of homogentisic acid. Other pioneers, such as L. Cuénot and R. Scott-Moncrieff, studied the associations between genotype and pigmentation in mice and in plants, recognizing that, at least in these cases, gene action had a biochemical consequence. Sewall Wright (1917), in summarizing the work to that date, recognized a direct relationship between genes and enzymes.

The beginning of the modern era of biochemical genetics can be identified with the studies of G. W. Beadle, E. L. Tatum, B. Ephrussi, and others in the 1930s. Working first with *Drosophila* and later especially with *Neurospora,* they demonstrated that each mutation could be associated with loss of a single enzyme. This led to the hypothesis that the function of genes is to direct synthesis of enzymes and that there is a one-to-one correspondence between genes and enzymes: the one–gene—one–enzyme hypothesis. More recent studies have shown the need to modify this to a one gene-one polypeptide form, and even in some circumstances two genes-one polypeptide, but the basic concept that each gene can be identified functionally with a particular enzyme is valid.

Since all enzymes are protein, either entirely or in large part, it follows that understanding gene action requires a knowledge of protein structure and function. In the following sections we will consider what proteins are and how they are synthesized under the direction of genes.

THE CHEMICAL STRUCTURE OF PROTEINS

Proteins are large organic molecules that occur in a variety of sizes, shapes, and functions. Molecular weights vary from 6000 for insulin and 13,500 for ribonuclease to 500,000 or higher. Proteins are composed of one or more chains of amino acids, folded and coiled into a three-dimensional structure. Twenty amino acids occur commonly in proteins (Table 8-1). The rare instances of other amino acids in proteins are thought to happen by conversion of one of the twenty to a related structure. Chemically, amino acids are compounds of the type structure $H_2N—CH(R)—COOH$, where R denotes a variety of organic groupings. Only proline and hydroxyproline deviate from this general structure.

In proteins, amino acids are linked by peptide bonds. This involves elimination of a hydroxyl from the carboxyl group of one amino acid and a hydrogen from the amino group of another:

$$H_2N-\underset{\displaystyle R'}{CH}-\overset{\displaystyle O}{\overset{\|}{C}}-OH + H\overset{\displaystyle H}{N}-\overset{\displaystyle R''}{CH}-\overset{\displaystyle O}{\overset{\|}{C}}-OH \longrightarrow H_2N-\overset{\displaystyle R'}{CH}-\overset{\displaystyle O}{\overset{\|}{C}}-\overset{\displaystyle H}{N}-\overset{\displaystyle R''}{CH}-COOH + HOH$$

The reaction is conveniently written as the elimination of water, but it is very much more complex than indicated. There remains a free amino group, conventionally written on the left, and a carboxyl group, written on the right. They are capable of reacting with other amino acids to form long *polypeptide chains*. Theoretically, there is no limit to the length of a polypeptide chain; many are known to consist of 600 or more amino acids. At one end there is always an amino group (the N-terminal end) and at the other a carboxyl (the C-terminal end). In some proteins, the amino group is acetylated so that it cannot react as a free amino group.

The sequence of amino acids in a polypeptide chain is known as its *primary structure*. Two additional levels of structure are distinguished in proteins: *secondary structure,* in which sections of polypeptide chains are coiled into a helix stabilized by hydrogen bonds between consecutive turns of the coil (Figure 8-1), and *tertiary structure,* in which polypeptide chains, including helical segments, are folded into a three-dimensional structure. Proper three-dimensional structure is necessary to provide a surface that functions biologically.

The tertiary structure is stabilized by various chemical forces between functional groups of the amino acid side chain. Of particular significance in most proteins is the amino acid cysteine. The side chain of cysteine contains a sulfhydryl group; two sulfhydryl groups can react to form a disulfide, producing thereby a covalent bond between adjacent polypeptide chains (or segments of a chain).

Heating and various other treatments disrupt the tertiary structure, destroying the biological activity of the molecule. This is called *denaturation.* In some cases, following the denaturation process, the active

Table 8-1 THE AMINO ACIDS CODED GENETICALLY IN THE SYNTHESIS OF PROTEINS

Most proteins have only these amino acids. A few derivatives, such as hydroxylysine and hydroxyproline, are formed after the protein is synthesized. With the exception of proline, all are α-amino acids, varying only in the side chain, symbolized by R. The structures of these side chains are given below. The capital letter preceeding each amino acid is the one-letter symbol for that amino acid. The three-letter symbol is given in parenthesis.

General formula for α-amino acids: $R-\underset{\displaystyle NH_2}{\underset{|}{CH}}-COOH$

Aliphatic amino acids:

- G. Glycine (Gly) $H-$
- A. Alanine (Ala) CH_3-
- V. Valine (Val) $CH_3-\underset{\displaystyle CH_3}{\underset{|}{CH}}-$
- L. Leucine (Leu) $CH_3-\underset{\displaystyle CH_3}{\underset{|}{CH}}-CH_2-$
- I. Isoleucine (Ile) $CH_3-CH_2-\underset{\displaystyle CH_3}{\underset{|}{CH}}-$

Hydroxyamino acids:

- S. Serine (Ser) $HO-CH_2-$
- T. Threonine (Thr) $HO-\underset{\displaystyle CH_3}{\underset{|}{CH}}-$

Dicarboxylic amino acids and amides:

- D. Aspartic acid (Asp) $HOOC-CH_2-$
- N. Asparagine (Asn) $H_2N-CO-CH_2-$
- E. Glutamic acid (Glu) $HOOC-CH_2-CH_2-$
- Q. Glutamine (Gln) $H_2N-CO-CH_2-CH_2-$

Basic amino acids:

- K. Lysine (Lys) $H_2N-CH_2-CH_2-CH_2-CH_2-$
- R. Arginine (Arg) $H_2N-\underset{\displaystyle NH}{\underset{\|}{C}}-NH-CH_2-CH_2-CH_2-$
- H. Histidine (His) [imidazole ring: N, NH] $-CH_2-$

Aromatic amino acids:

- F. Phenylalanine (Phe) [benzene ring]$-CH_2-$
- Y. Tyrosine (Tyr) $HO-$[benzene ring]$-CH_2-$
- W. Tryptophan (Trp) [indole ring: N]$-CH_2-$

Sulfur-containing amino acids:

- C. Cysteine (Cys) $HS-CH_2-$
- M. Methionine (Met) $CH_3-S-CH_2-CH_2-$

Imino acids:

- P. Proline (Pro) [pyrrolidine ring: N, H]$-COOH$

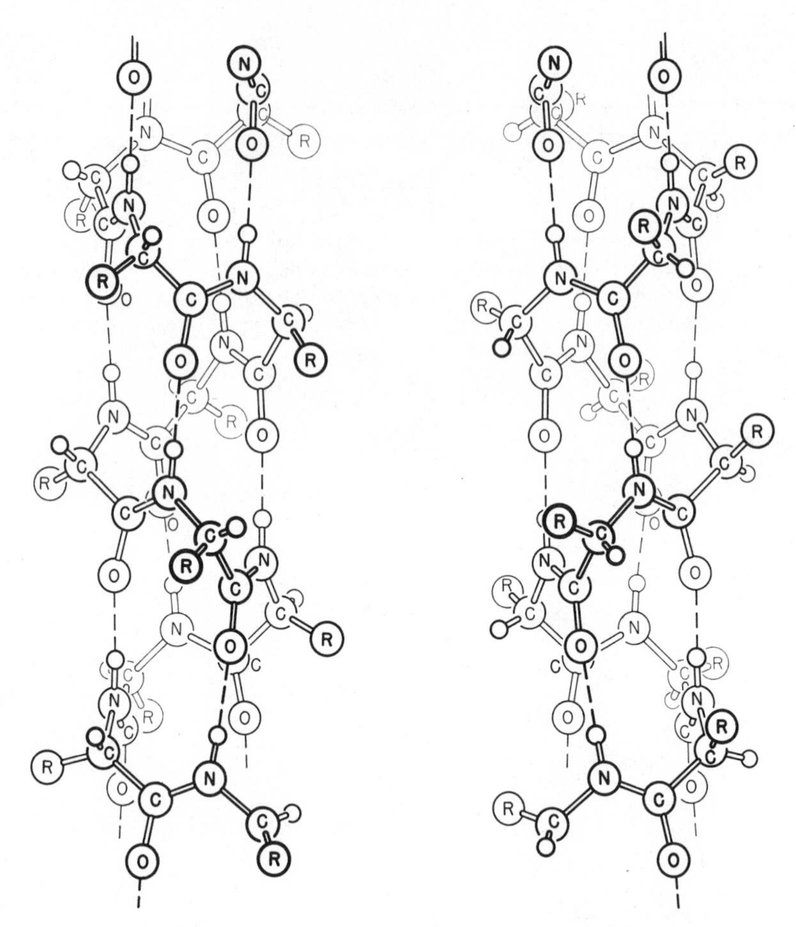

Figure 8-1 Diagram of protein α-helices, showing hydrogen bonds (dashed lines) between adjacent turns. Both left-handed and right-handed helices are shown. These are not mirror images, however. There are 3.6 amino acid residues per turn of the helix. The right-handed helix is the form most often found in proteins. (*Supplied by L. Pauling and R. B. Corey.*)

tertiary structure will re-form spontaneously after the denaturing agent is removed, demonstrating that the secondary and tertiary structures are largely, and perhaps entirely, a consequence of the primary structure.

HEREDITY AND PRIMARY STRUCTURE OF PROTEINS

The manner in which protein structure is influenced by genes has been elucidated through studies of abnormal proteins. Human hemoglobin was the protein first found to exist in an abnormal form as a result of genetic change.

The disease sickle cell anemia has been recognized for half a century. It is an inherited condition in which affected persons suffer from anemia resulting from excessive destruction of red cells in the circulatory system. In the presence of oxygen, the red cells are essentially normal. If oxygen pressure is low, as happens in many tissues, the red cells become distorted, in some instances resembling sickles (Figure 8-2). Formation of sickle-shaped cells in capillaries impedes circulation, leading to greater oxygen deficit and hence to more sickling. Interference with circulation leads to a variety of pathological consequences, depending on the tissues involved. Persons with the disease ordinarily do not survive to adulthood.

In addition to sickle cell anemia, there is a condition that occurs in normal persons in which red cells, although they do not sickle *in vivo,* can be induced to sickle *in vitro* in the presence of reducing agents. This condition is known as *sickle cell trait.* In 1949, Neel demonstrated that sickle cell disease is inherited as a simple Mendelian recessive condition. Persons with the severe disease are homozygous for the gene, and those with the trait are heterozygous.

The chemical nature of sickle cell anemia also was discovered in 1949 by Pauling, Itano, Singer, and Wells. Proteins carry electrical charges because of the ionization of amino, carboxyl, and other functional groups. These charges cause migration toward cathode or anode when proteins are placed in an electrical field (Figure 8-3). The speed of migration of a particular protein is influenced by several factors, including electrical charge of the molecule, shape of the molecule, and so on. The charge that a protein carries depends also on the *p*H of the surrounding solvent.

When hemoglobin from a normal person is examined for electrophoretic mobility, virtually all of the hemoglobin migrates as a single entity (Figure 8-4). Persons heterozygous for sickle cell disease possess two types of hemoglobin—normal hemoglobin and a slower moving va-

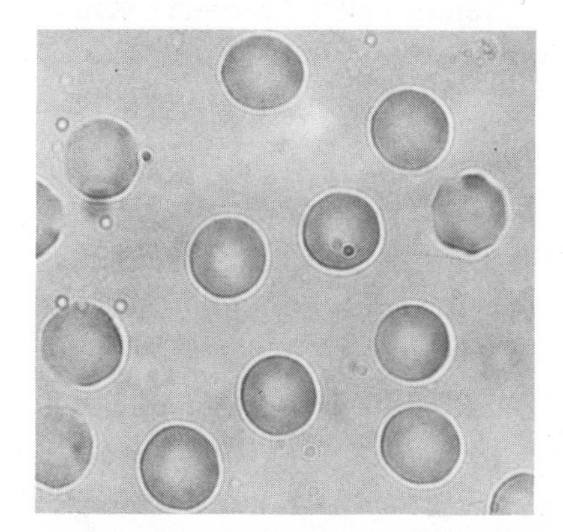
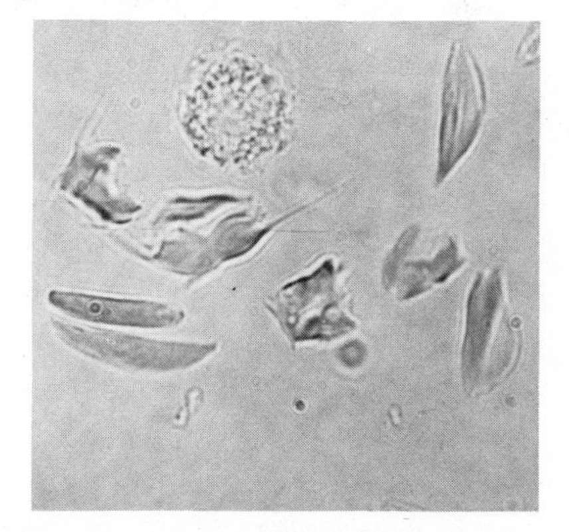

Figure 8-2 Formation of abnormal "sickle cells" under low oxygen pressure by red cells from persons homozygous for the sickle cell gene. On the left are normal red cells. On the right are sickle cells. (*Furnished by W. C. Levin.*)

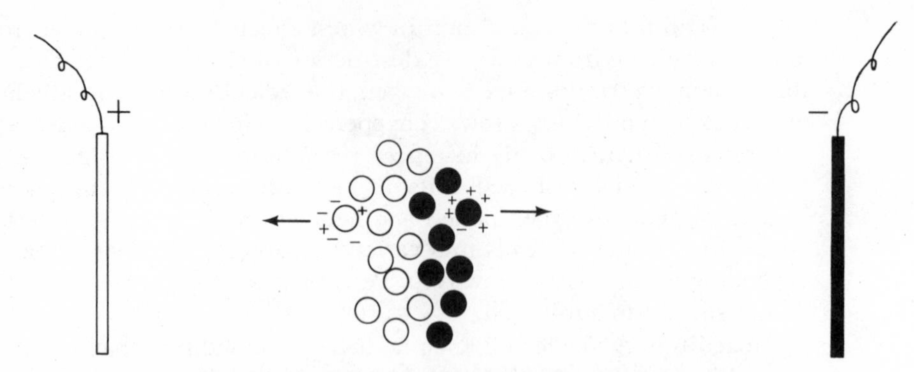

Figure 8-3 Diagram showing the movement of two different protein molecules in an electric field. One has an excess of negative charges and moves toward the positive pole. The other has more positive charges and moves toward the negative pole. At a more alkaline *p*H, both might have an excess of negative charges and move toward the positive pole, but the one with more negative charges would move faster, assuming the size and shape were not very different.

riety. Persons homozygous for the sickle cell gene possess only the slower moving variety. Pauling and co-workers suggested that the altered mobility of sickle cell hemoglobin reflects a change in molecular structure due to the effects of the sickle cell gene. The abnormal hemoglobin was designated hemoglobin S, and the normal adult hemoglobin was designated hemoglobin A. Sickle cell anemia thus became the first established example of a "molecular disease."

In addition to the abnormal hemoglobin associated with sickle cell anemia, a number of other hemoglobins have been detected (Chapter 10). Nearly all were recognized because they differ from hemoglobin A in electrophoretic mobility. Some of these hemoglobins are abnormal, usually resulting in anemia; others appear to function as normal hemoglobin.

The nature of the chemical difference characteristic of Hb S was established by Ingram in 1956. He digested the large hemoglobin molecule with the proteolytic enzyme trypsin, which breaks polypeptide chains wherever a lysine or arginine residue occurs, producing fragments consisting of short sequences of amino acids. Both Hb A and Hb S produced the same array of small peptide fragments, with a single exception. One of the peptides of Hb S differed slightly from the homologous peptide of Hb A. Detailed analysis of the amino acid sequences in these peptides gave the following results:

Hb A: Val-His-Leu-Thr-Pro-Glu-Glu-Lys
Hb S: Val-His-Leu-Thr-Pro-Val-Glu-Lys

These two peptides are identical except for the amino acid in position 6, which is glutamic acid in the case of Hb A and valine in the case of Hb S. Thus the effect of the Hb S mutation is to cause the substitution of a single amino acid out of a sequence of 146 amino acids in the β chain.

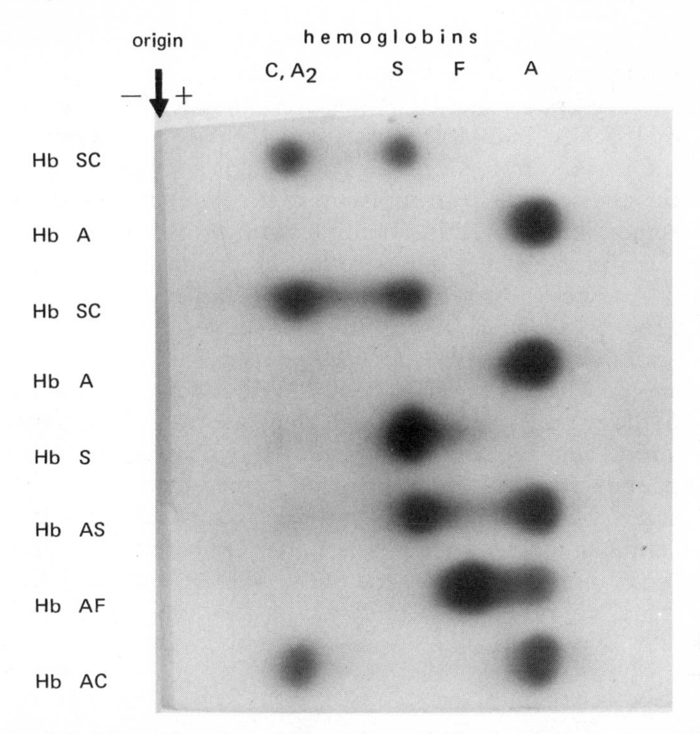

Figure 8-4 Electrophoretic mobility of hemoglobins from persons homozygous for Hb A, homozygous for Hb S, heterozygous for Hb's A and S, heterozygous for Hb's A and C, and heterozygous for Hb's S and C. The sample designated Hb AF is from umbilical cord blood, showing a high concentration of fetal hemoglobin. Hb A_2 is a normal hemoglobin present in small amounts (about 2%) in most persons. Mobilities of Hb C and Hb A_2 are similar under the conditions used. A slower mobility toward the anode at alkaline *p*H indicates a smaller negative charge. (*Furnished by Rose G. Schneider.*)

This first demonstration that different alleles code for different primary sequences in proteins has now been documented in hundreds of other examples. Studies of denaturation and renaturation of proteins also have indicated that secondary and tertiary structure are determined by the sequence of amino acids. It seems likely then that the role of genes in controlling protein function is in determining the primary structure only. Changes in molecular shape would arise as a secondary consequence of changes in amino acid sequences.

BIOSYNTHESIS OF PROTEINS

Proteins are synthesized in all cells, primarily in the cytoplasm but to some extent also in the nucleus. The sites of syntheses are ribosomes. Cells whose special function is secretion of protein (for example, salivary glands) are especially rich in ribosomes.

The principal features of protein synthesis have been established, although some of the details are lacking. Three kinds of RNA are involved—messenger RNA (mRNA), transfer RNA (tRNA), and ribosomal RNA (rRNA). Messenger RNA, the transmitter of genetic information from DNA to cytoplasm, is synthesized in the nucleus and then passes into the cytoplasm. In the cytoplasm, it combines with ribosomes. Transfer RNA combines with activated amino acids and aligns them in the proper sequence for the protein to be synthesized.

Figure 8-5 is a diagram of the major aspects of protein synthesis. Free amino acids react with adenosine triphosphate (ATP), forming amino acyl adenylates. An enzyme specific for each amino acid is necessary to catalyze these reactions. Without dissociating from the surface of the enzyme, the amino acyl adenylate reacts with transfer RNA, each amino acid requiring a specific tRNA. These amino acid-RNA compounds then align on a template formed by messenger RNA and ribosomes. The nature of the protein synthesized depends entirely on the source of messenger RNA, not on the source of ribosomes. A ribosome binds initially to the 5′ end of the mRNA, corresponding to the N-terminal region of the peptide chain. It aligns the tRNA molecule with

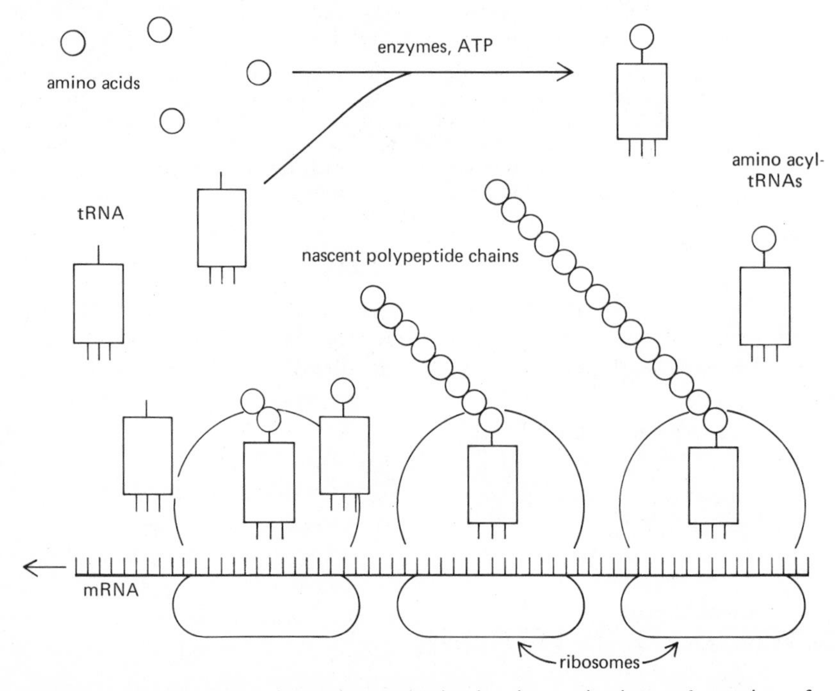

Figure 8-5 Diagram of protein synthesis showing activation and reaction of amino acids with transfer RNA, following by alignment on messenger RNA and formation of polypeptide. The ribosomes, mRNA, tRNA, and amino acids are not drawn to scale.

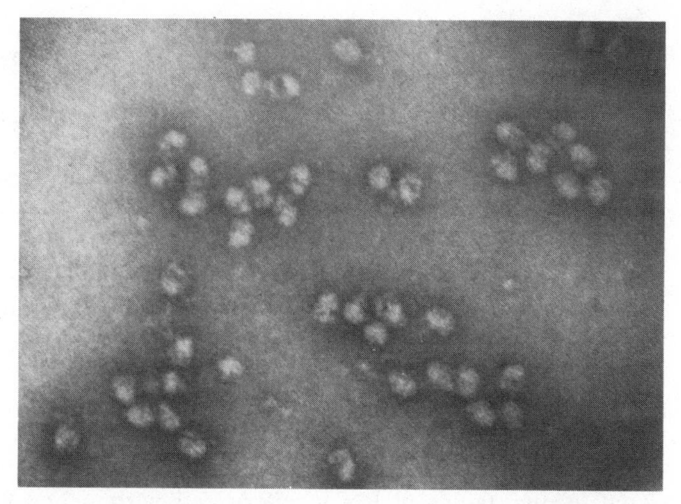

Figure 8-6 Electron micrograph of a polysome (complex of mRNA and ribosomes) from a preparation in which the mRNA is obtained from rabbit reticulocytes. The nascent polypeptide chains are not visible. (*Furnished by Boyd Hardesty.*)

mRNA, and associated enzymes catalyze the formation of peptide bonds between the amino acids attached to the tRNA molecules. The chemistry of this process is complex. When a peptide bond is formed, the ribosome then moves along the mRNA and the process is repeated. When the first ribosome moves away from the beginning of the mRNA, another ribosome can start the process anew. Eventually a number of ribosomes may be translating the same mRNA simultaneously (Figure 8-6). When the polypeptide chain is completed, it separates from the ribosomal complex and is free to fold into its biologically active tertiary structure.

The Genetic Code

Once the transfer of information from DNA to RNA to protein was established, attention was focused on the nature of the nucleic acid code. The most reasonable and economical assumptions proved to be correct. The linear nature of the DNA information as well as the linear nature of polypeptides suggested that information was stored colinearly; that is, the sequence of nucleotides in nucleic acids was translated into a corresponding sequence of amino acids. The number of nucleotides required to specify an amino acid had to be three or more. Since there are four nucleotides, one would give only four possibilities (A, G, C, or T), whereas two would give 4^2 or 16 (AA, AG, AC, and so on). There are 20 amino acids, so at least three are required. This gives 64 possible sequences, more than enough.

It turns out that a *codon*, the sequence of nucleotides required to specify one amino acid, consists of three nucleotides, and that 61 of the 64 possible sequences do code for an amino acid (Table 8-2). The remaining three code for chain termination. One may therefore "write" the primary structure of a protein entirely in the language of nucleic acids. The same code is used in all living organisms. Because a single amino acid is coded by more than one codon, the code is said to be degenerate.

In order for the DNA code to be *translated* into protein structure, it first must be *transcribed* into messenger RNA. As indicated earlier, this is accomplished by synthesis of an RNA sequence using DNA as the template. Since the RNA base sequence is complementary to the DNA template, it follows that the RNA codons are also complementary to DNA codons. The experimental methods for relating a specific codon to its corresponding amino acid make use of RNA; hence, it is conventional to give tables of RNA codons rather than DNA codons, as in Table 8-2.

The mRNA, usually after enzymatic addition of a "cap" (7-methyl guanosine) to the 5′ end and poly-A to the 3′ end, combines with ribo-

Table 8-2 **THE GENETIC CODE***

First position	Second position: U	C	A	G	Third position
U	UUU Phe	UCU	UAU Tyr	UGU Cys	U
	UUC	UCC	UAC	UGC	C
	UUA	UCA Ser	UAA Term	UGA Term	A
	UUG	UCG	UAG	UGG Trp	G
C	CUU Leu	CCU	CAU His	CGU	U
	CUC	CCC Pro	CAC	CGC Arg	C
	CUA	CCA	CAA Gln	CGA	A
	CUG	CCG	CAG	CGG	G
A	AUU	ACU	AAU Asn	AGU Ser	U
	AUC Ile	ACC Thr	AAC	AGC	C
	AUA	ACA	AAA Lys	AGA Arg	A
	AUG Met	ACG	AAG	AGG	G
G	GUU	GCU	GAU Asp	GGU	U
	GUC Val	GCC Ala	GAC	GGC Gly	C
	CUA	GCA	GAA Glu	GGA	A
	GUG	GCG	GAG	GGG	G

* RNA codons are shown, the 5′-phosphate to the left and the free 3′-hydroxyl to the right. The DNA codons in the transcribed strand would be complementary and would have reverse polarity, as would the tRNA anticodons.

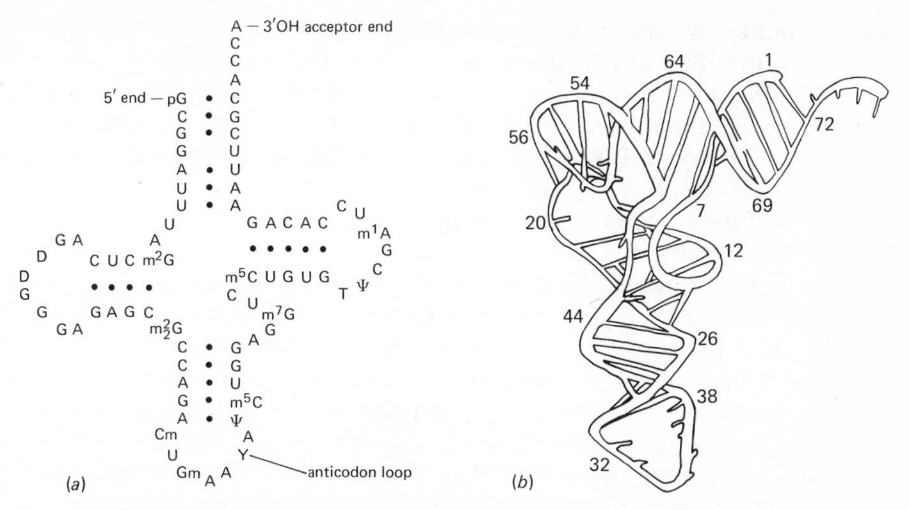

Figure 8-7 Diagram of yeast phenylalanine tRNA. There are 76 nucleotides in a single folded chain. (*a*) The primary structure, written in a "cloverleaf" form. Hydrogen bonds between bases are indicated by heavy dots. (*b*) Diagram showing the folding of tRNA into a three-dimensional structure. The ribose phosphate backbone is indicated by the continuous ribbon. Paired bases are shown as solid "rungs" and unpaired bases as projections. The numbers are the residue position in the molecule. The anticodon is at the bottom (34-36), and the amino acid is attached at position 76.

Abbreviations: A, adenosine; T, thymidine; G, guanosine; C, cytidine; U, uridine; D, dihydrouridine; ψ, pseudouridine; Y, a purine nucleoside; m_1, methyl; m_2, dimethyl. Superscripts indicate points of attachment. (*Redrawn from S. H. Kim and co-workers, 1974. Science 185: 435–440. © American Association for Advancement of Science.*)

somes and is translated into protein sequence. This requires recognition of the RNA codons by transfer RNA. There are many different transfer RNA molecules, each capable of combining with one specific amino acid. This specificity is conferred by the enzymes that can couple only one transfer RNA molecule with one amino acid. Each transfer RNA molecule, although largely twisted into double helical regions, has a loop of nucleotides that are unpaired (Figure 8-7). Three of these constitute the *anticodon* and pair with complementary codons of the mRNA, thus inserting the attached amino acid at the corresponding point in the growing polypeptide chain (Figure 8-5).

Studies in a number of experimental systems have indicated that translation of mRNA into protein starts at a specific point, proceeding from the N-terminal to the C-terminal end of the peptide chain. In *Escherichia coli,* the initial codon is AUG, and a special initiator tRNA always inserts formyl methionine as the N-terminal amino acid. Formyl methionine is a derivative of methionine in which a formyl group (HCO–) is attached to the α-amino group. The same codon in the interior of the message is translated by a different tRNA that inserts methionine. A similar situation exists in mammals but with methionine rather than formyl methionine as the initial amino acid residue. The N-terminal residues are removed enzymatically to expose a new N-terminus. It has been possible to investigate only a few systems, so that the generality of this mechanism is not fully established, but no exception is known.

Since more than one polypeptide may be coded on a single mRNA, it is necessary that termination points for protein translation also be coded. Three of the RNA codons, UAA, UAG, and UGA, code for termination of polypeptide chain synthesis. The termination codons differ from other codons only by single nucleotides and can arise by mutation in the middle of a structural gene. In this event, the polypeptide chain is terminated prematurely at that point. The product would usually be a nonsense protein.

Once translation has started, it proceeds by three nucleotides at a time, so that the *reading frame* is entirely dependent on the starting point. Insertion or deletion of a nucleotide causes the reading frame no longer to correspond to the original codons. Thus, if there has been an insertion or deletion, the reading frame will be in phase from the amino end of the chain up to the point of insertion or deletion. Beyond that point to the carboxyl end, the translation will be out of phase, and a completely different sequence of amino acids will be assembled.

Post-translational Modification of Proteins

Several things may happen to polypeptides before they become functional. Reference has already been made to the folding into secondary and tertiary structures, depending only on the primary amino acid sequence. The stability of such structures depends on the chemical environment. The traditional methods of denaturing proteins—acid, alkali, heat, low salt content, urea, organic solvents—reduce the stability of the native forms relative to other configurations. Such changes are often irreversible and occur to a limited extent under physiological conditions. Thus thermal denaturation of enzymes is a normal process in living cells. Some enzymes are very stable and show little evidence of denaturation in cells; others have a short life and must be replaced continually by synthesis.

Several other changes to polypeptides may occur before they become biochemically functional. Some of the more important are noted below.

Cleavage of N-Terminal Sequences. As described earlier, eukaryote structural genes appear to initiate synthesis regularly with methionine as the N-terminal residue. This has not been extensively studied in eukaryotes but has been demonstrated in several proteins, including human hemoglobins. In human hemoglobin, the additional

N-terminal methionine is cleaved off as synthesis of the nascent globin chain proceeds and is not present in completed chains.

Additional peptide segments often occur on the N-terminal region of a protein that also are cleaved during synthesis. In one of the better studied examples, rat albumin, an 18 amino acid segment beginning with methionine is on the N-terminus of the precursor or earliest synthesized molecule (Straus et al., 1977). This segment is ordinarily removed during synthesis to form proalbumin, which is itself a precursor of albumin.

Enzymatic Activation of Proteins. A number of examples are available of proteins (*proenzymes*) that must have portions removed by other enzymes before becoming active. The conversions of fibrinogen to fibrin, prothrombin to thrombin, proalbumin to albumin, and trypsinogen to trypsin involve enzymatic cleavage of a segment of the molecule. The conversion of proinsulin to insulin is especially interesting, since an internal segment is removed in the conversion (Figure 8-8). In this instance, therefore, a single structural gene gives rise to two poly-

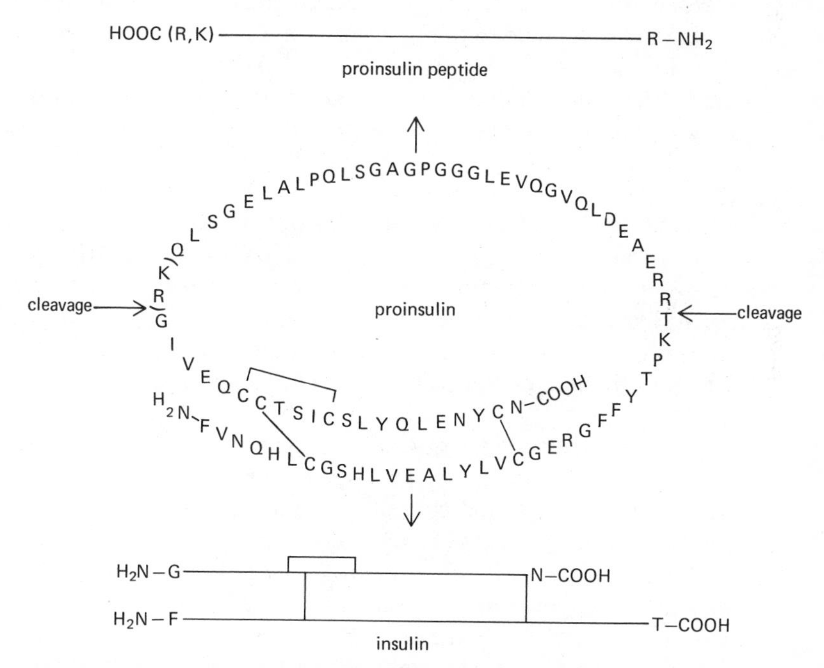

Figure 8-8 Synthesis of human insulin from proinsulin. Proinsulin is a single polypeptide chain, presumably reflecting a single structural gene. Insulin is formed by enzymatic cleavage of proinsulin to yield an inactive peptide plus insulin, the latter consisting of two polypeptide chains connected by disulfide bonds. The amino acids are represented by the single letter code. The heavy bars represent disulfide bonds.

peptide chains, although this variation in no sense violates the principle of one gene-one polypeptide chain.

Formation of Different Amino Acid Derivatives. A few proteins are noted for having unusual amino acids in their structures. For example, thyroid hormone, a polypeptide hormone, has thyroxine, an iodinated derivative of tyrosine. Associated with thyroid function also are mono- and di-iodotyrosine. The conversion of tyrosine to thyroxine occurs after tyrosine is incorporated into the hormone. One of the prominent constituents of collagen is hydroxylysine. In this case also the hydroxyl group is introduced after the lysine is incorporated into the polypeptide chain. Although only 20 amino acids are used as building blocks, over 140 have been isolated from naturally occurring proteins (Uy and Wold, 1977). Most have limited distribution.

Formation of Disulfide Bonds. The reaction of two sulfhydryl groups (—SH) to form disulfide bonds (—S—S—) between cysteine residues is a major factor in stabilizing the tertiary structure of proteins and hence increasing their thermal stability. This reaction is easily reversed by reduction under physiological conditions, so that disulfide bonds can be considered as being in equilibrium with the sulfhydryl form. In addition to stabilizing the folded form of a single polypeptide chain, disulfide bonds also form between polypeptide monomers in a polymer composed of protein subunits, thereby stabilizing the polymeric form.

Addition of Carbohydrate Side Chains. Many proteins have carbohydrate moieties attached at specific points along the polypeptide chain. These carbohydrate chains have structures specific for each protein and are attached during the "processing" of the newly synthesized protein in the Golgi structures. The function of the carbohydrate seems in general to relate to secretion across membranes.

Formation of Polymers. A large proportion of proteins have proved to consist of polymers composed of protein subunits. Hemoglobin is an example, consisting of four subunits, two of which are α-globin chains and two of which are β-globin chains. The molecular formula is therefore $\alpha_2\beta_2$. Proteins involving polymeric association of only one kind of polypeptide chain are *homopolymers*. Those involving more than one kind of chain are *heteropolymers*. Both types are common. In some cases, the monomeric subunits have some enzymatic activity, but usually the polymer is much more active than the subunits. The ability to form polymers is a function of the surface configuration of the monomers. Usually, subunits are held together by noncovalent forces, but occasionally disulfide bonds also add to the stability of the polymeric complex.

GENE STRUCTURE AND PROTEIN STRUCTURE

Reference has already been made to the fact that genes and polypeptides are colinear, that is, the linear order of mutations in DNA corresponds to a similar order of changes in primary structure of the respective proteins. This is easily explained by aligning the DNA codons in the same order as the amino acids. It was with great surprise that DNA was found to have additional untranscribed segments inserted in the middle of "structural genes." This finding was rapidly confirmed for a variety of proteins and seems now to be a general phenomenon.

Genes are classified as *structural* or *control*. Structural genes code for polypeptide chains and therefore have well defined beginnings and endings. Control genes are less well understood but probably consist both of genes that specify the structure of protein regulators (and hence are structural genes) and DNA regions that combine with protein (and other?) regulators.

The traditional model of a structural gene is one in which the DNA consists of a series of uninterrupted triplet codons that contain no information other than that necessary to specify the structure of the corresponding polypeptide chain. While the mRNA has leading and terminal sequences that are not translated, no examples are known in which untranslated sequences occur within the "structural" region of an mRNA. When techniques become available for determining nucleotide sequences of DNA, quite a different picture emerged for eukaryotes (Darnell, 1978; Crick, 1979). Interspersed among segments that are transcribed and translated are other segments that are not translated. In the several examples available, only two or three nontranslated segments usually are inserted, but in chicken ovalbumin seven such segments have been found (Figure 8-9). To distinguish translated from untranslated DNA, the term *exon* has been suggested to refer to a segment of translated DNA. An *intron* is an untranslated segment within a structural gene.

There is evidence that the entire region of DNA—both exons and introns—is transcribed but that post-transcriptional processing excises

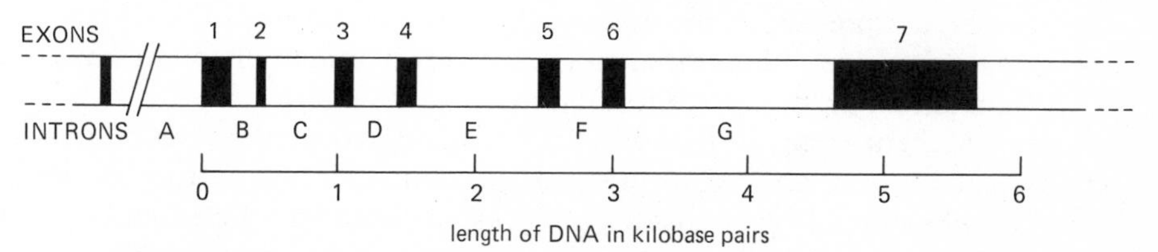

Figure 8-9 Organization of the chicken ovalbumin gene. The scale below the diagram is base-pairs of DNA. Introns are designated by A–G and exons by 1–7. The unlettered exon to the left of A is the initial segment of mRNA. The exact length of A is not known. The initiation point for protein translation (AUG in the mRNA code) is near the beginning of exon 1, and the termination point (UAA) is about a third of the way into exon 7. (*Based on Breathnach et al., 1978.*)

the segments that are not to be translated. The important conclusion is that a polypeptide sequence need not correspond to a continuous segment of DNA. The function of introns is unknown. It has been suggested that they may be concerned with gene regulation, but no evidence is available.

QUANTITATIVE ASPECTS OF PROTEIN SYNTHESIS

In the previous sections, the qualitative effects of genes were considered. It is known, however, that genes do not function at all times. Many proteins that play an important role in an adult animal are not synthesized during embryonic development, and probably very few are synthesized during early development of the zygote. Other proteins may be synthesized primarily during embryonic development but not during adulthood.

The factors responsible for turning genes on and off as part of the process of differentiation are unknown. A number of physiological factors, such as estrogens and androgens, are known to influence the activity of genes in specific tissues. The molecular mechanisms are not fully understood, but the effect seems to depend mostly on changing the rate of transcription of mRNA. Many external agents, including some drugs, influence the activity of specific enzymes. Increased activity in some cases appears due to increased synthesis, but most have not been completely investigated.

Control Genes

In bacteria, and in higher organisms as well, enzymes are known that are produced only in special environmental situations. An example in the bacterium *Escherichia coli* is the enzyme β-galactosidase. This enzyme splits the sugar lactose into galactose and glucose, which then can be metabolized for energy. If lactose is not present in culture media, *E. coli* does not produce galactosidase. If lactose (or certain other galactosides) are present in the media, then large quantities of the enzyme are manufactured. This responsiveness to environmental conditions is designated *induced* enzyme formation.

Mutant forms of *E. coli* have been isolated that cannot be induced. Instead, they are *constitutive,* manufacturing the enzyme continually whether or not galactosides are present. The enzyme produced is apparently normal enzyme, but a mutation has given rise to a defect in the control system for enzyme synthesis. It is possible to show by genetic means that the mutations that lead to constitutive enzyme production are located at definite positions on the chromosome, some adjacent to the locus that determines structure of β-galactosidase and others separated from this locus. Such genes might be considered *control genes* as opposed to *structural genes,* whose function is to specify the structure of a polypeptide.

Gene regulation in the bacterial galactosidase system and in several other bacterial systems investigated is according to the *operon* theory, first proposed by Jacob and Monod (1961) and illustrated in Figure 8-10. Fewer than a hundred repressor molecules for a particular operon are present in a bacterial cell, indicating the great effectiveness of the operon control as well as the technical difficulty in isolating repressors for study. Operon regulation is a negative control system, structural genes being repressed or "derepressed." The repressor, being diffusible, acts on sensitive operons on other chromosomes (*trans*) as well as on the chromosome from which it originated (*cis*).

In spite of extensive search, no clear cut example of operons has been demonstrated in eukaryotes. Yet it is clear that genes are regulated, as witness the turning off and on of genes during differentiation. It is further clear that most genes are not functional in any one tissue. However, nonfunctioning genes can sometimes be turned on under artificial circumstances. The requirements for gene control as part of differentiation may be very different from enzyme induction in bacteria and may have led to evolution of different control mechanisms.

One method of looking for control systems in mammalian cells has

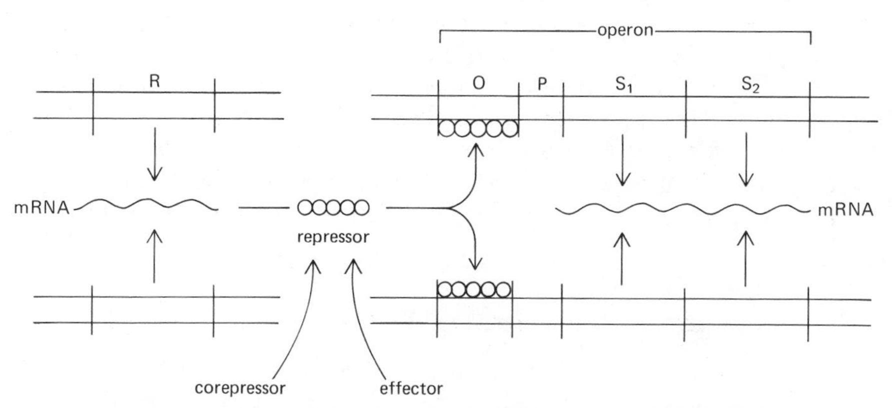

Figure 8-10 The *operon* theory of gene regulation (Jacob and Monod, 1961, modified.) *Regulator* genes (R) are responsible for synthesis of a protein *repressor* which can form complexes with specific *operators* (O). Presence of a repressor causes structural genes in the same operon not to synthesize the corresponding mRNA by interfering with binding of polymerase to the *promoter* (P). Mutation of an operator can lead to loss of ability to form complexes with repressors, with continued synthesis of mRNA by that operon. Cytoplasmic factors can either enhance or diminish the activity of repressors. The mRNA is shown as a single molecule carrying the information of two structural genes (S_1 and S_2). Studies of microbial systems indicate that all structural genes in one operon are transcribed onto one mRNA molecule.

been through cell hybridization (Chapter 14). Cells in culture can be induced to fuse, forming hybrids between two cell lines from different tissues or species. In one such study, rat hepatoma cells that produce normal liver enzymes were crossed with mouse fibroblast cells that do not (Weiss, 1975). The hybrids failed to produce the liver enzymes, indicating extinction of this gene function. The extinction was not due to irreversible changes, however, since continued culture led on occasion to loss of mouse chromosomes and re-expression of the liver enzyme genes. In this system, differentiation seems to be by means of diffusible repressors.

Quite different results were obtained using mouse hepatoma cells that secrete albumin and that were hybridized with human leukocytes, which do not. In this instance, the hybrids secreted both mouse and human albumin (Darlington et al., 1974). The human albumin gene is completely suppressed in leukocytes but can be activated by the differentiated mouse liver cells. The simplest explanation would be production of a diffusible activator by the liver cells, but more complex schemes involving repressors and derepressors can also be constructed.

Evidence for mammalian control genes other than those involved in differentiation comes from hybridization studies of mutant rat hepatoma cells that have lost the ability to synthesize hypoxanthine phosphoribosyltransferase (HPRT). These $HPRT^-$ cells were hybridized with $HPRT^+$ human fibroblasts. Many of the hybrids synthesized rat HPRT, others synthesized human HPRT, and one produced both (Croce et al., 1973). The presence of human HPRT depended on presence of the human X chromosome, which carries the structural gene for HPRT. Other human chromosomes were necessary for appearance of the rat HPRT, whose structural gene also is on the X chromosome. The mutation therefore involves a locus other than the HPRT structural gene, a control gene whose product can be supplied by human chromosomes, leading to activation of the intact rat structural gene.

Structure-Rate Control of Protein Synthesis

In view of the variety of codons and associated tRNA molecules, it would be surprising if all codons could be translated with equal speed. Some codons may require less abundant species of tRNA. Other tRNAs, for purely chemical or physical reasons, may enter into the ribosomal complex less rapidly. This potential variation in rate of translation led Itano (1957, 1965) to suggest that the rate of synthesis of a protein is determined in part by the codon structure of the mRNA, and this, of course, usually is reflected in structure of the protein. Because of the degeneracy of the code, some different codons will code for the same amino acid, but they may be translated at different rates. It was further proposed that evolution has favored those codon sequences that can be translated at a favorable rate, which should be neither too fast nor too slow. Thus the gene structure itself may be an important element in the control of protein synthesis.

The major evidence in support of the structure-rate hypothesis is provided by inherited variant hemoglobins. Many variants are known, and with few exceptions they are synthesized at rates less than the normal hemoglobin A (Chapter 10). For example, heterozygotes for sickle cell hemoglobin have approximately one-third Hb S and two-thirds Hb A. Since the difference between the two genes is thought to be limited to a single nucleotide, it would be difficult to explain substantially different rates of transcription of mRNA from DNA. Furthermore, there is no evidence that Hb S is destroyed more rapidly than Hb A. It is concluded therefore that Hb S is synthesized only half as fast as Hb A. This issue is further explored, especially with the Lepore hemoglobins, in Chapter 10.

The apparent sensitivity of rate of hemoglobin synthesis to codon structure does not necessarily apply to other proteins. Hemoglobin is synthesized at a high rate in cells that have differentiated in such a way as to divert virtually all the cell resources to this one protein. There may well be a great advantage to rapid translation before the resources become exhausted and before the nonrepairable synthetic machinery breaks down. In cells with still-functioning nuclei, where the particular protein is one of many products, small differences in mRNA structure may have less pronounced effects. The issue in part hinges on stability of mRNA and the risk of destruction at different parts of the synthetic cycle. mRNA complexed with ribosomes is thought to be more resistent to attack by nucleases than is free RNA. Variations in the movement of ribosomes along a strand of mRNA may lead therefore to variations in rate of destruction of the mRNA, influencing the number of times a particular molecule is translated before it is destroyed.

There is relatively little information on rate of synthesis of genetic variants other than hemoglobin. For the most part, protein variants appear to be synthesized at rates comparable to the wild type. A notable exception is the glucose-6-phosphate dehydrogenase variant G6PD Hektoen. This variant occurs at approximately four times the level of the common G6PD B. Yoshida (1970) showed that the variant involves a single amino acid substitution of tyrosine for histidine in the interior of the polypeptide chain. It is possible that in this instance, the normal histidine codon is rate limiting. If this were the case, then mutation at other positions might have relatively little effect on the net rate of synthesis.

The structure-rate hypothesis explains why a particular gene product is produced at a characteristic rate. It does not provide a mechanism for turning genes on and off. Nor does it provide for response to environmental agents. Thus control genes and the structure-rate hypothesis are complementary aspects of gene activity.

X-Chromosome Inactivation

One of the interesting problems posed by a chromosomal pattern of sex determination is the relatively small differences between males, with one X chromosome, and females, with two X chromosomes. Further, as was discussed in Chapter 7, maleness of femaleness results from presence or absence of the Y chromosome. How is it that genic imbalance does not occur with the X chromosome, which is one of the larger chromosomes carrying many genes? Indeed, persons with four X chromosomes are little different from those with two.

The hypothesis of X-chromosome inactivation is supported by several lines of evidence. The hypothesis states that in any one mammalian cell, only one X chromosome is active. In males, the single X chromosome is always active; in females, one of the two X chromosomes becomes inactive at an early stage of development. In any given cell, the choice of which X chromosome is to become inactive is entirely a matter of chance. Once inactivated, a chromosome continues to replicate at mitosis but remains inactive through subsequent cell cycles. This hypothesis has been most fully stated by Mary F. Lyon and is known as the *Lyon hypothesis*.

Several lines of evidence support the hypothesis. B. M. Cattanach's observations on coat color in mice initially provided support. He found that certain color genes in mice behave abnormally when translocated to X chromosomes. Instead of the usual dominant and recessive relationship of alleles characteristic of autosomal position, a variegated phenotype was produced, with expression of either the dominant or the recessive allele in different patches.

Autoradiographic studies of replicating chromosomes show that in females one of the X chromosomes replicates later than the other chromosome (Chapter 3). If tritium-labeled thymidine is added just before the late X begins replication, most of the label incorporated will be in that chromosome (Figure 3-8). If there are more than two X chromosomes, all but one is late-labeling.

The most impressive evidence illustrating X-chromosome inactivation involves the enzyme glucose-6-phosphate dehydrogenase (G6PD). This enzyme is found in virtually all forms of life, including man. It occurs in essentially all tissues, including red blood cells, where it acts to maintain the proper balance of reduced glutathione. Variant forms of the enzyme occur in man and are transmitted genetically by a gene on the X chromosome. When G6PD is placed in an electric field, mobility differences are observed among some of the variants. Males never have more than one form of G6PD, but females can be heterozygous and frequently have two forms.

Beutler, Yeh, and Fairbanks (1962) obtained evidence from studies of enzyme activities that red cells of heterozygotes are a mixture of two types of cells, each containing only one of the two forms of G6PD.

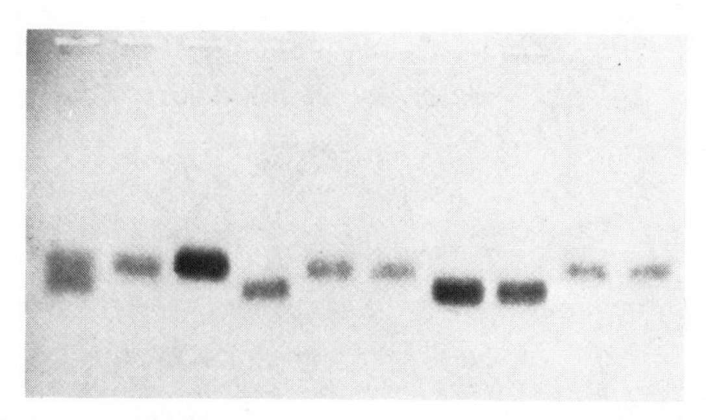

Figure 8-11 Electrophoretic separation of glucose-6-phosphate dehydrogenase variants from a woman heterozygous for types A (fast) and B (slow). On the left is an extract from a skin culture, showing both forms of the enzyme. The remaining preparations were extracted from clones derived from *single* cells of the skin culture. In individual cells, only one X chromosome is active and only one form of the enzyme is produced. Once an X chromosome becomes inactive, it remains so through subsequent cell divisions, even though it replicates. *(From Davidson, Nitowsky, and Childs, 1963.)*

Davidson, Nitowski, and Childs (1963) cultured cells from a woman heterozygous for two forms of G6PD. Electrophoresis of enzyme prepared from a mixture of cells showed both forms to be present (Figure 8-11). However, if single cells were isolated and permitted to replicate to form a clone of cells adequate for extraction, quite different results were obtained. In this case, each clone (or culture) contained only one form of G6PD. It might be either the fast or slow form but not both.

The results for the G6PD locus have been substantiated by studies of many other loci. These include the loci for hypoxanthine phosphoribosyl transferase (HPRT), Hunter's syndrome, α-galactosidase deficiency (Fabry's disease), and phosphoglycerate kinase (PGK). A patient doubly heterozygous for G6PD and PGK proved especially interesting. When her fibroblasts were cloned, 33 of 56 were G6PD A and PGK 1. The remaining 23 were G6PD B and PGK 2 (Gartler and co-workers, 1972). Inactivation therefore extended over a large segment of the chromosome, and loci in the *cis* configuration were either both active or both inactive.

Not all X-linked loci are inactivated, however. It has long been suspected that the *Xg* locus, responsible for Xg red cell types, might not be inactivated. Recently the locus for the skin condition X-linked ichthyosis, resulting from deficiency of steroid sulfatase, was shown not to undergo inactivation (Shapiro et al., 1979). Since this locus is located near the *Xg* locus, it suggests that the entire region remains active on both X chromosomes of females.

The exact period in development when inactivation occurs is not well established, although it is early in embryonic development. By five weeks from conception, most cells from a variety of tissues show inactivation (Migeon and Kennedy, 1975). Whether it occurs simultaneously

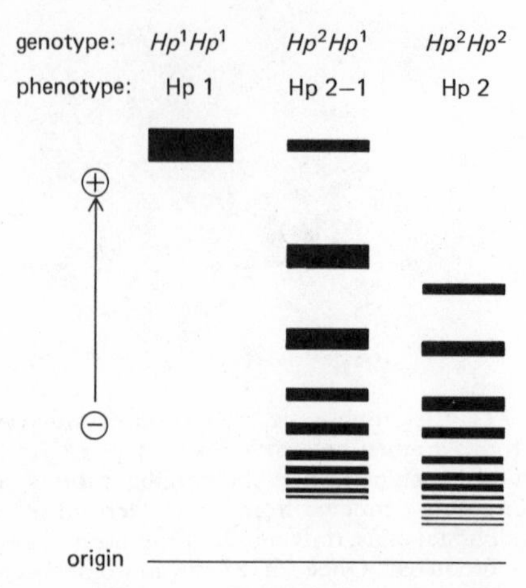

Figure 8-12 Diagram of haptoglobin (Hp) types as observed on gel electrophoresis. In free electrophoresis, the Hp molecules all migrate at very similar rates. However, on starch gel or acrylamide gel electrophoresis, the gel structure acts as a sieve, allowing the smaller molecules to migrate more readily than large molecules of otherwise equal electrophoretic mobility. The complex patterns of Hp 2-1 and Hp 2 are formed by polymers, the larger molecules migrating more slowly.

in all tissues and whether it extends to all tissues are unknown. Gartler, Liskay, and Gant (1973) demonstrated that in mature oöcytes both X chromosomes are active at the G6PD locus. They proposed that in the germ line, X-chromosome inactivation never occurs. If this is so, then inactivation may be irreversible *in vivo*.

Many aspects of the expression of genes located on the X chromosome can be explained by X-chromosome inactivation. Among the important questions remaining is the basic mechanism. By what pro-

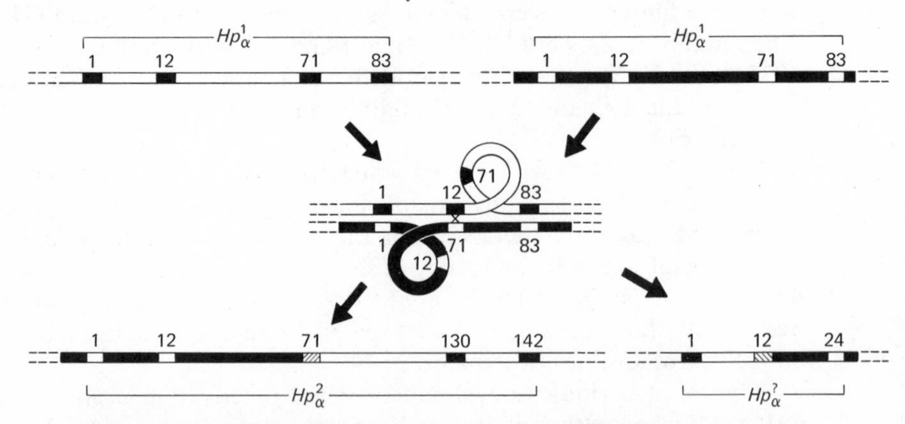

Figure 8-13 Origin of Hp_α^2 by mispairing of two Hp_α^1 alleles and crossing over. The numbers indicate the codons of the structural genes. Crossing over must have occurred between codons 12 and 71.

cess is such a large portion of a chromosome inactivated? How is it that all but one X chromosome is inactivated? Can inactivation be reversed? Is inactivation related to differentiation of other chromosome regions?

PROTEINS AND EVOLUTION

Proteins provide a record of gene structure and are useful in tracing evolutionary changes in genes. There is much debate on the extent to which amino acid substitutions (and nucleotide substitutions) may occur that are neutral with respect to natural selection (Chapter 22). But neutral or not, proteins with structural similarity are presumed to represent genes with structural similarity. The greater the similarity of two proteins, the more recent their evolutionary divergence. This applies not only to homologous proteins in different species but also to related proteins within an organism. This principle has been used effectively to trace the evolution of hemoglobin chains (Chapter 10).

One especially interesting example in man is provided by haptoglobin. Haptoglobin (Hp) is a serum protein that can form complexes with free hemoglobin. The molecule consists of two kinds of chains, α and β, each under control of a separate gene locus. The Hp_α locus is polymorphic in human populations, there being three common alleles, Hp_α^{1F}, Hp_α^{1S}, and Hp_α^{2}. With routine gel electrophoresis of plasma or serum, the first two alleles cannot be distinguished, so that often the system is treated as having two alleles, Hp^1 and Hp^2, producing three genotypes. As is noted in Figure 8-12, in two of the phenotypes—those involving the Hp_α^{2} allele—Hp consists of a series of polymers.

Analysis of the primary structure of the Hp α chains indicated the α1S and α1F chains to consist of 83 amino acid residues, identical except for one substitution at position 54. However, the α2 chain consists of 142 amino acid residues. The Hp_α^{2} gene would have to be nearly twice as long as its two alleles. The structural relationships among the three chains are shown in Figure 8-13. Apparently during the course of evolution, the Hp_α gene mispaired during meiosis, and crossing over occurred as indicated in Figure 8-13. The product of the new allele, Hp_α^{2}, was able to enter into the haptoglobin molecule and conferred on it the ability to form polymers of various sizes. This allele must have had some selective advantage in view of its widespread occurrence. Since the Hp_α^{2} allele is known only in man, the duplication probably occurred after divergence of the hominid line.

Haptoglobins provide an example of partial gene duplication revealed entirely by structural analysis of the protein product. There is strong evidence in other systems, especially with hemoglobins, of complete gene duplication followed by evolutionary divergence. Indeed, this must have been a major mechanism in achieving genetic complexity. Once a gene has duplicated, one copy is released from the previous constraints of natural selection and is free to mutate with less penalty for the organism. By studying the structural relationship of proteins in existing species, it should be possible to reconstruct many of the details of gene evolution.

SUMMARY

1. Studies of inherited defects of metabolism in man and a number of experimental organisms indicate that each gene can be associated with an enzyme or, more specifically, with a single polypeptide chain.

2. Proteins are chains of amino acids connected by peptide bonds. Only 20 amino acids are incorporated directly into peptide chains, although derivative amino acids occasionally are made after incorporation.

3. Studies of hemoglobin from sickle cell patients showed this inherited trait to be associated with substitution of valine for glutamic acid at position 6 of the β-globin chain. This and many similar observations indicate the function of genes to be determination of the primary amino acid sequence of polypeptides.

4. Synthesis of proteins occurs on the ribosomes. Messenger RNA carries the information on sequence of amino acid residues from the nuclear DNA to ribosomes. Specific transfer RNAs, with amino acids attached, align on the mRNA. Peptide bonds then form between the growing peptide chain and each aligned amino acid. The mRNA is translated into a peptide chain starting from the N-terminal residue.

5. A sequence of three nucleotides constitutes a codon for one amino acid. Of the 64 possible sequences of four nucleotides (A, G, C, U in the RNA code), 61 code for amino acids and 3 for termination of chain synthesis. The initial codon appears ordinarily to be for methionine (formyl methionine in bacteria), but this is subsequently cleaved off. The message is translated three nucleotides at a time. Addition or deletion of a nucleotide changes the reading frame and generates a different sequence of amino acids beyond that point.

6. After mRNA has been translated into a peptide chain, several kinds of modifications may occur. In most cases the N-terminal methionine (plus additional residues) is cleaved off. Additional cleavage may be necessary to achieve full activity. Other possible modifications are disulfide bond formation, addition of carbohydrate side chains, and formation of polymers.

7. The colinearity of DNA structural genes, messenger RNA, and polypeptide chains has been amply demonstrated. However, structural genes have segments of untranscribed nucleotide sequences (*introns*) inserted among the transcribed sequences (*exons*). The function of these introns is not known. Messenger RNA also has leading and terminal sequences that are transcribed but not translated.

8. In bacteria, control genes, concerned with quantitative activity of structural genes, act through an operon system in which transcription of mRNA depends on accessibility of the enzyme polymerase to the DNA. This system has not been detected in eukaryotes. Control genes, some of which are related to differentiation, can be demonstrated by hybridization of mammaliam cells from different tissues and species. Their mechanism of action is not known. In addition, some mammalian genes, such as human hemoglobin, appear to be synthesized at a rate that is influenced by the mRNA sequence.

9. X-Chromosome inactivation is an additional example of gene, or chromosome, regulation. Only one X chromosome in any cell is active, and only the genes on that chromosome are expressed. Females are mosaics of cells, some with one X active, some with the other.

10. Proteins are useful as records of evolution, since the primary structure of a protein reflects the gene structure. The evolutionary affinities of structural genes can be assessed by comparing the structures of the corresponding proteins. Human haptoglobin is an example of a locus at which partial gene duplication occurred, as indicated by the products of the present-day alleles.

REFERENCES AND SUGGESTED READING

BEUTLER, E., YEH, M., AND FAIRBANKS, V. F. 1962. The normal human female as a mosaic of X-chromosome activity: Studies using the gene for G-6-PD-deficiency as a marker. *Proc. Natl. Acad. Sci.* (U.S.A.) 48: 9–16.

BREATHNACH, R., BENOIST, C., O'HARE, K., GANNON, F., AND CHAMBON, P. 1978. Ovalbumin gene: Evidence for a leader sequence in mRNA and DNA sequences at the exon-intron boundaries. *Proc. Natl. Acad. Sci.* (U.S.A.) 75: 4853–4857.

CRICK, F. 1979. Split genes and RNA splicing. *Science* 204: 264–271.

CRICK, F. H. C., BARNETT, L., BRENNER, S., AND WATTS-TOBIN, R. J. 1961. General nature of the genetic code for proteins. *Nature* 192: 1227–1232.

CROCE, C. M., BAKAY, B., NYHAN, W. L., AND KOPROWSKI, H. 1973. Reexpression of the rat hypoxanthine phosphoribosyltransferase gene in rat-human hybrids. *Proc. Natl. Acad. Sci.* (U.S.A.) 70: 2590–2594.

DARLINGTON, G. J., BERNHARD, H. P., AND RUDDLE, F. H. 1974. Human serum albumin phenotype activation in mouse hepatoma-human leukocyte cell hybrids. *Science* 185: 859–862.

DARNELL, J. E. Jr. 1978. Implications of RNA. RNA splicing in evolution of eukaryotic cells. *Science* 202: 1257–1260.

DAVIDSON, R. G., NITOWSKI, H. M., AND CHILDS, B. 1963. Demonstration of two populations of cells in the human female heterozygous for glucose-6-phosphate dehydrogenase variants. *Proc. Natl. Acad. Sci.* (U.S.A.) 50: 481–485.

DINTZIS, H. M. 1961. Assembly of the peptide chains of hemoglobin. *Proc. Natl. Acad. Sci.* (U.S.A.) 47: 247–261.

GARTLER, S. M., AND ANDINA, R. J. 1976. Mammalian X-chromosome inactivation. *Adv. Hum. Genet.* 7: 99–140.

GARTLER, S. M., CHEN, S-H., FIALKOW, P. J., GIBLETT, E. R., AND SINGH, S. 1972. X-Chromosome inactivation in cells from an individual heterozygous for two *X*-linked genes. *Nature New Biol.* 236: 149–150.

GARTLER, S. M., LISKAY, R. M., AND GANT, N. 1973. Two functional X chromosomes in human oöcytes. Annual meeting, American Society of Human Genetics, Program p. 28A.

"The Genetic Code." *Cold Spring Harbor Symp. Quant. Biol.* 31 (1966).

INGRAM, V. M. 1956. A specific chemical difference between the globins of normal human and sickle-cell anaemia haemoglobin. *Nature* 178: 792–794.

ITANO, H. A. 1957. The human hemoglobins: their properties and genetic control. *Adv. Protein Chem.* 12: 216–268.

ITANO, H. A. 1965. The synthesis and structure of normal and abnormal hemoglobins. In *Abnormal Haemoglobins in Africa*, Jonxis, J. H. P. (Ed.). Oxford: Blackwell, pp. 3–16.

JACOB, F., AND MONOD, J. 1961. Genetic regulatory mechanisms in the synthesis of proteins. *J. Molec. Biol.* 3: 318–356.

JONES, O. W., AND NIRENBERG, M. W. 1962. Qualitative survey of RNA codewords. *Proc. Natl. Acad. Sci.* (U.S.A.) 48: 2115–2123.

KIM, S. H., SUDDATH, F. L., QUIGLEY, G. J., MC PHERSON, A., SUSSMAN, J. L., WANG, A. H. J., SEEMAN, N. C., AND RICH, A. 1974. Three-dimensional tertiary structure of yeast phenylalanine transfer RNA. *Science* 185: 435–440.

LYON, M. F. 1962. Sex chromatin and gene action in the mammalian X-chromosome. *Am. J. Hum. Genet.* 14: 135–148.

"The Mechanism of Protein Synthesis." *Cold Spring Harbor Symp. Quant. Biol.* 34 (1969).

MIGEON, B. R., AND KENNEDY, J. F. 1975. Evidence for the inactivation of an X chromosome early in the development of the human female. *Am. J. Hum. Genet.* 27: 233–239.

NEEL, J. V. 1949. The inheritance of sickle-cell anemia. *Science* 110: 64–66.

NIRENBERG, M. W., AND MATTHAEI, J. H. 1961. The dependence of cell-free protein synthesis in *E. coli* upon naturally occurring or synthetic polyribonucleotides. *Proc. Natl. Acad. Sci.* (U.S.A.) 47: 1588–1602.

OHNO, S. 1970. *Evolution by Gene Duplication.* Springer-Verlag, Berlin, New York. 160 pp.

PAULING, L., ITANO, H. A., SINGER, S. J., AND WELLS, I. C. 1949. Sickle-cell anemia, a molecular disease. *Science* 110: 543–548.

REZNIKOFF, W. S. 1972. The operon revisited. *Annu. Rev. Genet.* 6: 133–156.

SHAPIRO, L. J., MOHANDAS, T., WEISS, R., AND ROMEO, G. 1979. Non-inactivation of an X-chromosome locus in man. *Science* 204: 1224–1226.
STENT, G. S., AND CALENDAR, R. 1978. *Molecular Genetics,* 2nd Ed. San Francisco: W. H. Freeman, 773 pp.
STRAUS, A. W., DONOHUE, A. M., BENNETT, C. D., RODKEY, J. A., AND ALBERTS, A. W. 1977. Rat liver preproalbumin: *In vitro* synthesis and partial amino acid sequence. *Proc. Natl. Acad. Sci.* (U.S.A.) 74: 1358–1362.
STURTEVANT, A. H. 1965. Biochemical genetics. Chapter 16 in *A History of Genetics.* New York: Harper and Row, pp. 100–106.
SUTTON, H. E. 1970. The haptoglobins. *Prog. Med. Genet.* 7: 163–216.
UY, R., AND WOLD, F. 1977. Posttranslational covalent modification of proteins. *Science* 198: 890–896.
WARNER, J. R., KNOPF, P. M., AND RICH, A. 1963. A multiple ribosomal structure in protein synthesis. *Proc. Natl. Acad. Sci.* (U.S.A.) 49: 122–129.
WATSON, J. D. 1976. *Molecular Biology of the Gene,* 3rd Ed. Menlo Park, Calif.: W. A. Benjamin, 739 pp.
WEISS, M. C. 1975. Extinction, re-expression and induction of liver specific functions in hepatoma cell hybrids. In *Gene Expression and Carcinogenesis in Cultured Liver* (L. E. Gerschenson and E. B. Thompson, Eds.). New York: Academic Press, pp. 346–357.
WRIGHT, S. 1917. Color inheritance in mammals. *J. Hered.* 8: 224–235.
YOSHIDA, A. 1970. Amino acid substitution (histidine to tyrosine) in a glucose-6-phosphate dehydrogenase variant (G6PD Hektoen) associated with overproduction. *J. Molec. Biol.* 52: 483–489.
ZUBAY, G. L., AND MARMUR, J. (Eds.). 1973. *Papers in Biochemical Genetics,* 2nd Ed. New York: Holt, Rinehart & Winston, 622 pp.

REVIEW QUESTIONS

1. Define:

amino acid
peptide
protein
N-terminal
C-terminal
primary structure
secondary structure
tertiary structure
disulfide bond
denaturation
codon
anticodon
messenger RNA
transcription
translation
polymerase
intron
exon
ribosome
polysome
transfer RNA
reading frame
induced enzyme
constitutive enzyme
control gene
regulator gene
structural gene
operon
operator
promoter
repressor
homopolymers
heteropolymers
proenzyme
electrophoretic mobility
X-chromosome inactivation

2. a. Write an RNA sequence that would code for the Hb A peptide given on page 156. What change in the sequence would be required to code for the Hb S peptide?

b. If the sixth nucleotide were deleted from your nucleotide sequence, what would be the new amino acid sequence?

3. Most "mutant" proteins are recognized by differences in electrophoretic mobility as compared to the wild type protein. What two groups of mutations might not be detected by this procedure?

4. Fibrinogen is the plasma protein that, in the process of blood clotting, is converted to fibrin, the gel-like substance of the clot. The conversion consists in part in enzymatic removal of two peptide fragments called fibrinopeptides A and B. The structures of a series of fibrinopeptides A are given below. Based only on this information, construct an evolutionary tree for these species. (Dashes indicate a deletion, although the "deletion" may be the more ancestral form with additional residues having been added later.)

cow	E D G S D P P S G D F L T E G G G V R
cat	– – – G D V Q E G E F I A E G G G V R
dog	– – – T N S K E G E F I A E G G G V R
human	– – – A D S G E G D F L A E G G G V R
bear	– – – T D G K E G E F I A E G G G V R
horse	– – – – – T E E G E F L H E G G G V R
sheep	A D D S D P V G G E F L A E G G G V R
camel	– T D P D A D E G E F L A E G G G V R
donkey	– – – T K T E E G E F I S E G G G V R
chimpanzee	– – – A D S G E G D F L A E G G G V R
zebra	– – – T K T E E G E F I G E G G G V R
rhesus monkey	– – – A D T G E G D F L A E G G G V R

5. a. In a bacterial operon system, a mutation occurs in a regulator gene so that transcription of the regulator gene no longer occurs. What would be the effect on mRNA produced by the operon?

b. If mutation occurred in the operator so that it could no longer combine with repressor, what would be the effect?

6. How would you demonstrate that mRNA rather than ribosomes or tRNA carry information for protein structure?

7. What evidence is there that X-chromosome inactivation does not cause inactivation of the entire X chromosome?

8. Devise a scheme to explain X-chromosome inactivation in molecular terms. Your scheme should allow for only one active X chromosome in a cell, no matter what the total number is.

CHAPTER NINE

MUTATIONS

A mutation is defined as a heritable change. As observed in the human time scale, mutations are sudden. No transition from one form to another can be demonstrated, although induced mutations in some organisms do appear to have a transition state during which the gene may "repair" or change to a different form.

A key part of the definition of mutation is the word *heritable.* A new variation must be transmitted to offspring before it can be accepted as a mutation. A dominant mutation that causes loss of fertility cannot be shown to be heritable and can as well be attributed to an environmental factor. In rare instances, somatic mutation can be demonstrated. Mutation in a somatic cell gives rise to a clone of cells that differ from surrounding cell. The difference only rarely is detectable, usually only when visible features such as pigmentation or external morphological characteristics are altered. Without the possibility of transmission to offspring, it is difficult to establish the genetic nature of somatic mutations. The development of methods of studying somatic cells *in vitro* (Chapter 14) has removed this limitation somewhat.

Any change in the structure of genetic material is a mutation. Changes may involve substitution of a DNA base, deletion of one or more bases, or insertion of a new base. Alterations of chromosomes, in either number or structure, are mutations, although other terms often are more useful in describing these changes. Rearrangements of DNA sequences, such as inversions, change the genetic information. It is rarely possible to be certain that a mutation involves a single nucleotide pair (point mutation) or a larger segment such as a small chromosomal deletion. Either may be lethal.

Most mutations that influence function involve a change from function to nonfunction and usually are recessive. These can thus be observed only when they combine with a similar mutant allele, possibly many generations removed from the time of the original mutational event. In contrast, dominant mutations can be observed within one generation. Mutations on the X chromosome can be observed as soon as the mutant allele is transmitted to a male. One of the famous examples of a mutation whose occurrence can be pinpointed with some certainty is given in Figure 2-6.

THE CHEMICAL BASIS OF MUTATION

As a chemical substance, DNA is subject to chemical reactions that change its structure, often changing the genetic information. For example, base analogs, such as bromodeoxyuridine (BUDR) or 2-aminopurine, may be incorporated into DNA in place of the normal base. These often result in mutations as a result of mispairing during replication. BUDR pairs with adenine just as thymine does. However, it is more likely than thymine to pair with guanine because of the increased stability of the enol form (Figure 9-1). If this happens during replication, guanine will be incorporated into the newly synthesized strand in place of adenine.

Alkylating agents are a particularly effective class of mutagens. These are agents such as nitrogen mustard that attach organic groups to bases. The N^7 position of guanine is particularly susceptible. The attachment of such a group changes the configuration of the base so that mispairing again is stabilized. Bifunctional alkylating agents may also form stable bridges between two different bases. Other chemical reactions of DNA include the replacement of amino groups with hydroxyl groups on treatment with HNO_2. This has the effect of converting adenine to hypoxanthine, guanine to xanthine, and cytosine to uracil.

All of the preceding lead primarily to base substitution. Another group of compounds, exemplified by proflavine (Figure 9-2), have strong affinity for nucleic acid bases and intercalate into the normal base stacking of the double helix. These compounds lead especially to frameshift mutations, involving addition or deletion of base pairs.

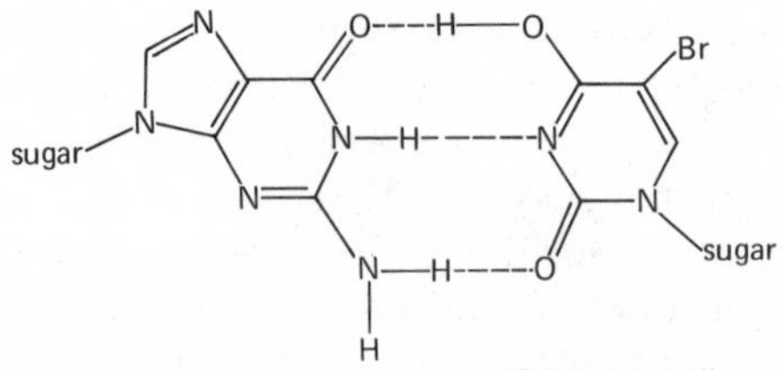

Figure 9-1 Base-pairing of the enolic form of BUDR to guanine.

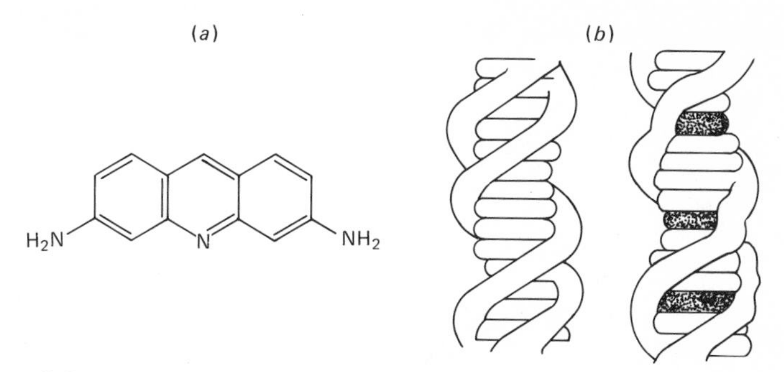

Figure 9-2 (*a*) Structure of proflavine. (*b*) Diagram of DNA helix without (left) and with (right) intercalated proflavine molecules (stippled). (*From Fishbein, Flamm, and Falk, 1970.*)

Many other compounds are effective mutagens, although the mechanisms are not always known. Free radicals have been shown to produce mutations, and this appears to be at least one mechanism involved in radiation-induced mutation. Nonionizing radiation (ultraviolet) causes adjacent thymine residues to form covalent bonds, so-called thymine dimers.

Most of the above examples lead to a change in base-pair sequences, but some mutagens, such as ionizing radiation, act in a substantial way by breaking the covalent structure of the DNA strands, leading ultimately to chromosome breakage. Some chemicals also do this and are therefore sometimes called radiomimetic. The precise mechanisms undoubtedly vary.

Repair of Mutations

Not every potential mutation becomes fixed as such. Chromosomes that break often rejoin, sometimes properly and sometimes improperly if multiple breaks have occurred. At the chemical level, repair of mutational damage also is possible (Hanawalt and Setlow, 1975). For example, only a few percent of the bases that are alkylated lead to mutation. Many of the remaining alkylated bases are replaced by the correct normal base, restoring the DNA to its original form.

The best example of DNA repair is found in the unscheduled DNA synthesis. Although studied originally in bacteria, it has also been found in man. If cells are irradiated with ultraviolet, thymine dimers are formed that can lead to stable mutation. Cells have a system for identifying such forbidden structures and replacing them with the correct nucleotides (Figure 9-3). Four separate steps can be identified: (1) recognizing the abnormal structure and cleaving the strand with an endonuclease; (2) removing a sequence of nucleotides with an exonu-

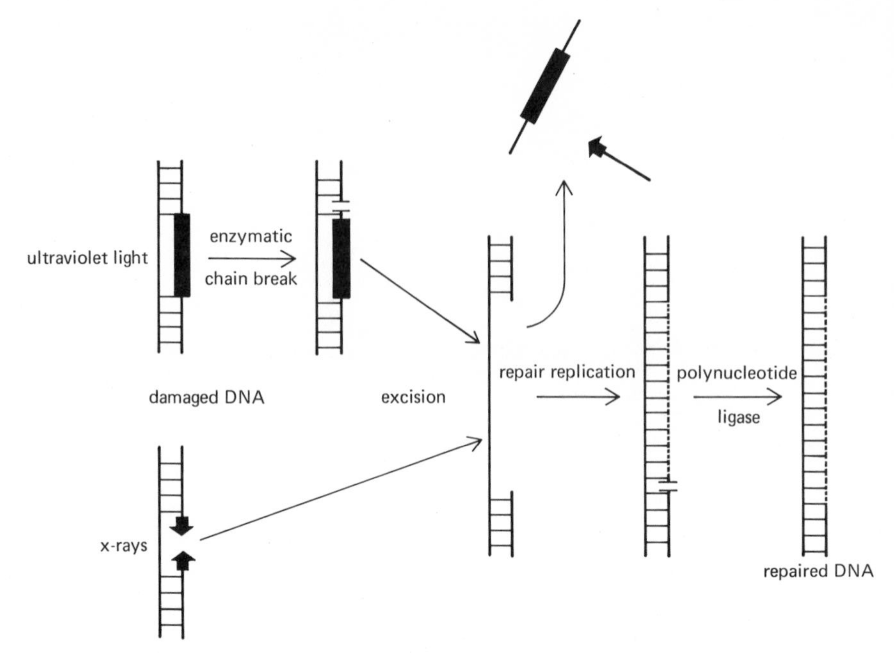

Figure 9-3 Possible scheme for DNA repair in mammalian cells indicating common pathways for repair of base damage and single strand breaks. This scheme illustrates merely the necessary stages of repair and not every biochemical reaction. (*Redrawn from Cleaver, 1969.*)

clease; (3) replacing the nucleotides, using the intact strand as a template, with a polymerase; and (4) rejoining the broken strand with a ligase. The result is an intact DNA double helix with the original structure.

That this repair system occurs in man is demonstrated by the autosomal recessive disease xeroderma pigmentosum. This disorder is characterized especially by sensitivity of the skin to sunlight, leading to a high incidence of tumors in exposed areas such as the face. Cleaver (1969) found that fibroblast cultures from patients lack unscheduled DNA synthesis. Normal fibroblasts, when irradiated with ultraviolet and then supplied with tritium-labeled thymidine, incorporate the radioactivity into the nucleus (Figure 9-4). Fibroblasts from xeroderma patients do not. They lack the repair mechanisms of normal cells and are therefore subject to very high somatic (and perhaps germinal) mutation rates. This is reflected also in increased chromosome breakage. In most patients, the missing enzyme appears to be the first step—the endonuclease—since some repair does occur after ionizing radiation that can break the covalent structure of the strand. At least six different complementation groups are known in this disease.

Other repair systems have been identified but are less thoroughly in-

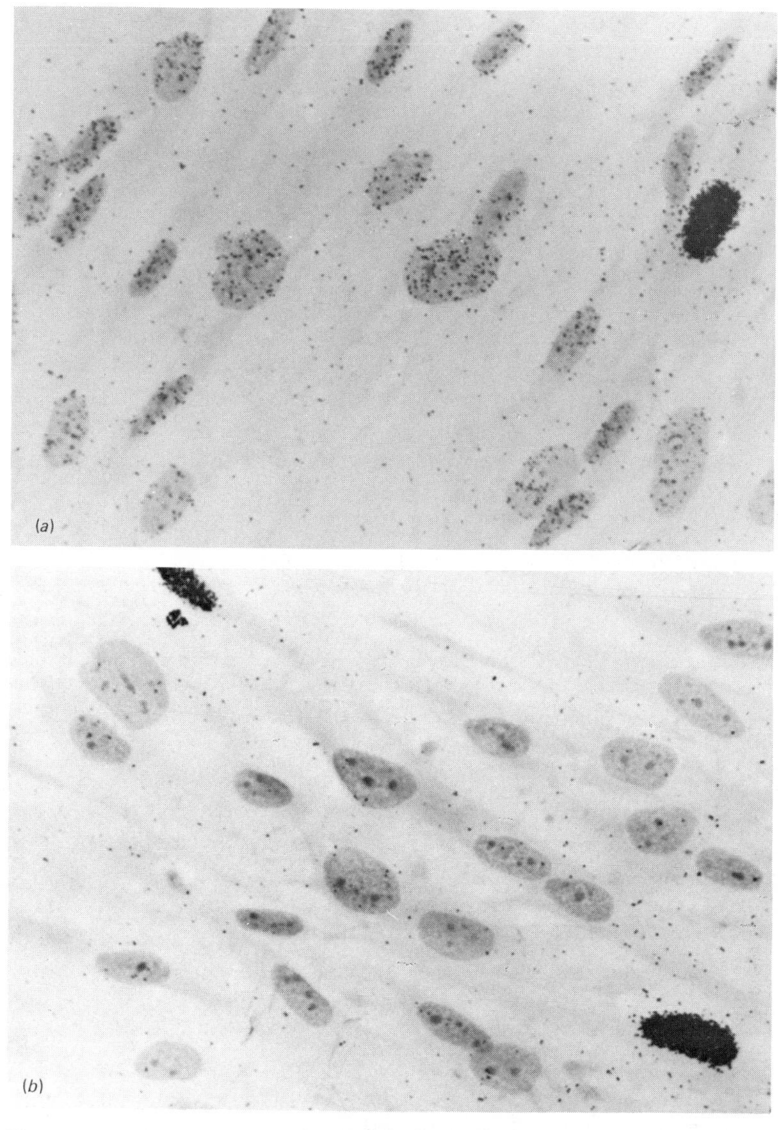

Figure 9-4 Autoradiographs of human fibroblasts in tissue culture irradiated with ultraviolet light (260 ergs/mm^2) and labeled with tritiated thymidine (10 μCi/ml, 23 Ci/millimole) for 1 hour. (*a*) Fibroblasts from a normal person showing heavily labeled S phase cells and lightly labeled cells undergoing repair. Unscheduled DNA synthesis is indicated by the concentration of thymidine in the nuclei, seen as a high frequency of disintegrations of tritium. (*b*) Fibroblasts from a xeroderma pigmentosum patient showing heavily labeled S phase cells but an absence of unscheduled synthesis due to the defective excision repair exhibited by most cases of this disease. (*Furnished by J. E. Cleaver.*)

vestigated than that described. There is ample evidence that only a small portion of altered nucleotides lead to mutation. The efficiency of repair mechanisms seems generally to be high.

FUNCTIONAL CONSEQUENCES OF MUTATION

The consequences of different kinds of changes in DNA information can be predicted and have been observed experimentally. If a base is substituted in a structural gene but the total number of bases remains the same, then a different amino acid codon has been introduced. Some amino acids have more than one codon, but chances are that new codon corresponds to a different amino acid. This would lead to an amino acid substitution in the polypeptide chain, such as apparently happened in many of the mutations of hemoglobin.

If a base pair is deleted in the DNA, a corresponding base would be deleted in mRNA. The effect of this would be to shift the reading frame for amino acid codons (Figure 9-5) so that each base to the "right" of the deletion shifts its position in its codon one space to the left. The first base in a codon would become the last base of the previous codon. Amino acids between the deletion and the C-terminal end of the polypeptide chain would be altered. If the deletion were very close to the part of the gene corresponding to the C-terminal end, the polypeptide chain might have only a few amino acids altered. If the deletion were close to the N-terminal end, where polypeptide synthesis begins, most of the codons of the chain would be displaced and the polypeptide chain would be largely nonsense. Insertion of an extra base in the structural gene would have the same effect as deletion, except that the reading frame would be shifted to the "left" rather than to the "right."

Crick and his associates, using bacterial viruses, have confirmed

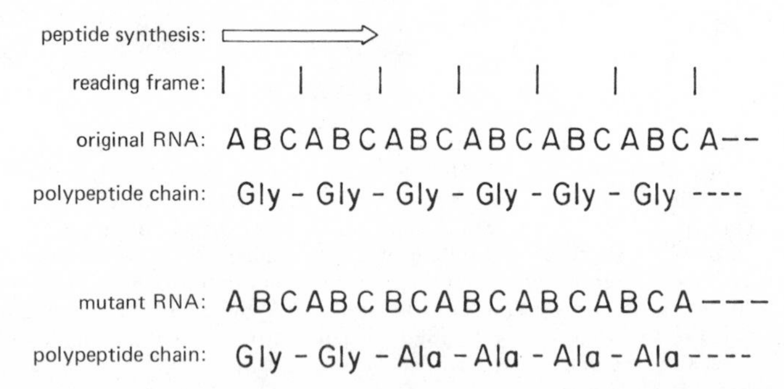

Figure 9-5 Deletion of one nucleotide, causing a shift in reading frame of the RNA. The seventh nucleotide has been eliminated, and all codons to the right have shifted to the left by one nucleotide.

these ideas by demonstrating that one insertion and one deletion very close together in a gene tend to counteract each other. If the two changes are far apart, they do not. Presumably, the shift in reading frame caused by deletion (or insertion) is restored by an insertion (or deletion). If only a few amino acids are altered between the two mutations, then the polypeptide may function. If many are altered, the polypeptide is largely nonsense. Insertion or deletion of three nucleotides would correspond exactly to one codon. If the changes are very close together, then only a small portion of the polypeptide chain would be affected and it might be able to carry out its biological function, although possibly with decreased activity compared with the original polypeptide chain.

Many mutations, especially those induced by radiation, involve more than single base changes. The effects on phenotype are likely to be correspondingly greater. Deletions of more than a few nucleotides lead to loss of amino acids from the polypeptide chain. This is unlikely to be compatible with biological function. Furthermore, the quantitative regulation of protein synthesis may be seriously impaired, although little is known on this subject at present.

Mutations also occur in control genes and other untranslated DNA. Elucidation of the operon system of control was based on the study of such mutants. Comparable studies have not been possible with genes concerned with mammalian gene control or differentiation or with the untranslated DNA sequences inserted into structural genes (introns).

MEASUREMENT OF MUTATION

Many different measures of mutation have been used with experimental organisms and with man. For the most part, these can be classified as lethal effects, specific locus mutations, and chromosomal aberrations. These are not mutually exclusive events, since chromosome aberrations and some specific locus mutations are lethal. There are many variations depending on the organism under study and the laboratory techniques available. Specific locus mutations may be designed to pick up small deletions, nucleotide substitutions, or frameshift mutations. They may depend on specific phenotypic effects that are produced by some point mutations but not by all. Mutation rates then are relative to a particular exposure, organism, and phenotypic response.

If it were possible to survey amino acid sequences in proteins on a mass screening basis, a majority of changes in structural gene sequence would be detected as mutations. Where protein function is the indicator, many mutations are missed, since mutant proteins sometimes are functional. If a mutation must be lethal in order to be detected, many nonlethal mutations will be missed. Similarly, so-called "visible" mutations include only those mutations compatible with life but causing externally visible changes. In discussions of mutation rate, it is necessary to keep in mind that the findings with one set of specific endpoints may differ from findings based on different endpoints.

Mutation rate is usually expressed as the number of mutations per gamete per locus. Sometimes the lethal mutations per X chromosome are measured rather than the mutations per locus. Or other portions of the chromosome complement may be studied and classes of mutations other than lethals (for example, visible morphological variants) may be counted. The particular kinds of mutations counted depend on the experimental material available and on the interests of the observer.

Mutation Studies in Experimental Organisms

The number of experimental systems that have been studied is so large that no effort to review them will be made. Only a few of special interest will be mentioned here. There are many discussions of mutational studies that may be consulted, and the journal *Mutation Research* carries many current research articles.

X-linked Recessive Lethals in Drosophila. In his original studies, Muller devised a test for measuring any X-linked recessive lethal that might be induced in *Drosophila*. Several derivative tests, such as the Muller-5 or the ClB, have been developed. Basically they depend on irradiating a male fly and mating him with a nonirradiated female. The F_1 females are heterozygous for an irradiated and a nonirradiated X chromosome. If these are then mated, any X chromosome on which a recessive lethal has been induced at any locus will fail to yield viable male offspring. Various markers and inversions are used to increase the efficiency of the experiment.

Dominant Lethals in Mammals. If a male mouse or rat is exposed to a mutagen prior to mating, any decrease in fertility associated with the exposure is assumed to be due to dominant lethal mutations. This is a widely used system for testing chemicals for mutagenicity. It has the virtue of requiring relatively few animals, since the number of loci under test is very large. A difficulty is the inability to analyze the mutations genetically, since affected animals die *in utero*. Dominant lethal mutations may be due mostly to chromosome aberrations.

Specific Locus Mutations in Microorganisms. Specific locus mutations also have been widely studied. In bacteria, where enormous numbers of cells can be examined by screening methods, very rare mutational events can be observed. Many of the bacterial mutations are nutritional. Through loss of function of a gene that controls synthesis of an enzyme on the biosynthetic pathway of a required substance, a bacterium mutates from nutritional independence (prototrophy) to a state in which there is a requirement for some specific vitamin, amino acid, or other substance (auxotrophy). The mutant bacteria cannot grow in the absence of the nutritional factor.

A series of *Salmonella* strains has been developed that are especially useful for screening potential mutagens. These strains are themselves mutant, being unable to grow in the absence of histidine. The test involves detecting back mutations to wild type. The test substance is placed at one point on an agar plate seeded with the his^- mutants. Occasional colonies appear spontaneously that are able to grow on histidine-free medium. If the test substance is mutagenic, a zone of such colonies appears around the point of application, from which the mutagen has diffused through the agar. This so-called *Ames test* is often carried out in the presence of extracts of mammalian liver to duplicate insofar as possible in a simple *in vitro* test the complex metabolism of mammals.

Single-locus Mutations in Mice. The need to study mutation directly in mammals has led to two single-locus systems in mice, one developed at Oak Ridge (U.S.), the other at Harwell (U.K.). The strains were originally developed to study radiation-induced mutations but are useful for chemical mutagens as well. The test system used at Oak Ridge is illustrated in Figure 9-6. The seven loci influence various externally visible features, such as coat color. Males with normal chromo-

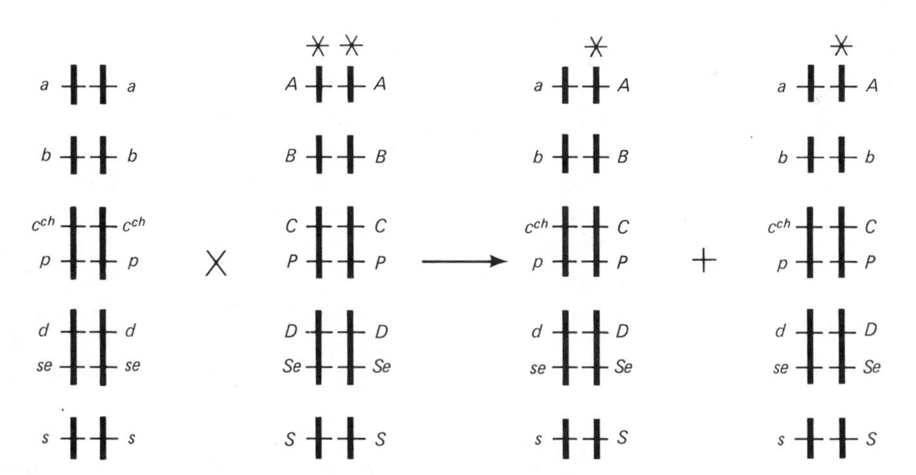

Figure 9-6 Diagram of Russell test stock for induced mutations in mice. The gene complements marked by asterisks are subjected to radiation. *C* and *P* are on one chromosome, as are *D* and *Se*. However, this is not relevant to this part of the experiment. (*From Russell, 1951.*)

somes are irradiated and mated to the test females. In the absence of newly induced mutations, offspring will have a normal phenotype. If a mutation has been induced at one of the seven loci, the offspring receiving that gene will show the mutant phenotype characteristic for the locus. Enormous numbers of mice have been tested in order to measure the frequency of mutation in the presence of low levels of radiation.

Measurement of Human Mutation Rates

Autosomal Dominant Mutations. In man, the only mutations whose rates are known with some certainty are those which are readily observed in heterozygous combination. Furthermore, the resulting condition must be such as to come to medical attention with high probability; it must not be lethal at an early age, otherwise it could not be proved to be genetic; and the frequency of phenocopies must be very low, preferably nonexistent. From a practical point of view, the condition must be uncommon so that new mutations will comprise an appreciable portion of cases. Otherwise, the investigator would have to examine hundreds, perhaps thousands, of cases in order to locate several mutant cases.

Some of the conditions that meet these requirements are listed in Table 9-1. Aniridia will serve as an example of the procedure by which a mutation rate is established. Aniridia (absence of iris of eyes) is inherited as a simple dominant trait. Persons heterozygous for the gene appear always to express the trait. Because these persons are blind, or nearly so, they come to the attention of ophthalmologists and of various agencies concerned with the blind. Nearly complete location of patients can be made through these sources.

In the state of Michigan in the period 1919 to 1959, 4,664,799 persons were reported born, of whom approximately 41 are estimated by M. W. Shaw to have been aniridic offspring with normal parents. (Twenty-eight were actually located; because of the methods of locating patients, an additional 13 were estimated to have been missed.) If these 41 persons are considered to represent new mutations, all at a single locus, then the mutation rate would be $41/(2 \times 4{,}664{,}799) = 4.4 \times 10^{-6}$ mutations per gamete per locus.

Although the procedures used in measuring mutation rate of dominant genes are simple conceptually, there are few diseases that satisfy all the conditions stated initially. It is rare that all the cases of a disease in a prescribed population can be ascertained. Few dominantly inherited conditions are completely penetrant and yet without phenocopies; and few are so readily diagnosed and referred to special agencies.

The conditions that do satisfy these requirements are consistent in their results. All have rates near 1×10^{-5} mutations per gamete per

Table 9-1 **SELECTED ESTIMATES OF MUTATION RATES FOR AUTOSOMAL DOMINANT HUMAN GENES***

Trait	Population	Mutants per 10^6 gametes	Authors
Achondroplasia	Denmark	10	Mørch, corrected by Slatis
	Northern Ireland	13	Stevenson
	Germany	6–9	Schiemann
Aniridia	Denmark	2.9(–5)	Møllenbach, corrected by Penrose
	Michigan	2.6	Shaw et al.
Dystrophia myotonica	Northern Ireland	8	Lynas
	Switzerland	11	Klein, corrected by Todorov et al.
Retinoblastoma	England, Michigan, Switzerland, Germany	6–7	Vogel
	Hungary	6	Czeizel et al.
	The Netherlands	12.3	Schappert-Kimmijser et al.
	Japan	8	Matsunaga
	France	5	Briart-Guillemot et al.
Acrocephalo-syndactyly	England	3	Blank
	Germany	4	Tünte and Lenz
Osteogenesis imperfecta	Sweden	7–13	Smårs
	Germany	10	Schröder
Tuberous sclerosis	England	10.5	Nevin and Pearce
	Chinese	6	Singer
Neurofibromatosis	Michigan	100	Crowe et al.
	USSR	44–49	Sergeyev
Polyposis intestini	Michigan	13	Reed and Neel
Marfan's syndrome	Northern Ireland	4.2–5.8	Lynas
Polycystic disease of the kidneys	Denmark	65–120	Dalgaard
Diaphyseal aclasis	Germany	6.3–9.1	Murken
von Hippel–Lindau syndrome	Germany	0.18	Burhorn

* From Vogel and Rathenberg (1975), where references to published figures can be found.

locus. Some are consistently less and others consistently more. This probably reflects inherent differences in stability of the respective genes. It may also be that, for some diseases, mutations at any of several loci are counted and the rate is therefore the sum of two or more separate rates.

It has been suggested that these rates are unrepresentative of all loci, since at least one mutation must have occurred in order for the locus to be recognized (Cavalli-Sforza and Bodmer, 1971). A more representative value might be 1×10^{-6} mutations per locus per gamete. An additional difficulty in interpretation stems from the fact that the primary gene defect is not known for any of these dominant disorders. It is possible that none represents diminished activity of an enzyme due to mutation at a structural locus.

Mutations at structural loci can, of course, be detected if the altered protein has a different electrophoretic mobility as compared to wild type (Chapter 8). Such mutants would be expressed in heterozygous combinations and therefore are dominant. Since every codon presumably could mutate to give an electrophoretic variant (*electromorph*), the number of mutant forms could be very large, although many would probably interfere with function of the protein. The frequency of electromorphs can be used to estimate indirectly the rates of this class of mutation if one makes certain assumptions about rate of chance elimination, selective neutrality, etc. Neel and Rothman (1978) calculated the rate as 1.6×10^{-5} per locus per generation.

Autosomal Recessive Mutations. Most mutations are from function to nonfunction or decreased function. For the most part, such mutations are recessive and are not evident unless they happen to be combined with another recessive allele. In human beings, it is not possible to know without pedigree studies that a person who is homozygous for a recessive gene has a newly mutated gene. Direct counting of mutant genes is therefore impossible in practice.

Estimates of mutation rates of recessives have been made on the assumption that, for a population in equilibrium, the number of recessive alleles removed from the population because of lack of fitness equals the number of recessive alleles added by mutation. If μ is the mutation rate and N the number of persons in a prescribed population, the number of new mutant genes is $2N\mu$. For autosomal genes, the number of genes removed because of lack of fitness is $2Nx(1 - f)$, where f is the fitness of homozygous recessives (Chapter 21), and x is the frequency of homozygous recessives. Equating these two quantities:

$$2N\mu = 2Nx(1 - f),$$
$$\mu = x(1 - f).$$

For conditions with a genetic fitness of zero, the mutation rate equals the frequency of the condition. The frequency in turn is equal to q^2, where q is the gene frequency.

$$\therefore \mu = q^2.$$

This procedure assumes that persons heterozygous for the mutant alleles are exactly equal in fitness to persons homozygous for the normal alleles. It is increasingly apparent that this requirement is rarely met. Furthermore, a very slight increase in fitness of heterozygotes will lead to grossly erroneous results in the calculated value of μ. This method therefore has been largely abandoned. No satisfactory method for measuring mutation rates of recessive genes has been developed.

Sex-linked Recessive Mutations. The first estimates of human mutation rates were made with X-linked recessive mutations (Haldane, 1935). At equilibrium, the number of new mutations is equal to the number lost through males (assuming the frequency to be sufficiently low so that homozygous recessive females can be ignored). Since only one-third of the X chromosomes occurs in males, the equilibrium formula becomes

$$(1 - f)q = 3\mu,$$

where f is the fitness in males. If the reproductive fitness of males with the mutant allele is near zero, the formula is

$$\mu = \frac{q}{3}.$$

In other words, the mutation rate should be one-third the gene frequency, which is equal to one-third the freguency of the affected males. Some examples of mutation rates for X-linked genes are given in Table 9-2. The accuracy of these estimates depends very much on the complete ascertainment of all cases within a defined population and on the absence of reduced fitness in heterozygous females.

Sex-linked Lethal Mutations. As in *Drosophila*, it is possible in man to detect classes of mutations without identifying specific loci. If a woman is exposed to a mutagenic agent, such as high-energy radiation, mutations will be induced at random throughout the genome. For the most part, the mutations will be recessive and expressed only in the rare event that a similar mutant allele comes from the father of the woman's offspring. Dominant mutations, although expressed in the offspring, will be very rare. Many mutations, both dominant and recessive, exert their deleterious effect early in gestation or at prezygotic stages so that direct observation is impossible.

If mutations are induced on the X chromosome, they will be expressed in sons of the woman. Many of the mutations will be lethal

Table 9-2 **SELECTED ESTIMATES OF MUTATION RATES FOR X-LINKED RECESSIVE HUMAN GENES***

Trait	Population	Mutants per 10^6 gametes	Authors
Hemophilia (A + B)	Denmark	32	Andreassen, corrected by Haldane
	Switzerland	22	Vogel
	Germany	23	Reith
Hemophilia A	Germany	57	Bitter et al.
	Finland	32	Ikkala
Hemophilia B	Germany	3	Bitter et al.
	Finland	2	Ikkala
Duchenne-type muscular dystrophy	Utah	95	Stephens and Tyler
	Great Britain	43	Walton
	Germany	48	Becker and Lenz
	Northern Ireland	60	Stevenson
	Great Britain	47	Blyth and Pugh
	Wisconsin	92	Morton and Chung
	Switzerland	73	Moser et al.
	Japan	65	Kuroiwa and Miyazaki
	England	105	Gardner-Medwin
	Poland	46	Prot
Incontinentia pigmenti	Germany	6–20	Essig
Oculofaciodigital syndrome	Germany	5	Majewski

* From Vogel and Rathenberg (1975), where references to published figures can be found.

prior to birth, but these can be recognized by their influence on the sex ratio. A deficiency of males among the offspring of women exposed to potential mutagenic agents can be accepted as evidence that such mutations were induced (Figure 9-7). Irradiation of males would give rise to female offspring with one irradiated and one untreated X chromosome. To the extent that induced mutations are dominant and deleterious, there should be a loss of female offspring, so that the sex ratio would increase.

Although the theory for predicting an effect of mutation on sex ratio is straightforward, efforts to use sex ratio for this purpose have been disappointing. There are unidentified factors that cause the sex ratio to fluctuate (Chapter 7), making it impossible to relate sex ratio to mutagenic exposure. Therefore this approach, once thought to be very promising, has been largely abandoned in man. Studies in mice also indicate that females are very much more resistant to induction of mutations by radiation (and perhaps other agents). Therefore in man the major effect

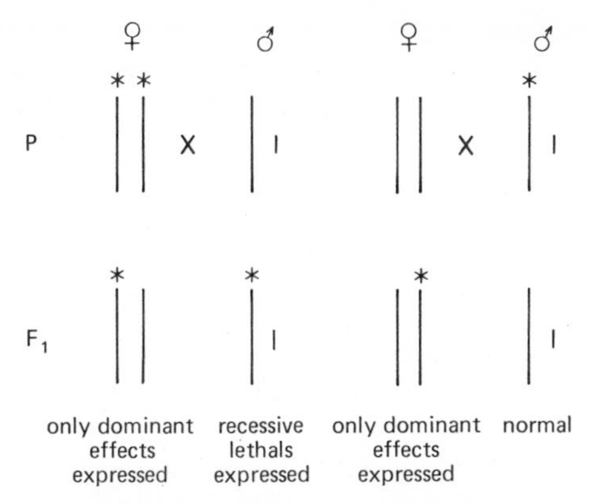

Figure 9-7 Effects of irradiation of maternal and paternal X chromosomes. If the mothers are irradiated (indicated by asterisks), there may be a small decrease in daughters as a result of dominant effects, but sons will show a greater effect, causing the sex ratio to decrease. If the fathers are irradiated, the sex ratio should increase slightly.

on sex ratio may be seen in the offspring of female offspring of treated males.

Chromosome Abnormalities. Chromosome abnormalities may be either numerical or structural and either germinal or somatic. The first dichotomy was discussed in Chapters 5 and 6. The second also was noted briefly. Either numerical or structural abnormalities may be considered to result from a mutational event, although numerical abnormalities need not interfere with the integrity of the chromosome. Therefore, they should be treated separately. Structural abnormalities are thought to be initiated by a chromosome break. Hence, they are more comparable to gene mutations.

For operational purposes, germinal mutations may be defined as those present in all cells of the body (or at least in those sampled) and presumed to have been present in the zygote. Somatic mutations are those that arise postzygotically. For point mutations, the inability to detect somatic mutations in general has limited discussion to germinal mutations. In the case of chromosome aberrations, it is possible to detect low frequencies of structurally abnormal chromosomes in circulating lymphocytes.

Jacobs (1972) summarized the results of 24,468 newborns (15,012 males and 9456 females) on whom chromosome studies had been reported. These were consecutive liveborn hospital births studied in five centers. The results are summarized in Table 9-3. Some 0.5% of all newborns have a detectable chromosome abnormality. Jacobs further considered those abnormalities that were transmitted from a parent (75% of autosomal structural abnormalities) and calculated the muta-

Table 9-3 **INCIDENCE OF CHROMOSOME ABNORMALITIES AMONG CONSECUTIVE NEWBORNS***

Total births	24,468
Abnormal sex chromosome complements (XO, XYY, XXY, XXX, other)	48
Autosomal trisomies (D, E, G, other)	32
Autosomal structural abnormalities	41
Total abnormalities	121
Incidence of abnormalities = 0.0049	

* Based on Jacobs (1972). Five mosaics were omitted from the tabulation.

tion rates shown in Table 9-4. These rates are much higher than the rate per locus for point mutations. This is expected since the processes counted are quite different, and chromosome abnormalities are expressed per cell rather than per locus.

In theory, one might also study the rate of postzygotic aberrant chromosomes detected in somatic tissue such as circulating lymphocytes. This has indeed been done in special exposure situations, but the sensitivity of chromosome breaks and gaps to variations in laboratory handling has generally made the results rather inconclusive. A minimum of several hundred cells must be fully analyzed in order to have meaningful estimates of events that occur with a frequency of a few percent. Structural rearrangements are more reliably scored, and these have been useful in detecting somatic effects of radiation exposure, for example.

Table 9-4 **MUTATION RATES (24,468 CONSECUTIVE HOSPITAL BIRTHS)***

	Sex chromosome abnormalities	Autosomal trisomics	Autosomal structural abnormalities	Total
Percent liveborn population	0.20	0.13	0.18	0.51
Proportion new mutants	100%	100%	25%	—
No. new mutants/100 liveborn	0.20	0.13	0.04	0.37
Mutation rate (per gamete per generation)	1.0×10^{-3}	0.65×10^{-3}	0.2×10^{-3}	1.85×10^{-3}

* From Jacobs, 1972.

Mutations in Cultured Cells. Mammalian cells grown in culture can mutate, but detection depends on an effective selective medium and on the ability of single mutant cells to form clones with high efficiency. Further, point mutations at a specific locus are sufficiently rare, so that in a diploid cell, the likelihood that the same gene will mutate on both chromosomes is exceedingly small. For example, if mutation occurred with a frequency of 1×10^{-6} per locus per cell cycle, homozygosity would be achieved by mutations with a frequency of $(1 \times 10^{-6})^2$ or 1×10^{-12}. This is an important consideration, since most selective media are designed to detect cells that have lost some specific enzyme function, and very often this requires homozygosity for the nonfunctional allele.

In spite of these obstacles, progress has been made in using mammalian cells as mutational test systems (Chu and Malling, 1968; Chu, 1970). Low cloning efficiency is still a problem with normal diploid human cells. Nevertheless, estimates of mutation rates in human diploid cultures have been made. The technique consists in screening for mutants at the HPRT locus, as described in Chapter 14, Mutation from $HPRT^+$ to $HPRT^-$ appears to occur at a rate of approximately 1×10^{-6} per locus per cell cycle, both in fibroblasts (DeMars and Held, 1972) and in lymphoblastoid culture (Sato and co-workers, 1972). In both systems, the rate is increased by known mutagens.

While cultured human cells have the advantage in mutational studies of possessing human genes, the system by no means duplicates an intact human being with the more complex metabolism provided by the liver and other specialized tissues. Cultured cells should be useful for many studies of mutagenesis, however.

BIOLOGICAL FACTORS THAT INFLUENCE HUMAN MUTATION

Several variables have been identified in man or in mammals that influence the rate of mutation. The following are selected to illustrate some of these factors. Many other factors are known also to influence mutation in some systems.

Paternal Age

In many organisms, mutation appears to be associated with cell replication, and certain types of mutations almost certainly occur as a result of errors in DNA replication. In females, there is no replication of oöcytes. Therefore, DNA replication should not make a significant contribution to the mutation rate. In males, spermatogenesis occurs over many years, and one might speculate that children conceived of older fathers would have a higher mutational risk.

The existence of such a paternal age effect was first demonstrated for achondroplasia (Penrose, 1955). Since then it has been demonstrated for several diseases but not for all (Figure 9-8). It is important to recall that these data are for fathers of cases resulting from new mutations and do not include fathers of transmitted cases. Why there should be a paternal age effect for some loci but not others is unknown. Where an increase exists, the increase in rate with age rises more steeply than would be predicted on the basis of a linear correlation with age.

Sex

Mutation should be influenced by the physiological state of the cells, and some of the major influences are associated with sex of the individual. Male and female mice are known to differ in their sensitivities to radiation-induced mutation, females having extremely low rates for low level radiation. The situation for spontaneous rates in man is not clear. There is strong evidence that the mutation rate for hemophilia A is higher in males than females, but the few other loci studied have generally failed to show a difference. Thus, in spite of the very different cell cycles in male and female germ cells, the mutation rates are not strikingly different.

Genotype

Genetic stability is influenced by specific genes in the genome as well as by external agents. This has been demonstrated for several organisms, notably bacteria and *Drosophila* (Mohn and Würgler, 1972). The influence can be to make the entire gene complement either more mutable or more stable. That genes themselves should influence mutational stability is to be expected, since they determine to a large extent the metabolic environment of the chromosomes. The ability to handle introduced chemical substances or metabolically produced mutagens, such as peroxides, is dependent on the array of enzymes whose synthesis is directed by the genes.

Several inherited disorders have been cited that are associated with chromosomal instability: Bloom's syndrome, Fanconi's anemia, xeroderma pigmentosum, and ataxia-telangiectasia. The repair defect in xeroderma pigmentosum is almost certainly associated with increased point mutations as well as chromosomal defects. Ataxia telangiectasia patients and cultured cells from them are more sensitive to x-rays than normal, with greater production of chromosome abnormalities. These particular disorders are sufficiently lethal so that an increase in germinal mutation rates would be hard to detect. Much of their clinical prob-

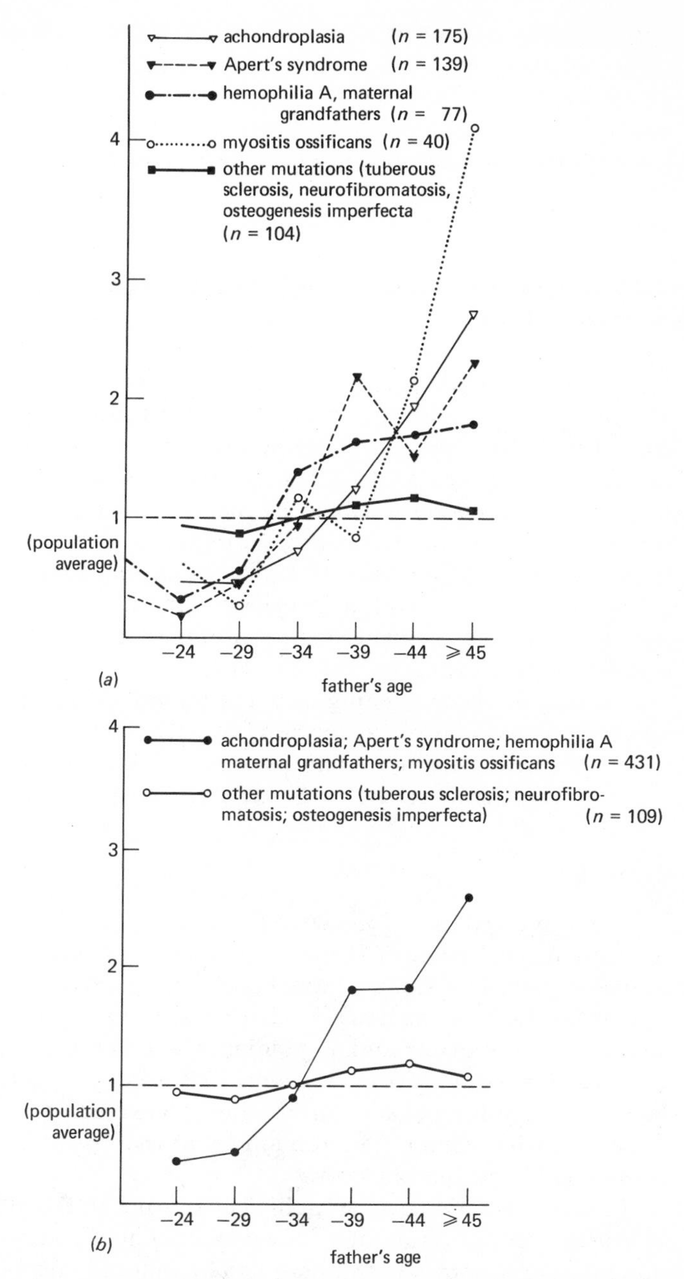

Figure 9-8 Relative incidence of mutations, depending on paternal age. (*Redrawn from Vogel, 1970.*)

lem probably originates from somatic mutations. There almost certainly are genes in the population that are associated with more modest increases in mutation rate. It would be almost impossible to demonstrate this with germinal mutations, but it may be possible to identify such persons through somatic mutations someday.

ENVIRONMENTAL FACTORS THAT INFLUENCE HUMAN MUTATION

Many environmental agents are known to increase mutation rates in experimental organisms. By contrast, the opportunities to observe such effects in man have been virtually nonexistent, because of the enormous populations required to study such rare events. Since man responds to most agents in much the same way that mice and other mammals do, it may be presumed that he undergoes mutation in response to the same agents, except where a species difference in metabolism or physiology is known to exist. Therefore the risk to man is ordinarily estimated on the basis of the exposure of man, assuming that he is as sensitive to that exposure as are mice.

No comprehensive listing of potential mutagens will be attempted, since the list would be far too long and the knowledge as to whether they in fact are causing mutations is largely lacking. The problems in assessing mutagenicity are no better illustrated than with radiation.

Radiation

The first induced mutations were reported in 1927 by H. J. Muller. He found that irradiation of *Drosophila* with x-rays greatly increases the mutation rate. His studies have been greatly extended not only with *Drosophila* but also with many other organisms. Irradiation with ultraviolet light or with any ionizing radiation—for example, x-rays, α particles, neutrons—produces mutations. The effectiveness is lowest for ultraviolet, since it must act entirely through excitation of the molecules rather than ionization. The remainder of the discussion will be concerned with ionizing radiation.

Ionizing radiation can be highly penetrating and therefore is capable of causing mutations in man. Effects are produced in part by direct action on chromosomes but in part also by induced chemical changes in the chromosome environment. For example, sterile irradiated culture media can be mutagenic for bacteria subsequently introduced. When an ionizing particle or photon passes near a chromosome, the large amount of energy released is adequate to break the chromosome. Breaks occasionally repair, restoring the intact chromosome, possibly with a mutation at the point of the break. If two breaks occur in the same cell, the broken ends can mismatch to give a chromosome rearrangement. All of the structural aberrations discussed in Chapter 6 can arise in this manner. A characteristic of these double-break rearrangements is the dose response. Single breaks respond linearly to dose, but, at higher dose rates, double breaks increase as a power of the dose. Point mutations may also result from irradiation.

The two major sources of radiation to man are background and medical x-rays. It is estimated that background radiation, from rocks, cosmic rays, and other natural sources, amounts to about 3 rems per 30 years (Table 9-5). (A rem is a unit of radiation adjusted for absorption by tissue and for biological efficiency. It is derived from the roentgen.) Thirty years is chosen to represent the average generation time of human beings in most Western countries. Medical x-rays account for an additional 2 rems, with other sources far behind. The radiation is not evenly distributed, of course. A person at the higher altitude of Colorado may receive as much as 100 millirems per year more radiation than someone living in Louisiana. The genetically significant dose is that delivered to the gonads. Proper shielding can reduce the genetically significant dose substantially.

The relationship of mutation rate to dose in man has not been ade-

Table 9-5 **SOURCES OF GENETICALLY SIGNIFICANT RADIATION***

	Millirems per year	
Source	Whole body exposure	Genetically significant exposure
Natural radiation		
Cosmic radiation	44	
Radionuclides in the body	18	
External gamma radiation	40	
Total	102	90
Man-made radiation		
Medical and dental	73	30–60
Fallout	4	
Occupational exposure	0.8	
Nuclear power (1970)	0.003	
Nuclear power (2000)	<1	
Radiation Protection Guide for man-made radiation (medical excluded) to the general population (for reference)	170	

* From BEIR Report, 1972.

quately established. This is because very large numbers of persons must be observed at various levels of radiation before the dose relationship can be ascertained. The doses commonly used in *Drosophila* are very much higher than those to which man is exposed. Human beings are killed by radiation that does little harm to *Drosophila.* The straight-line relationship between single-hit or point mutations and dose, amply demonstrated at high levels of radiation in *Drosophila,* has not been thoroughly studied at very low radiation intensity. Russell's results in mice suggest that the response may not be linear at very low intensity. Instead, the yield of mutations per unit of radiation is less for certain types of cells.

Radiation Effects on Man. A number of studies have been designed to test for an effect of radiation on human populations. Various irradiated groups have been used—patients treated for medical problems, x-ray technicians and radiologists, and survivors of the atomic bombs. The genetic studies have concentrated on diminished function and increased mortality among the offspring of these groups (see, for example, Neel and co-workers, 1974). In all studies, the numbers of offspring were too small to expect an observable effect at specific loci, assuming man not to be enormously more sensitive to radiation-induced mutation than are mice. The effects of radiation on the mutation rate have been so small that no single study shows a convincing change in sex ratio. However, some studies do show a small decrease in the proportion of males from irradiated mothers, and, taken together, the studies are consistent with the idea that radiation has been a cause of mutations in man. This has been assumed since the early demonstration of mutagenicity of x-rays, but quantitative data were lacking. The data still

Table 9-6 **ESTIMATED EFFECTS OF RADIATION FOR SPECIFIC GENETIC DAMAGE***

			Effect of 5 rems per generation	
	Current incidence per million live births	Number that are new mutants	First generation	Equilibrium
Autosomal dominant traits	10,000	2,000	50–500	250–2,500
X-chromosome-linked traits	400	65	0–15	10–100
Recessive traits	1,500	?	Very few	Very slow increase

* The range of estimates is based on doubling doses of 20 and 200 rems. The values given are the expected numbers per million live births. (From BEIR Report, 1972.)

Table 9-7 **ESTIMATES OF CYTOGENETIC EFFECTS FROM 5 REMS PER GENERATION***

	Current incidence	Effect of 5 rems per generation: First generation	Effect of 5 rems per generation: Equilibrium
Congenital anomalies			
Unbalanced rearrangements	1,000	60	75
Aneuploidy	4,000	5	5
Recognized abortions			
Aneuploidy and polyploidy	35,000	55	55
XO	9,000	15	15
Unbalanced rearrangements	11,000	360	450

* Values are based on a population of one million live births. Unbalanced rearrangements are based on male irradiation only. (From BEIR Report, 1972.)

are very poor, but they indicate that man probably is not notably more prone to radiation-induced mutation than are mice.

The evidence for the genetic risk to man of ionizing radiation has been summarized by a United Nations Committee (UNSCEAR, 1972) and by a committee of the National Academy of Sciences (BEIR Report, 1972). The BEIR report concluded that the doubling dose of radiation (the amount required to double the mutation rate above the spontaneous level) is between 20 and 200 rems. Using these figures, estimates of genetic damage from 5 rems per generation were calculated as shown in Tables 9-6 and 9-7.

In a comparison of radiation sensitivity versus haploid DNA content, Abrahamson and co-workers (1973) noted a linear relationship (Figure 9-9). From this they concluded that man, if he fits with other organisms, would have approximately 2.6×10^{-7} mutations per locus per rad. This conclusion is based on indirect evidence, but it is interesting that the calculated value falls into the broad range indicated by other approaches. For example, using 1×10^{-5} as the average spontaneous rate per locus in man, the doubling dose of radiation would be $(1 \times 10^{-5})/(2.6 \times 10^{-7})$ or 38 rads. If the "unbiased" estimate of 1×10^{-6} is used for the spontaneous rate, the doubling dose would be only 3.8 rads, a figure that seems too low.

Although no evidence of genetic damage to offspring of the atomic bomb survivors has been noted, chromosome studies on the peripheral leukocytes of the survivors themselves, done 20 years after exposure, still show increased structural rearrangements of chromosomes (Figure 9-10). This was somewhat unexpected, since some of the rearrangements would seem to present severe mechanical problems at mitosis.

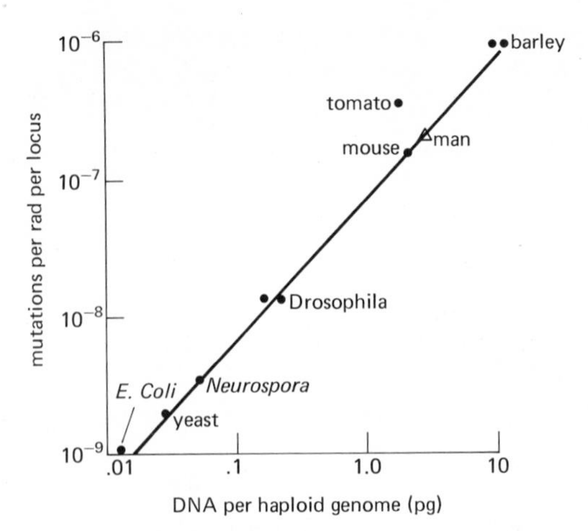

Figure 9-9 Relation between forward mutation rate per locus per rad and the DNA content per haploid genome. The line was drawn with a slope of 1.0 through the mouse point. The point for man was estimated from the DNA content. (*From Abrahamson et al., 1973.*)

Some of the other delayed effects seen in the survivors, such as increased leukemia and cancer, may be due to somatic mutation, but this is uncertain.

Studies of other exposed groups, such as uranium miners (Brandom et al., 1978) and nuclear dockyard workers (Evans et al., 1979), confirm the increase in chromosome aberrations in peripheral lymphocytes as a

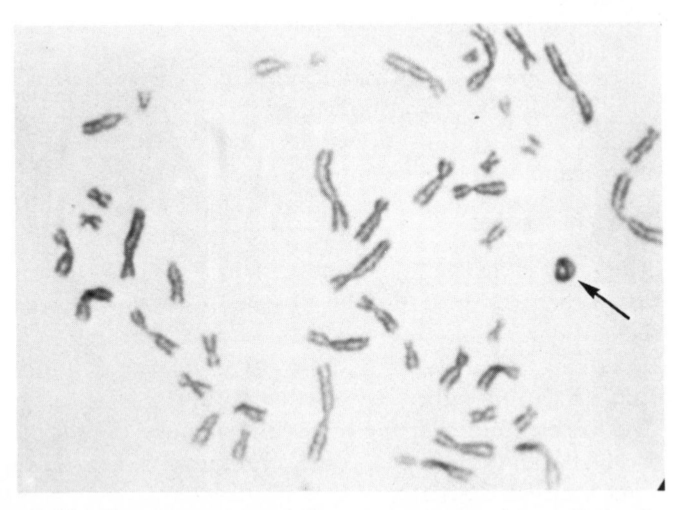

Figure 9-10 Chromosome spread from a man exposed to radiation from the atomic bomb in Hiroshima. The arrow indicates a ring chromosome. Such cells with abnormal chromosome complements are still found circulating in peripheral blood 25 years after the exposure. (*Furnished by Arthur D. Bloom.*)

function of cumulative exposure to radiation. Such aberrations were detectable at exposure levels substantially below the 5 rem per year maximum permissible allowed for industrial exposure. Indeed, the production of such chromosome aberrations has proved a very sensitive "dosimeter" to measure radiation exposure.

Chemical Mutagens

The ability of chemicals to induce mutations was first demonstrated in 1940 by C. Auerbach, using mustard gas and related compounds in experiments with *Drosophila.* Since then, many compounds ordinarily considered to be nontoxic have been shown to be mutagenic in specific situations. Any agent that affects the chemical environment of chromosomes is likely to influence, at least indirectly, the stability of DNA and its ability to replicate without error. However, a key factor in chemical mutagenesis is the necessity for the agent to enter the cell, usually the nucleus itself, in order for mutation to occur. The membranes of cells are quite selective in permitting substances to pass. The blood-testis barrier in mammals is an effective block to entry of some substances into the seminiferous tubules. Thus high concentration of a substance outside a cell is no guarantee that the substance is present in the nucleus. For this reason, results of studies of chemical mutagens cannot be transferred readily from one species to other species, especially if the species are not closely related.

Studies of the mutational risk experienced by man from exposure to chemicals are fraught with all the problems of radiation studies, plus the fact that there is an enormous variety of chemicals to be concerned about. These problems have been discussed by a number of authors (Fishbein and co-workers, 1970; Vogel and Röhrborn, 1970; Hollaender, 1970, 1974, 1976, 1978; Sutton and Harris, 1972).

Temperature

The rates of all chemical reactions are influenced by temperature. It is not surprising then that temperature can be mutagenic. While the effects of temperature are unpredictable for specific organisms and situations, in general the mutation rate increases with increasing temperature. In *Drosophila,* an increase of 10°C increases the mutation rate two- or threefold. Temperature probably affects both thermal stability of the DNA and the rate of reaction of other substances with DNA.

A study of Swedish nudists indicated that the scrotal temperature of human males in ordinary clothing is about 3°C higher than that of nude males (Ehrenberg and co-workers, 1957). The higher temperatures could well increase the mutation rate nearly twofold, leading the investigators to suggest that the wearing of pants has possibly been much more dysgenic than fallout from testing of nuclear devices threatens to be. They suggested the wearing of kilts as one solution.

SOMATIC MUTATION AND HUMAN HEALTH

Although somatic mutations have been demonstrated in a variety of experimental materials, their importance to human well-being is uncertain. This uncertainty arises in large part from the inability to measure mutation rates in somatic cells. Indeed, the identification of variant somatic cells as due to mutation rather than to developmental aberration cannot readily be tested. Because all genetic material is thought to be mutable (although not necessarily equally mutable under all circumstances), it is assumed that somatic mutations do occur and that some may accumulate with age of the organism, although this latter assumption may often be violated if the mutant cells regularly fail to survive.

Somatic mutation has been suggested to contribute to several physiological processes, of which ageing and cancer are especially noteworthy. In the case of ageing, there are only limited observations that support the hypothesis (Holliday and Tarrant, 1972). However, in the case of cancer, the evidence is much stronger.

Mutation and Cancer

Although the mutational theory of cancer was formulated many years ago, the first strong evidence was the similarity in mutagens and carcinogens. Radiation was early established as causing both mutations and cancer. Skin cancers occur primarily in areas exposed to the ultraviolet rays of the sun. Persons subjected to x-rays or other ionizing radiation through medical therapy or occupational exposure were identified as having a high risk of developing malignancy in the exposed organs. Such malignancies may develop long after exposure, however (Figure 9-11). Thus it is awkward to explain the entire process of malignant transformation in terms of mutation.

More support has come from the study of chemical mutagens. At first, the correlation between mutagenicity and carcinogenicity of chemicals was only moderate. Many carcinogens were not mutagenic in available test systems, primarily bacteria; and some strong mutagens were not carcinogens. As tests (and testors) have become more sophisticated, the differences have diminished. It is now known that many hydrocarbon carcinogens must first be metabolized to epoxides in order to cause tumors. This happens in mice, used to test for carcinogenicity, but not bacteria, used to test for mutagenicity. Quite different results

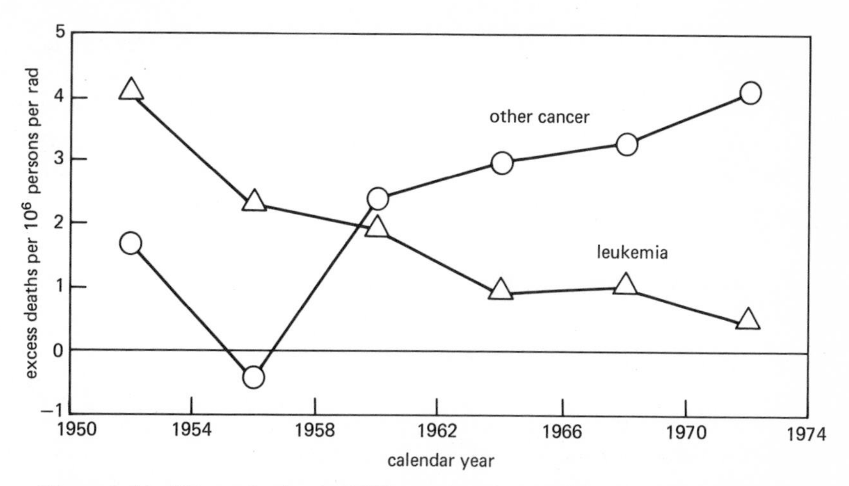

Figure 9-11 Excess deaths per million persons exposed per rad in Hiroshima and Nagasaki. The exposure occurred in 1945. The earliest risk of malignancy was from leukemia, which had already begun to decline by the time the studies began in 1950. However, there was still an increased risk 25 years after the bomb. This risk was primarily in Hiroshima, possibly due to the high neutron content of the Hiroshima radiation. As the risk for leukemia has dropped, the risk for solid tumors has increased and is still increasing. These other forms of malignancy have a much longer "induction" period. The reasons are not known. (*From Beebe et al., 1978.*)

are sometimes obtained when mammalian metabolism, in the form of liver homogenates, is coupled with the sensitivity of bacterial mutagenicity test systems (Ames et al., 1973). It appears that most, probably all, proximal carcinogens are mutagens, although some mutagens are not carcinogens (McCann and Ames, 1976; De Serres, 1976).

One of the more effective arguments in support of the mutational origin of cancer is provided by the hereditary cancers. The propensity for retinoblastoma is inherited as a simple dominant trait, although sporadic nontransmitted cases occur. Most persons with the gene develop one or more primary tumors of the retina, although a few develop no tumors. Knudson (1971, 1977) has analyzed the occurrence of tumors in persons with the transmitted and the sporadic forms and found the risk and distribution of tumors to be consistent with a one-hit event in the transmitted and a two-hit event in the sporadic cases. He proposed that a tumor arises when a cell has two genes for retinoblastoma. One may be transmitted, in which case the other arises as a rare somatic mutation. However, because of the large number of cells in the retina, most persons who inherit one gene will develop one or more mutant cells (and one or more tumors). A person who does not inherit one of the genes develops a tumor when two mutations occur in the same cell. This would be exceedingly rare, and the mutant genes would not be transmitted to offspring.

That these mutant cells are truly gene mutations has not been proved. They might represent some rare virus transformation. However, the rate at which the change from normal to tumor cell occurs is estimated to be 1×10^{-6} per cell generation, a figure comparable to true gene mutation. These studies have been extended with similar findings to Wilms tumor and neuroblastoma, two other childhood tumors.

The Clonal Origin of Tumors

Strongly supporting the mutational origin of many cancers is the demonstration that they originate as clones from a single cell. Those who favor somatic mutation as an integral part of the oncogenic transformation would expect a tumor generally to originate from a single mutant cell that has escaped normal regulation of growth. On the other hand, viruses have been implicated in the development of many tumors, and equally plausible reasons for expecting multicellular origin of tumors can be developed.

Resolution of the question in the case of the benign uterine tumors known as leiomyomas was given by D. Linder and S. M. Gartler. In several patients who were heterozygous for G6PD types A and B and who had uterine leiomyomas, cell cultures started from each tumor were of a single type, either A or B. Only in one of the 29 tumors examined were both G6PD A and B noted, and this was thought to be due to inclusion of nontumor tissue in the sample.

Similar results were obtained by Benditt and Benditt (1973) in their study of atherosclerosis and by Baylin et al., (1978) in the case of inherited medullary thyroid carcinoma.

Especially interesting results were obtained for Burkitt lymphoma by P. J. Fialkow et al. This malignancy is endemic among children in parts of Africa but occurs very rarely in other parts of the world. In Africa, it is always associated with infection with Epstein–Barr (EB) virus. EB virus that is indistinguishable is the causative agent in the benign infectious mononucleosis common in other countries. Studies of G6PD types as well as of surface antigens of the lymphocytes indicate the Burkitt lymphoma cells to be of monoclonal origin. Thus, whatever the nature of the viral action, it appears to lead very rarely to malignancy. Indeed the rate of conversion of normal to tumor cells is consistent with an event as rare as mutation. Why this should occur in Africa is an interesting unsolved problem.

In contrast to these studies, the results with hereditary multiple trichoepithelioma and hereditary neurofibromas indicate that most tumors have both types of G6PD and that they therefore ordinarily originate from more than one cell. This might be the case if the very high likelihood of tumor formation in these diseases led often to two or more transformed cells in sufficient proximity to form a single tumor.

It is conceivable that most tumors have mutation as one important factor in etiology. Even though viruses have been implicated in several tumors, the nature of the viral event that causes only a rare cell to become malignant is unknown. Many viruses appear to become intimately associated with chromosomes during infection, and possibly the viral effect is through induction of structural changes in the chromosomes, that is, mutation. Or viruses may activate cancer genes that are already present in inactive form, genes that also may be activated by mutation. The two hypotheses may therefore merge once the molecular basis of the viral effect is known.

SUMMARY

1. Any change in DNA or chromosomes that is heritable is a mutation. Substitution of base analogs or chemical modification of a base often causes mispairing of bases during replication, leading to substitution of the wrong nucleotide in the new DNA chain. Some compounds intercalate between adjacent base pairs, causing addition or deletion of nucleotides (frameshift mutations). Compounds that form free radicals are effective mutagens, and some of the strong mutagenic effect of radiation, both ionizing and nonionizing, is through the action of free radicals.

2. Mutations can often be repaired. The best studied repair mechanism in man is that responsible for removing thymine dimers caused by ultraviolet irradiation. The first step of repair requires cleaving the DNA strand containing the abnormal structure. In the autosomal recessive disease xeroderma pigmentosum, the enzyme required for this step is missing. As a consequence, persons with this disorder are unable to repair damage from ultraviolet irradiation. They develop many skin tumors, apparently from the greatly increased mutation risk.

3. The functional consequences of mutation depend on the nature of the polypeptide chain produced by the altered DNA. Some may function very well; others may function poorly, if at all, even though the structural change seems minor. If the mutation involves a change in reading frame, the polypeptide is likely to be completely nonfunctional. Mutations in mammalian control genes have not been studied in detail.

4. Mutation is quantified as the number of mutants induced per gamete per locus for point mutations. Often, interest is in the total number of lethal mutations induced per chromosome rather than per locus. A system of scoring mutants based on phenotype is certain to miss many mutations that do not yield the phenotype.

5. There are many different systems for measuring mutation rates in experimental organisms. Frequently used methods include induction of X-linked lethals in *Drosophila,* dominant lethals in mammals, and specific locus mutations in microorganisms and in mice.

6. The best estimates of human mutation rates are based on the occurrence of new mutations for autosomal dominant disorders. These rates average approximately 1×10^{-5} mutations per locus per gamete. It has been argued that these results are biased and that the correct rate may be closer to 1×10^{-6} per locus per gamete. Estimates of mutation rates of X-linked recessive and autosomal recessive genes have been made, but the latter especially are of doubtful validity. The sex ratio, which should reflect mutation on the X chromosome, appears to fluctuate in response to factors other than mutation.

7. Chromosome abnormalities occur much more frequently as new mutations than do point mutations. Studies of consecutive liveborn births gave a frequency of nearly 2 new aberrations per 1000 births. Chromosome aberrations also occur somatically, but it is difficult to obtain quantitative estimates of the magnitude.

8. Several biological factors influence mutation rate. For several dominant traits, but not for all, there is a strong positive association between mutation rate and age of the father. Mutation rates probably differ in males and females, although data in human populations are scant. The genetic background influences gene stability, and several simple Mendelian traits are known in man to lead to high mutation rates.

9. Radiation is the most thoroughly investigated environmental agent that increases mutation. Studies in man have yielded marginal evidence at best that man has been affected by exposure to radiation, although extensive experimental work has clearly shown ionizing radiation to be mutagenic. Apparently man is not especially more sensitive to radiation as compared to mice.

10. Many chemicals are known to be mutagenic for experimental organisms. Testing of chemicals is especially difficult, because of the metabolic differences among test organisms, including man. There is no direct evidence of chemicals causing mutation in man, but many of the chemicals to which man is exposed are highly mutagenic in other test systems.

11. The importance of somatic mutation has not been assessed. It has been suggested as contributing to several processes, including ageing and tumor formation. Most carcinogens are mutagens, especially if metabolism of the carcinogen is considered. The familial distribution of several childhood tumors—retinoblastoma, Wilms tumor, and neuroblastoma—is consistent with origin from a cell with two mutations, one of which may have been inherited.

12. In several types of tumor, it has been possible, using X-linked genetic markers to show that each tumor is a clone arising from a single cell. This supports the hypothesis that malignant transformation occurs as a very rare event in a single cell, which as a result escapes regulation. This rare event may be a mutation.

REFERENCES AND SUGGESTED READING

ABRAHAMSON, S., BENDER, M. A., CONGER, A. D., AND WOLFF, S. 1973. Uniformity of radiation-induced mutation rates among different species. *Nature* 245: 460–462.

AMES, B. N., DURSTON, W. E., YAMASAKI, E., AND LEE, F. D. 1973. Carcinogens are mutagens: a simple test system combining liver homogenates for activation and bacteria for detection. *Proc. Natl. Acad. Sci.* (U.S.A.) 70: 2281–2285.

AUERBACH, C., AND KILBEY, B. J. 1971. Mutation in eukaryotes. *Annu. Rev. Genet.* 5: 163–218.

BAYLIN, S. B., HSU, S. H., GANN, D. S., SMALLRIDGE, R. C., AND WELLS, S. A. JR. 1978. Inherited medullary thyroid carcinoma: A final monoclonal mutation in one of multiple clones of susceptible cells. *Science* 199: 429–431.

BEEBE, G. W., KATO, H., AND LAND, C. E. 1978. Studies of the mortality of A-bomb survivors. 6. Mortality and radiation dose, 1950 -1974. *Radiat. Res.* 75: 138–201.

BEIR (Biological Effects of Ionizing Radiation) Report. 1972. *The Effects on Populations of Exposure to Low Levels of Ionizing Radiation.* National Academy of Sciences-National Research Council, Washington, D.C. 217 pp.

BENDITT, E. P., AND BENDITT, J. M. 1973. Evidence of monoclonal origin of human atherosclerotic plaques. *Proc. Natl. Acad. Sci.* (U.S.A.) 70: 1753–1756.

BLOOM, A. D. 1972. Induced chromosomal aberrations in man. *Adv. Hum. Genet.* 3: 99–172.

BRANDOM, W. F., SACCOMANNO, G., ARCHER, V. E., ARCHER, P. G., AND BLOOM, A. D. 1978. Chromosome aberrations as a biological dose-response indicator of radiation exposure in uranium miners. *Radiat. Res.* 76: 159–171.

CAVALLI-SFORZA, L. L., AND BODMER, W. F. 1971. *The Genetics of Human Populations.* San Franciso: W. H. Freeman, 965 pp.

CHU, E. H. Y. 1970. Point mutations in mammalian cell cultures as measures for mutagenicity testing. In *Chemical Mutagenesis in Mammals and Man* (F. Vogel and G. Röhrborn, Eds.). New York-Berlin-Heidelberg: Springer-Verlag, pp. 241–250.

CHU, E. H. Y., AND MALLING, H. 1968. Mammalian cell genetics. II. Chemical induction of specific locus mutations in Chinese hamster cells *in vitro. Proc. Natl. Acad. Sci.* (U.S.A.) 61: 1302–1312.

CLEAVER, J. E. 1969. Xeroderma pigmentosum: a human disease in which an initial stage of DNA repair is defective. *Proc. Natl. Acad. Sci.* (U.S.A.) 63: 428–435.

CRICK, F. H. C., BARNETT, L., BRENNER, S., AND WATTS-TOBIN, R. J. 1961. General nature of the genetic code for proteins. *Nature* 192: 1227–1232.

CROW, J. F. 1961. Mutation in man. *Prog. Med. Genet.* 1: 1–26.

DE MARS, R., AND HELD, K. R. 1972. The spontaneous azaguanine-resistant mutants of diploid human fibroblasts. *Humangenetik* 16: 87–110.

DE SERRES, F. J. 1976. Mutagenicity of chemical carcinogens. *Mutat. Res.* 41: 43–50.

DRAKE, J. W. 1970. *The Molecular Basis of Mutation.* San Francisco: Holden-Day, 273 pp.

EHRENBERG, L., EHRENSTEIN, G. V., AND HEDGRAN, A. 1957. Gonad temperature and spontaneous mutation-rate in man. *Nature* 180: 1433–1434.

EVANS, H. J., BUCKTON, K. E., HAMILTON, G. E., AND CAROTHERS, A. 1979. Radiation-induced chromosome aberrations in nuclear-dockyard workers. *Nature* (Lond.) 277: 531–534.

FISHBEIN, L., FLAMM, W. G., AND FALK, H. L. 1970. *Chemical Mutagens. Environmental Effects on Biological Systems.* Academic Press, New York, 364 pp.

HALDANE, J. B. S. 1935. The rate of spontaneous mutation of a human gene. *J. Genetics* 31: 317–326.

HANAWALT, P. C., AND SETLOW, R. B. (Eds.). 1975. *Molecular Mechanisms for Repair of DNA.* Plenum Press, New York and London. 843 pp.

HOLLAENDER, A. (Ed.). 1970. *Chemical Mutagens,* Vols. 1 and 2. 1974, Vol. 3, 1976, Vol. 4. 1978, Vol. 5. New York: Plenum Press.

HOLLIDAY, R., AND TARRANT, G. M. 1972. Altered enzymes in ageing human fibroblasts. *Nature* 238: 26–30.

JACOBS, P. A. 1972. Chromosome mutations: frequency at birth in humans. *Humangenetik* 16: 137–140.

KNUDSON, A. G. JR. 1971. Mutation and cancer: Statistical study of retinoblastoma. *Proc. Natl. Acad. Sci.* (U.S.A.) 68: 820–823.

KNUDSON, A. G. JR. 1977. Genetics and etiology of human cancer. *Adv. Hum. Genet.* 8: 1–66.

MC CANN, J., AND AMES, B. N. 1976. Detection of carcinogens as mutagens in the *Salmonella*/microsome test: Assay of 300 chemicals: Discussion. *Proc. Natl. Acad. Sci.* (U.S.A.) 73: 950–954.

MOHN, G., AND WÜRGLER, F. E. 1972. Mutator genes in different species. *Humangenetik* 16: 49–58.

MULLER, H. J. 1927. Artificial transmutation of the gene. *Science* 66: 84–87.

NEEL, J. V., KATO, H., AND SCHULL, W. J. 1974. Mortality in the children of atomic bomb survivors and controls. *Genetics* 76: 311–326.

NEEL, J. V., AND ROTHMAN, E. D. 1978. Indirect estimates of mutation rates in tribal Amerindians. *Proc. Natl. Acad. Sci.* (U.S.A.) 75: 5585–5588.

NEEL, J. V., AND SCHULL, W. J. 1956. *The Effect of Exposure to the Atomic Bombs on Pregnancy Termination in Hiroshima and Nagasaki.* National Academy of Sciences-National Research Council, Washington, D.C. Publication 461. 241 pp.

OLIVER, C. P. 1930. The effect of varying the duration of X-ray treatment upon the frequency of mutation. *Science* 71: 44–46.
PENROSE, L. S. 1955. Parental age and mutation. *Lancet* 2: 312.
RUSSELL, W. L. 1951. X-ray-induced mutation in mice. *Cold Spring Harbor Symp. Quant. Biol.* 16: 327–336.
RUSSELL, W. L., RUSSELL, L. B., AND CUPP, M. B. 1959. Dependence of mutation frequency on radiation dose rate in female mice. *Proc. Natl. Acad. Sci.* (U.S.A.) 45: 18–23.
SATO, K., SLESINSKI, R. S., AND LITTLEFIELD, J. W. 1972. Chemical mutagenesis at the phosphoribosyltransferase locus in cultured human lymphoblasts. *Proc. Natl. Acad. Sci.* (U.S.A.) 69: 1244–1248.
SCHOLTE, P. J. L., AND SOBELS, F. H. 1964. Sex ratio shifts among progeny from patients having received therapeutic X-radiation. *Am. J. Hum. Genet.* 16: 26–37.
SCHULL, W. J., AND NEEL, J. V. 1958. Radiation and the sex ratio in man. *Science* 128: 343–348.
SHAW, M. W., FALLS, H. F., AND NEEL, J. V. 1960. Congenital aniridia. *Am. J. Hum. Genet.* 12: 389–415.
STRAUSS, B. S. 1964. Chemical mutagens and the genetic code. *Prog. Med. Genet.* 3: 1–48.
SUTTON, H. E., AND HARRIS, M. I. (Eds.). 1972. *Mutagenic Effects of Environmental Contaminants.* New York: Academic Press, 195 pp.
TROSKO, J. E., AND CHU, E. H. Y. 1975. The role of DNA repair and somatic mutation in carcinogenesis. *Adv. Cancer Res.* 21: 391–425.
UNSCEAR (United Nations Scientific Committee on the Effects of Atomic Radiation). 1972. *Ionizing Radiation: Levels and Effects.* (In two volumes.) United Nations, New York. 447 pp.
VOGEL, F. 1970. Spontaneous mutations in man. In *Chemical Mutagenesis in Mammals and Man* (F. Vogel and G. Röhrborn, Eds.). New York-Berlin-Heidelberg: Springer-Verlag, pp. 16–68.
VOGEL, F., AND RATHENBERG, R. 1975. Spontaneous mutation in man. *Adv. Hum. Genet.* 5: 223–318.
VOGEL, F., AND RÖHRBORN, G. (Eds.). 1970. *Chemical Mutagenesis in Mammals and Man.* New York-Berlin-Heidelberg: Springer-Verlag, 519 pp.

REVIEW QUESTIONS

1. Define:

alkylating agent	electromorph	mutation
aniridia	endonuclease	point mutation
auxotroph	frameshift	prototroph
base analog	germinal mutation	rem
carcinogen	host-mediated assay	somatic mutation
dominant lethal	intercalation	unscheduled DNA synthesis
doubling dose	ligase	

2. There has been considerable debate on whether most evolutionary change is by natural (Darwinian) selection or by chance fixation of neutral mutations (non-Darwinian selection). Considering the various kinds of mutations and their effects on protein structure, would you expect neutral mutations to be a frequent occurrence?

3. A new drug effective against upper respiratory infections is found to be highly active in causing frameshifts in *Salmonella*. What additional tests should be done to assure the safety of this drug before release? How would your answer change if the drug were effective against cancer?

4. What would be the approximate frequency of a recessive lethal that is maintained only by mutation pressure?

5. Some chemicals used in industrial processes are known to be mutagenic. Although exposure of workers to such chemicals may be kept very low, it is virtually impossible to avoid all exposure. Should there be restrictions on the age or sex of persons working in areas where exposure is likely? What other factors should be considered?

6. Would you expect birth control to increase or decrease the mutation rate? Why?

CHAPTER TEN

THE HUMAN HEMOGLOBINS

Human hemoglobins, in addition to having great clinical significance, have contributed remarkably to understanding of basic genetic principles. Reference was made in Chapter 8 to the insight into gene action gained by studies of sickle cell anemia. Other areas of genetics have also benefited from the rich variety of hemoglobin mutations available for study. There are numerous reasons for this key role of hemoglobin, including the fact that it occurs in relatively accessible cells, in relatively pure form, and in relatively large amounts. Exquisitely sensitive procedures are not required for many of the studies of hemoglobin.

A remarkable variety of genetic processes can be documented at the molecular level with human hemoglobin. It is therefore instructive to consider in detail some of the inherited variations in hemoglobin structure and function.

THE STRUCTURE AND FUNCTION OF NORMAL HEMOGLOBIN

Hemoglobin (Hb) is a tetrameric protein of molecular weight 64,500. Each of the four monomers is a single polypeptide chain (the *globin* portion) and has a single heme group (Figure 10-1). There are two each of two types of monomers, designated α and β in adult human Hb, similar in three-dimensional conformation to each other and to myoglobin (Mb), the oxygen-transporting protein of muscle cells (Figure 10-2). Myoglobin exists only as a monomer. It is thought that, during the course of evolution, a myoglobinlike precursor of Hb acquired the ability through gene mutation of forming stable dimers and then tetramers that satisfied the physiological requirements of oxygen transport in a manner superior to monomers. The interaction between Hb subunits is based on electrostatic and hydrophobic attractions rather than covalent bonds. Therefore, it is relatively easy to dissociate Hb into dimers and monomers, and such dissociation and reassociation probably occurs at a

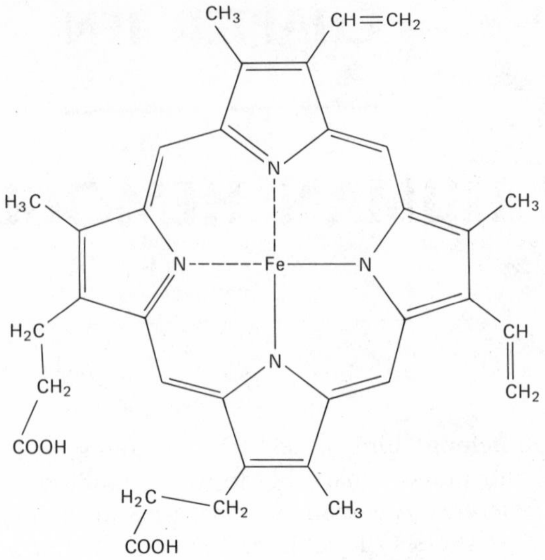

Figure 10-1 The structure of heme.

low level *in vivo*. The predominant dissociation is first into identical half-molecules:

$$\alpha_2\beta_2 \rightleftarrows 2\alpha\beta \rightleftarrows 2\alpha + 2\beta$$

The physiological advantage of the tetrameric Hb as compared to monomeric Mb is readily seen in the oxygenation curves. The degree of oxygen saturation of Mb is a simple function of the oxygen pressure to which it is exposed (Figure 10-3). However, once a Hb molecule picks up a single oxygen molecule, the three-dimensional structure changes so that the affinity for a second, third, and fourth oxygen molecule is increased. This "heme-heme interaction" causes the oxygenation curve to be sigmoid and means that Hb can pick up oxygen in the lungs and release it to tissues in which the pressure of oxygen is only slightly less than in the lungs. Hb function is also influenced by the chemical environment, responding especially to 2,3-diphosphoglyceric acid levels, to *p*H, and to carbon dioxide levels.

The Normal Hemoglobins

The Hb in a red cell is actually a mixture of hemoglobins. The predominant form in adults is Hb A, comprising some 90% of the total. A smaller component, Hb A_2, accounts for 2.5%. The remainder, sometimes called Hb A_3, is not fully defined but probably consists mostly of Hb A that has become altered following synthesis. Fetal Hb, Hb F, is the predominant form during fetal development. It is present in

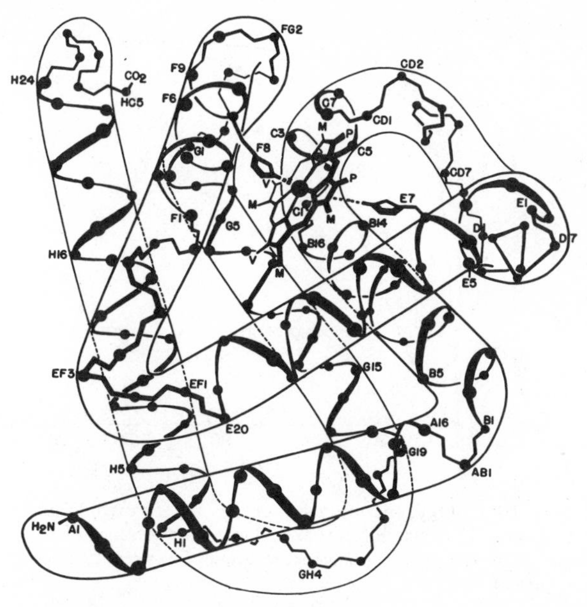

Figure 10-2 Drawing of myoglobin showing three-dimensional folding of the polypeptide chain. Other globins have the same configuration with only minor differences due to deletions or additions of amino acids. The large dots represent the α-carbon atoms of each amino acid residue. There are eight segments of the chain that form α helices (labeled A through H from the N-terminal end). Amino acid residues within a helix are assigned numbers beginning with the N-terminal residue of each helical segment. Thus, residue E7 is the seventh residue in the E helix. Residues in nonhelical segments are designated by the letters of the neighboring helical regions; thus, EF1, EF2, and so on. (*From R. E. Dickerson, 1964.* The Proteins, *Vol. II, 2nd Ed. New York: Academic Press, p. 603.*)

the earliest fetuses examined and is synthesized at high rate approximately until the time of birth, when synthesis is shut off or at least repressed to a very low level. By six months of age, the level of Hb F is less than 1% and remains so through adulthood. In embryos, an embryonic form of Hb occurs, known as Hb Gower 2. Also in early embryos, a form of Hb known as Portland 1 has been observed, but little is known of it.

All the human hemoglobins consist of four subunits. Two of the subunits are of the α type, identical for most hemoglobins, and the other two differ for each kind of hemoglobin (Table 10-1). Thus Hb A may be written as $\alpha_2\beta_2$, Hb F as $\alpha_2\gamma_2$, Hb A_2 as $\alpha_2\delta_2$, and embryonic Hb as $\alpha_2\epsilon_2$. Portland 1 apparently is $\zeta_2\gamma_2$, and Gower 1 is $\zeta_2\epsilon_2$.

Figure 10-4 illustrates synthesis of the various types of globin chains during development. Synthesis of β chains is shown to start near the time of birth. Using very sensitive techniques, Kazazian and Woodhead (1973) were able to show β-chain synthesis as early as the eighth

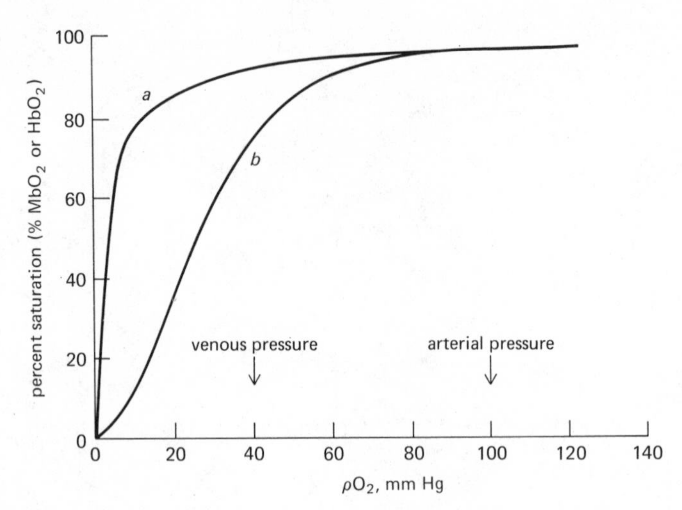

Figure 10-3 Oxygen saturation curves for (*a*) myoglobin (no heme-heme interaction) and (*b*) hemoglobin (heme-heme interaction). The presence of heme-heme interaction permits a much greater range of oxygenation for small changes in oxygen pressure.

week of gestation. Quantitatively the amount is very small, but it offers the possibility of recognizing β-chain defects early in pregnancy.

The α chains consist of 141 amino acids, whereas the β, γ, and δ chains have 146 amino acids. The complete sequences are shown in Figure 10-5. The ϵ and ζ chains have not been fully characterized chemically. The great similarity of the β and δ chains, differing only in 10 of

Table 10-1 **GENETIC AND MOLECULAR RELATIONSHIPS OF HUMAN HEMOGLOBINS**

Hb Portland 1 is omitted but presumably would represent an additional locus, Hb_ζ, whose product reacts with the γ chains. It is probable that some synthesis of all Hb chains occurs throughout life, but the non-α genes operate primarily during one period of ontogeny.

Genes		Hb's	Principal period of synthesis
	Hb_ϵ	$\alpha_2\epsilon_2$	Embryonic
	$Hb_{G\gamma}$	$\alpha_2{}^G\gamma_2$	Fetal
Hb_α	$Hb_{A\gamma}$	$\alpha_2{}^A\gamma_2$	
	Hb_β	$\alpha_2\beta_2$	Postnatal
	Hb_δ	$\alpha_2\delta_2$	

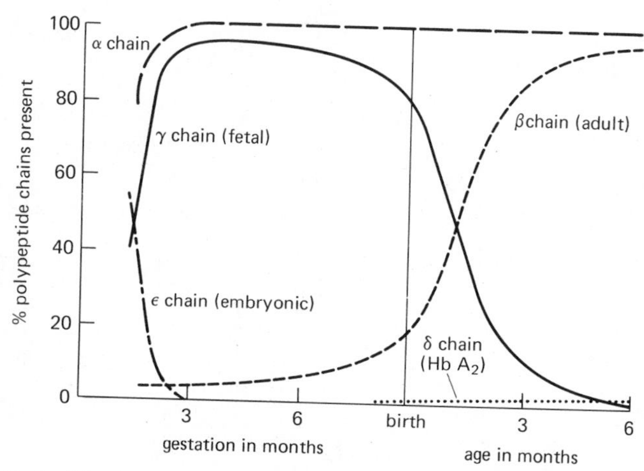

Figure 10-4 The formation of the α, β, γ, δ, and ϵ hemoglobin polypeptide chains in early life. (*Huehns and co-workers, 1964.*)

146 amino acids, suggests a recent divergence of the structural genes for these loci, recent at least in the evolutionary time scale. One additional distinction has been made in Hb types. Huisman and co-workers (1969) found that position 136 of the γ chain may be occupied either by glycine or alanine and that both forms are present in all normal persons. This suggests that Hb F should be represented by two formulas, $\alpha_2\gamma_2^{136\ \mathrm{Gly}}$ and $\alpha_2\gamma_2^{136\ \mathrm{Ala}}$, each γ chain, which can be symbolized as $G\gamma$ and $A\gamma$, apparently the product of a different structural gene.

In summary, six hemoglobins are found in most persons during the course of development from zygote to adult. Five have an α chain in common but differ in the non-α chain. Since each different polypeptide chain is coded by a different structural gene, a minimum of seven genes for Hb would be required to code for human hemoglobins.

FORMAL GENETICS OF HUMAN HEMOGLOBINS

Genetic variation is the primary tool of geneticists, and its use is nowhere better illustrated than in studies of human hemoglobins. As has been described in Chapter 8, sickle cell anemia involves a substitution of valine for glutamic acid at a specific position in the adult Hb molecule. With the complete sequencing of the Hb A molecule, this position was shown to be the sixth from the N-terminus on the β chain, since the tryptic peptide containing the substitution proved to be the N-terminal peptide of the β chain.

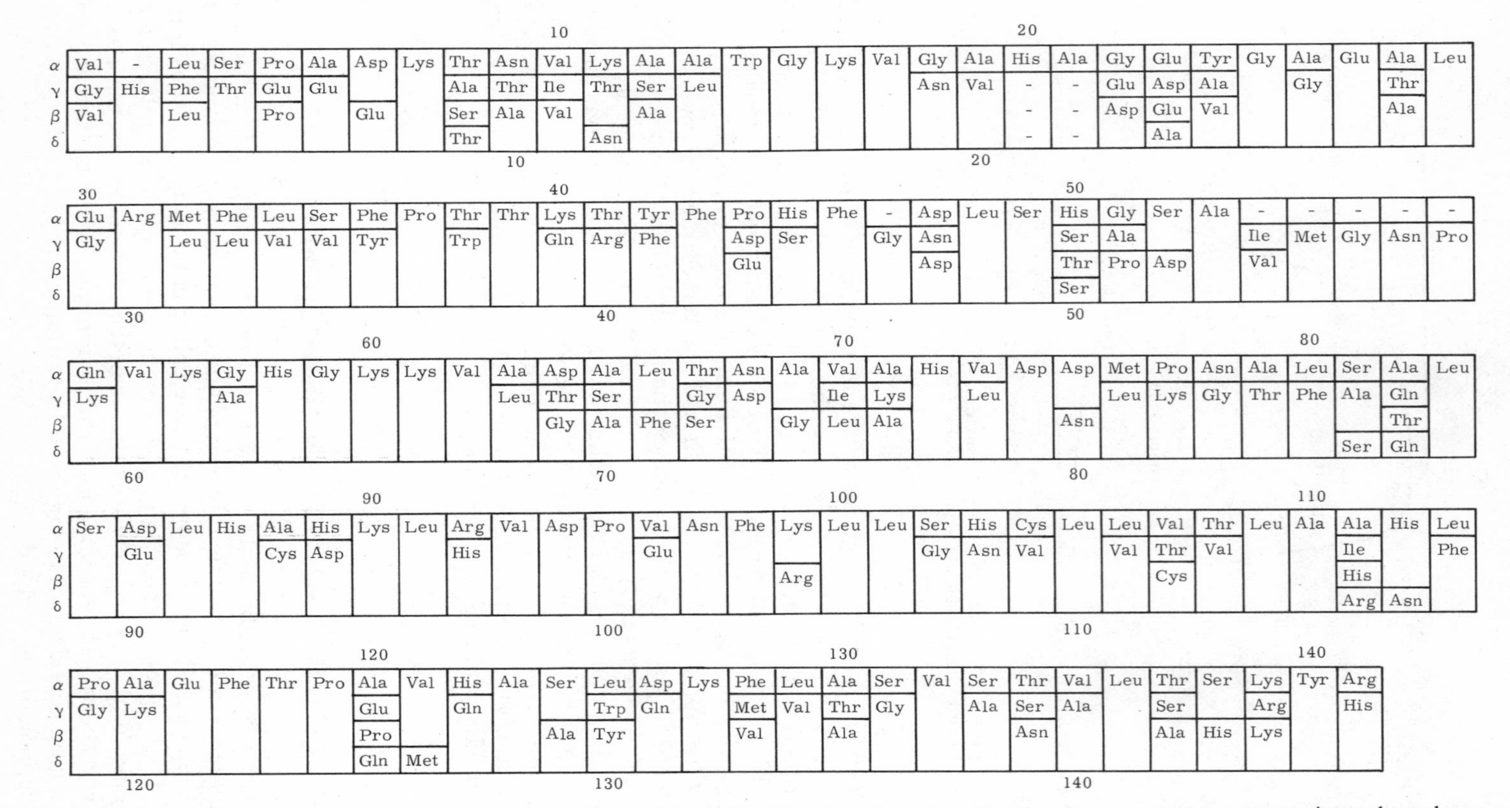

											10										20									
α	Val	-	Leu	Ser	Pro	Ala	Asp	Lys	Thr	Asn	Val	Lys	Ala	Ala	Trp	Gly	Lys	Val	Gly	Ala	His	Ala	Gly	Glu	Tyr	Gly	Ala	Glu	Ala	Leu
γ	Gly	His	Phe	Thr	Glu	Glu			Ala	Thr	Ile	Thr	Ser	Leu					Asn	Val	-	-	Glu	Asp	Ala		Gly		Thr	
β	Val		Leu		Pro		Glu		Ser	Ala	Val		Ala								-	-	Asp	Glu	Val				Ala	
δ									Thr			Asn									-	-		Ala						
										10										20										

	30										40											50								
α	Glu	Arg	Met	Phe	Leu	Ser	Phe	Pro	Thr	Thr	Lys	Thr	Tyr	Phe	Pro	His	Phe	-	Asp	Leu	Ser	His	Gly	Ser	Ala	-	-	-	-	-
γ	Gly		Leu	Leu	Val	Val	Tyr		Trp		Gln	Arg	Phe		Asp	Ser		Gly	Asn			Ser	Ala			Ile	Met	Gly	Asn	Pro
β															Glu				Asp			Thr	Pro	Asp		Val				
δ																						Ser								
		30										40										50								

							60										70										80			
α	Gln	Val	Lys	Gly	His	Gly	Lys	Lys	Val	Ala	Asp	Ala	Leu	Thr	Asn	Ala	Val	Ala	His	Val	Asp	Asp	Met	Pro	Asn	Ala	Leu	Ser	Ala	Leu
γ	Lys			Ala						Leu	Thr	Ser		Gly	Asp		Ile	Lys		Leu			Leu	Lys	Gly	Thr	Phe	Ala	Gln	
β											Gly	Ala	Phe	Ser		Gly	Leu	Ala				Asn							Thr	
δ																												Ser	Gln	
		60										70										80								

							90										100										110			
α	Ser	Asp	Leu	His	Ala	His	Lys	Leu	Arg	Val	Asp	Pro	Val	Asn	Phe	Lys	Leu	Leu	Ser	His	Cys	Leu	Leu	Val	Thr	Leu	Ala	Ala	His	Leu
γ		Glu			Cys	Asp			His				Glu						Gly	Asn	Val		Val	Thr	Val			Ile		Phe
β																Arg								Cys				His		
δ																												Arg	Asn	
		90										100										110								

							120										130										140	
α	Pro	Ala	Glu	Phe	Thr	Pro	Ala	Val	His	Ala	Ser	Leu	Asp	Lys	Phe	Leu	Ala	Ser	Val	Ser	Thr	Val	Leu	Thr	Ser	Lys	Tyr	Arg
γ	Gly	Lys					Glu		Gln			Trp	Gln		Met	Val	Thr	Gly		Ala	Ser	Ala		Ser		Arg		His
β							Pro				Ala	Tyr			Val		Ala				Asn			Ala	His	Lys		
δ							Gln	Met																				
		120										130										140						

Figure 10-5 Amino acid sequences of human α, β, γ, and δ globin chains. Gaps have been introduced as necessary to generate maximum homology. Numbers above and below the lines indicate residue positions in the α and non-α chains, respectively. (*Redrawn from Lehmann and Carrell, 1969.*)

Allelism of β-Chain Variants

Of the many additional variants of Hb structure that have been discovered, the next was designated Hb C (Itano and Neel, 1950). This variant also involved a charge difference in adult Hb. The immediate question was whether the Hb S and the Hb C variations were allelic. Such a question can be answered by studying the simultaneous segregation of Hb S and Hb C in the same families. If Hb S and Hb C are alleles, then a person with both alleles should have no additional alleles at that locus. Symbolically, a person whose genotype is Hb^S/Hb^C should have no Hb A, but, more important to the genetic reasoning, that person must have received the two alleles from different parents and must transmit one but not both to every child. On the other hand, if Hb S and Hb C represent variation at different loci, then these could be symbolized as $Hb_1^S/Hb_1^A, Hb_2^C/Hb_2^A$ for a person with both Hb S and Hb C. Theoretically, he might have Hb A in addition to Hb's S and C. Also he could have received both the Hb_1^S and the Hb_2^C alleles from the same parent and could transmit both, only one, or neither to any offspring.

The results of the studies of families with both Hb S and Hb C are shown in Table 10-2. It is clear that the results are not in agreement with the predictions of nonallelism and that they agree with the idea that Hb's S and C are alleles. When the chemical basis of the difference between Hb A and Hb C was subsequently worked out, it turned out that Hb C also involves a substitution in the β chain. As it happens, the same position, β^6, is involved, but lysine is substituted for glutamic acid in this instance. The involvement of this one position is of no special significance, since substitutions have been found at many other positions of the β chain. The fact that the β chain was involved and not the α chain is consistent with the idea that alleles should be concerned with the same structural gene and hence the same polypeptide chain.

Table 10-2 **SEGREGATION OF Hb S AND Hb C IN FAMILIES WITH ONE OR BOTH TRAITS***

Mating type	No. of families	Number of offspring					
		A	S	C	AS	AC	SC
A × C	2	0	0	0	0	5	0
A × CS	3	0	0	0	3	2	0
AS × AC	4	1	0	0	5	4	9
AS × CS	1	0	0	0	0	1	1
AC × CS	1	0	0	1	0	0	0

* From Ranney, 1954.

Nonallelism of β- and α-Chain Variants

It was not long before many additional Hb variants were discovered. The initial practice of naming variants by letters of the alphabet had to be discarded, and the geographical location of the Hb was used to identify variants. (Many of the variants were designated by the letter of the variant that they most resemble in electrophoretic mobility plus a place name.) The variant known as Hopkins-2 (the second variant described at Johns Hopkins University) is especially interesting as the first example of a variant nonallelic with Hb S. A crucial pedigree is given in Figure 10-6. The pattern of transmission is incompatible with the hypothesis of allelism for Ho-2 and S. This was consistent with the chemical evidence that the Ho-2 variant has a substitution in the α chain rather than in the β chain. Thus the gene that codes for the α chain is distinct from the gene that codes for the β chain, and, indeed, the two loci segregate quite independently, indicating that they are not closely linked. Following convention, we can symbolize the two loci as Hb_α and Hb_β, using superscripts to distinguish among alleles. A person with only normal adult Hb would be genotype $Hb_\alpha{}^A/Hb_\alpha{}^A; Hb_\beta{}^A/Hb_\beta{}^A$. A person heterozygous for sickle cell anemia would be $Hb_\alpha{}^A/Hb_\alpha{}^A; Hb_\beta{}^A/Hb_\beta{}^S$.

Genetic variants also are known for Hb F and Hb A_2, although they are rare (and easily overlooked). Using family data, it has been shown that the locus for the δ chain does not segregate from that for the β chain. Since the two chains differ by ten amino acid residues, different structural genes must code the two chains, but the genes are closely linked. This might be expected if the δ-chain gene arose through dupli-

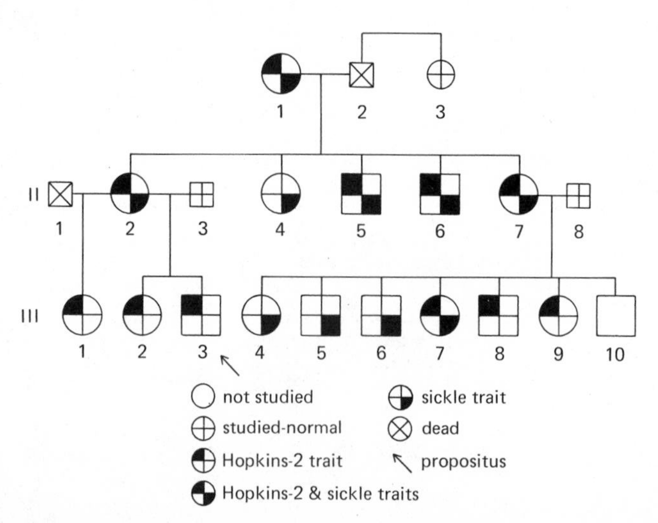

Figure 10-6 A pedigree showing segregation both of Hb S and Hb Hopkins-2. (*From Study of two abnormal hemoglobins with evidence for a new genetic locus for hemoglobin formation, by E. W. Smith and J. V. Torbert, 1958.* Bull. Johns Hopkins Hosp. *102: 38. © The Johns Hopkins University Press.*)

cation of the β-chain gene, with subsequent evolution leading to divergence between the two. It is noteworthy also that these two loci function at the same part of the life cycle. On the basis of Hb Kenya (page 229), the Hb_{γ} locus appears to be closely linked to the loci for β and δ chains, forming a complex locus consisting of the structural genes for the Aγ, Gγ, δ, and β globins. No genetic variants of embryonic Hb have been discovered; hence, the relationship of the Hb_{ϵ} locus to the Hb_{α} and Hb_{β} loci is unknown.

The Number of α- and β-Chain Loci

It has already been noted that there are at least two distinct loci for the γ chains, Gγ with glycine at position 136 and Aγ with alanine. The initial evidence was chemical, but structural mutations are limited to one or the other chain and are therefore consistent with the idea. The question must also be asked if there are possibly tandem duplications of other loci. If such duplications exist, the products of the loci must be identical, since most persons show no heterogeneity in α or β chains.

The complexity of the α-thalassemias has led to the suggestion that there are two Hb_{α} loci in man, as there are in some other mammals. In support of this are the observations that most α-chain variants do not constitute greater than 25% of the total Hb. These are quantitative arguments rather than genetic. A more convincing observation would be a person with three different α chains. Such a person should not exist if there is only one locus, since it would require three different copies of the α gene, not possible in a diploid organism if there is only one locus.

A Hungarian family has been reported with three such persons, each with the α variants Hb J Buda and Hb G Pest, as well as Hb A (Figure 10-7). Similarly, a family in India was reported in which two persons had the two α chain variants Rampa and Koya Dora in addition to A chains (De Jong et al., 1975), and Lie Injo et al. (1974) reported that persons homozygous for the α chain variant Constant Spring also have Hb A. Such persons must therefore have at least two Hb_{α} loci.

On the other hand, Abramson et al. (1970) reported two presumed homozygotes for Hb J Tongariki who lacked α^{A} chains. This would support the idea of a single Hb_{α} locus. It may be, as Abramson and coworkers suggested, that some human populations have two loci while others have only one. This suggestion is supported by quantitative studies of the α-chain variant G Philadelphia in blacks (Baine et al., 1976).

Hereditary Persistence of Fetal Hemoglobin

In the course of normal development, synthesis of fetal Hb virtually ceases at approximately the time of birth, with the Hb F level dropping to 1% or less by five months of age (Figure 10-4). At the same time, synthesis of β and δ chains increases. Concurrent with the decrease

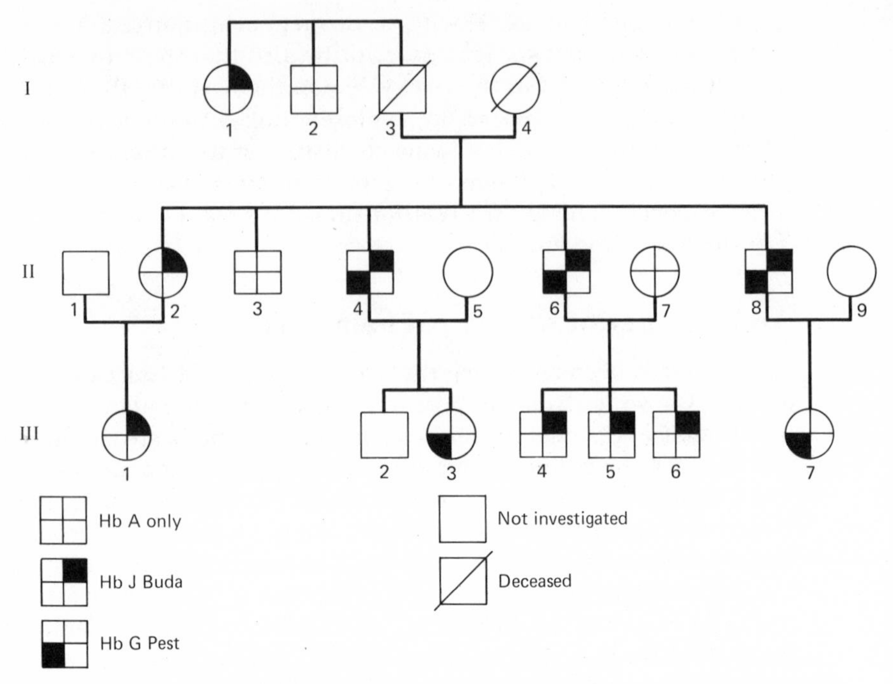

Figure 10-7 A portion of the Hungarian pedigree with three persons (II-4, 6, 8) who have both Hb J Buda and Hb G Pest in addition to Hb A. (*Taken from a larger pedigree in Hollán et al., 1972.*)

in γ-chain production is a shift from approximately 75% Gγ chains in cord blood to 40% in blood from persons 5 months and older, including adults.

A number of persons have been observed to have hereditary persistence of fetal Hb (HPFH). The condition was first described in American blacks and is more frequent among persons of African descent, although HPFH has been observed in several other ethnic groups. It is now apparent that there are several kinds of HPFH. All are closely linked to the Hb_β complex. No clinical defect is associated with HPFH, and even the two persons who are known to be homozygous and who lack any adult Hb are normal.

Heterozygotes for the Negro type HPFH have 10 to 35% Hb F, with decreased production of Hb A and Hb A_2. When the HPFH gene is in heterozygous combination with a β-chain variant such as β^S or β^C, no β^A chains are synthesized. This indicates that the HPFH gene suppresses β-chain synthesis for Hb_β genes in the *cis* configuration. The absence of both Hb A and Hb A_2 in HPFH homozygotes indicates that δ-chain synthesis also is suppressed. The failure to repress γ-chain synthesis is not simply a response to defective β-chain synthesis, since this does not occur in β-chain substitutions associated with low synthesis.

The relative amounts of Gγ and Aγ chains further differentiate Negro HPFH types. In one family, only Gγ chains were found. In two families, the Gγ chains accounted for 20% of the total. An additional group of families produces Gγ and Aγ chains in approximately equal amounts, although this group may also be heterogeneous on further study.

HPFH occurs in Greeks but is distinguished from the Negro type by lower levels of Hb F (10 to 18%) in heterozygotes. Hb A_2 is decreased, but less so than in the Negro type. The γ chains all are of the Aγ type. A white family not of Greek descent had a similar form of HPFH but with only 5% Hb F.

The existence of HPFH has stimulated much thought on the regulation of the *Hb* loci. Evidence has been presented that the structural genes for β, δ, and γ chains form a linkage complex. This would suggest mechanisms in which closely linked genes can be repressed and derepressed, perhaps by a modified operon system. In this instance, though, the complex, when active, exists in one of two alternate states. Either the Hb_{γ} genes are expressed or the Hb_{β} and Hb_{δ} genes are expressed, but both sets are not expressed simultaneously. The variety of HPFH types could represent deletions covering various regions of the β-δ-γ complex, including, perhaps, some regulatory regions.

Thalassemias

The thalassemias are a group of defects resulting from reduced synthesis of globin chains. Thalassemia major (Cooley's anemia) was originally described in persons of Mediterranean descent, with the name Mediterranean anemia also used as a synonym. It is characterized by severe hemolytic anemia due to shortened red cell life, by target cells (red cells with abnormally low amounts of Hb), and by abnormally shaped cells. Such persons are homozygous for a gene that, in heterozygous combination, causes thalassemia minor. This latter disorder has an array of abnormalities related to those of thalassemia major but with much less severe defect.

The basic defect in the thalassemias is reduced production of one or more specific Hb chains. Where some synthesis does occur, there is ordinarily no evidence that the structural gene is altered. This would not rule out a silent codon substitution, that is, one that requires a different tRNA but codes for the same amino acid. But, as will be noted later, many of the cases of thalassemia result from deletions. A few well recognized structural alterations, such as Hb Constant Spring and the Lepore hemoglobins, have very reduced synthesis and can be classified functionally with the thalassemias.

The following are some of the principal variations of thalassemia that have been described.

β-Thalassemia. This is the form that is found especially in Italy and Greece, where the frequency of heterozygotes may approximate 15%. The gene is closely linked to (or part of) the Hb_β locus. There appear to be several alleles differing in the amount of β^A chain synthesized. In some, no β chain is synthesized from the Hb_β gene that is coupled with the β^{thal} gene. In others, some normal β^A chains are synthesized. The genes are designated $\beta^{thal\ 0}$ and $\beta^{thal\ +}$, respectively. The Hb_β gene in the trans position is unaffected. In heterozygotes, the effect is to diminish β-chain synthesis and hence the amount of Hb A. In homozygotes, the Hb A may be very low (β^+-thalassemia) or absent (β^0-thalassemia). The diminished synthesis of β chains results in an excess of α chains. Free α chains are unstable and precipitate, forming inclusion bodies. These contribute to instability of the red cells. The Hb_δ gene coupled to the β^{thal} gene is unaffected, and indeed the level of Hb A_2 increases to 4 to 7% in β^{thal} heterozygotes. In homozygotes, most of the Hb is Hb F, but this is not adequate to compensate for the deficiency of Hb A. The Hb F is distributed unevenly among red cells.

Persons heterozygous for both β-thalassemia and an abnormal β-chain Hb show clearly that the effect of the thalassemia gene is limited to the chromosome on which it occurs. For example, a $\beta^{thal}\beta^S$ heterozygote may be severely affected with sickle cell anemia if the β^{thal} gene results in little or no normal β^A-chain synthesis. If some β^A-chain synthesis occurs, the defect may be mild. Heterozygotes for $\beta^{thal}\beta^C$ have very little clinical problem.

δ-β-Thalassemia. In this less common type of thalassemia, also linked to the β-δ complex, both β- and δ-chain synthesis is completely suppressed in the coupled loci. Hb F is produced. In heterozygotes, Hb F may constitute 5 to 20% of the total Hb, with Hb A_2 at normal (rather than elevated) levels.

α-Thalassemia. These forms of thalassemia are allelic with or closely linked to the Hb_α locus. They are especially prevalent in Southeast Asia and account for a substantial portion of fetal loss. Unlike the β^{thal} gene whose effect is limited to the postnatal period when β-chain synthesis is important, α^{thal} genes affect synthesis of α chains, which are essential for embryonic and fetal hemoglobins as well as adult. A deficiency of α chains leads to an excess of γ, δ, and β chains. The γ and β (and presumably the δ) chains are able to form pure tetramers, so that α thalassemias are characterized by the presence in the fetal and neonatal period of Hb Bart's (γ_4) and later of Hb H (β_4) molecules.

Several types of α-thalassemia can be distinguished on the basis of the degree of α-chain deficiency. The most severe is α-thalassemia-1, in which there is no α-chain production. The α-thal-1 gene involves inactivation of both the α loci on one chromosome, due apparently to a deletion covering both structural genes. Since this mutation is fairly common in Southeast Asia, homozygotes often occur, producing hydrops fetalis, with spontaneous abortion at about 34 weeks gestational age. Such fetuses have a large amount of Hb Bart's (γ_4) in their red cells. Heterozygotes for $Hb_{\alpha}^{\text{thal-1}}$ have a form of thalassemia minor, with 5 to 10% Hb Bart's at birth.

α-Thalassemia-2 involves loss of activity of only one of the two α genes on a chromosome. Therefore, homozygotes still have two functional α genes, one on each chromosome, and are only mildly affected with thalassemia minor. Heterozygotes may be still more mildly affected or normal. The heterozygous combination of an α-thal-2 with an α-thal-1 gene produces a more severe problem, hemoglobin H disease, in which the deficiency of α-chain synthesis causes an excess of β chains with formation of Hb H (β_4) inclusion bodies. This condition can also result from combination of Hb Constant Spring with α-thal-1.

Further heterogeneity of α-thalassemia is indicated by the situation in Africa, where α-thalassemia is known, but hydrops fetalis has not been observed and Hb H disease is rare. It is further not known why those blacks homozygous for a chromosome with only one normal Hb_{α} locus should be different from Asians homozygous for α-thal-2.

MUTATIONS OF THE HEMOGLOBIN LOCI

The readiness with which changes in Hb structure are observed and the high likelihood that changes in structure will have clinical consequences have made mutations at the Hb loci the best known of all human genes. They illustrate most of the mutational processes and in some instances provide examples that are unique. Studies of the Hb mutations complement very well those with viruses and bacteria. For microorganisms, the ability to make experimental crosses in a short time with large numbers of parents and progeny permits genetic analysis with very high resolution. For human Hb mutations, genetic analysis depends on the location of suitable matings in the population, many of which are rare or nonexistent. However, chemical analysis of Hb can be carried out with an ease and precision that is rare for microbial gene products.

Amino Acid Substitutions

By far the greatest number of known mutations involve substitutions of one amino acid for another at a particular position in the chain. A recent listing of substitutions contains over 275 for the α and β chains

(McKusick, 1978). These substitutions occur in all parts of the α and β chains, although the physiological consequences may be very different depending on the location.

Electrophoretic Variants. In nearly every instance, the substitution involves a difference in electrical charge; for example, glutamic acid or aspartic acid residues, both with negatively charged side chains, are replaced with neutral amino acids or with the positively charged amino acids lysine, arginine, or histidine. This high proportion of electrophoretic mutants is due to the fact that the principal means of detecting Hb variants is by electrophoresis, and many variants undoubtedly go unrecognized because the substitutions lead to no charge difference and hence no mobility difference in an electric field. It has been estimated that only a third of substitutions involve a change in charge.

The occurrence of variants of fetal Hb provides further evidence for the existence of two γ-chain structural genes, coding for Aγ and Gγ chains. A mutation involving substitution of a single amino acid residue should affect only one of the two loci. On the other hand, if the variation in the γ136 position occurs through ambiguous translation of a single structural gene, the mutation should affect both Aγ or Gγ chains. In each case analyzed, either Aγ or Gγ, but not both, is associated with the fetal variant. For example, Hb F Hull, Hb F Texas I, Hb F Kuala Lumpur, and Hb F Jamaica have only alanine at the γ136 position. Hb F Malta has only glycine. Consistent with this model is the very low proportion of variant Hb in heterozygotes for fetal variants, although other explanations could be advanced for this.

Three variant hemoglobins have substitutions at two different positions in the β chain. Hb C Harlem has β6 (Glu $\rightarrow$ Val) and β73 (Asp $\rightarrow$ Asn). These are the substitutions found in Hb S and Hb Korle-Bu, respectively, and Hb C Harlem is thought to have arisen by crossing over in a person heterozygous for the two variants. An additional mutation in a variant allele could also explain the origin. Other Hbs with two substitutions are Hb C Ziguinchor (β6 Glu $\rightarrow$ Val and β58 Pro $\rightarrow$ Arg) and Hb Arlington Park (β6 Glu $\rightarrow$ Lys and β95 Lys $\rightarrow$ Glu).

Amino Acid Deletions

A number of Hb variants are known in which one or more amino acids have been deleted (Table 10-3). The first described was Hb Freiburg, in which the valine of the β23 residue is missing. The mechanism by which this is thought to happen is shown in Figure 10-8. Mispairing of segments of DNA may occur because of chance similarities of nucleotide sequences in different parts of the gene. If crossing over occurs in these mispaired segments, the products are a duplication and a deficiency.

Table 10-3 **VARIANT HEMOGLOBINS WITH DELETIONS OF ONE OR MORE AMINO ACID RESIDUES**

Designation	Residues deleted	Reference*
Hb Leiden	β6(7) Glu	De Jong et al., 1968
Hb Lyon	β17–18 Lys-Val	Cohen Solal et al., 1974
Hb Freiburg	β23 Val	Jones et al., 1966
Hb Niteroi	β42–44(43–45) Phe-Gly-Ser	Praxedes et al., 1972
Hb Tochigi	β56–59 Gly-Asn-Pro-Lys	Shibata et al., 1970
Hb St. Antoine	β74–76 Gly-Leu	Wajcman et al., 1973
Hb Tours	β87 Thr	Wajcman et al., 1973
Hb Gun Hill	β91–95(92–96; 93–97) Cys-Asp-Lys,Leu,His	Bradley et al., 1967
Hb Leslie	β131 Gln	Lutcher et al., 1976

* References may be found in Weatherall and Clegg (1976).

The most striking deletion known occurs in Hb Gun Hill, with deletion of five amino acids. The usual sequence of the β chain, starting with position 90, is Glu-Leu-His-Cys-Asp-Lys-Leu-His-Val. Hb Gun Hill has Glu-Leu-His-Val for this region. Apparently the mispairing involved the Leu-His sequences, with crossing over somewhere within the mispaired region.

Amino Acid Insertions

A single example of an insertion of amino acids is known. Hb Grady is an α-chain variant in which residues 116-118 (Glu-Phe-Thr) are duplicated (Huisman et al., 1974). This presumably arose through mispairing and unequal crossing over.

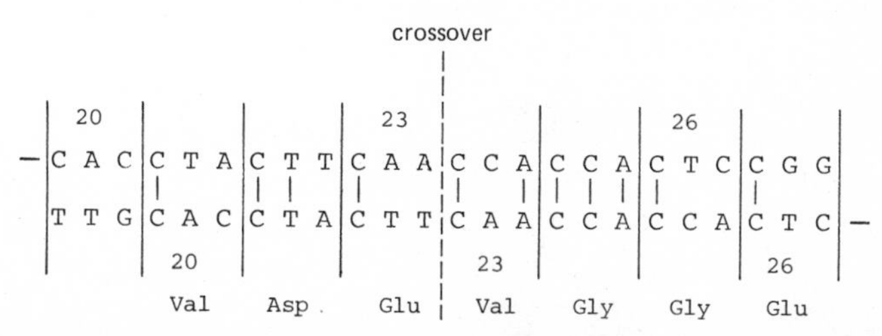

Figure 10-8 DNA code for region around β23 position of the Hb β chain. The two segments are shown one codon out of phase. Vertical solid lines indicate the identical nucleotides. The vertical dotted line indicates the position of the crossover that would account for the deletion of the β23 Val in one product (Hb Freiburg) and the duplication of the β23 Val in the other.

Fusion of Hemoglobin Genes

The Lepore Hemoglobins. Gerald and Diamond (1958) reported the first example of a group of hemoglobins known collectively as Lepore hemoglobins. Persons heterozygous for Hb Lepore have 10 to 15% of their Hb as a variant that has a normal α chain but a non-α chain that has some characteristics of the β chain and some of the δ chain. Structurally, the Lepore chain starts at the N-terminal end like a δ chain and ends at the C-terminus like a β chain (Baglioni, 1962). The δ and β chains differ by only ten residues. Consequently, they might be expected to mispair as in Figure 10-9. Crossing over would yield a series of Lepore genes, depending on where the crossover occurs. Three Lepore Hbs have been identified. In Lepore Hollandia, the crossover is between positions 22 and 50. In Lepore Baltimore, it is between 50 and 86. In Lepore Boston (Lepore Washington), it is between 87 and 116. Two Lepore genes might arise by crossing over at two different points, but if the region between the two points does not include one of the codon differences between the δ and β genes, there is no way to distinguish the crossover products. There should then be a maximum of nine distinguishable Lepore hemoglobins.

Anti-Lepore Hemoglobins. Three examples of the opposite crossover products are known. These are known collectively as anti-Lepore hemoglobins and arise as indicated in Figure 10-9. Hb P Congo

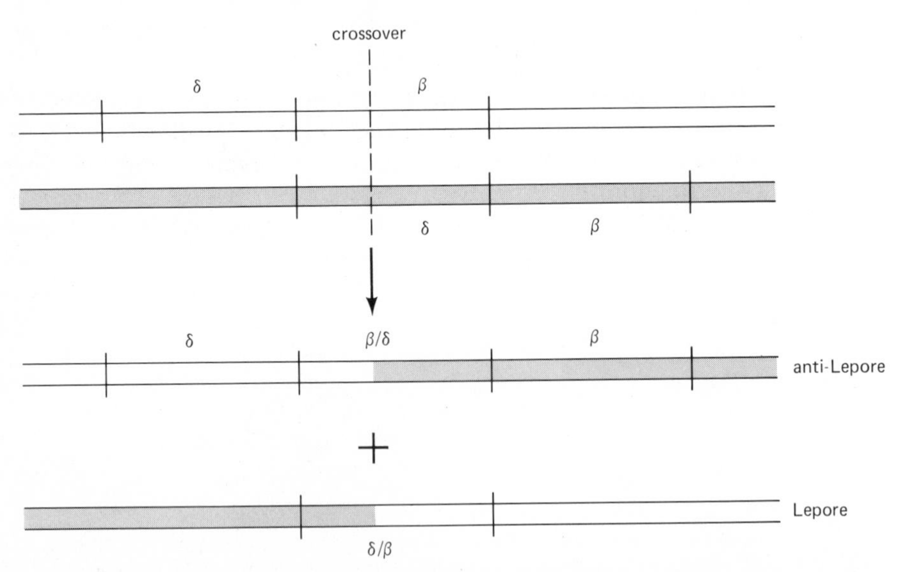

Figure 10-9 Diagram of mispairing with crossing over to produce Lepore and anti-Lepore hemoglobin genes. Details of the gene structure, such as introns and spacers between genes are not shown.

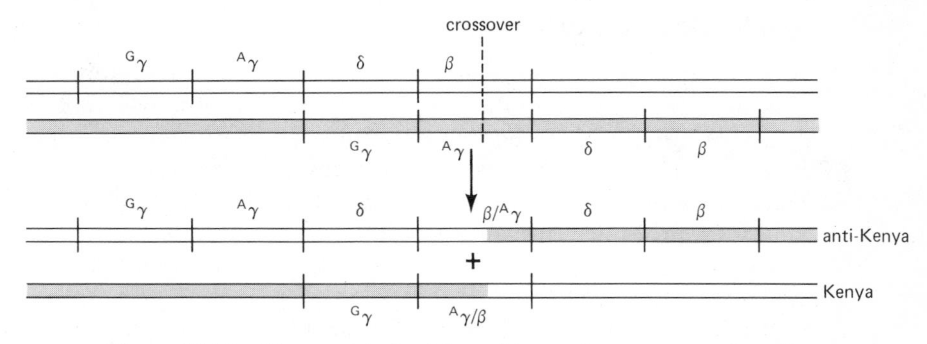

Figure 10-10 Diagram of mispairing with crossing over to produce Kenya and anti-Kenya hemoglobin genes.

(Lehmann and Charlesworth, 1970) and Hb Miyada (Ohta et al., 1971) have the expected 146 amino acid residues. Hb Lincoln Park, which involves a crossover between residues 22 and 50, also has a deletion of the valine in position 137 (Honig et al., 1978). In the case of anti-Lepore hemoglobins, each person has a full set of normal β genes, in addition to the β-δ fusion gene. Therefore, β-chain synthesis is normal, as are Hb levels.

Hb Kenya. In Hb Kenya, the non-α chain is the product of a γ-β fusion gene (Kendall et al., 1973). The N-terminal sequence is the same as a γ chain through residue 80. From residue 87 to the C-terminus, the sequence is the same as a β chain. Crossover between the γ and β genes apparently occurred in the region corresponding to residues 81-86, which are identical for the two chains. When in combination with $Hb_\beta{}^S$ on the homologous chromosome, no Hb A is produced. Hence there appears to be no intact β locus on the Hb Kenya chromosome. This suggests mispairing of two chromosomes with the gene sequence Hb_γ–Hb_δ–Hb_β, resulting in a deletion involving Hb_β (Figure 10-10). This clearly requires close proximity of Hb_γ to the δ-β complex, suggested by other observations.

Persons with Hb Kenya produce elevated levels of Hb F, all of the Gγ type. This suggests the deletion to have involved the Aγ gene. Both the Hb F (Gγ) and Hb Kenya occur at approximately 5% in heterozygotes. Thus the deletion seems to have increased the activity of the $Hb_{G\gamma}$ locus *cis* to it. The Hb F (Aγ) from the normal homologous chromosome could not be detected in the presence of this high level of Hb F (Gγ). These results with Hb Kenya established the most probable order of *Hb* loci as $Hb_{G\gamma}$–$Hb_{A\gamma}$–Hb_δ–Hb_β, an order that has now been confirmed by direct studies of DNA (Little et al., 1979).

Chain Elongation

A group of especially interesting Hb variants have additional amino acid residues attached to the C-terminal end of the chains (Table 10-4). It was originally supposed that mutation of a chain termination codon to a "sense" codon world account for the lengthening of the chain, if there is a sequence of nucleotides beyond the termination codon available for translation. In the case of Hb Tak, the authors note, however, that no single nucleotide substitution can convert a termination codon into a threonine codon.

In Hb Wayne, the change in sequence starts before the end of the chain. Therefore, this clearly was not a question of adding more amino acids by mutation of a termination codon. Rather, as suggested by the authors, it must have originated by a frameshift mutation. In support of the latter hypothesis was the fact that all the new codons of Hb Wayne can be generated from Hb Constant Spring by shifting the reading frame by one nucleotide, assuming Hb Constant Spring to have arisen by nucleotide substitution in a termination colon (UAA in this instance). This demonstration that mRNA has additional translatable sequences beyond the termination codon has been fully supported by subsequent determinations of α- and β-globin mRNA structures (see the following).

The normal α-chain termination codon should be able to mutate to nine other codons. Two would also be termination codons, but seven would be sense codons, coding for Glu, Gln, Ser, Lys, Tyr, and Leu. Four of the six possible amino acids have been observed (Table 10-4). The shortened chain of Hb Koya Dora as compared to Constant Spring suggests that an additional mutation to a termination codon may have occurred at or near codon 160.

Chain Shortening

A single example of a variant Hb with a shortened β chain has been reported (Winslow et al., 1975). In Hb McKees Rocks, the tyrosine codon in position 145 of the β chain apparently has changed to a terminating codon, causing the β chain to be two residues short. Such mutations could only be tolerated very near the C-terminus without seriously affecting the molecule.

MOLECULAR GENETICS OF HEMOGLOBINS

The relative ease of isolating hemoglobin mRNA from immature red cells has made possible the study of the structure not only of mRNA but also the corresponding DNA. The mRNA can serve as a template to

Table 10-4 **PARTIAL STRUCTURES OF HUMAN HEMOGLOBINS IN WHICH LONGER THAN NORMAL CHAINS OCCUR**

Hb chain	Sequence of C-terminal region
α-chains	137 141
normal A	-Thr-Ser-Lys-Tyr-Arg-COOH
	150 160
Constant Spring	-Thr-Ser-Lys-Tyr-Arg-Gln-Ala-Gly-Ala-Ser-Val-Ala-Val-Pro-Pro-Ala-Arg-Trp-Ala-Ser-Gln-Arg-Ala-Leu-Leu-Pro-Ser-Leu-His-Arg-Pro-Phe-Leu-Val-Phe-Glu-COOH
	172
Koya Dora*	-Thr-Ser-Lys-Tyr-Arg-Ser-Ala-Gly-Ala-Ser-Val-Ala-Val-Pro-Pro-Ala-Arg-Trp-Ala-Ser-Gln-Arg-COOH
Wayne	-Thr-Ser-Asn-Thr-Val-Lys-Leu-Glu-Pro-Arg-COOH
Seal Rock	-Thr-Ser-Asn-Thr-Arg-Glu-Ala- -(The remainder is identical to Constant Spring.)
Icaria	-Thr-Ser-Asn-Thr-Arg-Lys-Ala- -(The remainder is identical to Constant Spring.)
β-chains	142 146 157
normal A	-Ala-His-Lys-Tyr-His-COOH
Tak	-Ala-His-Lys-Tyr-His-Thr-Lys-Leu-Ala-Phe-Leu-Leu-Ser-Asn-Phe-Tyr-COOH
Cranston	-Ala-His-Lys-Ser-Ile-Thr-Lys-Leu-Ala-Phe-Leu-Leu-Ser-Asn-Phe-Tyr-COOH

* The Koya Dora chain may extend beyond position 158 by a few residues.

References: Hb Constant Spring, Clegg et al. (1971).
Hb Koya Dora, De Jong et al. (1975).
Hb Wayne, Seid-Akhavan et al. (1976).
Hb Seal Rock, Weatherall and Clegg (1976).
Hb Icaria, Clegg et al. (1974).
Hb Tak, Flatz et al. (1971).
Hb Cranston, Bunn et al. (1975).

make a DNA copy (cDNA) using RNA-dependant DNA polymerase (reverse transcriptase). The cDNA in turn can be inserted into bacterial plasmids by recombinant DNA techniques, producing large numbers of copies, and is a useful probe to hybridize with DNA genes. The discovery of restriction enzymes, an essential technique of recombinant DNA studies, also permits selective cleavage of DNA for studies of gene structure. The additional development of efficient techniques for sequencing nucleic acids has made it possible to establish the structure of DNA genes and mRNA.

The nucleotide sequences of human α- and β-globin mRNA have been completely established (Figure 10-11). The structures include all the features anticipated from the structures of the polypeptide chains, but there are additional points of interest. Beyond the terminating codons are the additional sequences of nucleotides that correspond to the amino acids seen in chain elongation mutants. No function is known for the additional sequences. The different *Hb* genes within a complex, for example the δ and β genes, are not transcribed onto the same mRNA, at least not as it exists after conversion from HnRNA to processed mRNA.

The structures of the genes also has yielded some unexpected findings. As shown in Figure 10-12, both the β and δ genes, located on chromosome 11, are interrupted by untranslated intervening sequences (introns) approximately 1000 base-pairs in length. This segment is apparently transcribed but is excised before translation. The β and δ genes are approximately 5400 base-pairs apart in the orientation predicted by the Lepore hemoglobins. The γ genes lie 13,500 bp to the "left" of the δ gene and are separated from each other by 3500 bp. The α genes are on chromosome 16 and lie in tandem, separated by a maximum of 3100 base pairs (Orkin, 1978). The presence of intervening sequences in the human α genes has not been established, but, by analogy to other globin genes, it is very likely that they are present. In the case of both α- and β-globin genes of mice, there are two intervening sequences at very similar positions (Leder et al., 1978), suggesting great antiquity for these structures and further suggesting a similar arrangement in the human genes.

The technique of nucleic acid hybridization has been especially useful in the understanding of thalassemias. Using radioactive-labeled cDNA prepared with β-globin mRNA as a template, several groups of investigators have shown that β mRNA is absent in many homozygotes for β^0-thalassemia and is untranslatable in others. In one case, a mutation of codon 17 of the β chain from AAG (lysine) to UAG (termination) was the basis for failure to produce β globin (Chang and Kan, 1979). With the discovery of restriction enzymes, the hybridization technique could be applied directly to DNA digests. Such studies indicate that β^0-thalassemics often are missing segments of DNA that contain

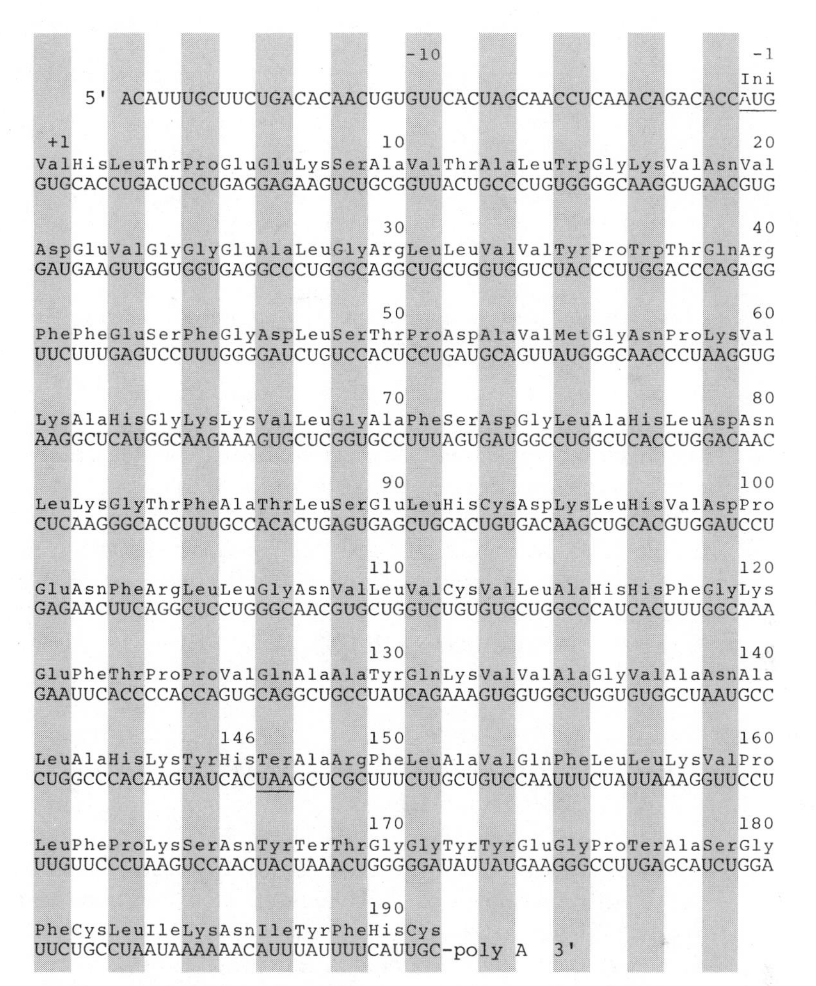

Figure 10-11 The nucleotide structure of human β-globin mRNA. The amino acids corresponding to each codon are shown, including the additional sequences on the 3′ end beyond the termination codon. (*The nucleotide sequence is from Kafatos et al., 1977.*)

the β-globin structural gene in the case of β-thalassemia and both β and δ genes in $\beta\delta$-thalassemia. Deletions have also been discovered in the Lepore hemoglobins, as expected, and in some cases of hereditary persistence of fetal hemoglobin (HPFH). The ability to clone such mutants and to determine DNA sequences with great efficiency should provide rapid advances in knowledge of the structure of normal and variant globin genes within the next several years.

One application of restriction enzymes deserves special mention. Kan and Dozy (1978) found a polymorphism in the sites sensitive to the restriction enzyme *Hpa* I. In most persons, this enzyme yields a DNA

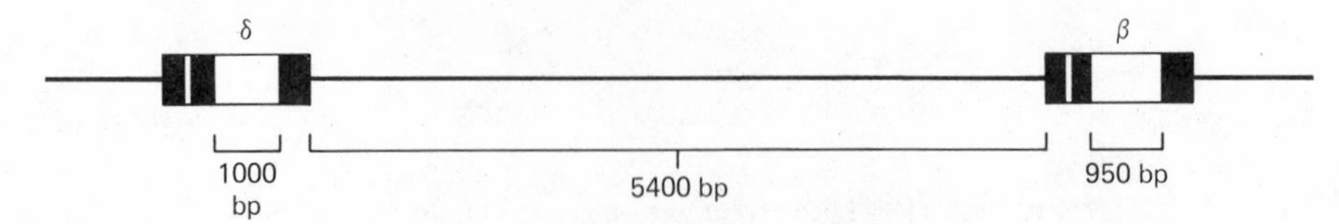

Figure 10-12 Diagram of the Hb β complex, showing number of nucleotides in the untranslated introns (open) and between the genes. The solid areas are the translated regions. The large intron occurs between codons 104 and 105 in both δ and β genes and presumably in the same position in the γ genes. The location of the small intron is not known with certainty. Small introns have not been demonstrated in the γ genes, which are 13,500 bp to the left, but it is likely that they occur. (*Based on Lawn et al., 1978; Little et al., 1979.*)

fragment containing the β-globin gene that is 7.6 kb (kilobase-pairs) long (Figure 10-13). Two variants occur in American blacks, however. In one, the fragment is only 7.0 kb. In the other, it is 13.0 kb. The 13.0 kb fragment was found in only 3% of blacks homozygous for Hb A but in 87% of blacks homozygous for Hb S. Therefore this DNA variation might have some use as a means of predicting the fetal Hb genotypes based on analysis of DNA from amniotic fluid cells that do not themselves produce hemoglobin.

PHYSIOLOGICAL PROBLEMS ASSOCIATED WITH VARIANT HEMOGLOBINS

Some variant hemoglobins are normal, not leading to detectable interference with red cell function. Usually, these have been located through screening of hematologically normal persons. Most variants are associated with some degree of defect, however, and this is notably true for the high frequency variants (Hb's S, C, and E). The following are some of the more frequently encountered problems. These deal primarily

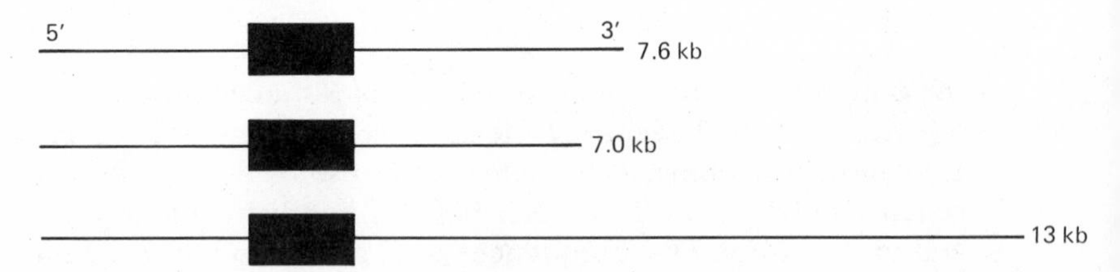

Figure 10-13 Diagram showing segments of human DNA of three different lengths that contain β-globin genes. Most persons produce only 7.6 kb segments, but some Blacks produce 7.0 kb or 13.0 kb segments, indicating a variation in DNA structure at those points. The 13.0 kb fragment is especially associated with the Hb S allele, which involves a nucleotide substitution in the structural gene. (*From Kan and Dozy, 1978.*)

with adult variants, since opportunities to study seriously defective embryonic or fetal variants are limited.

Rate of Synthesis

With few exceptions, variant Hb chains are not synthesized at rates comparable to normal chains. In Hb AS heterozygotes, typical proportions are 35–40% Hb S and 60–65% Hb A. The evidence suggests this is entirely due to differences in rates of synthesis rather than destruction. Therefore, the β^A mRNA is manufactured or is translated almost twice as fast as the β^S mRNA, even though the difference is limited to a single nucleotide among hundreds. The reason for such a large effect has not been established experimentally. Many other Hb variants are synthesized at even slower rates, resulting in one of the frequent consequences of Hb variants—hypochromic anemia. Such anemia is, of course, not responsive to iron therapy, since the problem is not the result of heme shortage.

The Lepore Hb's provide an unusual opportunity to study the regulation of rates of synthesis. The δ and β genes are in tandem, with the δ C-terminal toward the β N-terminal (Figure 10-9). Ordinarily, the β chain is produced in 40-fold greater quantities than the δ, in spite of a difference of only ten amino acid residues. Conceivably there could also be other codons that differ, although not coding for a different amino acid. Do the different rates reflect control regions external to the structural regions of the genes, or do they reflect differences in transcription or translation of structural regions? The three known Lepore Hb's all occur to the extent of 10–15% of total Hb in heterozygous combination. Conversely, the anti-Lepore Hb Miyada, which has the β structure only to the crossover point between positions 12 and 22, is found at levels of 17% in heterozygotes. Since this is in combination with two normal β genes, the predicted level would be about 30% .

At present there is no simple answer to what causes variant hemoglobins to be synthesized at such different rates. Very likely, the regions that lie outside the structural genes and that are responsible for initiation of transcription play a major role, one that must be better understood if we are to construct a coherent picture for structural variants.

Stability

Many of the structural changes in globin that result from mutation cause distortion of the tertiary structure of the globin. This may upset the globin-heme interaction, and one consequence is decreased stability. Free globin is much less stable than hemoglobin and precipitates in the red cells, forming inclusion bodies (Heinz bodies). Cells with inclusion bodies have increased tendency to lyse, and hemolytic anemia is often associated with unstable hemoglobins.

Lehmann and Carrell (1969) have discussed the mechanisms involved in Hb instability. They note that many of the unstable Hb's result from neutral amino acid substitutions in the interior of the globin molecule. Small differences in the size and shape of side chains interfere with folding of the globin chain. The variants that involve deletions (Hb's Freiburg, Leiden, Gun Hill) cause substantial distortion of the molecule and are unstable. This group of hemoglobins emphasizes the importance of molecular configuration in maintaining a stable, continuously functional protein. Especially in the case of Hb, where the molecules must function for the life of the red cell (100+ days) and no mechanism is available to replace damaged molecules, it is important that the Hb be very stable.

Methemoglobinemia

For hemoglobin to function as an oxygen carrier, the iron attached to the heme group must be in the ferrous (Fe^{2+}) form. However, ferrous iron readily oxidizes to the ferric (Fe^{3+}) form, producing methemoglobin (MetHb). The Fe^{3+} of MetHb must be reduced continuously by enzymes in red cells to ferrohemoglobin. A major function of globin is to stabilize the iron in the Fe^{2+} form. There are a number of variant Hb's noted primarily because they stabilize the iron in the Fe^{3+} form. Since MetHb cannot transport oxygen, such persons may have a relative deprivation of oxygen after severe exercise. The variants that cause MetHb are known collectively as hemoglobins M, with a geographical designation to differentiate among the types, for example, Hb M Boston.

Figure 10-14 shows some of the geometrical relationships of the heme group to the globin. The E7 and F8 histidine residues form complexes with the Fe in the heme group, stabilizing the Fe in the ferrous form. However, if tyrosine is substituted for either histidine residue, the phenolic group on the side chain forms a stable bond with Fe^{3+}, preventing its reduction to Fe^{2+}. Similarly, substitution of negatively charged glutamic acid at position E11 leads to stability of the ferric iron.

Since the reduction of Fe^{3+} to Fe^{2+} is catalyzed by the enzyme methemoglobin reductase, methemoglobinemia can also result from low activity of the enzymatic reduction as well as from alterations in Hb structure. MetHb absorbs light somewhat differently than does ferrohemoglobin, and persons with methemoglobinemia often have a gray color.

Sickling

The mechanism of red cell sickling associated with Hb S has been of great concern, since understanding it is the key to development of a

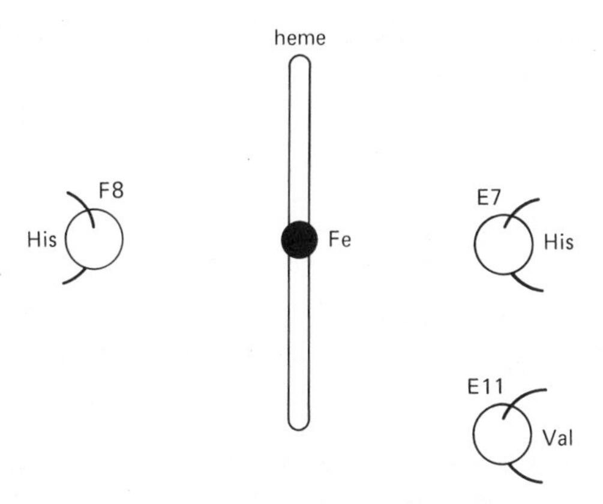

Figure 10-14 Diagram of the regions of globin molecules especially involved in heme binding. Residue E7 refers to the seventh residue in the E helix, and so on. Substitutions that give rise to methemoglobin are as follows:

Residue (helix no.)	Residue (chain no.)	Hb A residue	Hb M variant	
			residue	name
α chain				
E7	58	His	Tyr	Boston
F8	87	His	Tyr	Iwate
β chain				
E7	63	His	Tyr	Saskatoon
E11	67	Val	Glu	Milwaukee-1
F8	92	His	Tyr	Hyde Park

means of successful therapy. The studies of Murayama (1966) suggest that the substitution of valine with its uncharged hydrophobic side chain in place of negatively charged glutamic acid in the $\beta 6$ position permits formation of an intramolecular hydrophobic bond between the $\beta 1$ and $\beta 6$ positions. This in turn permits molecules to "stack," forming long gel-like structures that distort the cell membrane, leading to cell destruction. It only happens with deoxygenated molecules, apparently because the conformational shift known to occur when Hb is oxygenated interferes with the intermolecular interaction. Of the many variants of Hb now known, only Hb S and its derivative Hb C Harlem, which also has the $\beta 6$ Glu $\rightarrow$ Val substitution, form sickle cells.

Sickling occurs only if the Hb S comprises a very high proportion of the hemoglobin in a red cell. Thus AS heterozygotes do not have problems due to sickling *in vivo*, although their cells can be made to sickle under laboratory conditions. Even under conditions of low oxygen pressure, such as high altitude flying or prolonged athletic activity, persons with sickle cell trait show no defect. Occasional persons homozygous for the Hb S gene escape most of the sequelae associated with this genotype. In some instances, unusually high levels of Hb F appear to account for the protection.

DISTRIBUTION AND TREATMENT OF ABNORMAL HEMOGLOBINS

Frequency of Hemoglobin Abnormalities. Reference has been made to the high frequency variants of Hb as opposed to the rare or exotic variants. There are three electrophoretic variants that occur with substantial frequencies: Hb S in Africa and in persons of African descent, as well as in a few Mediterranean groups and in tribes of India; Hb C in West Africa and in descendants of West Africans; and Hb E in Asia. Each has been implicated in resistance to malaria, and the interplay of selection and malaria will be discussed further in a later chapter. The thalassemias also are widely distributed, β-thalassemia being especially prevalent in the Mediterranean areas, but occurring also in Africa, α-thalassemias occurring especially in Asia. Again, malaria has been implicated as a selective agent.

American blacks have moderate to high frequencies of Hb's S and C, β-thalassemia, and hereditary persistence of fetal Hb. Motulsky (1973) has estimated the frequencies to be as shown in Table 10-5. Six %

Table 10-5 **ESTIMATED FREQUENCIES OF HEMOGLOBINOPATHIES AMONG UNITED STATES BLACKS AT BIRTH***

Phenotype frequencies		Gene frequencies	
Hb AS	8%	Hb_{β}^{S}	0.04
Hb AC	3%	Hb_{β}^{C}	0.015
β thal heterozygote	1.5%	Hb β thal	0.0075
HPFH	0.1%	HPFH	0.005
Hb S-S	1:525		
Hb S-C	1:833		
Hb S/HPFH	1:25,000		
Hb S-β thal	1:1,667		
Hb C-C	1:4,444		
β thal homozygote	1:17,778		

* From Motulsky (1973).

of black newborns would be heterozygous for one of these traits. An additional three per thousand births would have sickle cell anemia, S/C disease, or S/β-thalassemia. The hemoglobinopathies thus constitute a major part of the health load for black children.

Therapy. Since there is as yet no way to replace bad genes with good, there is no no way to "cure" an inherited hemoglobinopathy. Present therapy is designed primarily to deal with symptoms, which, for the most part, result from chronic anemia. Sickled cells are a special problem, since they clog up capillaries and block circulation, leading to local hypoxia that in turn promotes more sickling. The spleen is active in removing sickled cells from circulation, and splenectomy is often performed on sickle cell anemia patients to prevent rupture of the spleen and to prevent excessive removal of red cells that, in spite of the abnormal Hb, function quite well in oxygen transport.

Indeed, Hb S functions normally as an oxygen carrier, as do most other variant hemoglobins. If a way could be found to prevent or reverse sickling, there should be relatively little medical problem associated with homozygosity for Hb S. A number of ideas have been tested, generally based on disruption of the hydrophobic bond formed between the β1 and β6 valines. In some of the earlier studies, urea or cyanate were used, the latter forming a covalent bond with free amino groups. While both work well *in vitro,* they have not proved useful *in vivo*. In the case of cyanate or any compound that reacts generally with proteins, there is a problem in restricting the site of action only to the molecule of interest, Hb. Recent attention has focused on small peptides, which are generally less toxic than many other agents but which unfortunately do not enter red cells readily. In spite of the frustration to date, there is still hope that substances can be found that will specifically interact with Hb without major side effects.

Prenatal Diagnosis of Hemoglobinopathies. The major variant Hb's, S, C, and E, involve alterations of the β chain, which is not produced in quantity until late in gestation. It was feared that prenatal diagnosis during the second trimester would be impossible since the genes, if active at all, would have very low activity. They do indeed have very low activity, but it is sufficient to detect products of the β locus if very sensitive techniques are used.

The initial efforts were based on *in vitro* synthesis of hemoglobins by immature red cells taken from fetuses. Such studies are technically difficult, but it was possible to demonstrate trace amounts of β chain synthesis even in 55 day embryos (Kazazian and Woodhead, 1973). Additional studies verified the ability to diagnose sickle cell anemia and thalassemia in fetuses, although a sample of fetal blood was required.

With development of the technique of cDNA hybridization, it became possible to test for the thalassemias that result from loss of structural genes. Kan, Golbus, and Dozy (1976) cultured fibroblasts from amniotic fluid from a pregnancy at risk for α-thalassemia and demonstrated that the fetus had α-thalassemia-1 but was not homozygous for the α-thal-1 gene. The ability to work with amniotic cells rather than blood cells considerably simplifies the procedure and reduces risk. Reference was made earlier to the possible use of restriction enzymes in predicting the presence of the Hb S gene in amniotic fibroblasts. While early diagnosis of adult hemoglobinopathies is not yet a routine procedure, the advances in molecular technology hold promise that it will become so.

EVOLUTION OF HEMOGLOBIN LOCI

From the foregoing, much of the evolutionary history of human Hb structural loci is apparent. On the assumption that those loci whose products are most closely related are most recently divergent in evolution, Ingram proposed the evolutionary scheme in Figure 10-15. Originally there must have been a gene whose product was somewhat like myoglobin. It was a monomer and could show no heme-heme interaction. This gene duplicated, allowing one copy to evolve while pre-

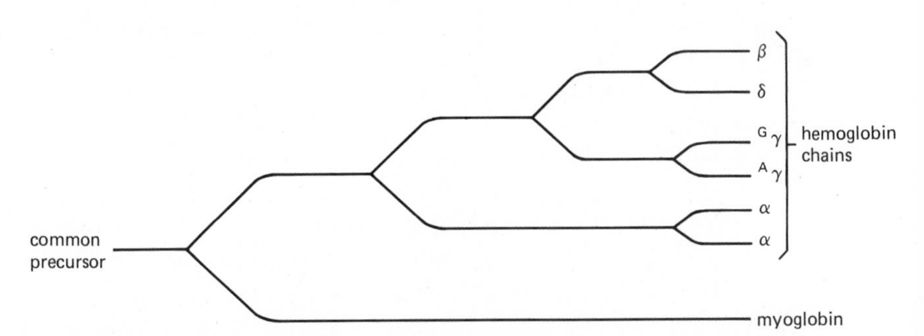

Figure 10-15 Evolution of the genes for the human globins. The probable order of the various gene duplications for which evidence is now available is depicted by the tree. The most ancient duplication and divergence produced the primitive myoglobin and hemoglobin genes. Duplication of the hemoglobin gene early in vertebrate evolution gave rise to primitive α and β chains, which, with further evolution (especially of the β gene), developed the ability to form tetramers. The β gene line underwent two further duplications. The first yielded the γ gene, expressed during fetal life, and the β gene, expressed in children and adults; the second yielded the two closely linked adult type genes which produce the major β component and the minor δ component. Other very recent duplications have occurred in both the α and γ lines; the two α loci produce identical proteins, while the two γ loci produce chains differing in one residue. Insufficient evidence is available to place the fetal ϵ and ζ genes on the tree with certainty.

serving the monomeric nature of the other. The copy that evolved mutated to a form that could form dimers. Further duplication and evolution of this gene gave rise to an α-like gene and a non-α gene. The latter further duplicated and evolved into a γ-like and a β-δ-like gene. Finally the β-δ gene duplicated and evolved into the present-day β and δ genes. Consistent with this idea is the generally conservative evolution of the α chain and the greater variations in non-α chains. Most of the species differences in Hb function are the result of β-chain differences.

One may compare the amino acid structures of homologous chains from different species to obtain ideas on the closeness of those species in evolution. For example, human beings and chimpanzees have identical β chains and differ by a single residue in the δ chains. Gorillas differ from man by one residue in both the β and δ chains. Clearly these genes have not changed much since divergence of these species from a common ancestor. More remote primates differ by greater amounts, as expected, and other mammals are still more different. It is possible to construct an evolutionary tree, such as that in Figure 10-16, purely by

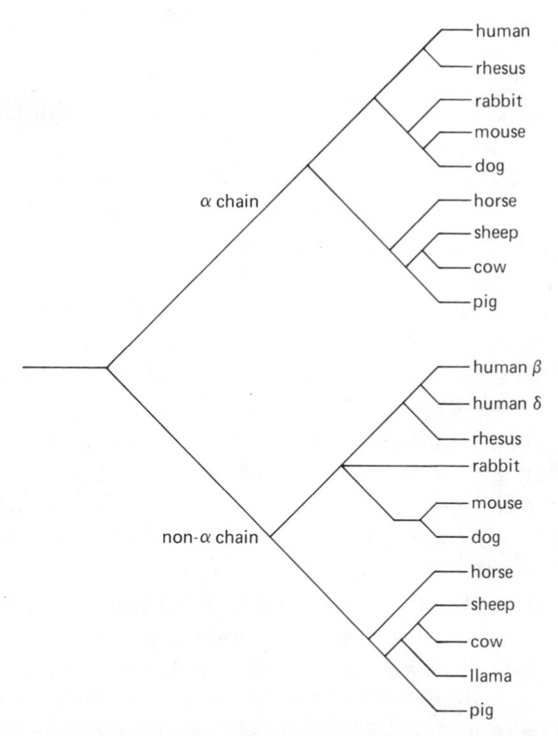

Figure 10-16 Evolutionary tree of hemoglobin chains. No effort is made to depict the relative length of time between the points of divergence. *(Data for this figure are summarized in Dayhoff, 1972.)*

comparing the structures of a limited group of genes, such as the Hb genes. Where differences are very small, chance may cause inappropriate assignments. However, with sufficient genes, it should be possible to reconstruct evolutionary trees of living species purely from protein sequences.

The two γ-chain loci in man, giving rise to $\gamma^{136\ \mathrm{Gly}}$ and $\gamma^{136\ \mathrm{Ala}}$, have also been reported in chimpanzees and gorillas (Huisman et al., 1973). Orangutans appear to have two γ genes, but they differ in the codons at position 135. These results confirm the close affinity of man, chimpanzees, and gorillas.

A more discriminating comparison can be made using nucleic acid sequences. Few are available for this purpose, but Kafatos et al. (1977) were able to compare the complete human and rabbit β-globin mRNA. Such comparisons, when extended to other species, should be very informative, since they can also take into consideration silent mutations and untranslated regions.

SUMMARY

1. Hemoglobin (Hb) is a tetramer consisting of two each of two kinds of polypeptide chains to which heme is attached. Five Hb's have been identified in human beings as follows: an early embryonic Hb, Portland 1, symbolized $\zeta_2\gamma_2$; embryonic Hb, $\alpha_2\epsilon_2$; fetal hemoglobin, $\alpha_2\gamma_2$; adult Hb, $\alpha_2\beta_2$; and a minor adult constituent, Hb A_2, $\alpha_2\delta_2$.

2. The α, β, δ, ϵ, and ζ polypeptide chains each represent the action of separate structural genes, symbolized Hb_α, Hb_β, and so on. It is suggested that in some but not all human populations, there may be two Hb_α loci in tandem producing identical α chains. There are two kinds of γ chains differing by a single amino acid residue. These are produced by two γ genes in tandem, $Hb_{A\gamma}$ and $Hb_{G\gamma}$.

3. The major adult *Hb* loci, Hb_α and Hb_β, are not linked. The loci producing δ and γ chains are closely linked to the Hb_β locus, probably in the order $Hb_{G\gamma}$-$Hb_{A\gamma}$-Hb_δ- Hb_β, the regions coding for the N-terminals of the polypeptide chains being to the left. Locations of the Hb_ϵ and Hb_ζ loci are unknown.

4. Normally the synthesis of fetal Hb drops to very low levels at birth, concomitant with a rise in synthesis of adult Hb. The switchover mechanism is unknown, but several mutations have occurred that interfere with the switchover. These are linked to the Hb_β complex. The persistence of high levels of fetal Hb is not just in response to anemia, but rather represents "normal" synthesis. Homozygotes are normal with

100% fetal hemoglobin. It is possible that the different persistent fetal Hb genes are deletions of various segments of the Hb_{β} complex.

5. Thalassemia is a quantitative defect in globin chain synthesis that may affect either the α or β chains. The nature of some of the mutations is unknown, but in some instances there is a deletion of the genes involved. Defects in synthesis of a particular chain are transmitted as closely linked to the structural gene.

6. Many mutations of the *Hb* loci are known. Most involve amino acid substitutions consistent with a single nucleotide substitution in the corresponding codon. Less common mutations are known to involve deletions of one or more codons within a structural gene and deletions that overlap two or more *Hb* loci. The deletions are thought to arise by mispairing and unequal crossing over. Mutations are also known that lead to elongation of the polypeptide chain. They arose through unequal crossing over, through mutation of a termination codon to a sense codon, or through a frameshift mutation.

7. The complete sequences of human α and β globin mRNA have been established. They both contain untranslated sequences on the 5′ and 3′ ends. The latter is translated in part in chain termination mutants. Hybridization of cDNA copied from globin mRNA to nuclear DNA indicates the presence of a nontranslated intervening sequence 1000 base pairs long in both δ and β genes. Restriction enzyme analysis shows the δ and β genes to be separated by 7000 base pairs. The α genes are separated by a maximum of 3100 base pairs.

8. Mutation affects several functional properties of Hb. Most common is the rate of synthesis. Other properties that are sometimes altered are stability and the equilibrium between ferrohemoglobin and methemoglobin. Substitution of valine at the $\beta 6$ position causes a gel-like formation that leads to sickling and is the cause of sickle cell anemia.

9. Several of the Hb mutations are polymorphic in certain populations in which malaria has been an important agent of natural selection. These include Hb's S and C and β-thalassemia in Africans and their descendants, Hb S and β-thalassemia in Mediterranean groups, and Hb E and α-thalassemia in Asia. Efforts to treat hemoglobinopathies have largely been limited to treating symptoms, although attempts have been made to alter the sickling process by chemical means.

10. Prenatal diagnosis of defects in "adult" hemoglobins have been shown to be possible, since fetal cells synthesize β chains at a very low level. Hybridization of cDNA to DNA from fibroblasts can reveal deletions of Hb genes, and this has also been used in prenatal diagnosis.

11. The primary structures of Hb chains in living species can be used to trace their evolutionary relationships. Such studies affirm the close affinities of man, chimpanzees, and gorillas.

REFERENCES AND SUGGESTED READING

ABRAMSON, R. K., RUCKNAGEL, D. L., SHREFFLER, D. C., AND SAAVE, J. 1970. Homozygous Hb J Tongariki: Evidence for only one alpha chain structural locus in Melanesians. *Science* 169: 194–196.

BAGLIONI, C. 1962. The fusion of two peptide chains in hemoglobin Lepore and its interpretation as a genetic deletion. *Proc. Natl. Acad. Sci.* (U.S.A.) 48: 1880–1886.

BAINE, R. M., RUCKNAGEL, D. L., DUBLIN, P. A. JR., AND ADAMS, J. G. III. 1976. Trimodality in the proportion of hemoglobin G Philadelphia in heterozygotes: Evidence for heterogeneity in the number of human alpha chain loci. *Proc. Natl. Acad. Sci.* (U.S.A.) 73: 3633–3636.

BOYER, S. H., NOYES, A. N., TIMMONS, C. F., AND YOUNG, R. A. 1972. Primate hemoglobins: polymorphisms and evolutionary patterns. *J. Hum. Evol.* 1: 515–543.

BRADLEY, T. B., WEHL, R. C., AND SMITH, G. J. 1975. Elongation of the alpha globin chain in a Black family: Interaction with Hb G Philadelphia. *Clin. Res.* 23: 131A.

BUNN, H. F., SCHMIDT, G. J., HANEY, D. N., AND DLUHY, R. G. 1975. Hemoglobin Cranston, an unstable variant having an elongated β chain due to nonhomologous crossover between two normal β chain genes. *Proc. Natl. Acad. Sci.* (U.S.A.) 72: 3609–3613.

CAPP, G. L., RIGAS, D. A., AND JONES, R. T. 1970. Evidence for a new haemoglobin chain (ζ-chain). *Nature* (Lond.) 228: 278–280.

CHANG, J. C., AND KAN, Y. W. 1979. β^0 thalassemia, a nonsense mutation in man. *Proc. Natl. Acad. Sci.* (U.S.A.) 76: 2886–2889.

CLEGG, J. B., AND WEATHERALL, D. J. 1976. Molecular basis of thalassaemia. *Br. Med. Bull.* 32: 262–269.

CLEGG, J. B., WEATHERALL, D. J., CONTOPOLOU-GRIVA, I., CAROUTSOS, K., POUNGOURAS, P., AND TSEVRENIS, H. 1974. Haemoglobin Icaria, a new chain-termination mutant which causes α thalassaemia. *Nature* (Lond.) 251: 245–247.

CLEGG, J. B., WEATHERALL, D. J., AND MILNER, P. F. 1971. Haemoglobin Constant Spring—a chain termination mutant? *Nature* (Lond.) 234: 337–340.

CONLEY, C. L., WEATHERALL, D. J., RICHARDSON, S. N., SHEPARD, M. K., AND CHARACHE, S. 1963. Hereditary persistence of fetal hemoglobin: a study of 79 affected persons in 15 Negro families in Baltimore. *Blood* 21: 261–281.

DAYHOFF, M. O., HUNT, L. T., MC LAUGHLIN, P. J., AND JONES, D. D. 1972. Gene duplications in evolution: the globins. In *Atlas of Protein Sequence and Structure*, Vol. 5 (M. O. Dayhoff, Ed.). National Biomedical Research Foundation, Silver Spring, Md. pp. 17–30.

DE JONG, W. W., MEERA KHAN, P., AND BERNINI, L. F. 1975. Hemoglobin Koya Dora: High frequency of a chain termination mutant. *Am. J. Hum. Genet.* 27: 81–90.

DICKERSON, R. E. 1964. X-ray analysis and protein structure. In *The Proteins*, Vol. II, 2d Ed. New York: Academic Press, pp. 603–778.

FLATZ, G., KINDERLERER, J. L., KILMARTIN, J. V., AND LEHMANN

H. 1971. Haemoglobin Tak: a variant with additional residues at the end of the β-chains. *Lancet* 1: 732–733.

FLAVELL, R. A., KOOTER, J. M., DE BOER, E., LITTLE, P. F. R., AND WILLIAMSON, R. 1978. Analysis of the β-δ-globin gene loci in normal and Hb Lepore DNA: Direct determination of gene linkage and intergene distance. *Cell* 15: 15–41.

GERALD, P. S., AND DIAMOND, L. K. 1958. A new hereditary hemoglobinopathy (the Lepore trait) and its interaction with thalassemia trait. *Blood* 13: 835–844.

HOLLÁN, S. R., SZELENYI, J. G., BRIMHALL, B., DUERST, M., JONES, R. T., KOLER, R. D., AND STOCKLEN, Z. 1972. Multiple alpha chain loci for human haemoglobins: Hb J-Buda and Hb G-Pest. *Nature* (Lond.) 235: 47–50.

HONIG, G. R., SHAMSUDDIN, M., MASON, R. G., AND VIDA, L. N. 1978. Hemoglobin Lincoln Park: A βδ fusion (anti-Lepore) variant with an amino acid deletion in the δ chain-derived segment. *Proc. Natl. Acad. Sci.* (U.S.A.) 75: 1475–1479.

HUEHNS, E. R., DANCE, N., BEAVEN, G. H., HECHT, F., AND MOTULSKY, A. G. 1964. Human embryonic hemoglobins. *Cold Spring Harbor Symp. Quant. Biol.* 29: 327–331.

HUISMAN, T. H. J., SCHROEDER, W. A., DOZY, A. M., SHELTON, J. R., SHELTON, J. B., BOYD, E. M., AND APELL, G. 1969. Evidence for multiple structural genes for the gamma-chain of human fetal hemoglobin in hereditary persistence of fetal hemoglobin. *Ann. N.Y. Acad. Sci.* 165: 320–331.

HUISMAN, T. H. J., SCHROEDER, W. A., KEELING, M. E., GENGOZIAN, N., MILLER, A., BRODIE, A. R., SHELTON, J. R., SHELTON, J. B., AND APELL, G. 1973. Search for nonallelic structural genes for γ-chains of fetal hemoglobin in some primates. *Biochem. Genet.* 10: 309–318.

HUISMAN, T. H. J., WILSON, J. B., GRAVELY, M., AND HUBBARD, M. 1974. Hemoglobin Grady: The first example of a variant with elongated chains due to an insertion of residues. *Proc. Natl. Acad. Sci.* (U.S.A.) 71: 3270–3273.

ITANO, H. A., AND NEEL, J. V. 1950. A new inherited abnormality of human hemoglobin. *Proc. Natl. Acad. Sci.* (U.S.A.) 36: 613–617.

JONES, R. T., BRIMHALL, B., HUISMAN, T. H. J., KLEIHAUER, E., AND BETKE, K. 1966. Hemoglobin Freiburg: abnormal hemoglobin due to deletion of a single amino acid residue. *Science* 154: 1024–1027.

KABAT, D., AND KOLER, R. D. 1975. The thalassemias: Models for analysis of quantitative gene control. *Adv. Hum. Genet.* 5: 157–222.

KAFATOS, F. C., EFSTRATIADIS, A., FORGET, B. G., AND WEISSMAN, S. M. 1977. Molecular evolution of human and rabbit β-globin mRNAs. *Proc. Natl. Acad. Sci.* (U.S.A.) 74: 5618–5622.

KAN, Y. W., AND DOZY, A. M. 1978. Polymorphism of DNA sequence adjacent to human β-globin structural gene: relationship to sickle mutation. *Proc. Natl. Acad. Sci.* (U.S.A.) 75: 5631–5635.

KAN, Y. W., GOLBUS, M. S., AND DOZY, A. M. 1976. Prenatal diagnosis of α-thalassemia. *N. Engl. J. Med.* 295: 1165–1167.

KAZAZIAN, H. H. JR., CHO, S., AND PHILLIPS, J. A. III. 1977. The mutational basis of the thalassemia syndromes. *Prog. Med. Genet.* n.s. 2: 165–204.

KAZAZIAN, H. H. JR., AND WOODHEAD, A. P. 1973. Hemoglobin A synthesis in the developing fetus. *N. Engl. J. Med.* 289: 58–62.

KENDALL, A. G., OJWANG, P. J., SCHROEDER, W. A., AND HUISMAN, T. H. J. 1973. Hemoglobin Kenya, the product of a γ-β fusion gene: studies of the family. *Am. J. Hum. Genet.* 25: 548–563.

LAWN, R. M., FRITSCH, E. F., PARKER, R. C., BLAKE, G., AND MANIATIS, T. 1978. The isolation and characterization of linked δ- and β-globin genes from a cloned library of human DNA. *Cell* 15: 1157–1174.

LEDER, A., MILLER, H. I., HAMER, D. H., SEIDMAN, J. G., NORMAN, B., SULLIVAN, M., AND LEDER, P. 1978. Comparison of cloned mouse α- and β-globin genes: Conservation of intervening sequence locations and extragenic homology. *Proc. Natl. Acad. Sci.* (U.S.A.) 75: 6187–6191.

LEHMANN, H., AND CARRELL, R. W. 1969. Variations in the structure of human haemoglobin. *Br. Med. Bull.* 25: 14–23.

LEHMANN, H., AND CHARLESWORTH, D. 1970. Observations on haemoglobin P (Congo type). *Biochem. J.* 119: 43P.

LEHMANN, H., AND HUNTSMAN, R. G. 1975. *Man's Haemoglobins,* 2d Ed. Philadelphia: J. B. Lippincott, 478 pp.

LIE-INJO L. E., GANESAN, J., CLEGG, J. B., AND WEATHERALL, D. J. 1974. Homozygous state for Hb Constant Spring (slow-moving Hb X components). *Blood* 43: 251–259.

LITTLE, P. F. R., FLAVELL, R. A., KOOTER, J. M., ANNISON, G., AND WILLIAMSON, R. 1979. Structure of the human fetal globin gene locus. *Nature* (Lond.) 278: 227–231.

LORKIN, P. A. 1973. Fetal and embryonic haemoglobins. *J. Med. Genet.* 10: 50–64.

MC KUSICK, V. A. 1978. *Mendelian Inheritance in Man,* 5th Ed. Baltimore: Johns Hopkins Press, 976 pp.

MEARS, J. G., RAMIREZ, F., LEIBOWITZ, D., NAKAMURA, F., BLOOM, A., KONOTEY-AHULU, F., AND BANK, A. 1978. Changes in restricted human cellular DNA fragments containing globin gene sequences in thalassemias and related disorders. *Proc. Natl. Acad. Sci.* (U.S.A.) 75: 1222–1226.

MOTULSKY, A. G. 1973. Frequency of sickling disorders in U.S. blacks. *N. Engl. J. Med.* 288: 31–33.

MURAYAMA, M. 1966. Molecular mechanism of red cell "sickling." *Science* 153: 145–149.

OHTA, Y., YAMAOKA, K., SUMIDA, I., AND YANASE, T. 1971. Haemoglobin Miyada, a β-δ fusion peptide (anti-Lepore) type discovered in a Japanese family. *Nature New Biol.* 234: 218–220.

ORKIN, S. H. 1978. The duplicated human α globin genes lie close together in cellular DNA. *Proc. Natl. Acad. Sci.* (U.S.A.) 75: 5950–5954.

RANNEY, H. M. 1954. Observations on the inheritance of sickle-cell hemoglobin and hemoglobin C. *J. Clin. Invest.* 33: 1634–1641.

SEID-AKHAVAN, M., WINTER, W. P., ABRAMSON, R. K., AND RUCKNAGEL, D. L. 1976. Hemoglobin Wayne: A frameshift mutation detected in human hemoglobin α chains. *Proc. Natl. Acad. Sci.* (U.S.A.) 73: 882–886.

SHIBATA, S., MIYAJI, T., UEDA, S., MATSUOKA, M., IUCHI, I., YAMADA, K., AND SHINKAI, N. 1970. Hemoglobin Tochigi (beta 56-59 deleted). A new unstable hemoglobin discovered in a Japanese family. *Proc. Jap. Acad.* 46: 440–445.

SMITH, E. W., AND TORBERT, J. V. 1958. Study of two abnormal hemoglobins with evidence for a new genetic locus for hemoglobin formation. *Bull. Johns Hopkins Hosp.* 102: 38–45.

STAMATOYANNOPOULOS, G. 1972. The molecular basis of hemoglobin disease. *Annu. Rev. Genet.* 6: 47–70.

WEATHERALL, D. J., AND CLEGG, J. B. 1976. Molecular genetics of human hemoglobin. *Annu. Rev. Genet.* 10: 157–178.

WEATHERALL, D. J., AND CLEGG, J. B. 1979. Recent developments in the molecular genetics of human hemoglobin. *Cell* 16: 467–479.

WINSLOW, R. M., SWENBERG, M.-L., GROSS, E., CHERVENICK, P. A., BUCHMAN, R. R., AND ANDERSON, W. F. 1975. Hemoglobin McKees Rocks ($\alpha_2\beta_2^{145\ \mathrm{Tyr}\rightarrow\mathrm{Term}}$): a human "nonsense" mutation leading to a shortened β-chain. *J. Clin. Invest.* 57: 772–781.

REVIEW QUESTIONS

1. Define:

heme–heme interaction	Hb Bart's
myoglobin	Hb Kenya
fetal hemoglobin	thalassemia
embryonic hemoglobin	HPFH
Lepore hemoglobin	cDNA
anti-Lepore hemoglobin	hemolytic anemia
methemoglobin	hypochromic anemia

2. In Figure 10-6, which persons provide most critical evidence for nonallelism of Hb S and Hb Ho-2?

3. In Table 10-2, fill out the expected proportions of each phenotype for each mating on the assumption of allelism for Hb S and Hb C. Do the same on the hypothesis of nonallelism.

4. Draw a diagram showing the mispairing of the DNA sequences necessary to produce Hb Gun Hill. What is the number of identical nucleotides in this segment? If the same procedure is carried out on Hb Tours, how do the results compare?

5. What should be the maximum number of "sense" mutations that could occur by mutation of the β termination codon? What amino acids would be produced?

CHAPTER ELEVEN

METABOLIC VARIATION AND DISEASE

Metabolism may be viewed as the product of enzymes acting on the environment. Enzymes, being proteins, are subject to the great variety of structural changes known for hemoglobin. However, they do not occur in such high concentrations conveniently packaged for study. Understanding the molecular basis for the inherited functional changes in enzymes has therefore been difficult. There are some notable successes, however, and these confirm the idea that inherited disease can be approached effectively in the context of molecular genetics.

FUNCTIONAL PROPERTIES OF ENZYMES

Although the conventional definition of an enzyme emphasizes its ability to function as a catalyst, enzymes must be able to interact with a variety of cell constituents in order to function in a coordinated way in the complex metabolic network. Whether or not a mutation leads to metabolic defect cannot always be answered by *in vitro* measurements with artificial substrates. The variety of ways in which defects may arise can best be approached by considering some of the aspects of enzyme function.

Enzymes As Catalysts. The ultimate function of an enzyme is to catalyze a specific reaction or group of similar reactions. This is sometimes written as

$$En + S \rightarrow [En \cdot S] \rightarrow En + P$$

where S is the substrate and P the product. From purely chemical considerations, the conversion of S to P is one of many that S might undergo. In any chemical reaction, a stable molecule must become "activated" to a less stable state by absorption of energy, resulting in greater atomic vibrations, bond distortions, and so forth. This energy barrier provides stability to a molecule, and the energy barriers for different reactions that a molecule may undergo usually differ. The contribution of the enzyme is to lower the energy barrier between S and P for a particular reaction. For readily reversible reactions, at equilibrium, the enzyme does not change the relative amounts of the substrate and product (although equilibrium conditions must be rare in living organisms). Rather, it changes the speed with which interconversion occurs.

Coenzymes. The *active site* of an enzyme is that region that is directly involved in the substrate reaction. In most cases, the enzyme consists of a *coenzyme* and the *apoenzyme*. The coenzymes are nonprotein small molecular weight compounds, many of which are vitamins, that bind in very specific ways to the apoenzyme (the protein portion). The coenzyme enters into the chemical reaction with the substrate and is the principal component of the active site. Only in certain hydrolytic reactions is the active site derived entirely from amino acids of the protein chain.

Coenzymes are usually bound to the apoenzyme by noncovalent bonds. The surface contours and distribution of charges and nonpolar groups in the apoenzyme determine whether a coenzyme will be bound and with what strength. One frequent consequence of mutation is to alter the avidity with which coenzymes are bound. A decrease in enzyme activity may therefore reflect diminished coenzyme binding rather than decreased synthesis of apoenzyme. The decreased affinity sometimes may be overcome by higher concentrations of coenzyme.

Unbound coenzymes typically have some catalytic activity. As compared to bound coenzyme, the activity is much lower, and the substrate specificity is very broad. The low activity can be thought of as resulting from the absence of the activity of the apoenzyme in promoting alignment of the coenzyme and the substrate. The apoenzyme therefore greatly enhances the activity of the coenzyme, but the chemistry of the reaction is in the coenzyme.

Substrate Specificity. A major contribution of the apoenzyme is to limit the substrates that gain access to the nondiscriminating coenzymes. Some enzymes can act only on one substrate. Others may act on many related substrates. In general, those that function in biosynthetic pathways or in key metabolic and energy pools have narrow specificity. Those that are involved in degrading food, for example, have very broad specificity. Specificity is a function of the conformation of the apoprotein and of the surface charges and nonpolar groups around the active site. A highly specific enzyme is one that excludes molecules even closely related to the substrate. Mutation may distort the apoenzymes so that more molecules can reach the coenzyme and react with it, leading to broader specificity.

Enzyme Inhibitors. The binding of substrate to the enzyme is a key part of an enzymatically catalyzed reaction. Very often molecules with structures that resemble the substrate can bind at the same site as the substrate, even though they may not have the chemical groups necessary for reaction. In metabolism, the product of a reaction often is structurally similar to the substrate and can bind to the active site. If a high level of product accumulates, this may inhibit formation of additional product, acting as a feedback regulator of the net enzyme activity. The inhibitor need not be the product but can be any metabolite, drug, or other substance with sufficient structural similarity. So long as the inhibitor binds reversibly, the inhibition is a function of the relative concentrations of the substrate and inhibitor, and the inhibition is called *competitive* inhibition. Irreversible binding with an inhibitor causes *noncompetitive* inhibition.

Inhibition may also occur with metabolites that are structurally very different from the substrate. The binding of such metabolites is at sites other than the catalytically active site, but the binding either causes conformational changes in the enzyme or stabilizes the enzyme in an inactive conformation. The inhibitors in some instances are the end products of biosynthetic pathways, and the inhibited enzyme is the first for that one pathway. An example is given in Figure 11-1. The effect is to prevent excess synthesis of the end product by reducing the activity of the first enzyme. Such a negative feedback mechanism is called *feedback inhibition.* The site at which the inhibitor binds is the *allosteric* site. Enzymes with allosteric sites are referred to as allosteric enzymes. Mutation that alters the allosteric site may influence the rate of an enzyme reaction, even though the catalytic site has not been altered.

It is useful to distinguish between feedback inhibition, which reduces the activity of an enzyme, and *feedback repression,* in which synthesis of the enzyme is reduced. The same end product molecule may act both as a repressor and an inhibitor, but the mechanisms are different.

Enzyme Complexes. Some enzymes function in solution, relatively independent of other enzymes. Many others function as part of multienzyme complexes or bound to membranes. The ability to form complexes depends on the structures of the components. Mutation may alter the structure of an enzyme so that it can no longer enter into complex formation. It may still be potentially active as a catalyst, but it no longer is located where it can come in contact with the substrate. The metabolic effect is greatly diminished activity.

The function of enzymes in membranes has not been extensively studied. In addition to occurring in intracellular membranes, enzymes

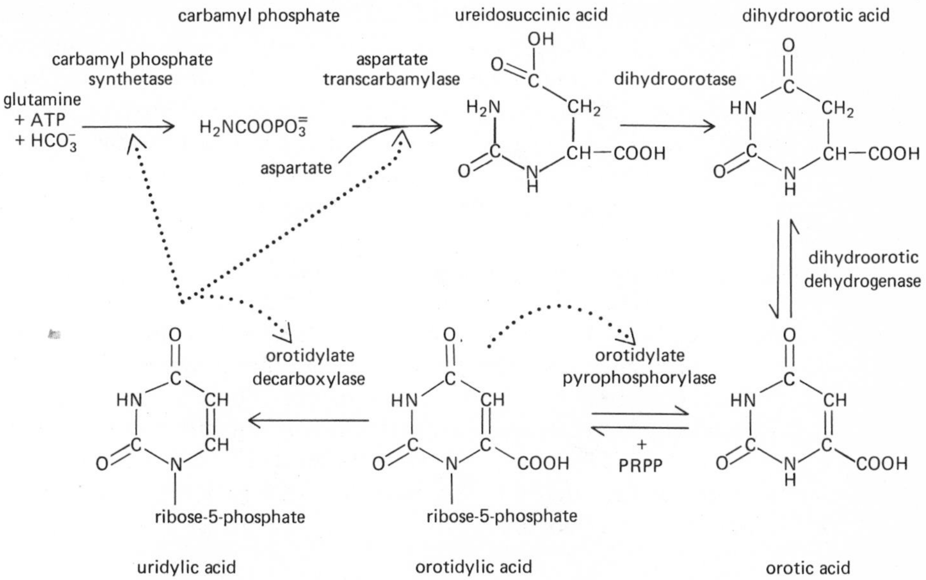

Figure 11-1 Biosynthetic pathway for pyrimidines. Solid arrows indicate pathways. Dotted arrows indicate inhibition. The first step uniquely on the pathway is catalyzed by the enzyme aspartate transcarbamylase. This and other enzymes are inhibited by the end product uridine-5′-monophosphate, preventing the excess synthesis. In the disease, oroticaciduria, the last two steps of the pathway are blocked, leading to accumulation of orotic acid. However, if pyrimidine nucleotides are supplied exogenously in high levels, the initial enzymes are inhibited, reducing thereby the synthesis and accumulation of orotic acid. (*Wuu and Krooth, 1968; Kelley and Smith, 1978.*)

also occur in the cell membrane, where they function in the transport of molecules to the interior of the cell. Many *permease* mutations are known in microorganisms, in which the defect is in the ability to transport substances to the interior of the cell rather than convert them to something else.

INHERITED DEFECTS: METABOLIC BLOCKS

A primary error in metabolism leads to many secondary metabolic defects through accumulation of normal or abnormal metabolites that can act as inhibitors. It is not always easy, therefore, to identify the primary error in a metabolic disease. A good case can be made if activity of one enzyme is totally absent, but for reasons discussed above, the mutation may be in a gene other than the structural gene for that enzyme. The following are examples of some of the inherited metabolic diseases in which the primary defect is known with some certainty.

Mutations that result in lack of enzyme activity are known as metabolic blocks. The effects of a metabolic block depend on a variety of factors in addition to the blocked metabolic reaction. The presence of alternate pathways of metabolism, of efficient methods of excretion, and so on, determine whether a metabolic block is lethal or benign. A block in a biosynthetic pathway may be lethal if there is no alternative source of the end product, or it may be innocuous if the end product is also available from the diet. For example, blocks in the biosynthesis of heme, a component of hemoglobin, cytochromes, catalase, and peroxidase, are never complete blocks. A zygote with a complete block could not develop without cytochromes. Several partial blocks are known in this pathway, giving rise to the group of diseases known as porphyrias.

Most metabolic blocks are recessive. It seems that in general one good copy of a structural gene is adequate to direct synthesis of sufficient enzyme to meet metabolic requirements. There are undoubtedly many metabolic "bottlenecks" for which this is not true, but the known examples are few. Direct assay of enzyme levels in heterozygotes often reveals only half normal activity, indicating the relative excess of enzyme activity that exists for most metabolic reactions.

Blocks in Biosynthetic Pathways: Albinism

A vast number of genes (and enzymes) have as their function the synthesis of body constituents. In some instances this synthetic potential may serve primarily as a backup system to produce substances that may be provided primarily by diet. In most instances, the synthetic pathways provide end products for which there are no alternate sources, at least no sources that can supply the end products in the quantities needed and to the intracellular locations of specific tissues. A complete block in an essential biosynthetic pathway is therefore likely to have profound effects on development. Indeed, most probably are embryonic lethals. Most of these that are available for study and hence are not lethal involve either the synthesis of nonessential (but useful) products or incomplete blocks that often are difficult to identify.

Albinism is one of the earliest inherited traits studied. Persons with albinism are easily identified, and the familial nature often is obvious. Garrod proposed albinism as an example of an inherited enzyme defect, and subsequent studies have proved him correct. The formation of melanin, the epidermal pigment, whether in a blond Scandinavian or a black African, occurs by the same mechanisms (Figure 11-2). Differences in color are due to differences in the amount of melanin as well as to differences in size and shape of the pigment-containing granules.

The three types of albinism most commonly seen are tyrosinase-positive oculocutaneous albinism, tyrosinase-negative oculocutaneous albinism, and ocular albinism. In the first two, pigment is absent both in the eyes and in skin; in ocular albinism, only eye pigment is missing. For both forms of oculocutaneous albinism, affected persons are homozygous for an autosomal recessive gene, whereas ocular albinism is

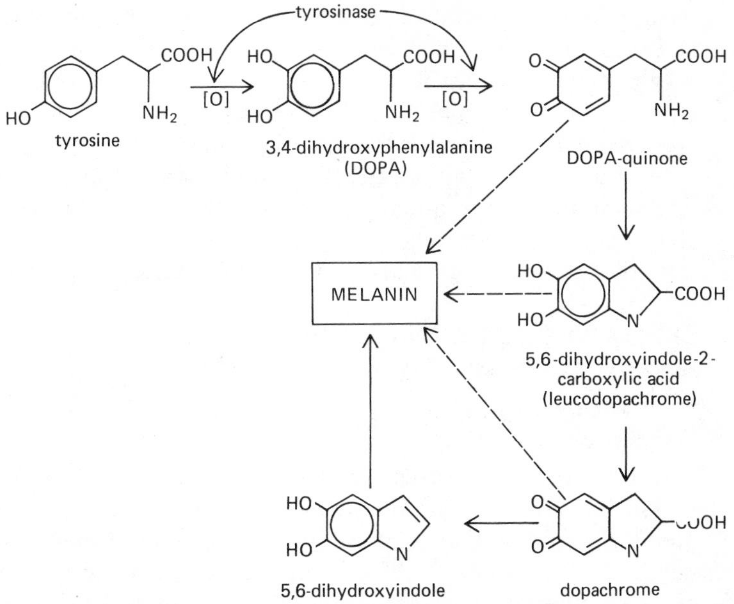

tyrosinase
[O]
[O]
tyrosine
3,4-dihydroxyphenylalanine (DOPA)
DOPA-quinone
MELANIN
5,6-dihydroxyindole-2-carboxylic acid (leucodopachrome)
5,6-dihydroxyindole
dopachrome

Figure 11-2 Conversion of tyrosine to melanin. The enzyme tyrosinase catalyzes the first two steps. The steps beyond that appear to occur nonenzymatically. Melanin is a polymer of 5,6-dihydroxyindole and probably includes also the intermediates indicated by dashed arrows. (*Redrawn from Duchon, J., Fitzpatrick, T. B., and Seiji, M. 1968.* Year Book of Dermatology. *Chicago: Year Book Med. Pub., pp. 6–33.*)

transmitted as an X-linked recessive. In addition to these three forms of albinism, there are several very rare forms as well as several other diseases that involve hypopigmentation (Witkop et al., 1978). These latter are not generally classified as albinism, however.

The genetic heterogeneity of oculocutaneous albinism was first clearly demonstrated by matings between albinos. For matings between persons homozygous at one locus, all children should be affected. Matings were observed, however, in which all children were normally pigmented. They must therefore have been doubly heterozygous at two recessive loci. The biochemical basis for this heterogeneity was discovered by Witkop and co-workers (1970), who reported that if hair bulbs are immersed in a tyrosine solution, those from some albinos turn black, indicating melanin formation, whereas those from other albinos remain colorless. The former group apparently have normal tyrosinase, which, for unknown reasons, cannot act on the tyrosine *in vivo*. It has been suggested that perhaps the deficit is in the transport of tyrosine into the melanocytes, but this is not established. The latter group of albinos lack tyrosinase, and the mutation is thought to be in the structural gene for this enzyme, although again this is not established. Tyrosinase-negative albinos are completely without pigment throughout their lives, whereas tyrosinase-positive albinos may have traces of pigment and acquire more as they age. A tyrosinase-positive albino African adult may have darker skin than a normally pigmented blond European. In those matings in which two albino parents have produced normal children, one parent is tyrosinase negative and the other tyrosinase positive.

Although albinism is generally a detrimental trait, at least in primitive societies, the frequency is high in some populations. Pigment is clearly not necessary for normal growth, thereby providing an opportunity to observe a nonlethal block in a biosynthetic pathway.

Blocks in Degradative Pathways: Alkaptonuria and Phenylketonuria

Alkaptonuria is one of several metabolic blocks known in the metabolism of phenylalanine in man (Figure 11-3). In persons affected with this autosomal recessive disease, one of the four diseases studied by Garrod, the enzyme homogentisic acid oxidase is missing. Consequently, homogentisic acid cannot be converted to its oxidation product, maleylacetoacetic acid. Homogentisic acid accumulates as a consequence, but it is not reabsorbed by the kidney from the glomerular filtrate and, therefore, cannot reach very high levels in the tissues and blood. Large amounts of homogentisic acid appear in the urine and on excretion undergo further oxidation spontaneously, forming a black pigment. The condition is present from birth and can be recognized because of the dark color appearing on wet diapers.

Alkaptonuria is a relatively benign disease. The amino acids phenylalanine and tyrosine are normal dietary constituents. However, the diet supplies much more than the body can use as such, and the excess is metabolized via homogentisic acid. No essential metabolites are formed subsequent to homogentisic acid. The fact that the kidney cannot retain excess homogentisic acid prevents major deviation from normal metabolism. The only pathological consequence of alkaptonuria is a slow deposition of pigment in some of the joints, leading ultimately to a form of arthritis.

Phenylketonuria (PKU), an autosomal recessive disease, shows quite a different effect of enzyme activity. The normal function of the enzyme phenylalanine hydroxylase is to convert excess phenylalanine to tyrosine, which is then further metabolized. The enzyme normally is found in the liver, with lesser amounts detected in the kidney. Persons who lack this enzyme have the disease known as phenylketonuria. In normal persons, the level of free phenylalanine in blood is approximately 1 mg per 100 ml plasma. In persons with PKU, the phenylalanine may be as high as 50 mg per 100 ml. Phenylalanine is efficiently reabsorbed by the kidney, and blood levels must be considerably elevated before appreciable amounts appear in urine.

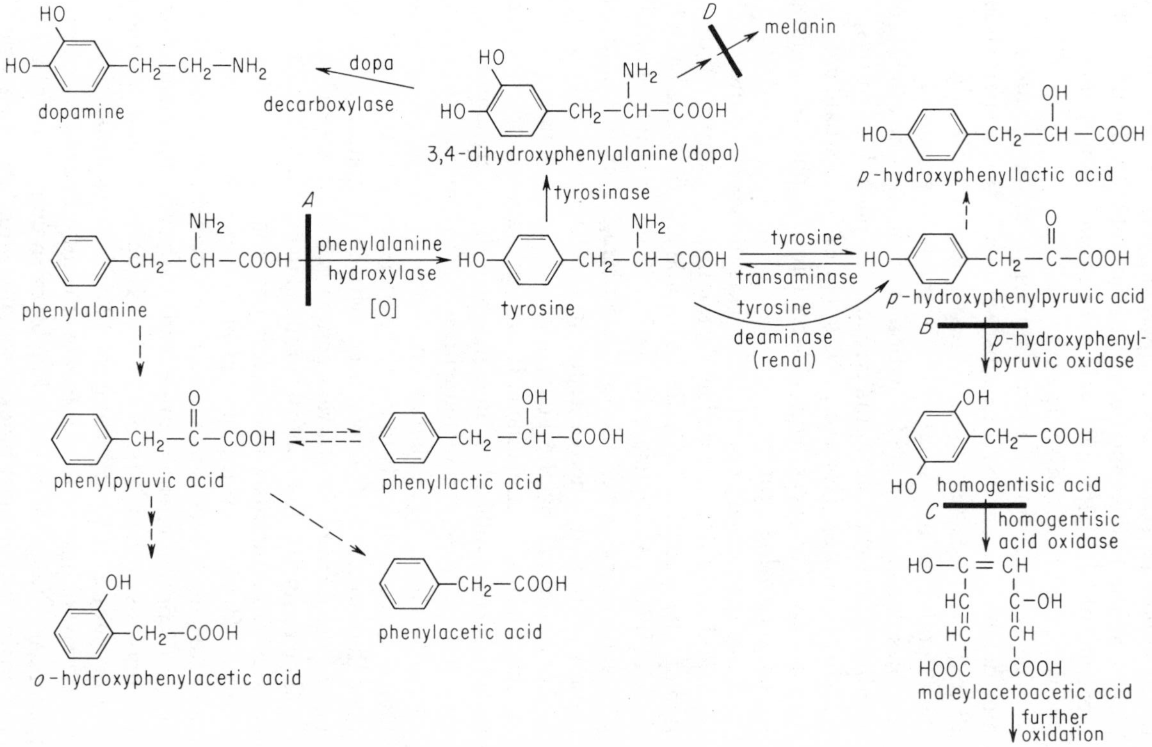

Figure 11-3 Metabolic pathways of phenylalanine metabolism, showing blocks in phenylketonuria (A), tyrosinosis (B), alkaptonuria (C), and albinism (D). Reactions shown by broken arrows are quantitatively unimportant in normal persons but become important in phenylketonuria.

Phenylalanine may form a variety of products other than tyrosine. Normally, however, these other reactions are quantitatively unimportant. When the phenylalanine levels are greatly elevated, these side reactions become major pathways for disposing of phenylalanine. Some of the products, such as phenylpyruvic acid, are not effectively reabsorbed by the kidney and are rapidly excreted in the urine. It was the presence of phenylpyruvic acid that initially enabled recognition of phenylketonuria as a disease entity.

The accumulation of abnormal phenylalanine products leads to various deleterious effects. The most serious is severe mental deficiency. Phenylketonurics usually have less pigment than other members of their family, because some of the by-products apparently interfere with enzymes that catalyze the formation of melanin. The product of phenylalanine hydroxylase action is tyrosine. In phenylketonurics, very little tyrosine is formed. However, it is supplied in adequate amounts in the diet, so none of the effects observed in the phenylketonuric can be traced to tyrosine deficiency.

Since the primary culprit in PKU is the grossly elevated phenylalanine, and since this arises from diet, it might be supposed that elimination of phenylalanine from the diet would eliminate the pathological effects of PKU. Complete elimination is not possible, however, since phenylalanine is essential for the building of proteins, and the body cannot synthesize phenylalanine to meet its own requirements. The amount required by the body, however, is very much less than the amount normally consumed. It is now standard practice for persons with PKU to be placed on diets in which the phenylalanine level is very low. If the disease is detected within the first few months of life, administration of low phenylalanine diets prevents development of the neurological defect that leads to mental deficiency. If diet control is delayed, the deleterious effects may not be avoided. Initiation of diet control as late as the third or fourth year generally has little beneficial effect.

The timing of diet control is related to maturation of the central nervous system. Infants who are homozygous for the PKU gene do not show abnormality at the time of birth, since the mother's metabolism can take care of excess phenylalanine. Once the child is on his own, the accumulation of phenylalanine leads to irreversible changes in the central nervous system. The need for early detection and treatment has led many states to require testing for PKU in newborns.

Phenylketonuria also illustrates the difference between primary and secondary metabolic blocks. The block in conversion of phenylalanine to tyrosine is a primary block associated with lack of enzyme. The deleterious effects, however, are due to secondary blocks resulting from inhibition of enzymes by the abnormal by-products of phenylalanine. The inhibition is particularly demonstrated in pigment formation. The enzymes responsible for pigment formation are normal in phenylketon-

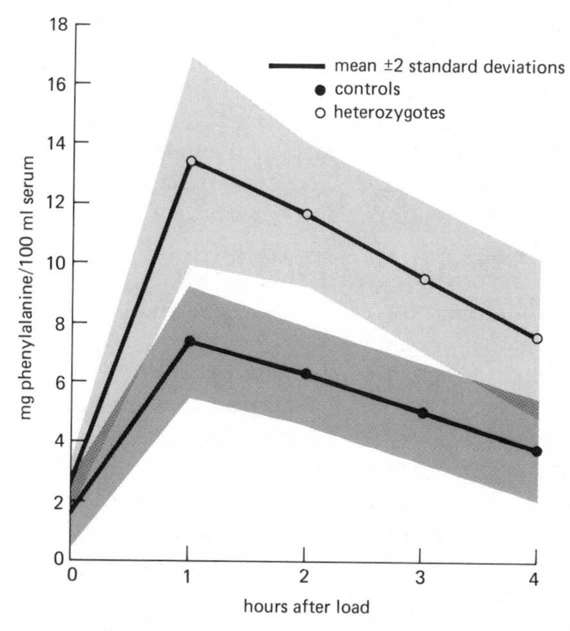

Figure 11-4 Phenylalanine tolerance curves for normal individuals and for individuals heterozygous for phenylketonuria (parents of phenylketonurics). L-phenylalanine (0.1 g per kg body weight) is ingested at zero hours. (*Based on data from Berry, Sutherland, and Guest, 1957.*)

urics, but, because of inhibition by phenylalanine metabolites, less pigment is formed. When phenylketonurics are placed on a low phenylalanine diet, the inhibitors are no longer present, and normal pigment formation occurs.

Phenylalanine hydroxylase occurs only in liver and kidney, and the enzyme can be assayed directly only by biopsy of these tissues. Prenatal diagnosis would require biopsy of the fetal tissues, a procedure that is not done at present. Heterozygotes usually can be recognized by a phenylalanine tolerance test (Figure 11-4). A large amount of phenylalanine is ingested (0.1 g per kg body weight), and the blood levels of phenylalanine are observed during a period of several hours. In heterozygotes, the blood level goes higher than in homozygous normal persons, and the elevation persists longer. However, there is some overlap between high normals and low heterozygotes, so that the results may be ambiguous on occasion. The tolerance studies do confirm, though, that heterozygotes are deficient in phenylalanine hydroxylase as compared to normal, but the amount they possess is much more than required for normal metabolism.

The severe degree of mental retardation observed in most untreated PKU patients has meant that most have been institutionalized. Some have intelligence near normal, and a few are in the normal range without treatment. These have provided opportunities to observe the effects of elevated phenylalanine levels in PKU mothers on fetuses heterozygous for PKU. Ordinarily, heterozygous fetuses would develop normally. However, if the mother is homozygous, the fetus is exposed to elevated phenylalanine and its metabolites. As a consequence, most such offspring are severely defective mentally and many have other congenital malformations. With the increased frequency of successful dietary management of PKU children, the number of potential PKU mothers also will increase. These will require careful management during pregnancy if the child is to be normal.

A condition known as hyperphenylalaninemia is very similar to PKU in the biochemical pattern, except that blood phenylalanine levels are only moderately elevated, and the persons are otherwise normal. Such persons are detected in PKU screening tests, but no dietary control is required for normal development. It has been suggested that these persons are homozygous for an allele of the PKU gene, giving rise to greatly diminished phenylalanine hydroxylase activity. Still, the activity is sufficient to avoid the very high levels of blood phenylalanine usually found in PKU.

Not all cases of PKU have proved to have deficiency of phenylalanine hydroxylase (PAH), and these exceptional cases illustrate the variety of defects that may produce similar phenotypes. PAH requires tetrahydrobiopterin (THB) as a cofactor (Figure 11-5). Ordinarily THB is

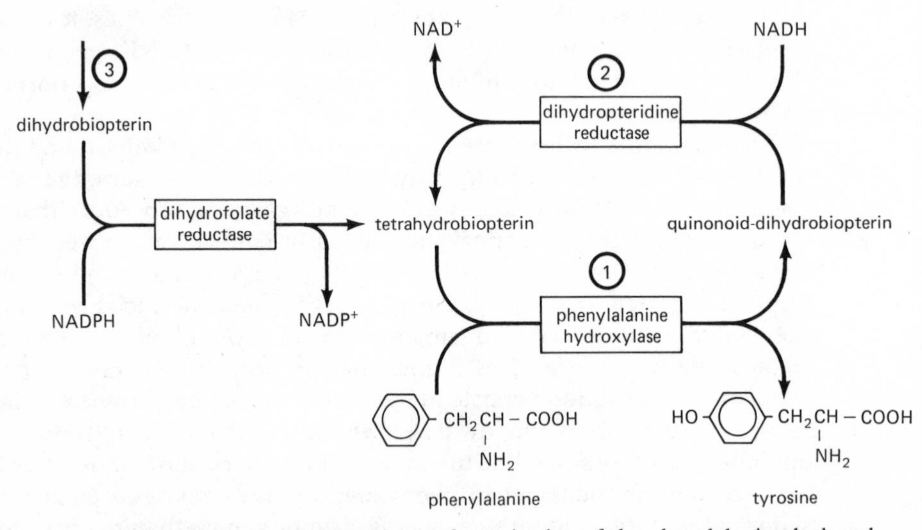

Figure 11-5 Metabolic pathways for activation of the phenylalanine hydroxylase system. The various blocks are indicated by circled numbers. In classical phenylketonuria (1), the apo-phenylalanine hydroxylase (PAH) is nonfunctional. In very rare variants, tetrahydrobiopterin, the cofactor for PAH, is deficient, due either to loss of activity of dihydropteridine reductase (2) or deficiency of biopterin for the oxidation/reduction cycle (3).

manufactured by the body from a precursor, quinonoid dihydrobiopterin, catalyzed by the enzyme dihydropteridine reductase. In two rare cases of PKU, one was found to be deficient in biopterin, necessary for operation of the oxidation/reduction cycle (Kaufman et al., 1978). The other was deficient in dihydropteridine reductase (Kaufman et al., 1975). The effect in either case was the same: phenylalanine hydroxylase *activity* was low, although due in this instance to lack of cofactor. These rare cases were noted because they failed to respond favorably to low phenylalanine diets. Apparently the tetrahydrobiopterin deficiency causes neurological defect that is not dependent on blood phenylalanine level.

Blocks in Degradative Pathways: The Mucopolysaccharidoses

Many tissues of the body, but especially connective tissue, have large molecular weight polysaccharides combined with protein as an integral part of their structure. These *mucopolysaccharides* occur in a variety of structures, but they share some chemical properties. They all have amino sugars and uronic acids as major constituents. Most also have a high concentration of sulfate. This confers an acidic character to them. Mucopolysaccharides are continuously synthesized and broken down, the latter step taking place in the scavenger intracellular organelles, the lysosomes.

A series of genetic disorders has been identified in which defective mucopolysaccharide breakdown is the common factor. These are characterized by accumulation of intermediates of mucopolysaccharide breakdown in lysosomes, with interference in cell function, and by excretion of elevated levels of mucopolysaccharide metabolites in urine. The first disorders identified were designated Hurler's syndrome and Hunter's syndrome. These two diseases are very similar in having skeletal defects, a characteristic facial appearance, and progressive mental retardation. The term gargoylism was applied to both. They were distinguished by (1) autosomal recessive inheritance for Hurler's syndrome as opposed to X-linked recessive inheritance for Hunter's syndrome; (2) corneal clouding in Hurler's but not in Hunter's syndrome; and (3) a generally more severe progressive defect in Hurler's as compared to Hunter's syndrome.

Initially, the distinction between Hurler's and Hunter's syndromes was made primarily on the different modes of inheritance. Recognition that the two disorders are biochemically complementary to each other confirmed that the primary defects are different and hence should involve different genes. For example, Neufeld and Fratantoni (1970) found that fibroblast cultures from patients with either disorder accumulate mucopolysaccharides in the media. However, if the cultures are

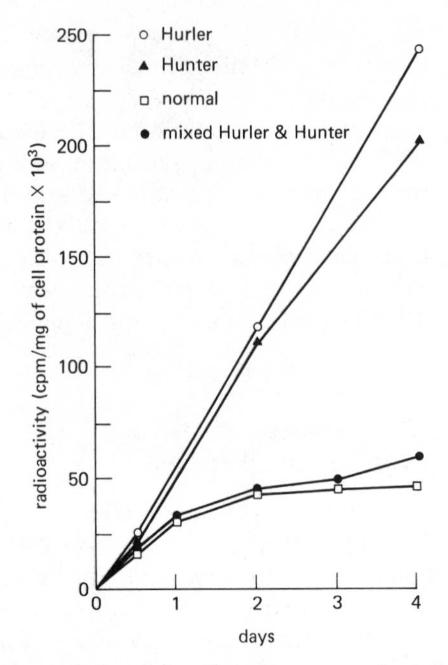

Figure 11-6 Accumulation of radioactive mucopolysaccharide by Hurler, Hunter, and normal fibroblasts, as well as by a mixture of Hurler and Hunter fibroblasts. (*From E. R. Neufeld and J. C. Fratantoni, 1970* Science *169:141–146.* © *American Association for the Advancement of Science.*)

grown together, no accumulation occurs (Figure 11-6). Each type of cell is able to compensate for the metabolic deficit of the other, indicating not only that the defects involve different factors but also that normally these factors are released from the cells. Indeed, the Hurler and Hunter factors are found in plasma of normal persons.

The "factors" missing in Hurler's and Hunter's syndromes, as well as in some of the other MPS syndromes, have been shown to be enzymes. In each syndrome, a lysosomal enzyme is missing (Table 11-1). Normal mucopolysaccharides are synthesized, but they cannot be completely broken down, leading to accumulation in the lysosomes and interference in other cell processes.

Although the Hurler and Scheie syndromes are distinguishable clinically, the latter being much less severe, they involve the same enzyme defect. It is possible but not yet proved that these are due to homozygosity for different alleles at a single locus, the Hurler allele giving a more profoundly defective enzyme. If so, this would require that some enzyme activity be present in persons with the Scheie syndrome, at least under *in vivo* conditions.

The fact that the enzyme defects of the mucopolysaccharidoses are characteristic of many tissues, perhaps all, makes these diseases suit-

Table 11-1 **MUCOPOLYSACCHARIDOSES CLASSIFIED ACCORDING TO MC KUSICK (1978)***

In some instances, there is a range of clinical defect within a group, presumably reflecting a range of enzyme deficits. For example, in MPS I, Hurler syndrome has a marked deficiency of enzyme and severe clinical defect. Scheie syndrome is milder with less enzyme deficit. All are inherited as autosomal recessive disorders except MPS II, which is an x-linked recessive.

MPS type	Eponym	Substances excreted	Enzyme deficiency
I	Hurler/Scheie	dermatan sulfate heparan sulfate	α-L-iduronidase
II	Hunter	dermatan sulfate heparan sulfate	L-iduronide-2-sulfate sulfatase
IIIA	Sanfilippo A	heparan sulfate	*N*-sulfoglucosaminide sulfamidase
IIIB	Sanfilippo B	heparan sulfate	*N*-acetyl-α-D-glucosaminidase
IIIC	Sanfilippo C	heparan sulfate	α-glucosaminide *N*-acetyltransferase
IVA	Morquio A	keratan sulfate	galactosamine-6-sulfate sulfatase
IVB	Morquio B	keratan sulfate	β-galactosidase
VI	Maroteaux-Lamy	dermatan sulfate	aryl sulfatase B
VII	Sly	dermatan sulfate heparan sulfate	β-glucuronidase
VIII	Di Ferrante	keratan sulfate heparan sulfate	*N*-acetylglucosamine-6-sulfate sulfatase

* See also McKusick et al. (1978).

able for prenatal diagnosis. Both Hurler's and Hunter's syndromes have been diagnosed with cells obtained by amniocentesis, and presumably all could be so diagnosed once the enzyme assays are generally available.

Blocks in Degradative Pathways: Tay–Sachs Disease

One of the more prevalent genetic diseases involving a failure to degrade a normal body constituent is Tay–Sachs disease (TSD). TSD is an autosomal recessive disorder, and, like the mucopolysaccharidoses, is a lysosomal storage disease. Homozygotes appear normal at birth, but before one year of age they begin to show signs of central nervous system deterioration. There is progressive mental retardation, blindness, and loss of neuromuscular control. Death occurs usually at 3 to 4 years of age.

The TSD gene occurs in all populations, but it is rare except in Ashkenazi Jews. This is the group that, in recent centuries, has lived primarily in Eastern Europe. Most American Jews are of Ashkenazi descent. Among American Jewish populations, the frequency of the TSD gene has been estimated to be as high as $\frac{1}{60}$, giving a heterozygote frequency of $\frac{1}{30}$ and a homozygote frequency of 1 in 3600 births. The frequency of homozygous affected children in other populations is only about 1% as high.

Proper nerve function, and probably functions of other cells, depends on the synthesis of complex structures known as sphingolipids (Figure 11-7). Like most other cell constituents, the sphingolipids are constantly being broken down by catabolic processes and replaced by newly synthesized molecules. The degradative pathway of gangliosides is shown in Figure 11-8. The pathology of TSD is a consequence of the accumulation of a particular sphingolipid known as G_{M2} ganglioside in nerve cells. The G_{M2} ganglioside normally occurs in small quantities in the brain. In the brains of TSD patients, the levels may be increased 70-fold, forming intracellular inclusions and interfering with nerve function.

The biochemical basis for the block in degradation of G_{M2} ganglioside is the absence of the enzyme hexosaminidase A (Okada and O'Brien, 1969). On the basis of electrophoretic mobility and using artificial substrates, it is possible to show the presence in a variety of tissues of two enzymes that can hydrolyze off a terminal hexosamine, such as occurs in G_{M2} ganglioside. These two enzymes were labeled hexosaminidase A and B, and patients with TSD are deficient only in hexosaminidase A (Figure 11-9). Apparently it is the A form that acts in degradation of the G_{M2} ganglioside. Since the degradative enzymes act in sequence, the G_{M2} ganglioside cannot be degraded further in the absence of hexosaminidase A. There is no mechanism for excreting it undegraded, and it accumulates.

That the two enzymes are related is indicated by the rare variants of TSD known as type 2 (Sandhoff–Jatzkewitz disease). These patients are homozygous for a gene that causes both A and B forms of hexosa-

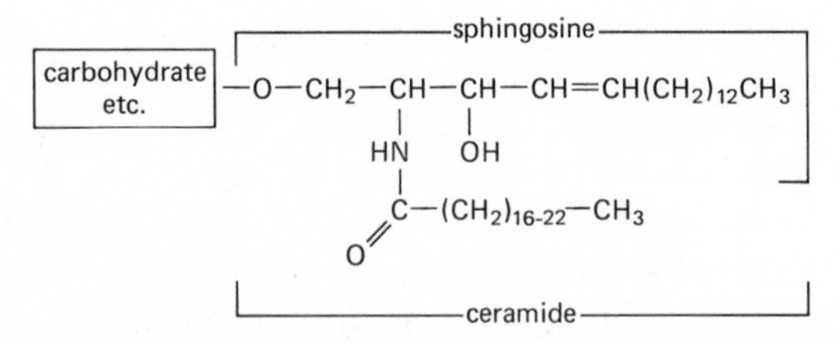

Figure 11-7 Structure of sphingolipids. All have a basic ceramide unit, which consists of a long chain fatty acid connected through an amide linkage to sphingosine. The various sphingolipids are distinguished primarily by the type of substances esterified to the 1-position of sphingosine. Gangliosides have carbohydrate side chains, including sialic acid.

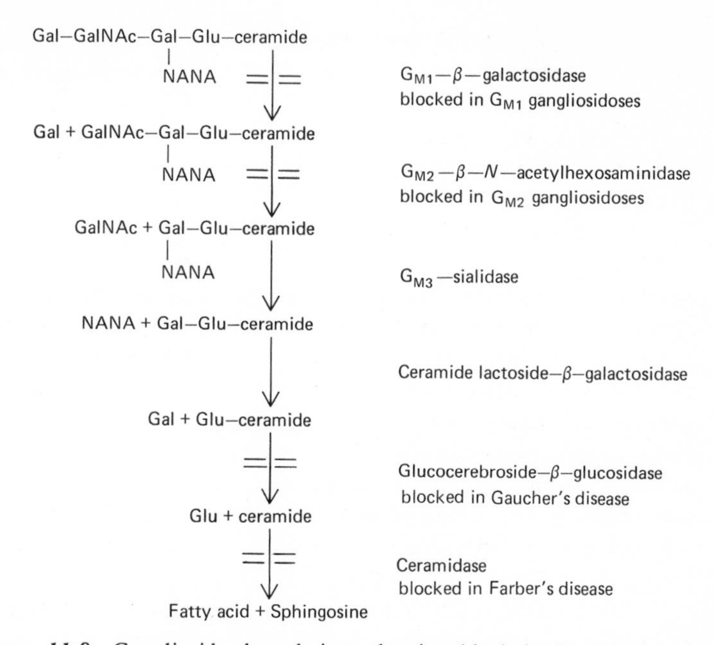

Figure 11-8 Ganglioside degradation, showing block in Tay–Sachs disease (G_{M2}-β-N-acetylhexosaminidase) and other related disorders.
Abbreviations: Gal, galactose; GalNAc, *N*-acetylgalactosamine; Glu, glucose; NANA, *N*-acetylneuraminic acid (sialic acid).

minidase to be absent. The accumulation of G_{M2} gangliosides is much more general, occurring especially in kidneys.

Immunological and chemical studies have shown the hexosaminidases A and B to consist of two kinds of polypeptide subunits, each under control of a separate genetic locus. The molecular formula for hexosaminidase A can be represented as $\alpha_2\beta_2$ and that for hexosaminidase B as β_4. Thus mutation at the α locus causes absence only of hexosaminidase A, but mutation at the β locus leads to absence of both. As

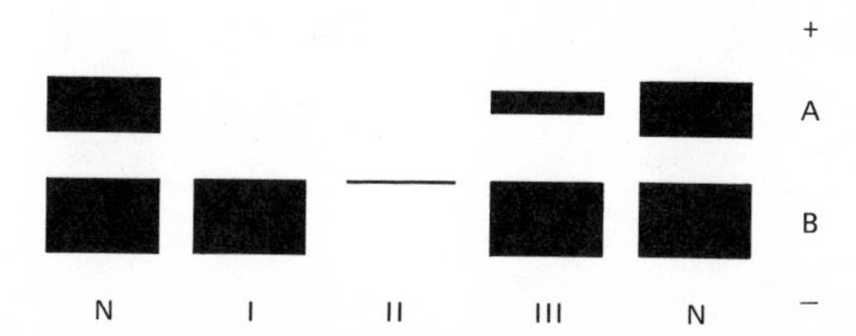

Figure 11-9 Starch gel electrophoretic patterns of hexosaminidase from liver homogenates of (N) normal controls, (I) Tay–Sachs disease, (II) Sandhoff's disease, (III) juvenile G_{M2} gangliosidosis. Hexosaminidase A is absent in I, both A and B are absent in II, and A is diminished in III. (*Redrawn from O'Brien, 1972.*)

in Hunter and Hurler syndromes, fibroblasts from TSD types 1 and 2 are able to complement each other when cultured together, since they involve two different defects.

A type 3 TSD has also been described in which hexosaminidase A activity is only partially reduced. Onset of symptoms is later than in the usual type 1 TSD (2 to 5 years of age as opposed to 8 to 10 months), and death occurs at 5 to 15 years of age.

Another rare mutation at the α locus functions normally *in vivo* but behaves abnormally in the laboratory. A family reported from Israel showed the phenotypically normal father of a Tay–Sachs patient to have zero activity of hexosaminidase A (Figure 11-10). He should therefore have been affected also. Further testing showed that he had one allele for TSD, as expected, and a second allele that gave no activity in the laboratory test but that functioned normally *in vivo*. Apparently the product of the second allele acts normally on natural substrates but has no activity on the artificial substrates used in the laboratory assay. Several similar examples have been observed, but fortunately these are very rare.

Although the pathology in TSD is especially noted in brain cells, the genetic defect is in all cells, and the deficiency of hexosaminidase A is readily demonstrated in other tissues, such as liver, kidney, plasma,

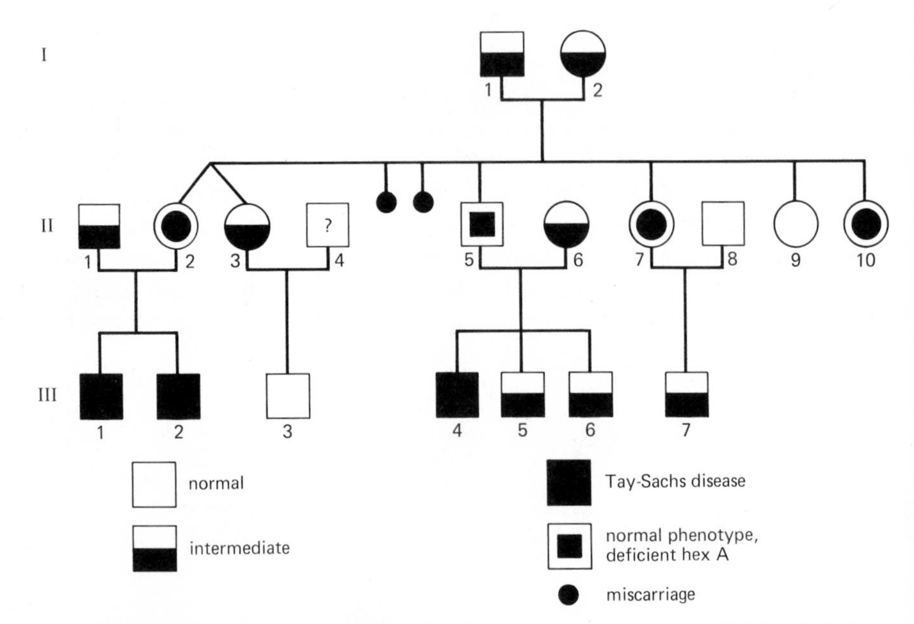

Figure 11-10 Pedigree of family with Tay–Sachs disease and an additional allele that produces a normal phenotype but that does not show hexosaminidase activity in laboratory tests. Presumably the gene product cannot react with the artificial substrate used in the laboratory assay, but it can act on the natural substrate. (*Redrawn from Navon et al., 1973.*)

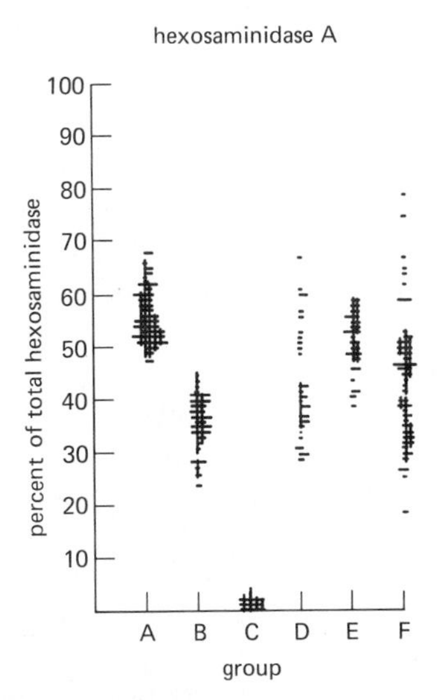

Figure 11-11 Hexosaminidase A activity in serum: (*a*) normal subjects; (*b*) heterozygotes for Tay–Sachs disease (parents of affected); (*c*) patients with Tay–Sachs disease; (*d*) relatives of Tay–Sachs patients; (*e*) patients with diabetes mellitus; (*f*) other hospitalized patients with various diseases. (*Reprinted by permission from J. S. O'Brien, S. Okada, A. Chen, D. L. Fillerup, 1970.* N. Engl. J. Med. *283:15.*)

leukocytes, and cultured fibroblasts. TSD can be diagnosed prenatally by the absence of hexosaminidase A activity in amniotic fluid and in cells cultured from amniotic fluid. Where both parents are known to be heterozygous for TSD, amniocentesis and testing for TSD in the fetus has become a frequent practice.

Most heterozygotes also can be recognized by their intermediate level of hexosaminidase A (Figure 11-11). The A form is more sensitive to heat denaturation than the B form, and the rate of loss of activity of a blood serum sample at high temperature is a measure of the amount of hexosaminidase A present. Such tests have been used in screening programs in Jewish populations to detect heterozygotes. In approximately 1 in 900 marriages between Jews, both persons would be heterozygous, with a risk of affected children.

An interesting question in the case of TSD is the reason for the high frequency of the gene in Ashkenazi Jews but not in other populations. A related observation is the relatively high frequency in the same group of some other similar degenerative diseases of the central nervous system—Gaucher's disease and Niemann–Pick disease—involving different primary gene defects. It has been suggested that persons heterozygous for the TSD gene may have some advantage over homozygous normal persons under certain conditions (Myrianthopoulos and Aronson, 1966). What this advantage might be—if it exists—is unknown.

Blocks Due to Defects in Coenzyme Binding: Homocystinuria and Cystathioninuria

A number of disorders are known that respond to large doses of a specific vitamin. Most of these have not been investigated at the molecular level, but, in a few cases, it seems likely that the primary defect is a structurally altered apoenzyme that cannot bind coenzyme with the same efficiency as wild-type apoenzyme. By administering large amounts of the specific vitamin (coenzyme), the binding is shifted to produce larger amounts of active apoenzyme-coenzyme complex.

Two of the better documented examples are the rare autosomal recessive traits homocystinuria and cystathioninuria. Cystathionine is an intermediate in the transfer of sulfur from methionine to cysteine (Figure 11-12). It is formed by the condensation of homocysteine and serine, catalyzed by the enzyme cystathionine synthase, and is cleaved by the enzyme cystathionase. Both enzymes require pyridoxal phosphate as a coenzyme.

In patients with deficient cystathionine synthase, homocysteine accumulates and is converted to homocystine and excreted. Similarly, in cystathionase deficiency, cystathionine accumulates and is excreted. In both disorders, some but not all patients become essentially normal if massive doses of vitamin B_6 (pyridoxine) is added to the diet. Pyridoxal phosphate (the active form of pyridoxine) is a coenzyme in many different reactions. The effect of deficiency of pyridoxine is well known. None of the other reactions that depend on pyridoxal phosphate are diminished in homocystinuria or cystathioninuria, so that normal supplies of pyridoxal phosphate appear to be available. Therefore it is con-

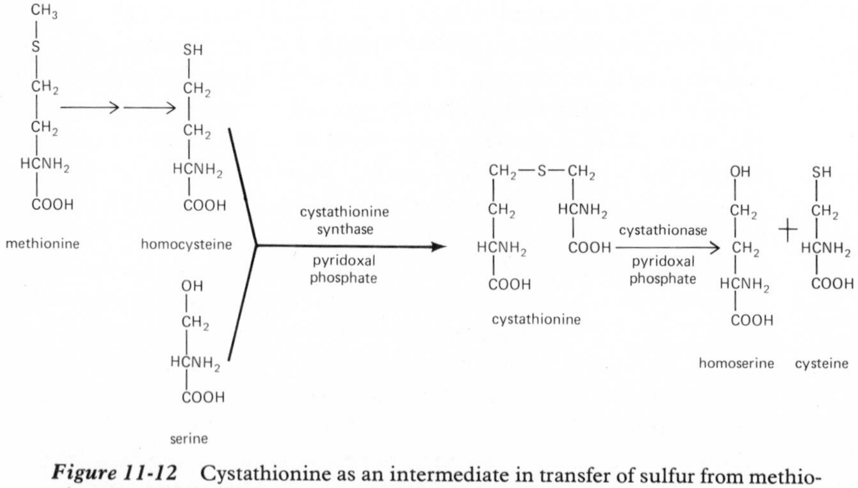

Figure 11-12 Cystathionine as an intermediate in transfer of sulfur from methionine to cysteine. The enzymes cystathionine synthase and cystathionase require pyridoxal phosphate as a coenzyme.

cluded that the apoenzymes are unable to function properly in the presence of normal amounts of pyridoxal phosphate but can if excess amounts are present. The most likely explanation is lowering of the coenzyme binding constant of cystathionine synthase or cystathionase in affected persons.

These two disorders are illustrative of what undoubtedly exists in less severe form for many enzymes. The concept that all normal persons have the same quantitative requirements for vitamins is not supported by experimental evidence. There likely are many persons whose needs for one or more vitamins are greater than average, owing to specific inherited enzymes with low coenzyme binding constants. Proper nutrition may be as individual as other traits.

RECEPTOR AND TRANSPORT DEFICIENCIES

Many proteins function, not as catalysts in the conversion of a substrate to a different product, but in the binding and transport of substances. Hemoglobin is the prime example of a transport protein, carrying O_2 from one location to another, where it is released unchanged. As was discussed in Chapter 10, mutations of the hemoglobin molecule influence this function through a variety of mechanisms. The deficiency of dihydrotestosterone-binding activity has been identified as the cause of testicular feminization (Chapter 7). Wilson's disease is associated with a deficiency of ceruloplasmin, the copper-binding plasma protein, although the primary nature of that mutation is not established.

The transport of substances across membranes often involves an initial binding to specific receptors on the membrane surface, followed by metabolic transport across the membrane, with release on the other side. Energy is typically consumed in these reactions, especially if the transfer is against a concentration gradient. Study of the membrane proteins involved in binding is technically difficult, and very little is known about inherited structural variations. The binding of substances to receptors is an important metabolic function, and defects in receptor activity are likely to become more important in metabolic disease (Brown and Goldstein, 1976).

Cystinuria

Cystinuria may well be the first inherited metabolic disorder to have been studied scientifically. As early as 1810, Wollaston studied cystine stones obtained from urinary bladders of patients, presumably with cystinuria. Indeed, the amino acid cystine was first discovered in such stones. Cystinuria is a rare autosomal recessive disease. Persons homozygous for the cystinuria genes have defective reabsorption of cystine, lysine, ornithine, and arginine in the renal tubules. As a result, these substances accumulate in urine. The condition is benign except for the low solubility of cystine, especially under acid conditions. Crystals of cystine precipitate out of solution in the bladder, and calculus formation may occur. Lysine, ornithine, and arginine are quite soluble and do not form renal calculi. One form of treatment is by manipulation of the diet to maintain an alkaline urine.

The enzymatic basis for cystinuria is unknown, but presumably the four amino acids involved share a common pathway for reabsorption from the tubular lumen. That the defect is not limited to kidney cells is indicated by a decrease in absorption of cystine from the intestine in some cases.

If one examines relatives of cystinurics, those in some families are found not to have elevated amino acid excretion, whereas in others the first degree relatives have elevated excretion of lysine and cystine (but not arginine and ornithine). Accordingly, there must be at least two genetic forms of cystinuria, one in which the disorder is mildly expressed in heterozygotes and the other in which there is no expression. The "homozygous" expression of cystinuria in children whose parents are of different types suggests that the two forms are allelic.

Familial Hypercholesterolemia

Regulation of the cholesterol level in blood is very complex. Fat intake is clearly a factor for most persons, although the relationship between quantity and quality of diet to cholesterol synthesis is not fully understood. In many studies, heredity has appeared to be a factor also. In some families, hypercholesterolemia is inherited as an autosomal dominant trait, affected persons usually developing coronary disease in the third to sixth decade of life. The rare persons homozygous for the high cholesterol gene may develop levels of cholesterol greater than 800 mg per 100 ml plasma, even as children. Death from myocardial infarction often occurs before the age of 30 and has been recorded in children less than 10 years old.

Using fibroblast cultures from patients homozygous for this disorder, Goldstein and Brown (1973) demonstrated that the defect lies in the regulation of 3-hydroxy-3-methylglutaryl coenzyme A reductase (HMG CoA reductase). Persons with familial hypercholesterolemia have elevated levels of this enzyme. More importantly, in contrast to normal controls, fibroblasts from such persons do not shut off synthesis of the enzyme in the presence of high levels of plasma cholesterol (Figure 11-13). The defect seems not to be in the loss of ability to be inhibited by cholesterol. Indeed, fibroblasts from homozygous persons do reduce synthesis of the enzyme in the presence of unbound cholesterol.

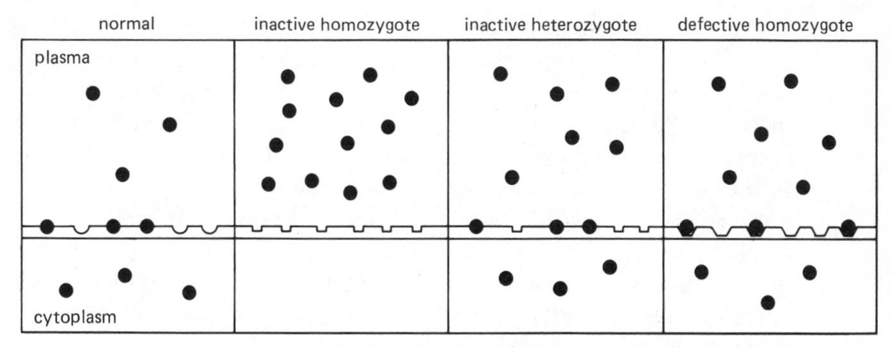

Figure 11-13 Diagram showing normal binding of low-density lipoprotein (LDL) to cell membrane compared to lack of binding by inactive receptors and diminished binding by defective receptors, observed in familial hypercholesterolemia. Esterified cholesterol is bound by LDL and is transported into the cell by binding of the LDL to specific receptors. Inside the cell, the cholesterol ester is hydrolyzed to free cholesterol, the level of which regulates the amount of cholesterol synthesized. Inactive receptors do not bind LDL; hence synthesis of cholesterol is not regulated by plasma levels. In heterozygotes for inactive receptors or in homozygotes for defective receptors, LDL is transported into the cell, but it takes a much higher plasma level than normal to achieve a given intracellular level.

Rather, it appears that a receptor at the surface of the fibroblast membrane fails to bind the normal plasma carrier of cholesterol, the low density lipoprotein (LDL). There are in fact at least two mutations that cause familial hypercholesterolemia, one in which the receptors are completely inactive and one in which they are defective but with low activity (reviewed in Fredrickson et al., 1978). These are thought to be alleles.

It is estimated that 0.1 to 0.2% of the population has a dominant gene for familial hypercholesterolemia. This accounts for only a portion of inherited hypercholesterolemia, and other forms are normal in the regulation of HMG CoA reductase. The recognition of the gene defect in this one entity permits it to be separated from the heterogeneous group of hypercholesterolemias. Since the gene defect is expressed in fibroblast cultures, it is potentially diagnosable in children and in fetuses.

INHERITED NORMAL VARIATION

Different alleles may produce enzymes that are functionally distinguishable but nevertheless normal. Many examples are known. Those cited here are chosen because they illustrate especially well some of the principles of genetic control of metabolism.

Geneticists prefer to deal with discontinuous traits. The ability to classify rather than to measure is not just a matter of convenience. Discontinuous traits often correspond to different genotypes, and environmental influence can be ignored. Continuous traits frequently are more complex and reflect environmental influences. The quantitative variation that is distributed according to the familiar "bell-shaped curve" is not necessarily devoid of genetic interest. People may well fall at a particular point on the curve because of their genotype. This can sometimes be shown by intrafamilial correlations. Occasionally one is fortunate enough to identify discontinuous attributes associated with different parts of the activity curve.

Acid Phophatase

The enzyme acid phosphatase (AcP) in red cells shows a wide range of activities among persons, but if the distribution of activities for a population is plotted, it is clearly continuous. The importance of genetic variation in determining the activity of AcP became clear with the discovery of electrophoretic variants (Figure 11-14). In the British population, six phenotypes were found, corresponding to the six genotypes produced by three alleles (Hopkinson and co-workers, 1963). Type A persons, homozygous for P^a, constitute 13% of the British population;

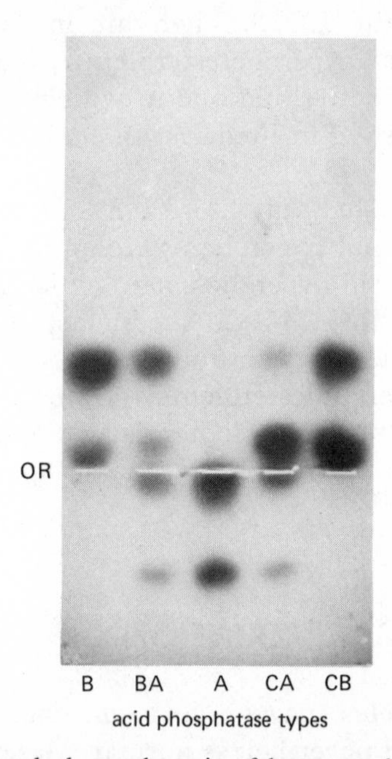

Figure 11-14 Starch gel electrophoresis of human red cell acid phosphatase showing the five common phenotypes found in most populations. Hemolysates are inserted into the gel at OR and migrate toward the anode (top) or cathode (bottom) depending on the net charge at the *p*H of the buffer. (*From Karp and Sutton, 1967.*)

Table 11-2 **RED CELL ACID PHOSPHATASE ACTIVITIES ASSOCIATED WITH EACH PHENOTYPE***

Activities are micromoles *p*-nitrophenol released from *p*-nitrophenyl phosphate per 30 minutes per gram Hb.

Type	Mean activity	Standard deviation
A	122.4	16.8
BA	153.9	17.3
B	188.3	19.5
CA	183.8	19.8
CB	212.3	23.1

* From Spencer and co-workers, 1964.

type B (P^bP^b), 36%; type C (P^cP^c), approximately 0.2%; type BA (P^bP^a), 43% type CA (P^cP^a), 3%; and type CB (P^cP^b), 5%. All other populations are polymorphic for acid phosphatase, although the frequencies vary, and P^c is very rare in some.

Spencer and co-workers (1964) measured the activity associated with each phenotype and found distinct differences (Table 11-2). If one assumes that each allele contributes a given amount of activity and that these are additive, results consistent with Table 11-2 are obtained, with P^a contributing approximately 60 units, P^b 94 units, and P^c 120 units. In this instance, there appears to be no interaction between alleles in producing the final amount of enzyme. In Figure 11-15, the means and standard deviations associated with each phenotype are plotted, the area of each plot being weighted according to the frequency of that phenotype in the population. If these are added together, a distribution for the entire population is obtained that matches closely the observed distribution. Thus, in the case of acid phosphatase, a continuous distribution is seen to be the sum of a series of distributions, each representing combinations of Mendelian genes. This is probably true for many other continuous variables, but only in a few instances has a qualitative difference been detected that permits resolution of the population into separate genotypes.

PHARMACOGENETICS

The widespread use of drugs in therapy has uncovered a number of genetic variations in response to drugs. In some instances these variations are of little significance, but in others appropriate therapeutic levels can

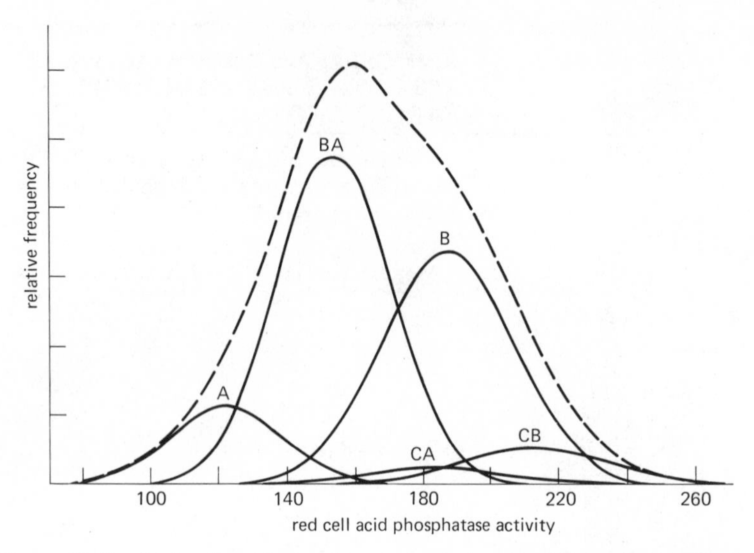

Figure 11-15 Red cell acid phosphatase levels in a British population (broken line) and in the various phenotypes. The latter are weighted according to their relative frequencies in the population. (*From H. Harris,* Proc. Roy. Soc., *B. 164: 298–310, 1966.*)

be maintained only if the physician is aware of individual differences in drug metabolism. The study of genetic variation in response to drugs—pharmacogenetics—is still in its infancy but has already been established as an important area of medicine (Motulsky, 1964; La Du and Kalow, 1968; Vesell, 1973).

Glucose-6-Phosphate Dehydrogenase Deficiency

The most widespread and widely studied variation in drug response is that due to glucose-6-phosphate dehydrogenase (G6PD) deficiency. During World War II, the antimalarial primaquine was administered to thousands of U.S. servicemen returning from malarial areas of the Pacific. Although adverse reactions had not been significant in prior testing of primaquine, severe hemolytic episodes were induced in a large number of persons, primarily in Negroes. Ultimately this primaquine sensitivity was traced to a deficiency of reduced glutathione in red cells, causing them to be unable to cope with the challenge of primaquine as well as many other drugs, including acetanilid, some sulfa drugs, and nitrofurantoin. Exposure to naphthalene also causes hemolysis.

The level of reduced glutathione in red cells depends on the activity of the enzyme glutathione reductase and on the supply of NADPH, its reduced coenzyme. As shown in Figure 11-16, the supply of NADPH is maintained by the dehydrogenation of glucose-6-phosphate. Prima-

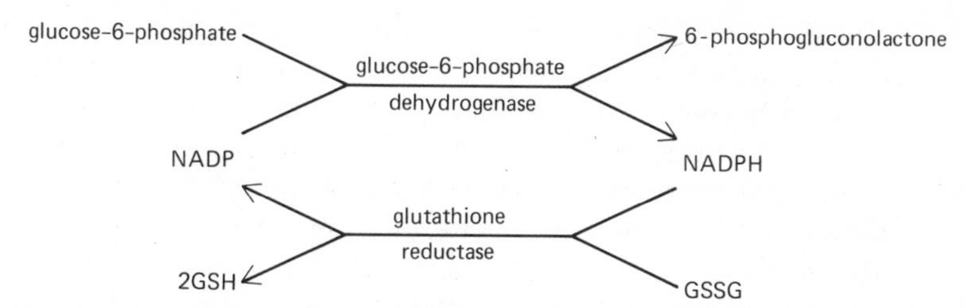

Figure 11-16 Initial step in metabolism of glucose-6-phosphate via the pentosephosphate pathway to give NADPH, which in turn is essential to convert glutathione (GSSG) into reduced glutathione (GSH).

quine sensitivity is due to a deficiency of this reaction because of decreased activity of G6PD. A condition common in some Mediterranean populations, favism, is also due to G6PD deficiency, but in these populations the precipitating factor usually is the ingestion of fava beans.

G6PD deficiency is inherited as an X-linked trait. Hemizygous males therefore are either affected or normal, depending on which allele they have. Heterozygous females are usually intermediate in the degree of deficiency and the response to drugs. However, X-chromosome inactivation appears to occur sufficiently early in development so that by chance a heterozygous female may have a preponderance of sensitive cells or conversely a preponderance of normal cells. Therefore it is not always possible to judge the genotype of a female solely on the basis of the phenotype. The high frequency of the G6PD deficiency allele in American blacks, approximately 15%, leads to a substantial number of homozygous deficient females.

The defect in G6PD deficiency is due largely to instability of the variant enzyme. Many different variants have been identified, primarily by electrophoretic mobility. Very often these first are investigated because of clinical problems resulting from lack of enzyme stability. Indeed some 70 structural variants are known, and additional variants have not been fully studied. Certain variants characterize particular populations and are useful as markers (Chapter 23). In tissues with rapid enzyme turnover, the instability of the variants is of lesser importance, since new enzyme is constantly being made. In red cells, where replacement does not occur during the 100+ days of existence, small differences in stability may have pronounced effects if the activity of G6PD drops too low before the end of the normal red cell life span.

Cholinesterase

One of the earliest examples of inherited deficit in drug metabolism was noted in the case of succinylcholine. This drug is a strong muscle relaxant, interfering with acetylcholinesterase, the enzyme that breaks down acetylcholine, the synaptic transmitter for many nerve cells. Succinylcholine is rapidly destroyed by plasma cholinesterase (sometimes called pseudocholinesterase), so that the effect is very short-lived, normally no more than three or four minutes. For this reason, succinylcholine has been used extensively in electroshock therapy, in order to protect patients from violent muscle contractions during the short period of the shock.

In 1952, soon after use of succinylcholine began, it was reported that rare patients, approximately 1 in 2500, fail to recover from the effects of succinylcholine in the expected time. The patients have prolonged apnea and must be supported by artificial respiration for an hour or more. The defect was shown to be due to homozygosity for a gene resulting in markedly deficient plasma cholinesterase activity (Kalow and Genest, 1957). The deficient, or atypical, form of cholinesterase has much lower substrate affinity. Consequently, at therapeutic doses of succinylcholine, almost no destruction occurs, and the drug continues to exert its inhibitory effect on acetylcholinesterase.

Although homozygotes for the atypical gene are rare, their existence emphasizes dramatically the variability in response to specific drugs that may occur. An additional allele at the cholinesterase locus has been detected, but this does not cause marked deficiency of cholinesterase activity. Rather, the enzyme is not inhibited by fluoride to the extent that the normal allele is. Homozygosity for this allele is associated with only moderate sensitivity to succinylcholine. Other alleles include a rare "silent" allele, which is associated with complete loss of cholinesterase activity.

Isoniazid Inactivation

The drug isoniazid has been used primarily for the treatment of tuberculosis. It is inactivated by acetylation, in which form it is rapidly excreted (Figure 11-17). Two phenotypes are commonly identified on the basis of the rapidity of isoniazid inactivation (Figure 11-18). These correspond to a dominant allele for rapid inactivation and a recessive for slow inactivation. Using a somewhat different technique, S. Sunahara

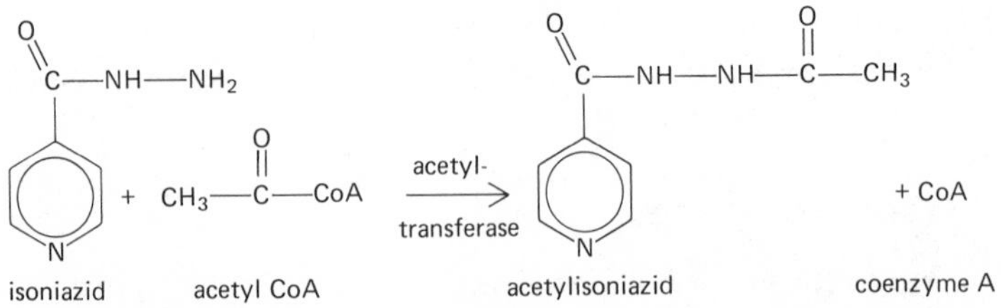

isoniazid (pyridine ring–C(=O)–NH–NH_2) + acetyl CoA (CH_3–C(=O)–CoA) —acetyl-transferase→ acetylisoniazid (pyridine ring–C(=O)–NH–NH–C(=O)–CH_3) + CoA (coenzyme A)

Figure 11-17 Acetylation of isoniazid (isonicotinic acid hydrazide) catalyzed by the liver enzyme acetyltransferase.

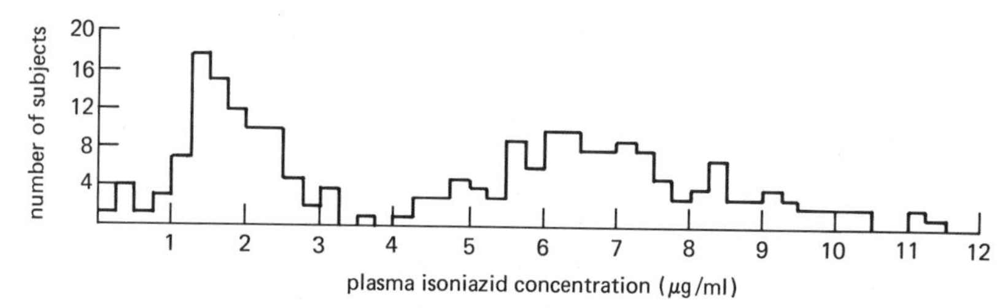

Figure 11-18 Distribution of plasma isoniazid levels 6 hours after oral administration of a standard dose (40 mg/kg "metabolically active mass"). Subjects were Caucasians. The distribution is clearly bimodal, the group on the left being rapid inactivators. (*From Evans and co-workers, 1961.*)

was able to distinguish between homozygous and heterozygous rapid inactivators. The frequency of the rapid inactivation allele is much higher in Orientals than in Europeans.

It is not clear to what extent proper management of isoniazid therapy should take into consideration whether a person is a fast or slow inactivator. If the objective is to maintain a given blood level of isoniazid, therapy may be less effective if a regimen is adopted for a fast inactivator that is appropriate for a slow inactivator. The drug procainamide, which induces antinuclear antibodies and lupus erythematosus after prolonged use, has been shown to cause these toxic effects much faster in slow acetylators (inactivators) than in fast (Woosley et al., 1978).

Increasingly, as additional variability in drug metabolism is identified, it may become necessary to do genetic studies on patients in order to design the most effective drug therapy.

SUMMARY

1. Enzymes function as catalysts, lowering the energy barrier for a particular reaction among the many that a substrate might be able to undergo. This usually involves a chemical reaction between the coenzyme, which is bound to the apoenzyme, and the substrate. The apoenzyme greatly enhances the activity of the coenzyme and narrows the substrate specificity.

2. Enzymes may be inhibited by many substances, including products of the metabolic pathway in which they function (feedback inhibition). Often binding of the metabolic product is at a site different from the catalytic site, in which case it is called an allosteric site.

3. Loss of activity of an enzyme results in a metabolic block. Albinism is an example of a block in a biosynthetic pathway.

4. Blocks in degradative pathways of small metabolites are exemplified by phenylketonuria (PKU) and alkaptonuria. Accumulation of phenylalanine in patients with PKU causes severe mental retardation which can be prevented if homozygotes are detected as newborns and placed on low phenylalanine diets. Alkaptonuria is benign since blocked metabolites do not accumulate.

5. The mucopolysaccharidoses are examples of blocks in breakdown of large molecular weight components, the mucopolysaccharides, which accumulate in the cells. A variety of lysosomal enzymes is involved in this breakdown, and inherited defects are known for many of them.

6. Tay–Sachs disease (TSD) results from a block in the breakdown of gangliosides due to deficiency of the enzyme hexosaminidase A. Related diseases are TSD 2 (Sandhoff disease), in which both A and B forms of hexosaminidase are absent, and TSD 3 (juvenile TSD), in which hexosaminidase A is present in reduced amount.

7. Inherited changes in apoenzyme structure may alter the affinity with which coenzyme is bound, therefore influencing enzyme activity. The enzyme cystathionase requires pyridoxal phosphate as coenzyme. In cystathioninuria, activity of this enzyme is low, but it can be increased with high levels of dietary pyridoxine, suggesting that coenzyme binding is diminished. Similar responses are sometimes observed in homocystinuria, in which there is deficiency of the enzyme cystathionine synthase.

8. Enzymes also function in transport of metabolites across cell membranes. In cystinuria, the transport system is defective, with reduced reabsorption of cystine, lysine, ornithine, and arginine from the glomerular filtrate. Familial hypercholesterolemia is due to defective receptors for plasma low-density lipoprotein (LDL), which cannot transport LDL into the cell. This causes lack of regulation of cholesterol synthesis.

9. Normal inherited variation in activity is known for many enzymes. An example is red cell acid phosphatase, for which there are three common electromorphic alleles. It has been shown that the quantitative variation in the population reflects the quantitative differences associated with each allele.

10. Pharmacogenetics is concerned with inherited variation in response to drugs. Persons with the X-linked trait glucose-6-phosphate dehydrogenase deficiency may have hemolytic crises after exposure to many drugs. Rare persons with deficiency of plasma cholinesterase do not destroy the muscle relaxant succinylcholine with normal rapidity and may have prolonged apnea. In this disorder, the enzyme is present but has a low affinity for succinylcholine. Isoniazid is an example of a drug that some persons acetylate rapidly and others slowly, the difference being under genetic control.

REFERENCES AND SUGGESTED READING

BEADLE, G. W. 1945. Biochemical genetics. *Chem. Rev.* 37: 15–96.

BERRY, H., SUTHERLAND, B., AND GUEST, G. M. 1957. Phenylalanine tolerance tests on relatives of phenylketonuric children. *Am. J. Hum. Genet.* 9: 310–316.

BEUTLER, E. 1978. Glucose-6-phosphate dehydrogenase deficiency. In Stanbury et al. (1978). pp. 1430–1451.

BEUTLER, E., KUHL, W., AND COMINGS, D. 1975. Hexosaminidase isozyme in type 0 G_{M2} gangliosidosis (Sandhoff–Jatzkewitz disease). *Am. J. Hum. Genet.* 27: 628–638.

BROWN, M. S., AND GOLDSTEIN, J. L. 1976. New directions in human biochemical genetics: Understanding the manifestations of receptor deficiency states. *Prog. Med. Genet.* n.s. 1: 103–119.

EVANS, D. A. P., MANLEY, K. E., AND MC KUSICK, V. A. 1960. Genetic control of isoniazid metabolism in man. *Br. Med. J.* 2: 485.

EVANS, D. A. P., STOREY, P. B., AND MC KUSICK, V. A. 1961. Further observations on the determination of the isoniazid inactivator phenotype. *Bull. Johns Hopkins Hosp.* 108: 60–66.

FREDRICKSON, D. S., GOLDSTEIN, J. L., AND BROWN, M. S. 1978. The familial hyperlipoproteinemias. In Stanbury et al. (1978). pp. 604–655.

GARROD, A. E. 1909. *Inborn Errors of Metabolism.* (Reprinted with a supplement by H. Harris, 1963.) London: Oxford Univ. Press, 207 pp.

GOLDSTEIN, J. L., AND BROWN, M. S. 1973. Familial hypercholesterolemia: Identification of a defect in regulation of 3-hydroxy-3-methylglutaryl coenzyme A reductase activity associated with overproduction of cholesterol. *Proc. Natl. Acad. Sci.* (U.S.A.) 70: 2804–2808.

HARRIS, H. 1975. *The Principles of Human Biochemical Genetics,* 2nd Ed. Amsterdam/Oxford: North-Holland, 473 pp.

HARRIS H., MITTWOCH, U., ROBSON, E. B., AND WARREN, F. L. 1955. Phenotypes and genotypes in cystinuria. *Ann. Hum. Genet.* 20: 57–91.

HIRSCHHORN, R., AND WEISSMAN, G. 1976. Genetic disorders of lysosomes. *Prog. Med. Genet.* n.s. 1: 49–101.

HOPKINSON, D. A., SPENCER, N., AND HARRIS, H. 1963. Red cell acid phosphatase variants; a new human polymorphism. *Nature* (Lond.) 199: 969.

KALOW, N., AND GENEST, K. 1957. A method for the detection of atypical forms of human cholinesterase: determination of dibucaine numbers. *Can. J. Biochem. Physiol.* 35: 339.

KARP, G. W. Jr, AND SUTTON, H. E. 1967. Some new phenotypes of human red cell acid phosphatase. *Am. J. Hum. Genet.* 19: 54–62.

KAUFMAN, S., BERLOW, S., SUMMER, G. K., MILSTIEN, S., SCHULMAN, J. D., ORLOFF, S., SPIELBERG, S., AND PUESCHEL, S. 1978. Hyperphenylalaninemia due to a deficiency of biopterin. *N. Engl. J. Med.* 299: 673–679.

KAUFMAN, S., HOLTZMAN, N. A., MILSTIEN, S., BUTLER, I. J., AND

KRUMHOLZ, A. 1975. Phenylketonuria due to a deficiency of dihydropteridine reductase. *N. Engl. J. Med.* 293: 785–790.

KELLEY, W. N., AND SMITH, L. J. JR. 1978. Hereditary orotic aciduria. In Stanbury et al. (1978). pp. 1045–1071.

KOLODNY, E. H. 1976. Lysosomal storage diseases. *N. Engl. J. Med.* 294: 1217–1220.

LA DU, B. N. 1972. The isoniazid and pseudocholinesterase polymorphisms. *Fed. Proc.* 31: 1276–1285.

LA DU, B. N., AND KALOW, W. (Eds.). 1968. Pharmacogenetics. *Ann. N.Y. Acad. Sci.* 151: 691–1001.

LEHMANN, H., AND LIDDELL, J. 1972. The cholinesterase variants. In *The Metabolic Basis of Inherited Disease,* 3rd Ed. (J. B. Stanbury, J. B. Wyngaarden, D. S. Fredrickson, Eds.) New York: McGraw-Hill, pp. 1730–1736.

LEHMANN, H., SILK, E., AND LIDDELL, J. 1961. Pseudocholinesterase. *Br. Med. Bull.* 17: 230.

MAC CREADY, R. A., AND LEVY, H. L. 1972. The problem of maternal phenylketonuria. *Am. J. Obstet. Gynecol.* 113: 121.

MC KUSICK, V. A. 1978. *Mendelian Inheritance in Man,* 5th Ed. Baltimore: Johns Hopkins U. Press, 976 pp.

MC KUSICK, V. A., NEUFELD, E. F., AND KELLY, T. E. 1978. The mucopolysaccharide storage diseases. In Stanbury et al. (1978). pp. 1282–1307.

MOTULSKY, A. G. 1964. Pharmacogenetics. *Prog. Med. Genet.* 3: 49–74.

MODIANO, G. 1976. Genetically determined quantitative protein variations in man, excluding immunoglobulins. *Atti della Accademia Naz. dei Lincei,* Memorie Sc. fisiche, ecc., Series 8, Vol. 13, Sess. 3, pp. 55–437.

MYRIANTHOPOULOS, N. C., AND ARONSON, S. M. 1966. Population dynamics of Tay–Sachs disease. I. Reproductive fitness and selection. *Am. J. Hum. Genet.* 18: 313–327.

NAVON, R., PADEH, B., AND ADAM, A. 1973. Apparent deficiency of hexosaminidase A in healthy members of a family with Tay–Sachs disease. *Am. J. Hum. Genet.* 25: 287–293.

NEUFELD, E. F., AND FRATANTONI, J. C. 1970. Inborn errors of mucopolysaccharide metabolism. *Science* 169: 141–146.

O'BRIEN, J. S. 1978. The gangliosidoses. In Stanbury et al. (1978). pp. 841–865.

O'BRIEN, J. S., OKADA, S., CHEN, A., AND FILLERUP, D. L. 1970. Tay–Sachs disease: detection of heterozygotes and homozygotes by serum hexosaminidase assay. *N. Engl. J. Med.* 283: 15–20.

OKADA, S., AND O'BRIEN, J. S. 1969. Tay–Sachs disease: generalized absence of a beta-D-*N*-acetylhexosaminidase component. *Science* 165: 698–700.

ROSENBERG, L. E. 1976. Vitamin-responsive inherited metabolic disorders. *Adv. Hum. Genet.* 6: 1–74.

SCRIVER, C. R., AND HECHTMAN, P. 1970. Human genetics of membrane transport with emphasis on amino acids. *Adv. Hum. Genet.* 1: 211–274.

SEGAL, S. 1976. Disorders of renal amino acid transport. *N. Engl. J. Med.* 294: 1044–1051.

SPENCER, N., HOPKINSON, D. A., AND HARRIS, H. 1964. Quantitative differences and gene dosage in the human red cell acid phosphatase polymorphism. *Nature* (Lond.) 201: 299.

STANBURY, J. B., WYNGAARDEN, J. B., AND FREDRICKSON, D. S. (Eds.). 1978. *The Metabolic Basis of Inherited Disease,* 4th Ed. New York: McGraw-Hill, 1862 pp.

SUNAHARA, S., URANO, M., AND OGAWA, M. 1961. Genetical and geographic studies on isoniazid inactivation. *Science* 134: 1530.

SUTTON, H. E., AND WAGNER, R. P. 1975. Mutation and enzyme function in humans. *Annu. Rev. Genet.* 9: 187–212.

TOURIAN, A. Y., AND SIDBURY, J. B. 1978. Phenylketonuria. In Stanbury et al. (1978). pp. 240–255.

VESELL, E. S. 1973. Advances in pharmacogenetics. *Prog. Med. Genet.* 9: 291–367.

WITKOP, C. J. JR., NANCE, W. E., RAWLS, R. F., AND WHITE, J. G. 1970. Autosomal recessive oculocutaneous albinism in man: evidence for genetic heterogeneity. *Am. J. Hum. Genet.* 22: 55–74.

WITKOP, C. J. JR., QUEVEDO, W. C. JR., AND FITZPATRICK, T. B. 1978. Albinism. In Stanbury et al. (1978). pp. 283–316.

WOOSLEY, R. L., DRAYER, D. E., REIDENBERG, M. M., NIES, A. S., CARR, K., AND OATES, J. A. 1978. Effect of acetylator phenotype on the rate at which procainamide induces antinuclear antibodies and the lupus syndrome. *N. Engl. J. Med.* 298: 1157–1159.

WUU, K. D., AND KROOTH, R. S. 1968. Dihydroorotic acid dehydrogenase activity of human diploid cell strains. *Science* 160: 539–541.

REVIEW QUESTIONS

1. Define:

catalyst	competitive inhibition	allosteric site
coenzyme	noncompetitive inhibition	metabolic block
apoenzyme	feedback inhibition	lysosomes
active site	feedback repression	enzyme induction
		pharmacogenetics

2. Discuss the ways in which two disorders can be distinguished genetically even though they are similar clinically.

3. Consider two persons affected with an inherited disease, in both cases due to deficiency of the same enzyme. Yet one responds to massive vitamin therapy and the other does not. How might this be explained?

4. Which of the following would you expect most often to be dominant or recessive: (a) metabolic blocks, (b) loss of feedback inhibition, (c) inactivation of a drug, (d) defect in membrane transport? Why?

5. How do you explain the existence of genetic variation for new drugs to which man has not been exposed previously?

6. A new drug is introduced that causes a hemolytic reaction in 10% of blacks, men and women being about equally affected. Why would you expect this not to be caused by G6PD deficiency?

7. Some of the mucopolysaccharidoses and sphingolipidoses are characterized by delayed onset. Yet the gene defect is present from birth. How is this explained?

CHAPTER TWELVE

IMMUNOGENETICS: ANTIBODY RESPONSE

The production of antibodies provides vertebrates with a powerful weapon against invasion by bacteria and other microorganisms. It may well be that acquisition of the immune system, with its associated increased likelihood of individual survival, permitted evolution to favor organisms with fewer offspring and longer periods of development rather than those organisms that must devote a major portion of their energy to producing vast numbers of young in order to circumvent the high probability that the young will not survive to reproduce. The details of how the present-day immune system evolved are still obscure, but enough is now known of the structure, function, and cellular control of immunoglobulins to suggest some fascinating possibilities.

THE NATURE OF ANTIBODIES

Functionally, antibodies are defined by their specific reactions with antigens. Chemically, they are classified as globulins, that is, they are part of the globulin fraction of plasma (being less soluble than albumin), and most of them move in an electric field with the major fraction of globulin designated γ. Because some antibodies are not, in fact, γ-globulins, as defined by electrophoretic mobility, the general term *immunoglobulins* (Ig) has come into widespread use as being more meaningful.

Antibodies are a prime example of adaptation in response to environmental stimuli. An organism not previously exposed to a foreign substance (antigen) does not have detectable levels of antibody that react with that antigen. Following exposure, by infection, injection, and so on, there is a build up of antibodies that react with the specific antigen but not with other unrelated antigens (see Figure 12-1). The

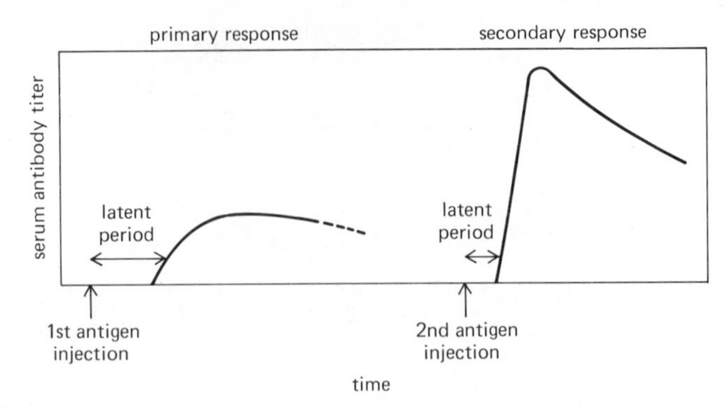

Figure 12-1 Appearance of antibody following a primary immunization compared to a secondary immunization. (*From Abramoff and LaVia, 1970.*)

level of antibody may taper off if exposure has not continued, but a second exposure (immunization) causes a prompt rise in antibody titer, usually to higher levels than previously. The immune system has thus become primed by its first exposure, and the machinery for producing antibodies is already assembled and ready to function.

Antigens are quite diverse in their structure. They may be cells of another person or animal, or they may be bacteria or viruses. They may be chemical substances of biological origin, or they may have been synthesized in a chemistry laboratory. Not all foreign substances function equally well as antigens, but synthetic materials that could not have been significant in vertebrate evolution may be just as antigenic as an invading organism that often must have posed a threat. Therefore, it was reasonable to suppose that the specificity of antibodies was derived from information supplied by the antigen, that somehow the antigen served as a template upon which the antibody was formed.

The idea became increasingly difficult to accommodate as evidence mounted from other systems that the function of genes is to specify the amino acid sequence of proteins. But antibodies are proteins, and it seemed unlikely that two systems of protein synthesis would exist, one deriving its information from genes, the other from any foreign substance that happened to be introduced. But how could the enormous diversity of antibody structures be encoded in genes, and would it be likely that evolution would perpetuate such a diversity, most of which probably would not be used by the organism? Although it is now clear that evolution *has* provided for the preexistence of a great diversity of antibody genes, the extent to which this diversity is of germinal or somatic origin is still in contention. But much has been learned about immunoglobulin structure, and from this as well as from genetic studies, some has been learned about the genes that direct immunoglobulin synthesis.

The Structure of Immunoglobulins

There are five classes of Ig molecules now known in man (Table 12-1). Separation of these classes is accomplished in part by solubility differences, but they can be distinguished also on the basis of their differences as antigens. The most prevalent in plasma is IgG. IgG passes readily across the placenta and therefore is especially important in passive protection of newborns, although it clearly is important to adults also. The next major Ig molecule is IgA, which appears to be expecially important in secretions. IgM has a molecular weight of 900,000. It is the first antibody manufactured in response to antigenic stimulation and therefore seems especially important in early immunity. Two minor classes are IgD and IgE, which occur at less than 0.5 mg/ml. The function of IgD is unknown, but IgE is involved in allergic reactions.

The basic structure of Ig molecules was proposed in 1959 by R. R. Porter. All the molecules consist of two kinds of polypeptide chains, designated light (L) and heavy (H). In all except IgM, there are two of each kind, giving the tetrameric formula L_2H_2. IgM has five such tetramers, with the formula $(L_2H_2)_5$. Under some conditions, IgA also forms higher polymers. The L chains are the same in all five classes, the differences residing in the H chains. L chains have a molecular weight of 22,000. H chains have molecular weights varying from 53,000 for γ chains to approximately 80,000 for ϵ chains.

There are two kinds of L chains, κ and λ, distinguished by antigenic differences associated with structural differences. Both κ and λ are present in all five Ig classes. There actually are three types of λ chains, KERN(+)/Oz(−), KERN(−)/Oz(+), and KERN(−)/Oz(−). These are detected because of different antigenicity. They depend on the presence of glycine at position 157 (KERN+) or serine (KERN−) and lysine at position 188 (Oz+) or arginine (Oz−). Each person has all three combinations, but not KERN(+)/Oz(+); therefore there must be at least three different λ loci.

Table 12-1 **THE IMMUNOGLOBULINS OF HUMAN PLASMA**

Group	Molecular weight	Plasma concentration	Molecular formula*	Molec. wt. of H chain
Ig G	150,000	8–16 mg/ml	$L_2\gamma_2$	53,000
Ig A	160,000	2–4	$L_2\alpha_2$	60,000
Ig M	900,000	0.5–2	$(L_2\mu_2)_5$	70,000
Ig D	180,000	0.5	$L_2\delta_2$	70,000
Ig E	200,000	0.1	$L_2\epsilon_2$	80,000

* The light chain L can be of either class κ or λ.

Table 12-2 **SUMMARY OF LIGHT AND HEAVY CHAIN TYPES AND SUBCLASSES AND THEIR COMBINATIONS INTO Ig MOLECULES**

Polypeptide chains		
Light*	Heavy	Ig molecules
κ	$\gamma1$	$\kappa_2(\gamma1)_2$; $(\lambda Oz+K-)_2(\gamma1)_2$; $(\lambda Oz-K-)_2(\gamma1)_2$; $(\lambda Oz-K+)_2(\gamma1)_2$
$\lambda Oz+K-$	$\gamma2$	$\kappa_2(\gamma2)_2$; and so on for all possible combinations of light and heavy chains
$\lambda Oz-K-$	$\gamma3$	
$\lambda Oz-K+$	$\gamma4$	
	$\alpha1$	
	$\alpha2$	
	$\mu1$	
	$\mu2$	
	δ	
	ϵ	

* K = KERN

The situation with H chains is somewhat more complex. IgG molecules have heavy chains designated γ chains, and there are at least four types of γ chains ($\gamma1$, $\gamma2$, $\gamma3$, and $\gamma4$), giving rise to four subclasses of IgG. The different subclasses are distinguished primarily by their antigenic differences, which are attributable to structural differences in the heavy chains. Similarly, the H chains of IgA, designated α chains, occur in two forms, $\alpha1$ and $\alpha2$. The μ chains that are found in IgM may be either $\mu1$ or $\mu2$. No subclasses of δ chains (IgD) or ϵ chains (IgE) are known. Each person then possesses the variety of Ig chains shown in Table 12-2. It is possible, indeed probable, that additional subclasses of Ig molecules exist. To demonstrate their presence ordinarily requires production of specific antisera against them, and this must be done in a species whose own immunoglobulins lack the particular antigenic configuration.

The light and heavy chains are associated by covalent disulfide bonds as shown in Figure 12-2. The exact position and number of disulfide bridges varies with the Ig class and subclass. In each case, however, there is association of L and H chains into a Y-shaped molecule (Figure 12-3). For IgM, five such units are associated by interaction of the bases of the Y shapes (the C-terminal ends) into a disc (Figure 12-4). On the basis of electron micrographs, it is clear that the variable regions of the antibodies are the antigen combining sites.

Additional polypeptide chains are found combined with the Ig molecules on occasion. A J chain of molecular weight ca. 15,000 is found in IgM and appears important to the polymerization process.

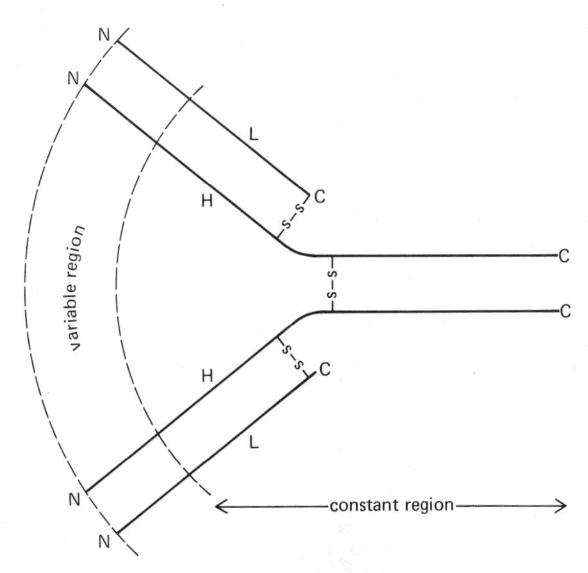

Figure 12-2 Representation of structure of immunoglobulin molecule. The overall shape is a Y-shape, with the N-terminal region forming the two arms of the Y. Additional disulfide bonds are present, the number and location varying with the subclass. Both the H (heavy) and L (light) chains occur with many different amino acid sequences in the variable region but with a limited number of sequences in the remaining "constant" region. The differences in the constant region are responsible for the classes and subclasses of Ig.

Heterogeneity of Immunoglobulins

Within the Ig classes described above, there is great heterogeneity of structure. A particular subclass of Ig molecules from a single person will contain a large variety of molecules, no one of which is in a concentration sufficient for characterization. This variety enables the organism to react with many different antigens and is of obvious benefit for survival. The diversity of immunoglobulins is at the crux of the genetic control of antibody structure. Does it represent the action of many genes? Have a relatively few genes diverged somatically, through mutation or some other mechanism, to produce a variety of genotypes not originally present in the zygote?

Better definition of the question would exist if the chemical differences among the Ig molecules in a single person were known. This has not proved possible, but insight is provided by studies of persons with multiple myeloma. Such persons produce large quantities of a single type of Ig. Apparently malignant transformation of a single Ig-producing cell occurs, and a large clone of Ig-producing cells results. Normally each cell appears to produce only one type of Ig. In multiple myeloma the entire clone produces the same Ig molecule. This permits

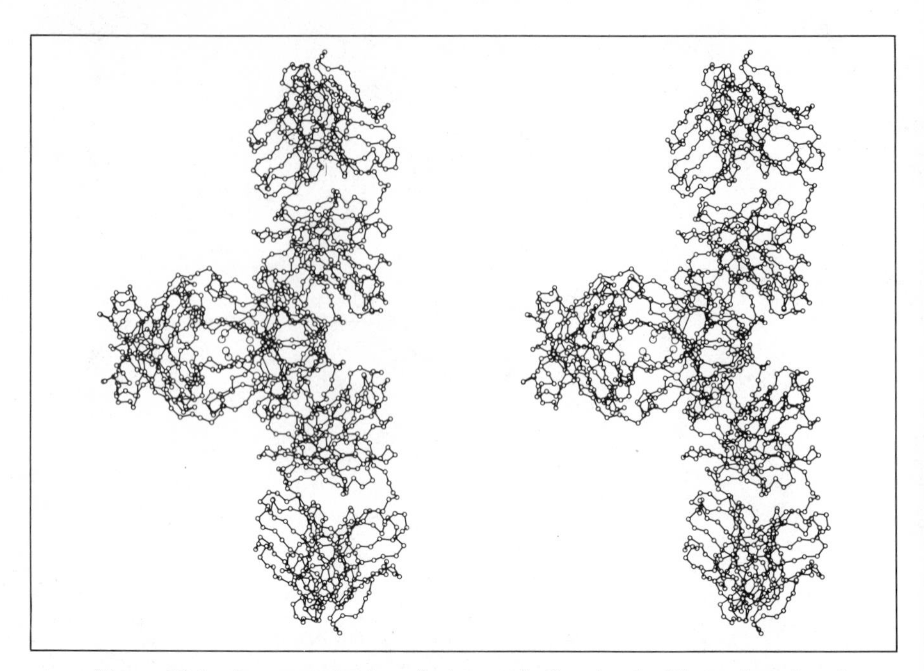

Figure 12-3 Stereoscopic view of a human IgG molecule. The various domains are seen as globular clusters, each composed of corresponding regions of an L and an H chain (vertically aligned regions), or two H chains (horizontal region at left). The variable regions are at the top and the bottom and are the surfaces that interact with antigen.

To see the structure in three dimensions, hold a piece of paper vertically above the figure so that the left eye sees only the left figure and the right eye only the right. Visually merge the two figures (stare at a distance), then adjust the focus of the eyes so that a sharp image is obtained. (*Reproduced from Silverton, Navia, and Davies, 1977.*)

isolation and chemical study of the one type of molecule. The L chains, being small enough to pass through the glomerulus, often are excreted in urine of myeloma patients, where they are designated *Bence–Jones protein.*

Such Ig myeloma proteins have been studied from many patients, and patterns of structural variation are now clear. As suggested in Figure 12-2, L chains differ from each other in the N-terminal region, but there is very little variation in the C-terminal regions from different patients. This has given rise to the concept of a variable portion of the L chain (residues 1–110, approximately) and a constant portion (the remainder of the molecule to the C-terminal). These are designated V regions and C regions, respectively. A similar situation exists with respect to H chains, the initial 100–110 residues being quite variable from one myeloma protein to another, the remainder very similar among H chains of a given class.

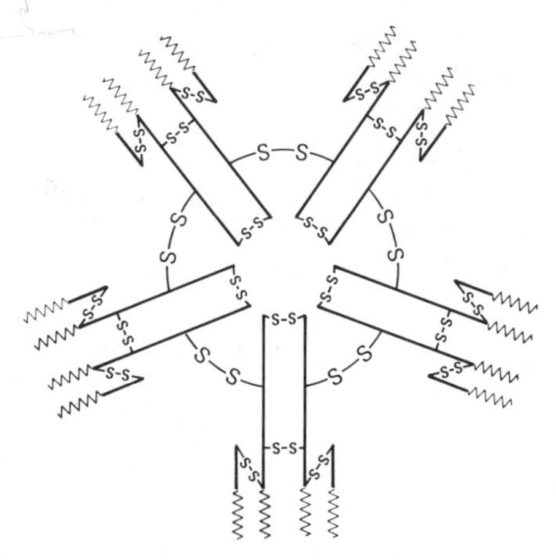

Figure 12-4 Diagrammatic representation of an IgM molecule. The zigzag regions are the variable N-terminal segments. The solid lines are the constant regions of the L and H chains. (*From Putnam and co-workers, 1971.*)

The picture that emerges is of a family of molecules composed of a great variety of L and H chains, the variety generated primarily by variability in the N-terminal region of both L and H chains. There is limited variability in the C portion of the different H chains that causes Ig molecules to fall into major classes and into subclasses. The variable ends are responsible for differences in antibody specificity.

Immunoglobulin Allotypes

The previous section has been devoted to variation among the population of Ig molecules found in any one person. There is additional Ig genetic variation that distinguishes different persons. This is referred to as *allotypic* variation. It was first detected in 1956 by R. Grubb, using a somewhat complicated system of red blood cell agglutination in the study of rheumatoid arthritis. In Grubb's system, human red cells are coated with "incomplete" Rh antibodies, that is, antibodies that react with Rh(+) red cells but that will not agglutinate them. Such antibodies belong to the IgG class, and persons with rheumatoid arthritis often have a plasma factor that causes such coated cells to agglutinate, the factor acting very much as an antibody against the Rh antibodies on the red cells. Grubb found that plasma from certain persons, if mixed with the rheumatoid agglutinating (Ragg) plasma prior to addition of coated red cells, prevents the latter from agglutinating the red cells. Plasma from others did not interfere with the agglutination. Since the interfering factor was shown to be in the γ-globulin (IgG) fraction, persons with the factor were designated Gm(a+), the Gm symbolizing gamma, and persons without were Gm(a−). These thus appeared to be simple Mendelian traits, with the allele Gm^a dominant to *Gm*.

Since the initial studies, many normal persons have been identified as having the antibodylike activity useful in agglutinating coated red cells. Such sera designated SNagg (serum normal agglutinating). Since these are found in persons who have received transfusions or in women who have been pregnant, it seems likely that the SNagg activity represents normal antibody production against "foreign" Ig molecules. Such sera would agglutinate red cells coated with incomplete antibodies having this same "foreign" antigen. Consistent with this concept is the discovery of many SNagg sera useful for detecting different Gm factors.

Table 12-3 **LIST OF HUMAN IMMUNOGLOBULIN ALLOTYPES, WITH NOMENCLATURE AS RECOMMENDED BY A STUDY GROUP OF THE WORLD HEALTH ORGANIZATION***

	Recommended designations		
Ig chain	Alphameric	Numeric	Earlier designation, if different
$\gamma 1$	G1m(a)	G1m(1)	
	(x)	(2)	
	(f)	(3)	(b^w), (b2), (4)
	(z)	(17)	
$\gamma 2$	G2m(n)	G2m(23)	
$\gamma 3$	G3m(b0)	G3m(11)	(b^β)
	(b1)	(5)	(b), (bγ), (12)
	(b3)	(13)	(Bet), (25)
	(b4)	(14)	
	(b5)	(10)	(bα)
	(c3)	(6)	Gm-like, (c)
	(c5)	(24)	Gm-like, (c)
	(g)	(21)	
	(s)	(15)	
	(t)	(16)	
	(u)	(26)	(Pa)
	(v)	(27)	(Ray)
$\alpha 2$		A2m(1)	
		(2)	
κ		Km(1)	Inv(1)
		(2)	(2)
		(3)	(3)

* (*Am. J. Human Genet.* 29: 117–120, 1977.) In earlier nomenclature, Gm was used to designate allotypes of IgG and Am of IgA. In this newer system, the particular subclass on which the variation occurs (1, 2, etc.) is inserted between the G and m. Earlier designations are given for allotypic equivalencies that are not obvious. Other allotypes have been reported, but antisera are no longer available for testing.

A particular combination of incomplete Rh antiserum and SNagg serum will define a particular factor.

The first additional factor recognized after Gm^a was designated Gm^b. This appeared to be allelic to Gm^a, so that the possible genotypes were Gm^a/Gm^a, Gm^a/Gm^b, or Gm^b/Gm^b. Several additional alleles were soon identified. When Baltimore blacks were tested, however, 95% turned out to be Gm(a+b+), apparently heterozygotes. This unexpected result is inconsistent with a two-allele hypothesis, and still another allele, Gm^{ab}, was proposed, an allele that is common in persons of African descent but rare in persons of European descent. As additional Gm factors were discovered, it became apparent that the *Gm* locus is very complex and that groups of factors are inherited as a unit, recombination among them being exceedingly rare. A list of known Gm allotypes is given in Table 12-3.

It was soon demonstrated that most of the allotypic factors are limited to specific subclasses of Ig molecules. For example, the originally described Gm(a) is found only in the IgG class and indeed, only in $\gamma 1$ chains. Gm(b) is found in $\gamma 3$ chains, and so on. Other variants have been assigned to $\gamma 2$ and to $\alpha 2$ chains. The symbols used (Table 12-3) reflect these assignments by inserting the subclass designation into the symbol, that is, G3m for $\gamma 3$ variants.

The above allotypic variants are properties of the heavy chains. Several variants are known that occur in all Ig classes. This would be the case for light chain variants, and these variants have been shown to occur on the κ chains.

In general, the bases for the antigenic variations have not been established. In the case of G1m(a), the antigen is associated with aspartic acid in position 356 of the $\gamma 1$ chain rather than glutamic acid in G1m(a−) persons and other primates. Additional allotypes have also been associated with single amino acid changes in the constant region of the Ig molecules, and this is presumably the usual explanation for allotypic variation.

Genetic Structure of Ig Loci

The allotypic variation, with rare exceptions, occurs in the constant regions of the H and L chains. The variable regions show many differences in amino acid sequences among the myeloma proteins studied. These variations fall into families of related sequences, but the total number of different sequences identified is large—several dozen at least with indications of many more not yet encountered. There appears to exist then within any one person quite a large library of variable regions but a restricted choice of constant regions.

The existence of allotypes has permitted formal genetic analysis of Ig loci and has led to recognition of three Ig gene complexes. The larg-

Table 12-4 **SOME COMMON IgG HAPLOTYPES***

Heavy chain location of factors			
G1m	G2m	G3m	Population
4	23	5,11,13,14	White
4	—	5,11,13,14	White
1,17	—	21	White, Ainu, Bushmen, Micronesian, Melanesian
1,2,17	—	21	White, Mongoloid, Melanesian
1,17	—	5,11,13,14	Negroid
1,17	—	5,6,11,24	Negroid
1,17	—	5,13,14	Bushmen, Pygmy, Micronesian, Melanesian

* This list is highly selected. A complete list would be very long. The populations are selected to be illustrative. The haplotypes are those present in high frequency. Other haplotypes are also present in lower frequencies.

est is the H-chain complex. All of the H-chain allotypes segregate as if closely linked, leading to the supposition that all H-chain genes are part of a single complex. By contrast the L-chain genes are in two separate groups, the κ and λ, which are not linked to each other or to the H-chain complex.

Because the allelic combinations of H-chain allotypes are transmitted as a group, it is common to refer to them as *haplotypes*. Although the number of known haplotype combinations is large, not all possible combinations have been observed. Some of the common haplotypes are given in Table 12-4.

The Origin of Antibody Variability

Within each Ig subclass, there is a great variety of antibodies produced by any one person. If each cell produces only a single type, as seems to be the case, it follows that there must be a great variety of cells, each producing its distinct antibody. According to the clonal selection theory of Burnett, this variety exists prior to exposure to an antigen, and the effect of antigen is to stimulate growth of those cells producing antibody to the antigen.

The existence of so many varieties of L and H chains has been awkward to explain. On any simple one gene-one polypeptide chain hypothesis, either (1) there must be a very large number of loci, all very similar in the constant regions but not in the variable regions, or (2) there must be a smaller number of loci with an unusual amount of somatic mutation in the variable regions but not the constant regions. Neither of these hypotheses is attractive.

Of the large number of L and H chains that have been sequenced, no two are identical (with one exception) in their variable regions. On the other hand, the constant regions are exceedingly constant within a class or subclass, except for the allotypic variation limited to a relatively few amino acid substitutions. If one imagines a series of γ1 genes, perhaps arranged in tandem, the variable portions could well have diverged through mutation. One would expect more divergence in the constant regions, but selection could be invoked to explain the constancy. However, all, or nearly all, of the IgG1 molecules from a particular chromosome will be of the same Gm types. This would require that each of the genes in the tandem array have the same codon changes for the allotype variations. There is no known mechanism that would generate such a series of genes with an acceptable likelihood.

A case of myeloma reported by Wang and co-workers (1970, 1973) is especially instructive from the point of view of gene control of Ig structure. This patient, unlike all others studied previously, manufactured both IgG2 and IgM in greatly excessive amounts. In both instances a κ light chain was involved, and both the γ2 and μ heavy chains were typical of their classes. However, the variable regions of the two H chains were identical (Wang et al., 1977). This suggests that the V regions were coded by the same genetic locus, but the C regions clearly were coded by two different loci. Other similar cases have also been reported.

These cases could only occur if the V and C regions are coded by separate genes whose products can join together into a single polypeptide chain. At the time of this report, there was no precedent for such an event. With the discovery of intervening sequences in structural genes, the picture has changed. Separated coding sections of a single polypeptide chain appear to be common, and the special requirements for Ig H and L chain production are perhaps not so unusual.

Recent studies of a mouse myeloma cell line show that the V and C regions of the λ chain are separated by a sequence of 1250 untranslated base-pairs (Brack and Tonegawa, 1977). Indeed, it appears that the V gene itself is divided into two segments, one 97-98 codons long, with the remaining segment of ca. 13 codons, designated the J region, separated by a sequence of untranslated nucleotides (Bernard et al., 1978). A fragment of mouse DNA containing the genes for the C region of a heavy chain consisted of a series of exons, each corresponding to a "domain" or to the "hinge" region (Sakano et al., 1979). Studies of transcription of a κ chain gene indicate the primary transcription unit to be ca. 10,000 nucleotides long with the V and C regions near the 3′-end and some 2000 nucleotides apart. The final mRNA derived from this transcript is much smaller, with the V and C regions fused (Gilmore-Hebert et al., 1978).

A current picture of the H chain complex is given in Figure 12-5. Many of the details are yet to be filled in. It seems probable, for example, that each of the constant genes shown consists of several (four?) exons separated by intervening sequences. It seems inescapable though

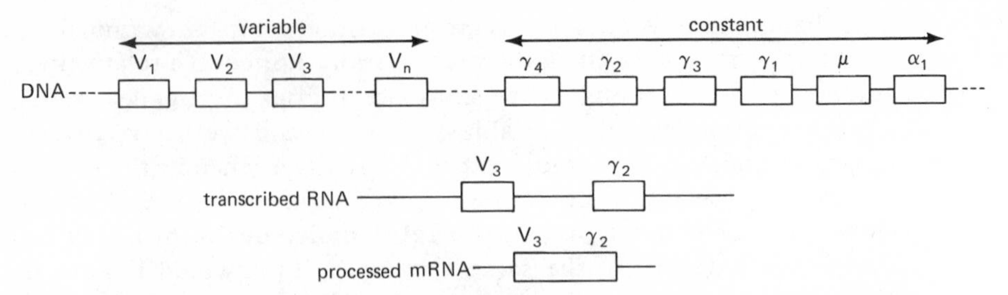

Figure 12-5 Diagram of H chain gene complex showing possible relationship of genes for variable regions in tandem with genes for constant regions. A particular V gene (V_3 in the example) and a C gene (e.g., γ2) are transcribed and, after processing of mRNA, are juxtaposed to provide a single translated unit. Whether this juxtaposition occurs by posttranscriptional removal of nucleotides, as with intervening sequences, or whether there are additional mechanisms for bringing the two together during transcription is not known. The arrangement of the genes as shown is hypothetical.

that each person has a large number of V genes and a smaller number of C genes. These somehow are transcribed and processed in such a way that a particular V gene and a C gene are represented by a continuous sequence of mRNA.

Apparent errors in this mechanism are sometimes seen in persons with "heavy-chain disease," a myeloma-like condition in which polypeptide chains are manufactured that are small enough to pass through the glomerulus into the urine but that have some of the structural features of heavy chains. Analysis of the structures of some of these "heavy chains" shows the N-terminal to be homologous to the variable region and the C-terminal to be homologous to the corresponding region of a heavy chain. The structure appears to result from deletion of the middle of the molecule, including portions of both the V and C regions.

Allelic Exclusion. An interesting phenomenon observed in antibody-producing cells is allelic exclusion. By this is meant that in a cell heterozygous for Ig allotypes, only one of the two possible genetic types is produced. This principle in fact is somewhat more far-reaching than the exclusion of alleles. When an animal is immunized, only a small number of cells respond with synthesis of specific antibody. These not only manufacture antibody but also are stimulated to grow and divide, building up thereby the population of cells with the potential for specific antibody response. Such cells ordinarily manufacture a single antibody, both in terms of specificity and in terms of Ig subclass. And, as noted above, if the cell is heterozygous for the subclass, only one genetic type is produced. Thus from its repertory of Ig genes, a cell selects a single type L chain and a single type H chain for synthesis. There is evidence that the shift from predominantly IgM synthesis in the initial response to immunization to predominantly IgG synthesis with the same antibody specificity may reflect a switchover in single cells. Whether this shift is abrupt or whether at a given time there may be two kinds of H chains being synthesized in one cell, albeit with the same specificity, remains to be determined.

Evolution of Immunoglobulin Structure

As with other protein molecules, the evolutionary divergence of Ig molecules can be reconstructed by comparison of present-day amino acid sequences. The greater the time since divergence, the more opportunity for differences in amino acid sequences to arise, or, stated conversely, the greater the number of differences between two sequences, the earlier the separation.

Homology among Ig chains is observed, not only between different chains but also between different segments of one chain. In fact, all Ig chains seem to have evolved by gene duplication (and reduplication) and amino acid substitution, deletion, or addition from a primordial molecule of approximately 10,000 molecular weight. As shown in Figure 12-6, each L chain contains two such basic units (*domains*), and the H chains contain four. Since the V and C regions should be treated as separate entities, the V regions of both L and H chains and the light C region should be similar in size to the primordial gene. The heavy C regions are approximately three times that size. The evolution of the C regions (excluding δ and ϵ chains, about which too little is known) is fairly clearly established. The V regions are less clear because of their extreme variability and because the total number and kind are not known.

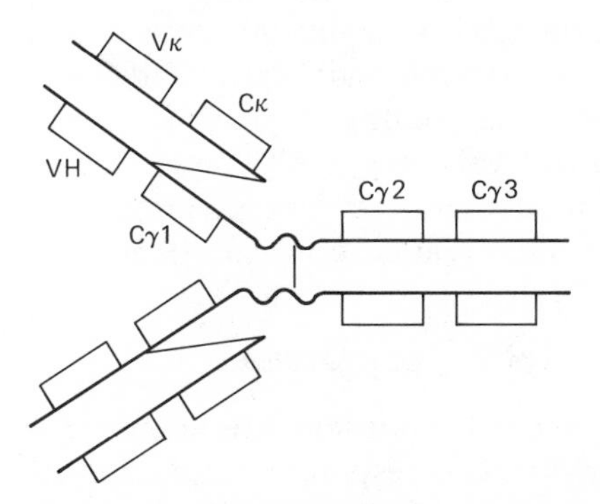

Figure 12-6 The basic structure of an IgG molecule showing the two homology regions of the L chain (Vκ and Cκ) and the four homology regions of the H chain (VH, Cγ1, Cγ2, and Cγ3). Each region of homology is known as a domain, and the region between Cγ1 and Cγ2 is known as the hinge region. (*Modified from Fudenberg and co-workers, 1978.*)

IMMUNE RESPONSIVENESS

The sequence of events leading to production of antibodies following exposure to an antigen is complex and only partially understood. At least three cell types are involved: T cells, B cells, and macrophages. These can be further subdivided on the basis of special properties and functions. In this presentation, we will be concerned only with genetic differences in the ability to respond.

Immune Tolerance. It is important for organisms not to make antibodies against their own tissues. This distinction between self and non-self is developed during the embryonic period. Animals do not respond to antigens present during embryonic development, even if these antigens are reintroduced much later after a long period during which they were absent (Billingham et al., 1956). In pregnancies involving twins, anastomoses of the vascular system occasionally lead to exchange of blood-forming tissue. This has been studied especially in cattle, where exchange of blood between male and female twin pairs causes the female to become a freemartin. Such animals often show a mixture of red cell types, one type representing the genotype of the animal under study and the other type the genotype of the co-twin (Owen, 1945). The permanent exchange of blood-forming cell lines between twins has been observed in man, but it is very rare. Such immune tolerance is limited to those antigens present in the fetal period, and the organism is fully capable of responding to other antigenic stimuli.

Tolerance may sometimes work against an organism. Rodents injected in the newborn period with lymphocytes may subsequently become the target of the antibodies secreted by those lymphocytes in a "graft-versus-host" reaction. Injected lymphocytes ordinarily are destroyed by the immune defenses of human infants. Exceptions have been reported in which intrauterine and exchange transfusions gave rise to "grafted" lymphoid tissue that eventually formed antibodies against the host, leading to death (Parkman et al., 1974). It has been suggested that this may rarely happen by transfer of maternal lymphocytes to a fetus that is relatively immunoincompetent, leading to death even after many years (Schwartz, 1974).

Immune Deficiency Diseases

Several inherited conditions are known in man in which there is generalized inability to respond to antigenic stimulus. These are grouped under the term *dysgammaglobulinemia* or *agammaglobulinemia*, but the pathogenetic mechanisms are varied.

Bruton's Agammaglobulinemia. This was the first immune deficiency disease to be clearly recognized (1952). It is inherited as a sex-linked recessive. It is rare, and only males have been observed to be affected. Circulating globulin levels are very low (less than 25 mg/100 ml), associated with an absence of plasma cells (antibody-producing cells). Other aspects of the immune mechanism are normal. The first symptoms usually appear between 5 and 12 months of age when the immunoglobulins passively transferred from the mother during the fetal period are exhausted. Recurrent bacterial infections occur that usually can be managed with antibiotics and γ-globulin injections. Resistance to viral infections is normal, since this depends on the thymus function, which is intact.

Severe Combined Immunodeficiency (Swiss-Type Agammaglobulinemia). This rare disorder occurs both as an autosomal recessive and a sex-linked recessive disease. It is a combined immunodeficiency disease, since circulating immunoglobulins are virtually absent and lymphoid tissues, including the thymus, are greatly underdeveloped. This leaves affected persons with no defense against bacterial or viral infections. Repeated infections usually lead to death before 2 years of age. No successful treatment has been developed.

The discovery of complete absence of adenosine deaminase (ADA) activity in several cases suggests that the primary gene defect may be directly related to this enzyme (Giblett et al., 1972; Chen et al., 1974). Supporting this is the discovery that some cases with normal ADA activity are deficient in nucleoside phosphorylase (Giblett et al., 1975). These metabolic blocks in nucleic acid metabolism should be very helpful in working out the chain of events that lead to lack of immune response.

Wiskott–Aldrich Syndrome. This disorder shows sex-linked recessive inheritance. Both circulating immunoglobulins as well as cellular immune response are partially deficient. The disease is characterized by recurrent infections, skin rash, and bleeding.

There are several additional immunodeficiency syndromes whose genetic status and cellular bases are less well defined. In addition, some forms of hypogammaglobulinemia can be acquired. The generalized immunodeficiency diseases result from absence of a cell type or absence of function of a cell type. In this sense, they are developmental disorders. They illustrate that defects in function of single genes can lead to failure of development of function of an entire cell type.

Specific Immune Response Variations

Both mice and guinea pigs have a series of specific *IR* (immune response) loci, and there is now evidence that a similar locus exists in man. The first such locus was clearly delineated in guinea pigs, controlling responsiveness to several artificial antigens, such as poly-L-lysine (PLL), poly-L-arginine (PLA), copolymers of L-glutamic acid and L-lysine (GL), and to substances conjugated to these structures (Levine and co-workers, 1963). Animals belonging to strain 2 produced antibodies to these antigens; those belonging to strain 13 did not. Crosses indicated a single locus to be responsible. Animals with the *PLL* gene (or allele) could respond to these antigens; those lacking the *PLL* gene could not. It is important to note that responders produce a variety of specific antibodies that can distinguish among this group of antigens. The *PLL* gene determines *whether* they respond but not the *kinds* of antibodies produced. Other genes have been found in guinea pigs controlling response to other groups of antigens. These genes are very closely linked to the *PLL* gene and are distinct from Ig genes.

Mice have proved especially informative for the *IR* systems. A series of genes has been described controlling response to a variety of antigens. Most are shown to be located in the midst of the H-2 region, a complex region that determines the major histocompatibility antigens (see Chapter 13). Transplantation studies from responder donors to nonresponder hosts show the action of the *IR* genes to be concerned with T-cell function, that is, in the recognition of antigens. The structures of the antibodies are determined by the various heavy and light chain loci, that are not linked to the *IR* genes. An organism may thus have the structural information necessary to manufacture specific antibodies, but these will not be made in the absence of recognition by the T cells.

Human IR Genes. The existence of variation in immune response has been reported for man. Allergies have long been known to be familial, suggesting genetic variation in the response to allergens. Several groups have reported that sensitivity to a component of ragweed pollen is probably hereditary, but the precise definition of alleles has not been made. Other examples may include an apparent X-linked immune deficiency to Epstein–Barr virus, normally a relatively benign virus that causes infectious mononucleosis but which, in this family, caused cancer and other major problems (Purtilo et al., 1977).

The study of genetic variation in immune responsiveness is a new field in which only a few beginning observations have been made. Considering the importance to human health of allergies and disease resistance, and, in view of their apparent dependence on antigen-recognizing systems, the forthcoming discoveries in this field may have substantial impact for many persons.

SUMMARY

1. Antibodies (immunoglobulins, Ig) are produced in response to introduction of foreign substance (antigens) into the body. Ig molecules are protein, belonging to the blood globulins. They combine very specifically with the appropriate antigens and not with others. Since they are not present before antigenic stimulation, the nature of the genetic control of antibody structure is of special interest.

2. There are five major Ig classes: IgG, IgA, IgM, IgD, and IgE. All consist of two types of polypeptide chains, L (light) and H (heavy), the basic formula being L_2H_2. IgM is a pentamer of this basic unit. The L chains are the same in all classes, but the H chains differ according to class; they are designated γ (in IgG), α (IgA), μ (IgM), δ (IgD), and ϵ (IgE). The γ, α, and μ chains are further divided into subclasses based on related but structurally distinct forms. There are two kinds of L chains, κ and λ.

3. Each genome codes for all of the above variations. In addition, there is allotypic variation transmitted as Mendelian traits. The allotypes are the result of structural variations in the C-terminal portions ("the constant regions") of the chains.

4. Most of the H-chain loci appear to be closely linked to each other. However, the κ- and λ-chain loci are not linked to the H -chain loci or to each other in rabbits and presumably not in man.

5. Ig-producing cells ordinarily produce a single antibody type with a single specificity. Antibody variability appears to arise by clonal selection. That is, when an antigen is introduced, those cells capable of making appropriate antibodies are stimulated to proliferate. This implies a large number of cells with different antibody potential in order to make the variety of antibodies produced by an organism.

6. Substantial evidence now exists for the idea that there are separate genes for the variable and constant regions of Ig chains. These genes are separated in the DNA and in the initially transcribed RNA but are fused into a single structural region in mRNA.

7. Immunoglobulins appear to have evolved from a primordial gene about half the size of the present day *L* genes. The *H* genes are about four times the size of the primordial gene. It appears that duplication and diversion have been extensive in the Ig genes, as judged by homologies in sequences of Ig molecules.

8. There are several inherited defects in Ig production in man. These include the X-linked Bruton's agammaglobulinemia, in which there is deficiency of circulating Ig but cellular immunity is intact. In severe combined immunodeficiency disorders, there commonly are blocks in nucleic acid metabolism. Other immunodeficiency disorders are known but are less well understood.

9. The ability to respond to antigenic stimulation requires not only the appropriate structural genes for Ig but also appropriate *IR* (immune response) genes. Little is known of *IR* genes in man, but some allergies, as well as other diseases, may be examples.

REFERENCES AND SUGGESTED READING

ABRAMOFF, P., AND LA VIA, M. F. 1970. *Biology of the Immune Response.* New York: McGraw-Hill, 492 pp.

BENACERRAF, B., AND MC DEVITT, H. O. 1972. Histocompatibility-linked immune response genes. *Science* 175: 273–274.

BERNARD, O., HOZUMI, N., AND TONEGAWA, S. 1978. Sequences of mouse immunoglobulin light chain genes before and after somatic changes. *Cell* 15: 1133–1144.

BILLINGHAM, R. E. 1968. The biology of graft-versus-host reactions. *Harvey Lectures* 62: 21–78.

BILLINGHAM, R. E., BRENT, L., AND MEDAWAR, P. B. 1956. Quantitative studies on tissue transplantation immunity. III. Actively acquired tolerance. *Phil. Trans.* 239: 357–414.

BLANC, M., et al. 1977. Report of standardized nomenclature: human immunoglobulin markers. *Am. J. Hum. Genet.* 29: 117–120.

BRACK, C., AND TONEGAWA, S. 1977. Variable and constant parts of the immunoglobulin light chain gene of a mouse myeloma cell are 1250 nontranslated bases apart. *Proc. Natl. Acad. Sci.* (U.S.A.) 74: 5652–5656.

CHEN, S-H., SCOTT, C. R., AND GIBLETT, E. R. 1974. Adenosine deaminase: demonstration of a "silent" gene associated with combined immunodeficiency disease. *Am. J. Hum. Genet.* 26: 103–107.

EDELMAN, G. M., CUNNINGHAM, B. A., GALL, W. E., GOTTLIEB, P. D., RUTISHAUSER U., AND WAXDALL, M. J. 1969. The covalent structure of an entire γG immunoglobulin molecule. *Proc. Natl. Acad. Sci.* (U.S.A.) 63: 78.

FRANGIONE, B., AND MILSTEIN, C. 1969. Partial deletion in the heavy chain disease protein ZUC. *Nature* (Lond.) 224: 597.

FRANKLIN, E. C., PRELLI, F., AND FRANGIONE, B. 1979. Human heavy chain disease protein WIS: Implications for the organization of immunoglobulin genes. *Proc. Natl. Acad. Sci.* (U.S.A.) 76: 452–456.

FUDENBERG, H. H., PINK, J. R. L., WANG, A-C., AND DOUGLAS, S. D. 1978. *Basic Immunogenetics,* 2nd Ed. New York: Oxford Univ. Press, 262 pp.

GIBLETT, E. R., AMMANN, A. J., SANDMAN, R., WARA, D. W., AND DIAMOND, L. K. 1975. Nucleoside phosphorylase deficiency in a child with severely deficient T-cell immunity and normal B-cell immunity. *Lancet* 1: 1010–1013.

GIBLETT, E. R., ANDERSON, J. E., COHEN, F., POLLARA, B., AND MEUWISSEN, H. F. 1972. Adenosine deaminase deficiency in two patients with severely impaired cellular immunity. *Lancet* 2: 1067–1069.

GILMORE-HEBERT, M., HERCULES, K., KOMAROMY, M., AND WALL, R. 1978. Variable and constant regions are separated in the 10-kbase transcription unit coding for immunoglobulin light chains. *Proc. Natl. Acad. Sci.* (U.S.A.) 75: 6044–6048.

GRUBB, R. 1970. *The Genetic Markers of Human Immunoglobulins.* Springer-Verlag, New York. 152 pp.

HILDEMANN, W. H. 1973. Genetics of immune responsiveness. *Annu. Rev. Genet.* 7: 19–36.

HILL, R. L., DELANEY, R., FELLOWS, R. E. JR., AND LEBOVITZ, H. E. 1966. The evolutionary origins of the immunoglobulins. *Proc. Natl. Acad. Sci.* (U.S.A.) 56: 1762–1769.

LEVINE, B. B., OJEDA, A., AND BENACERRAF, B. 1963. Studies on artificial antigens. 3. The genetic control of the immune response to hapten-poly-L-lysine conjugates in guinea pigs. *J. Exp. Med.* 118: 953–957.

MARSH, D. G., BIAS, W. B., HSU, S. H., AND GOODFRIEND, L. 1973. Association of the HL-A7 cross-reacting group with a specific reaginic antibody response in allergic man. *Science* 179: 691–693.

MILSTEIN, C. 1969. The basic sequences of immunoglobulin kappa chains: sequence studies of Bence Jones proteins Rad, FR 4, and B_6. *FEBS Lett.* 2: 301.

OWEN, R. D. 1945. Immunogenetic consequences of vascular anastomoses between bovine twins. *Science* 102: 400–401.

PARKMAN, R., MOSIER, D., UMANSKY, I., COCHRAN, W., CARPENTER, C. B., AND ROSEN, F. S. 1974. Graft-versus-host disease after intrauterine and exchange transfusions for hemolytic disease of the newborn. *N. Engl. J. Med.* 290: 359–363.

PURTILO, D. T., DE FLORIO, D. JR., HUTT, L. M., BHAWAN, J., YANG, J. P. S., OTTO, R., AND EDWARDS, W. 1977. Variable phenotypic expression of an X-linked recessive lymphoproliferative syndrome. *N. Engl. J. Med.* 297: 1077–1081.

PUTNAM, F. W., SHIMIZU, A., PAUL, C., SHINODA, T., AND KOHLER, H. 1971. The amino acid sequence of human macroglobulins. *Ann. New York Acad. Sci.* 190: 83–103.

PRAHL, J. W. 1967. N and C terminal sequences of a heavy chain disease protein and its genetic implications. *Nature* (Lond.) 215: 1386.

SAKANO, H., ROGERS, J. H., HÜPPI, K., BRACK, C., TRAUNECKER, A., MAKI, R., WALL, R., AND TONEGAWA, S. 1979. Domains and the hinge region of an immunoglobulin heavy chain are encoded in separate DNA segments. *Nature* (Lond.) 277: 627–633.

SCHWARTZ, R. S. 1974. Trojan-horse lymphocytes. *N. Engl. J. Med.* 290: 397–398.

SELIGMANN, M., DANON, F., HUREZ, D., MIHAESCO, E., AND PREUD'HOMME, J.-L. 1968. Alpha chain disease: a new immunoglobulin abnormality. *Science* 162: 1396–1397.

SILVERTON, E. W., NAVIA, M. A., AND DAVIES, D. R. 1977. Three-dimensional structures of an intact human immunoglobulin. *Proc. Natl. Acad. Sci.* (U.S.A.) 74: 5140–5144.

TERRY, W. D., HOOD, L. E., AND STEINBERG, A. G. 1969. Genetics of Ig κ chains: chemical analysis of normal human light chains of differing inV types. *Proc. Natl. Acad. Sci.* (U.S.A.) 63: 71–77.

TERRY, W. D., AND OHMS, J. 1970. Implications of heavy chain disease protein sequences for multiple gene theories of immunoglobulin synthesis. *Proc. Natl. Acad. Sci.* (U.S.A.) 66: 558–563.

TONEGAWA, S., MAXAM, A. M., TIZARD, R., BERNARD, O., AND GILBERT, W. 1978. Sequence of a mouse germ-line gene for a variable region of an immunoglobulin light chain. *Proc. Natl. Acad. Sci.* (U.S.A.) 75: 1485–1489.

WANG, A. C., GERGELY, J., AND FUDENBERG, H. H. 1973. Amino acid sequences at constant and variable regions of heavy chains of monotypic immunoglobulins G and M of a single patient. *Biochemistry* 12: 528–534.

WANG, A-C., WANG, I. Y., AND FUDENBERG, H. H. 1977. Immunoglobulin structure and genetics: Identity between variable regions of a μ and a $\gamma 2$ chain. *J. Biol. Chem.* 252: 7192–7199.

WANG, A.-C., WILSON, S. K., HOPPER, J. E., FUDENBERG, H. H., AND NISONOFF, A. 1970. Evidence for control of synthesis of the variable regions of the heavy chains of immunoglobulins G and M by the same gene, *Proc. Natl. Acad. Sci.* (U.S.A.) 66: 337–343.

REVIEW QUESTIONS

1. Define:

immunoglobulin	Ig heavy chain	agammaglobulinemia
immune response	Ig light chain	haplotype
antigen	B cell	myeloma
antibody	T cell	allelic exclusion
allotype	IR gene	Bence-Jones protein

2. A person is typed as Gm(a,z,g). The various Ig classes and subclasses are isolated. Which would you expect to find positive for these factors?

3. What are the major classes of Ig molecules?

4. What is the distinction between classes and subclasses?

5. How many Ig linkage groups are known?

6. For a particular kind of IgG, how many genes are thought to be required for structural information?

7. What genetic loci, in addition to those for Ig structure, are necessary for immune response?

CHAPTER THIRTEEN

IMMUNOGENETICS: ANTIGENIC VARIATION

RED BLOOD CELL ANTIGENS

The scientific study of red blood cell types began in 1900 when Landsteiner noted that certain blood samples, when mixed together, resulted in agglutination, but with other combinations no agglutination occurred. Agglutination was shown to be due to the plasma of one person reacting with the cells of another. Landsteiner and his associates soon were able to classify persons into four groups—now designated A, B, AB, and O—on the basis of their agglutination reactions. In subsequent years, inheritance of ABO groups was shown to follow simple Mendelian rules, and the ABO system became the best example of a polymorphic Mendelian system in human beings.

The Typing of Red Cell Antigens

The earliest classification of ABO types was dependent on the presence both of red cell antigens and of naturally occurring antibodies in plasma, as shown in Table 13-1. The origin of the antibodies is unknown, but they are uniformly present if the red cells lack the corresponding antigen. It is the presence of these antibodies that prevents transfusion to persons of different ABO type and that most often causes agglutination when blood samples are mixed together.

The ability to discriminate blood cells of different types depends entirely on the availability of antibody preparations ("reagents") that recognize the different antigens. In blood cell typing, the reaction most often is agglutination, although systems that involve lysis in the presence of complement sometimes are used. Not all antibody preparations are capable of causing cells to agglutinate. Some can only adsorb to cells that have the specific antigen. These "incomplete" antibodies form a coat ~~of hum~~an Ig molecules on the red cell. The cells can then

Table 13-1 **ANTIGENS AND ANTIBODIES ASSOCIATED WITH THE ABO BLOOD GROUPS**

Blood group	Antigens on red cells	Antibodies in plasma
A	A	Anti-B
B	B	Anti-A
AB	A, B	None
O	H	Anti-A, anti-B

be agglutinated by reaction with antibodies to human Ig molecules. This is the *indirect Coombs test.*

Most antibodies to red cells are not "natural" in the sense of the antibodies to A and B blood groups. Rather, they are manufactured in response to immunization, either from transfusions of blood or from transfer of red cells from fetal to maternal circulation in pregnancy. Experimental animals may also be immunized with human red cells. Such antibodies to red cells from a different species are more likely to be antihuman rather than directed against single antigenic factors. The more discriminating reagents are made in the same species in which the antigens occur, since the antibodies there are directed only against antigens possessed by the donor but not by the recipient.

Many, perhaps most, antibody preparations are complex mixtures. This may not be evident if the only source of test antigen is equally complex. Sometimes cells other than those used for immunizing may have fewer antigenic factors and react only with a portion of the antibodies, removing from the reagent those antibodies against antigens that are common but leaving unreacted those antibodies against antigens found only on the immunizing cells. By this and similar manipulations it is often possible to prepare highly specific antibody reagents useful in typing red cells. Such reagents may be referred to as *monospecific,* but this should be understood as an operational term based on the availability of appropriate test antigens.

There are many antigens on red cells. Yet a person may be transfused repeatedly with blood matched only for ABO type without forming antibodies. Or, if antibodies are formed, they may be against only one or a few of the antigens, even though the donor cells may have many antigens not shared by the recipient. Some antigens are very much more effective in inducing antibody response than are others. Some antigens are so ineffective that only a few examples of antibody formation are known in spite of tens of thousands of persons transfused with blood unmatched for the antigen. The reasons for such variations in antigenicity are unknown, but it makes the availability of antisera dependent on rare unpredictable responses.

The ABO System

Landsteiner's early studies showed the A and B antigens to be transmitted as dominant traits but did not distinguish between one-locus and two-locus hypotheses. In the one-locus hypothesis, three alleles would be required to account for the four phenotypes. In the two-locus hypothesis, two alleles at each locus would account for the four phenotypes. At the *A* locus, there would be an *A* allele and a *non-A* allele; *B* and *non-B* alleles would occur at the *B* locus. It was not until 1924 that Bernstein proved by statistical analysis of populations that the one-locus, three-allele hypothesis accounted for the observed frequencies of the four types. Using the symbol *I* to designate the *ABO* locus, there are three common alleles, I^A, I^B, and I^O. I^A and I^B are dominant to I^O and are codominant to each other, giving the corresponding genotypes and phenotypes of Table 13-2. The *ABO* locus is located on chromosome 9.

Although variation in the A antigen had been noted as early as 1911, it was not until 1930 that the two major subgroups, A_1 and A_2, were clearly delineated. If antisera to A are absorbed by A_1 cells (that is, reacted with an excess of red cells), no anti-A activity remains and no reaction occurs subsequently with A_1 or A_2 cells. However, if absorption is with A_2 cells, the antisera which can no longer react with A_2 cells still can agglutinate A_1 cells. It is as if A_2 lacked some antigens present in A_1. The true explanation is unknown but seems to be more complex than that. The inheritance of group A subtypes is due to variation at the ABO locus, the allele I^A actually being either I^{A1} or I^{A2}. I^{A1} is dominant to I^{A2}. Therefore A_2 cells correspond to the genotypic combinations $I^{A2}I^{A2}$ or $I^{A2}I^O$. The distinction between A_1 and A_2 cells generally has no significance for transfusion. Additional rare subgroups of A have been described, but these are poorly understood. Subgroups of B have also been described, but these are even more rare.

Persons of type O are generally so classified because their red cells fail to react with anti-A and anti-B sera. They do have a related

Table 13-2 **INHERITANCE OF THE ABO GROUPS**

Genotype	Phenotype
I^AI^A	A
I^AI^O	A
I^BI^B	B
I^BI^O	B
I^AI^B	AB
I^OI^O	O

antigen, designated H, that is also present in lower concentrations on A and B cells. Since persons of all ABO types possess the H antigen, none will form antibodies to it. Several plant-seed extracts *(lectins)* react with H antigen in much the way that antibodies react, and these will agglutinate cells with H antigen. The H antigen has proved important in understanding the chemical basis of gene action in the ABO system.

Frequencies of ABO Types. Because of their medical significance and early discovery, ABO types constitute the genetic system most widely surveyed in man. It is a rare population indeed that has escaped testing for distribution of ABO types. This interest stems in part from the large variation in genotype and phenotype frequencies in different populations (Table 13-3). Accordingly, the system is useful in tracing the origins of populations and the degree of relationship among populations. These points are explored more fully in later chapters. It will only be noted here that nearly all populations are polymorphic at the ABO locus.

Secretion of ABO Substances in Body Fluids. The ABO antigens occur on many cell types other than red cells and also occur in soluble form in some persons. Indeed, A-like antigenic activity has been

Table 13-3 **FREQUENCIES OF ABO TYPES IN SELECTED HUMAN POPULATIONS***

	Phenotype				Gene frequencies		
Population	**O**	**A**	**B**	**AB**	*p*	*q*	*r*
Armenians	.289	.499	.132	.080	.351	.112	.537
Eskimos (Greenland)	.472	.452	.059	.017	.272	.039	.689
Austrians	.427	.391	.115	.066	.262	.095	.644
Danes	.423	.434	.101	.042	.276	.074	.650
Irish	.542	.323	.106	.029	.195	.070	.735
French	.417	.453	.091	.039	.287	.067	.645
Russians (Moscow)	.340	.377	.206	.077	.261	.154	.585
Indians (Calcutta)	.304	.255	.375	.066	.178	.255	.567
Vietnamese	.418	.218	.305	.059	.150	.203	.648
Chinese	.439	.270	.233	.058	.181	.158	.662
Japanese (Niigata)	.296	.369	.233	.102	.273	.185	.543
Nigerians (Yoruba)	.515	.214	.232	.039	.136	.146	.718
Amerindians (Lima)	.860	.140	.000	.000	.073	.000	.928
Amerindians (Bolivia)	.931	.053	.016	.001	.027	.008	.965
U.S. whites (St. Louis)	.453	.413	.099	.035	.258	.070	.673
U.S. blacks (Iowa)	.491	.265	.201	.043	.168	.131	.701

* The symbols for gene frequencies are p (for I^A), q (for I^B), r (for I^O). Data were selected from Mourant et al. (1976).

noted in many species, including plants and bacteria. The antigenic specificity is due to certain carbohydrate structures that, in man, may be bound to lipid of cell membranes or to protein. The glycolipid form is often referred to as alcohol soluble, while the glycoprotein form is water soluble.

Whether the water soluble form occurs in body secretions is determined by genetic variation at another locus, the *secretor* locus. Saliva ordinarily is used to test for the presence of soluble antigens. Serial dilutions of saliva are mixed with standard amounts of anti-A, anti-B, or anti-H, depending on the ABO type of the subject. After a period of incubation, test red cells of the appropriate A, B, or O type are added to the mixture. If antigens secreted in the saliva have already formed complexes with the antibodies, no agglutination occurs. However, if there were no antigens in the saliva, the antibodies are still free to agglutinate the red cells.

The secretion of A, B, and H antigens in water-soluble form is inherited as a simple dominant trait. Secretors may be either *SeSe* or *Sese;* nonsecretors are *sese*. Only the two alleles are recognized. Among persons of European ancestry, approximately 23% are nonsecretors, giving gene frequencies of 0.52 for *Se* and 0.48 for *se*. Most other populations have frequencies similar to these.

The Bombay Phenotype. In 1952, several unusual persons from India were reported to have red blood cells that lacked H antigen as well as A and B antigens. In addition, they formed antibodies to H antigen. This unusual type was designated the Bombay type and symbolized as O_h.

Inheritance of the Bombay type was clarified by the family reported by Levine and co-workers (1955) shown in Figure 13-1. The proposita must have received an I^B allele from her mother and transmitted it to her first daughter. Yet she expresses no B antigen, nor do her cells react with anti-H. On the basis of this and several additional Indian families, it is concluded that the Bombay phenotype occurs when a person is homozygous for a recessive gene at a locus different from the ABO locus. Since the normal function of the locus appears to be concerned with the production of H antigen, the symbol *H* is used for the normal dominant allele and *h* for the recessive. The *h* allele is very rare. In the Marathi-speaking people around Bombay, homozygotes were found with a prevalence of 1 in 13,000 persons. In England, no example has been found in over a million persons tested.

The Chemistry of ABO Specificity. The availability of water-soluble forms of the ABH antigens has enabled structural studies of the differences among this group. These chemical studies have also provided understanding of the relationships between the ABO and the H

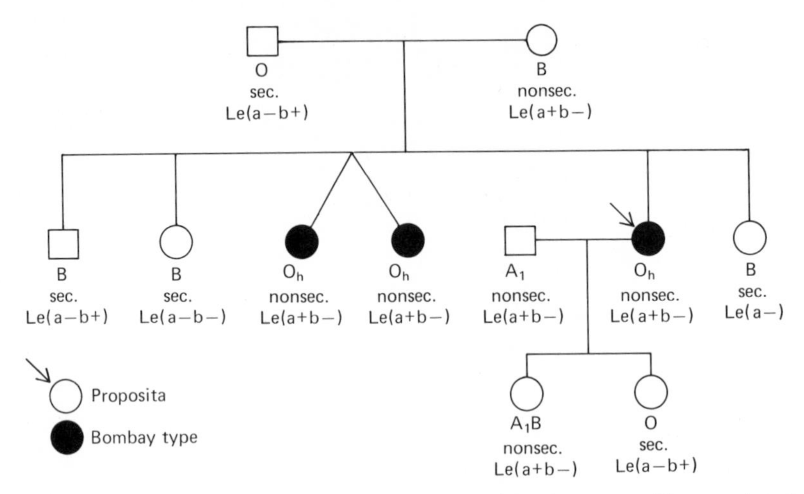

Figure 13-1 Pedigree showing presence of Bombay phenotypes. The results of ABO, secretor, and Lewis system typing are shown under each symbol. The symbol O_h refers to the absence of H as well as A and B antigens on the red cells of persons with the Bombay type. (*From P. Levine, E. Robinson, M. Celano, O. Briggs, L. Falkinburg, 1955. Gene interaction resulting in suppression of blood group substance B.* Blood *10:1100, by permission.*)

loci and have suggested how the secretor system and the Lewis system, also related to ABO, might work (Figure 13-2). The earliest identified precursors must acquire a fucosyl residue to form H substance. The *H* gene product is therefore thought to be an α-L-fucosyltransferase. In red cells no other locus is known to be required for this activity, but in secretory tissue the dominant *Se* allele is required at the secretor locus. Persons with the Bombay phenotype are homozygous for an allele *h* that is associated with lack of this enzyme activity. It is tempting to speculate that the *H* locus is the structural gene for this enzyme, but no evidence exists on this point.

The *ABO* locus is concerned with the addition of a galactosaminyl residue in the case of the I^A allele or of a galactosyl residue in the case of the I^B allele. In both instances the addition is to the H product. It is especially noteworthy that the enzymes associated with this locus have different specificities insofar as the sugar residue is concerned. The respective enzymes have been isolated and appear to be similar in structure (Nagai et al., 1978*a*, *b*). If the *ABO* locus is the structural locus for these enzymes, small structural differences must have arisen, changing the enzyme specificity in part.

These theories of gene action in the ABO and related systems do not suggest the basis of such allelic differences as A_1 versus A_2. Chemically, the products of I^{A1} and I^{A2} are identical. The main difference seems to be quantitative. In secretions, only quantitative differences are associated with these different alleles. However, on red cells, qualitative dif-

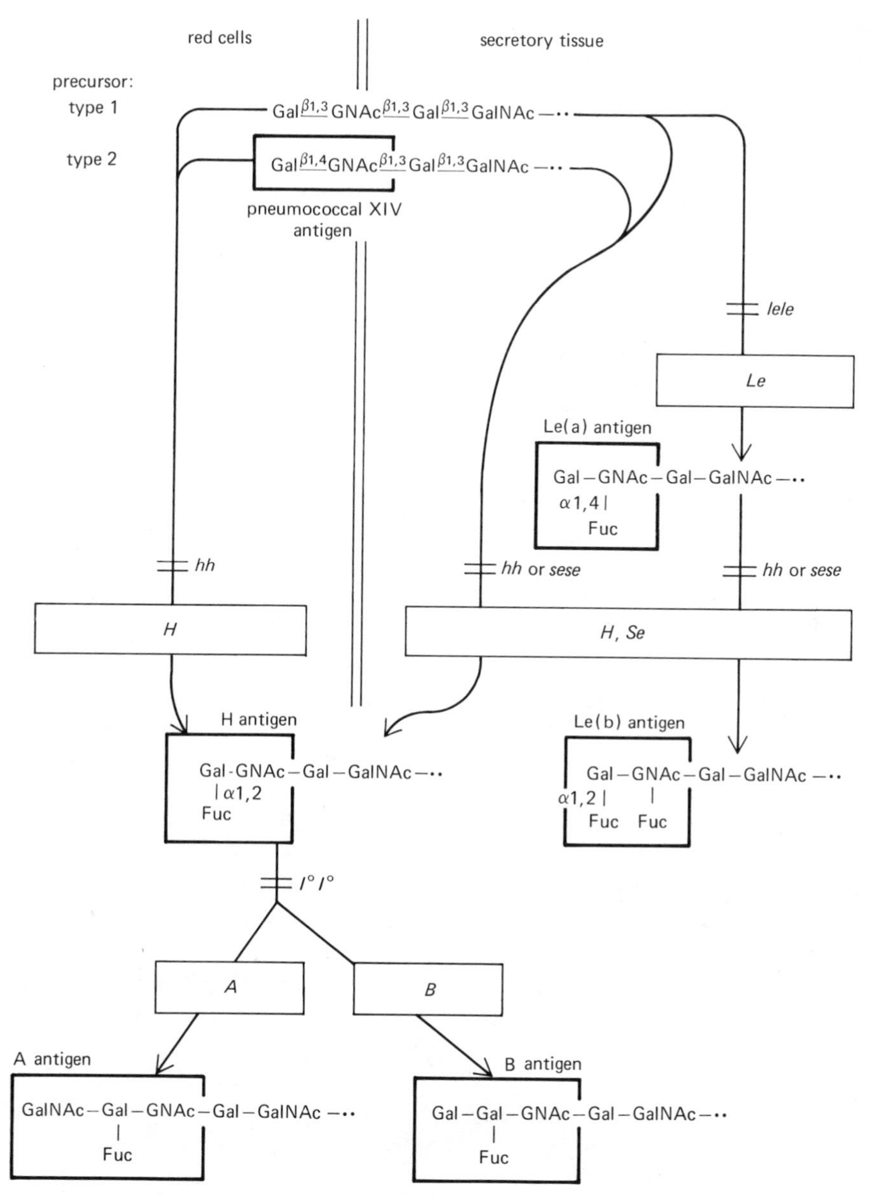

Figure 13-2 Diagram of reactions associated with *ABO, Hh, Sese,* and *Lele* loci. Alleles associated with enzyme activities are shown in boxes. Homozygous recessive combinations that block reactions are shown as double bars across the particular reaction. The Le enzyme can act only on type 1 precursor. The *H* gene acts both in secretory tissue and on red cell membranes, but in secretory tissue it can do so only if an *Se* allele is present at the secretor locus. The function of the *Se* allele is unknown. (*References may be found in Morgan and Watkins, 1969.*)

Abbreviations: Gal, D-galactopyranosyl; Fuc, L-fucopyranosyl; GNAc, *N*-acetyl-D-glucosaminopyranosyl; GalNAc, *N*-acetyl-D-galactosaminopyranosyl.

ferences also occur. Very little is known about the antigenic sites on red cells, except that the number is enormous—over a million for the ABO system alone. As these sites become more densely filled due to high activity I^A alleles—I^{A1} as compared to I^{A2} for example—the likelihood increases that adjacent sites will be filled with A chains, forming new antigenic surfaces consisting of multiple A chains.

The Lewis System. Some 23% of Europeans have red cells that are positive for an antigen designated Le^a. Shortly after discovery of Le^a in 1946, a related antigen, Le^b, was reported. Initially Le^a appeared to be recessive to Le^b, but clear exceptions soon appeared, and the inheritance was obscure for a number of years. Of special interest was the observation that all Le(a+) persons are nonsecretors of ABH substances.

Additional observations, and especially the chemical studies done on soluble Le^a and Le^b substances, have clarified the relationships of the Lewis system to the ABO, H, and secretor loci. Of special importance was the realization that the Lewis antigens occur only as secreted antigens and that their presence on red cells is due to adsorption of Le^a and Le^b from plasma. Persons may fail to show Le^a activity on red cells but nevertheless have Le^a in saliva and other secretions. Reference to Figure 13-2 will clarify the situation in the case of Lewis types. There are two alleles, *Le* and *le*. The dominant *Le* is responsible for attaching fucosyl residues to the penultimate *N*-acetylglucosaminyl residue of type 1 precursor. In type 2 precursor the 4 position is already blocked, so that it does not enter into this reaction. The product of the reaction has Le^a specificity. If the person has the dominant *Se* allele as well as the almost ubiquitous *H* allele, most of the Le^a substance is converted to Le^b by addition of another fucosyl residue. The saliva of such persons has sufficient Le^a to be Le(a+), but the red cells will be Le(a−). However, if the persons are nonsecretors, the *H* gene cannot act, and Le^a is the only product. Both saliva and red cells of such persons are Le(a+b−).

In contrast to the 23% of Europeans whose red cells are Le(a+), some 97% have Le(a+) saliva, giving an *Le* gene frequency of 0.82. The frequency of Le(a+) saliva in Africans is somewhat lower, approximately 55% with an *Le* frequency of 0.32.

The Rh System

The antigenic system subsequently designated Rh was first noted in 1939 by Levine and Stetson. These authors reported the presence of an antibody in the serum of a woman who had given birth to a stillborn fetus and who had suffered a severe hemolytic reaction when transfused with her husband's ABO-compatible blood. The antiserum from this woman agglutinated the cells of her husband as well as cells of approximately 80% of other persons who were ABO compatible. Levine and Stetson suggested that the mother lacked an antigen present on the cells of her husband and on the cells of the fetus, that fetal cells had entered the maternal circulation and immunized the mother, and that the antibodies produced by the mother had then entered the fetal circulation and caused destruction of the fetal red cells. This explanation proved correct for the rather common problem at birth, *erythroblastosis fetalis,* or hemolytic disease of the newborn.

Landsteiner and Wiener reported in 1940 their studies on immunization of rabbits with cells from the rhesus monkey, *Macaca mulatta.* The antisera agglutinated not only rhesus cells but also 85% of cells from human populations. The persons with this new antigen were designated Rh-positive and those without Rh-negative. It was soon recognized that the antigen reported by Levine and Stetson was the same as the Rh antigen, and that Rh incompatibility between Rh(−) mothers and Rh(+) fetuses was a major cause of hemolytic disease in newborns.

There followed a period of rapid discovery of antisera that defined antigens related to the *Rh* locus, and it soon became apparent that the *Rh* locus is very complex. There are many aspects yet to be clarified, and no chemical studies have been possible on isolated material.

The Genetics of Rh. There are five major antigenic classes associated with Rh, with many subtle variations in each class. In 1943, R. A. Fisher noted that two of the four classes then recognized behaved as alleles, and he designated them C and c. The remaining two classes were designated D and E. Subsequently, a class that is alternative to E was discovered and labeled e. No alternative to D has been found. Fisher considered the results to be best explained by three closely linked loci—*C, D,* and *E.* The alternative alleles *c, d,* and *e* occur, but only *c* and *e* lead to antigenic factors.

This point of view has been strongly challenged by Wiener, who correctly notes that interpretations other than linked loci are consistent with the observations. Indeed, there is no compelling reason to impose a linear concept derived from formal genetics onto a three-dimensional antigenic structure, and Wiener developed a system of nomenclature based on the assumption of a complex locus with many alleles. Nevertheless, the notation devised by Fisher (usually designated the Fisher–Race nomenclature) has proved a convenient means of communicating the results of Rh testing. Most workers use it without meaning to imply anything about the chemical activity or structure or the *Rh* locus, and we will follow this example. The *Rh* locus is now known to be on chromosome 1.

The five recognized antigenic classes—C, D, E, c, and e—form a stable genetic complex among which recombination, if it is possible, is exceedingly rare and has not been observed unambiguously. C is alternative to c and E to e. There is no antigenic alternative to D, but the genetic alternative to D is symbolized by *d*. These six genetic elements may occur in eight combinations, as shown in Table 13-4.

A number of variants of C and D are known. The variants C^w and D^u are especially common, the latter in African populations. The difficulty with such variants lies in their potential for misclassification. For example, the D^u type gives a weak reaction with anti-D and may give a negative reaction. Yet it is capable of stimulating an anti-D antibody response. A D^u person might therefore be considered compatible with an Rh(−) person (*cde/cde*), and his blood might be used for transfusion. Because of this it is generally wise to test apparent D(−) red cells with special antisera known to react with D^u cells.

Hemolytic Disease of the Newborn. Rh incompatibility has been a major contributor to hemolytic disease of the newborn. By incompatibility is meant any maternal-fetal combination in which the fetus possesses an antigen not possessed by the mother. The recognition of the disorder in Rh(−) women bearing Rh(+) fetuses can be translated into more modern terminology as *rr* (*cde/cde*) women bearing R^1r (*CDe/cde*), R^2r (*cDE/cde*), or R^0r (*cDe/cde*) children. The common antigenic factor in these children is D, and this has become the factor usually equated with Rh positiveness. The D factor is very antigenic;

Table 13-4 THE PRINCIPAL Rh GENE COMBINATIONS*

Reactions of the gene product with the five major antisera are shown. Heterozygotes involving two different complexes have the positive reactions of both.

Fisher–Race nomenclature	Modified Wiener nomenclature	Reactions with anti- C	D	E	c	e	Frequency of gene complex in England
CDe	R^1	+	+	−	−	+	0.408
cde	r	−	−	−	+	+	0.389
cDe	R^2	−	+	+	+	−	0.141
cDe	R^0	−	+	−	+	+	0.026
C^wDe	R^{1w}	†	+	−	−	+	0.013
Cde	r'	+	−	−	−	+	0.010
cdE	r''	−	−	+	+	−	0.012
CdE	r^y	+	−	+	−	−	rare
CDE	R^z	+	+	+	−	−	0.002

* From Race and Sanger, 1975.
† Most anti-C is also anti-C^w. Pure anti-C^w is required to recognize the C^w antigen.

that is, it is a potent stimulant of antibodies and was understandably involved in most early cases of incompatibility. C and E are less strongly antigenic but occasionally are responsible for induction of maternal antibodies. Neither occurs very often without D.

A mother who is R^1R^1 (*CDe*/*CDe*) could have a fetus who is R^1r (*CDe*/*cde*), in which case the fetus would possess the c antigen not present on the mother's cells. This would constitute incompatibility, even though both mother and fetus would be classifed as Rh(+). The c factor is not a strong antigen, so that in most such cases no problem would ensue. However, in a small proportion of cases, the mother would make antibodies against c, and hemolytic disease could result. It is such uncommon but possible situations that have provided the antisera necessary to recognize the very large number of antigenic specificities associated with the *Rh* locus and many other loci. Most cases though have their origin in a D(+) fetus carried by a D(−) mother.

It is widely observed that the first incompatible pregnancy in a mating usually has no problem. It is the second and later such pregnancies that suffer from hemolytic disease of the newborn. The explanation appears to be that most of the transfer of red cells from fetal to maternal circulation occurs at or near the time of birth. Thus there is insufficient time for a primary antibody response to occur during the first incompatible pregnancy. However, for later such pregnancies, the antibody response is much more rapid, since the immunization is a secondary immunization (see Figure 12-1). If the father is heterozygous, the series of incompatible pregnancies may be interspersed with compatible pregnancies that have no problems. Incompatibility at the ABO locus serves as a protective mechanism against Rh immunization. For example, if the fetal cells are type A, Rh(+) and the mother is O, Rh(−), the anti-A in the mother's plasma combines with the fetal cells, leading to their destruction before they elicit an immune reaction to Rh.

A preventive measure has been introduced in the management of Rh disease that has greatly reduced the impact of this very common problem (Clarke, 1972). Shortly after birth of an Rh(+) child to an Rh(−) mother, antibody preparations (anti-D) are injected into the mother. The antibodies combine with any fetal cells in the maternal circulation, removing them as antigenic stimuli. The mother therefore does not develop a primary antibody response. This very effective procedure has reduced the likelihood of maternal immunization to less than 1% as opposed to a 17% overall risk without treatment.

The Xg Locus

Of the numerous blood groups thus far identified, one of the more interesting is the sex-linked dominant Xg^a group. There are two alleles, Xg^a, giving rise to Xg(a+) cells, and its silent allele, *Xg*, which is known only as the nonantigenic alternative to Xg^a. The frequency of Xg^a in European-derived populations is 0.67. This is also the frequency of Xg(a+) males, and the frequency of Xg(a+) females is 0.89.

The *Xg* locus has no clinical significance, persons who form antibodies to it being exceedingly rare. It is, however, of great interest as a polymorphic marker on the X chromosome. Thus it can be used to recognize the origin of nondisjunction of sex chromosomes (see Chapter 5). In a few rare families, including some with XX males, Xg inheritance is abnormal. It is as if the *Xg* locus were translocated to the Y chromosome. It has been suggested that this may rarely occur during meiosis when the X and Y chromosomes show end-to-end pairing of their short arms. In that case, some X chromosomes would have male-determining genes from the Y short arm, and the Y would get the *Xg* region in exchange. A few observations are consistent with this hypothesis but more examples are needed. The vast majority of families show quite regular X-linked inheritance for Xg.

Other Red Cell Antigens

A large number of additional antigens are known, many of them polymorphic. Some, such as the MNSs system, appear to be as complex as the Rh system. Others are known only by two alleles. A reference such as Race and Sanger (1975) should be consulted for information on additional systems.

One especially interesting blood group is the Duffy system. There are two codominant alleles, Fy^a and Fy^b, recognized by their different antigens. A third allele, *Fy,* has no corresponding antigen. It was recently shown that the Duffy antigen is the receptor for the malarial parasites *Plasmodium vivax* and *P. knowlesi* (Miller et al., 1976). Persons homozygous for *Fy,* who are Fy(a−b−), are also resistant to these forms of malaria. The *Fy* allele occurs in high frequency only in Africans and appears to explain the resistance of Africans to these particular forms of malaria. The *Fy* allele (and resistance to vivax malaria) is uncommon in other populations.

Two groups of antigens that merit note are the nonpolymorphic antigens. Those that are very frequent are sometimes called "public" antigens, since nearly everyone possesses them. They are recognized when the rare person homozygous for an alternative allele is unintentionally immunized. At the other extreme are the "private" antigens. They have been detected in a very few families, perhaps only one. Typically they are transmitted as a dominant, with carrier males nearly always marrying negative females and giving rise to incompatible offspring. For both public and private antigens, the supplies of antisera for testing are likely to be scarce.

HISTOCOMPATIBILITY

When an organ or a tissue is transplanted from one species to another (*xenograft*), antibodies are formed against it and it is rejected. Similarly, when such a graft is made between different individuals within a species (*allograft* or *allogeneic* graft), rejection also occurs unless the two individuals are genetically identical. By contrast, grafts are accepted within genetically homogeneous strains of mice or between monozygotic twins (*isografts; isogeneic* or *syngeneic* grafts).

The basis for rejection is the presence on the donor cell surfaces of antigens that are foreign to the recipient. There are many such antigens under control of a number of gene loci. Both in mice and in man, the two species most extensively studied, the most intense immune response is to a series of antigens determined by the *major histocompatibility complex* (MHC), designated H-2 in mice and HLA (for human leukocyte A, where A refers to the first system) in man. These two loci appear to be homologous, and much has been learned about the probable organization of the *HLA* locus through study of the *H-2* locus.

Reference has already been made in Chapter 7 to the H-Y antigen, controlled by a locus on the Y chromosome and thought to be involved in sex determination.

Human Leukocyte Antigens

Skin grafting is the predominant histocompatibility test in mice but is hardly acceptable in man, although it has been used in rare instances to test for monozygosity in twins. Little progress was made in defining human histocompatibility antigens until *in vitro* tests were developed in the 1950's. The first such test involved agglutination of leukocytes in the presence of complement (Dausset, 1954). The histocompatibility antigens are found on the surfaces of virtually all nucleated cells, including leukocytes. Antibodies against these antigens bind to the leukocyte surface, but no visible reaction occurs in the absence of complement.

The predominant test in use now is the lymphocytotoxity test. In the presence of complement, cells that have adsorbed antibodies will lyse. Certain dyes are taken up only by dead cells, and a high proportion of stained cells is scored as a positive reaction. An additional test, useful for certain purposes, is a complement fixation test, using platelets or lymphocytes.

It was possible with these tests to show that antibodies are commonly formed against antigens found on cells other than red cells following transfusion or pregnancy. In both cases, white cells and platelets are the apparent source of antigenic stimulation. Such antibodies are not generally of importance for subsequent transfusions or pregnancies, but they may be on occasion. Definition of the antigens was initially very difficult because most sources of antisera contained antibodies against a number of antigens. Eventually, through cooperation of a number of laboratories, it was possible to identify a large number of antigens, controlled by alleles at two separate but closely linked loci. The first locus is known as the *A* locus (formerly the *LA* locus) and appears homologous to the *D* locus of the mouse H-2 region. The second human locus is the *B* locus (formerly the *4* locus), which is homologous to the mouse *K* locus. A third locus, the C locus, has also been identified, but fewer alleles are known as yet. A list of alleles at these three loci is given in Table 13-5.

Table 13-5 **LIST OF KNOWN ALLELES AT THE *HLA-A, -B, -C,* AND *-DR* LOCI WITH SELECTED GENE FREQUENCIES***

Allele	Europeans	Africans	Japanese	Allele	Europeans	Africans	Japanese
*A*1	.158	.039	.012	*B*5	.059	.030	.209
*A*2	.270	.094	.253	*B*7	.104	.073	.071
*A*3	.126	.064	.007	*B*8	.092	.071	.002
*A*w23	.024	.108	—	*B*12	.166	.127	.065
*A*w24	.088	.024	.367	*B*13	.032	.015	.008
*A*25	.020	.035	—	*B*14	.024	.036	.005
*A*26	.039	.045	.127	*B*18	.062	.020	—
*A*11	.051	—	.067	*B*27	.046	—	.003
*A*28	.044	.089	—	*B*15	.048	.030	.093
*A*29	.058	.064	.002	*B*w38	.020	—	.018
*A*w30	.039	.221	.005	*B*w39	.035	.015	.047
*A*w31	.023	.042	.087	*B*17	.057	.161	.006
*A*w32	.029	.015	.005	*B*w21	.022	.015	.015
*A*w33	.007	.010	.020	*B*w22	.036	—	.065
*A*w43	—	.040	—	*B*w35	.099	.072	.094
Blank	.022	.110	.042	*B*37	.011	—	.008
				*B*40	.081	.020	.218
				*B*w41	.012	.015	—
				*B*w42	—	.123	—
				Blank	.024	.179	.076
*C*w1	.048	—	.111	*DR*w1	.062	—	.045
*C*w2	.054	.114	.014	*DR*w2	.112	.087	.165
*C*w3	.094	.055	.263	*DR*w3	.089	.117	—
*C*w4	.126	.142	.043	*DR*w4	.078	.035	.144
*C*w5	.084	.010	.012	*DR*w5	.151	.074	.054
*C*w6	.126	.177	.021	*DR*w6	.086	.099	.067
Blank	.467	.502	.535	*DR*w7	.156	.066	—
				*WIA*8	.056	.072	.072
				Blank	.211	.450	.453

* From Bodmer and Bodmer, 1978.

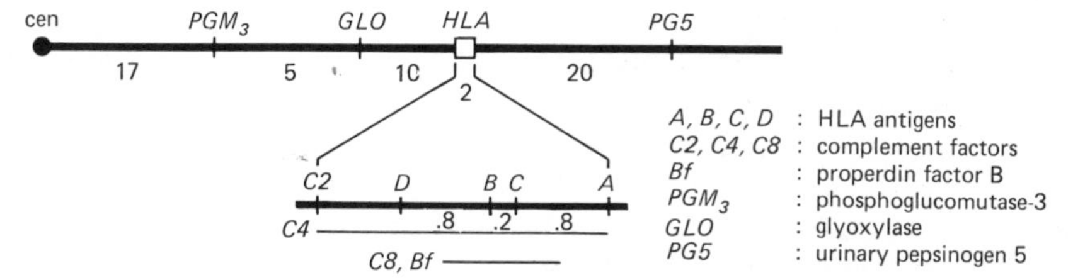

Figure 13-3 Diagram of the short arm of chromosome 6 showing approximate location of HLA with respect to other genetic markers. The HLA complex is expanded to show more detail. The numbers below the chromosome segments are the map distances in centimorgans.

The recombination distance between *D* and *K* of the mouse H-2 complex is 0.3 centimorgans. Estimates of recombination between *A* and *B* of the HLA complex give values of approximately 0.8 cM. These distances are sufficient to contain hundreds of genes. In the mouse, some of these additional loci have been found, including loci that appear to be functionally related to the H-2 complex but are just outside the region defined by *D* and *K*.

Mixed Lymphocyte Cultures

The growth of lymphocytes in culture is in part a response to antigenic stimulation by mitogens. Based on this idea, Bain and Lowenstein (1964) and Bach and Hirschhorn (1964) demonstrated that lymphocytes from two different sources mutually stimulate each other to grow, usually measured by DNA synthesis. The test was further refined by treating lymphocytes from one source with x-rays or chemical agents so that they are no longer capable of dividing but remain alive and serve as a stimulant to the untreated cells. The stimulation provided by the treated cells was shown to be divisible into a number of separate antigens, and these segregate with the HLA genes. However, it is now established that the mixed lymphocyte culture (MLC) antigens are distinct from previously recognized HLA systems but are transmitted as part of the same complex. Thus the MLC test identifies a new locus, the *D* locus, that has been shown to lie close to the *B* locus but outside the *A*-*B* segment (Figure 13-3). Alleles of the *D* locus are given in Table 13-5.

Organization and Genetics of the HLA Complex

The HLA complex is on chromosome 6. In addition to the four antigenic loci, there are several others related to immunity in the complex (Figure 13-3). These include complement factors 2, 4, and 8 and properdin factor B. The loci for the Chido blood group and the Rogers blood group are close to or within the HLA complex. Comparison with the mouse *H-2* locus suggests that additional loci should be detected. These include especially the *Ir* (immune response) genes, for which there is evidence in man but it is inadequate.

The location of so many genes with related functions in the same chromosome region has led to speculation that these loci originated by duplication and divergence and that natural selection has maintained this organization. There is no hint as to what the selective advantage of this particular organization may be. It should be remembered, though, that we probably know about only 1% of the loci in the HLA region.

HLA Haplotypes. Crossing over occurs between the HLA loci, but it is rare. Within families, then, the alleles that happen to be together on the same chromosome usually are transmitted together as a unit. The term *haplotype* is applied to such a combination. If the population were at equilibrium with respect to crossing over, the haplotype frequencies should be the product of the various allele frequencies at the segregating loci. However, this is far from the case. Certain combinations are frequent; others are exceedingly rare. The complex is therefore said to be in linkage disequilibrium.

The enormous number of haplotype combinations means that most persons are heterozygous and that any one person chosen at random almost certainly differs from any other person who might be chosen. This is important, not only in matching donors for organ transplantation, but also in establishing genetic relationships. Most matings involve four different parental haplotypes. Letting *a, b, c,* and *d* represent different haplotypes, matings typically are $ab \times cd$, with possible offspring *ac, ad, bc,* and *bd,* each with equal likelihood.

HLA typing has become a powerful tool in forensic medicine. In paternity cases, the required haplotype received from the father can usually be established with certainty, and only a small proportion of the male population is likely to have that haplotype. In one unusual case, fraternal twins were shown to have arisen from different fathers on the basis of HLA types (Terasaki et al., 1978).

The reason for the extreme polymorphism of the HLA genes are not known. It seems unlikely to be a matter of chance. No other locus is known to be so polymorphic. It is presumed therefore that the variability has a selective advantage. Transplantation could not have been a factor in mammalian evolution, leaving only maternal-fetal compatibility for a natural system involving antibodies to HLA antigens. This provides no obvious solution, since maternal-fetal incompatibility does not seem to lead to fetal wastage. (Virtually all pregnancies are incompatible.) Males do not produce antibodies unless they have received transfusions.

The Structure of the HLA Antigens

Very little information is available on detailed structure of HLA antigens. The antigens that are products of the mouse *H-2K* and *H-2D* genes and human *HLA-A* and *HLA-B* genes are embedded in the plasma membrane, with the antigenic portions exposed. They are glycoproteins with molecular weights of ca. 55,000 and consist of two subunits. The larger subunit is the HLA product and is responsible for variation among the loci. The smaller subunit is β_2-microglobulin, with molecular weight 11,600. The β_2-microglobulin is the same in all histocompatibility antigens, and its structure is determined by a locus that is not part of the MHC.

There is great similarity in structure between products of the different MHC loci. This suggests that they originated by duplication. Somewhat surprising is the observation that there is also some homology between the MHC antigens and immunoglobulin sequences. This would suggest that both had a common origin.

HLA and Transplantation

Transplantation of organs from one person to another requires matching as nearly as possible for histocompatibility antigens. If the graft is between monozygotic twins, the match is perfect. If it is between other persons, the results may be simply summarized: If the donor and recipient are matched for HLA type, the graft may be accepted. If they are not matched, the graft will be rejected. Matching does not assure acceptance, because there are other histocompatibility loci that may be different, and, indeed, there probably are loci within the MHC that are not yet recognized. But the HLA system is the *major* histocompatibility complex, and careful use of immunosuppressive agents will often permit other incompatibilities to be tolerated.

Because of the great number of HLA alleles, it is not easy to find an unrelated person who matches a potential graft recipient for HLA type. The best chance is among siblings of the recipient. One-fourth should be identical for HLA (but may differ for other loci). Kidneys are commonly transplanted from siblings if at all possible. If there are no siblings matched for HLA type, unrelated persons with the same type are used, but the success rate of such transplants is not as good. With much more research on the HLA and other histocompatibility loci, it should be possible to achieve a much higher success rate with organ transplants.

HLA and Disease

An embarrassingly large number of diseases have been reported to be associated with one or another HLA haplotype. If there are *Ir* loci in the

Table 13-6 **DISEASES ASSOCIATED WITH PARTICULAR HLA TYPES***

Disease	HLA type	Relative risk
Ankylosing spondylitis	B27	103.5
Celiac disease	Dw3	64.5
Reiter's disease	B27	40.4
Psoriasis (Japanese)	B37	26
Psoriasis (Japanese)	B13	22
Psoriasis (Japanese)	A1	21
Psoriasis (Japanese)	Cw6	15
Hemochromatosis	A3	9.7
Celiac disease	B8	8.1
Rheumatoid arthritis	DRw4	7.2
Chronic active hepatitis	B8	6.1
Psoriasis (Caucasoids)	B13	5.7
Graves' disease	Dw3	5.1

* Selected from Bodmer and Bodmer (1978).

MHC complex, it should not be surprising to find some infectious diseases more prevalent with certain haplotypes, if the HLA alleles of those haplotypes are in turn associated with differential abilities to form specific antibodies. But it is difficult to accept without reservation the dozens of associations reported so far.

The most impressive association, and the one about which there is least question, is between ankylosing spondylitis and antigen B27 (reviewed in Sachs and Brewerton, 1978; McMichael and McDevitt, 1977). This particular antigen is found in only about 5% of the general population but it is found in 95% of patients with this disorder. The observation has been repeated by different observers in different racial groups but with the same results. Not all persons with B27 will develop ankylosing spondylitis. Only about 3% do. But this is a greatly increased risk as compared to persons without B27. The relative risk is greater than 100, that is, a person with B27 has over a hundred times greater chance of developing the disease than someone without B27. This then could be one of the genetic factors that predisposes to a disease but is not sufficient to produce the disease. A list of diseases associated with HLA antigens in given in Table 13-6.

SUMMARY

1. Human red cells have many different antigens for which genetic variation is known. Antigenic types ordinarily are detected by agglutination of red cells with specific antisera. Such antisera are most often obtained when a person lacking a particular antigen is immunized by cells with the antigen, either through transfusion or through escape of red cells from fetal to maternal circulation.

2. The ABO antigens were the first discovered. There are three major alleles, giving rise to the four phenotypes A, B, AB, and O. The major A and B alleles can be divided further based on their determination of subtypes. Other loci associated with ABO types are the Bombay (*H*) locus, the secretor (*Se*) locus, and the Lewis (*Le*) locus. A and B antigens on red cells depend on activity of the *H* locus to produce the substrate for the enzymes controlled by the ABO locus. In secretory tissue, the *Se* allele is necessary for activity of the *H* locus. The *Le* locus acts in secretory tissue by adding fucose to one of the carbohydrate precursors of the ABO blood groups, making it Le(a+). If *H* and *Se* are both present, Le(a) is converted to Le(b).

3. The Rh antigens are especially important because of their association with hemolytic disease of the newborn (HDN). Mothers lacking an antigen often produce antibodies if that antigen is present on fetal red cells. This may lead to destruction of red cells in the fetus. A number of antigens are associated with the *Rh* locus, any particular combination being transmitted as a stable complex. The antigenic factor Rh D in the Fisher–Race nomenclature is most likely to be involved in HDN. Prevention of HDN by injecting antibodies to the D factor at the time of delivery of an Rh(D+) infant from an Rh(D−) mother has been successful in preventing most cases of HDN in subsequent children.

4. Other red cell antigens include Xg^a, determined by a locus on the X chromosome, and the MNSs system, also a complex system but without medical significance. The Duffy system, with three alleles (Fy^a, Fy^b, and *Fy*) has been shown to be the erythrocyte receptor system for *Plasmodium vivax* and *P. knowlesi*. Persons homozygous for *Fy*, primarily Africans, are resistant to these forms of malaria. Many additional inherited blood groups are known, including the "public" groups, possessed by nearly every person, and the "private" groups, possessed by very few.

5. The rejection of tissue and organ transplants is due to histocompatibility antigens. These are antigens found on most nucleated cells, but not on red cells. The strongest antigens are those associated with the major histocompatibility complex (MHC), designated HLA in man and H-2 in mice.

6. The *A*, *B*, and *C* loci in man are recognized by their stimulation of antibodies in transfusions and pregnancies. The *D* locus is recognized by the stimulation of lymphocyte growth in mixed lymphocyte culture (MLC). The four loci are part of the same complex and occur in the order *A-C-B-D*. Other loci in the MHC complex are for complement fractions C2, C4, C8, and properdin factor Bf. Loci for the Chido and Rogers blood groups also are in the MHC.

7. There are many alleles at the HLA loci. They are close together, crossing over infrequently. Each chromosomal combination (haplotype) is transmitted as a unit, and the great majority of persons are heterozygous for haplotype combinations. The reason for the great diversity is unknown.

8. HLA antigens consist of a polypeptide chain of approximately 45,000 molecular weight combined with a β_2-microglobulin molecule of MW 11,600. This complex is located in the plasma membrane. The similarity of molecular structure suggests that the different HLA alleles arose by gene duplication.

9. Successful transplantation of organs from one person to another requires matching of HLA types. Because of the great diversity of types, matching can be difficult except among siblings, where the likelihood of a match for any sib pair is 1/4. Even with matching of HLA types, there are minor histocompatibility loci that may create problems of rejection.

10. Many diseases are associated with particular HLA types. The most striking association is with ankylosing spondylitis, in which the relative risk of having the disease is increased over 100-fold if the person has the B27 allele.

REFERENCES AND SUGGESTED READING

BACH, F., AND HIRSCHHORN, K. 1964. Lymphocyte interaction: A potential histocompatibility test in vitro. *Science* 143: 813–814.

BACH, F. H., AND VAN ROOD, J. J. 1976. The major histocompatibility complex—genetics and biology. *N. Engl. J. Med.* 295: 806–813, 872–878, 927–936.

BAIN, B., AND LOWENSTEIN, L. 1964. Genetic studies on the mixed leukocyte reaction. *Science* 145: 1315–1316.

BODMER, W. F. (Ed.). 1978. The HLA system. *Br. Med. Bull.* 34: 213–320.

BODMER, W. F., AND BODMER, J. G. 1978. Evolution and function of the HLA system. *Br. Med. Bull.* 34: 309–316.

CLARKE, C. A. 1972. Prevention of Rh isoimmunization. *Prog. Med. Genet.* 8: 169–223.

DAUSSET, J. 1954. Leuco-agglutinins. IV. Leuco-agglutinins and blood transfusion. *Vox Sang.* 4: 190–198.

FERRARA, G. B. (Ed.). 1977. *HLA System—New Aspects.* Amsterdam: North-Holland, 170 pp.

FRANCKE, U., AND PELLEGRINO, M. A. 1977. Assignment of the major histocompatibility complex to a region of the short arm of human chromosome 6. *Proc. Natl. Acad. Sci.* (U.S.A.) 74: 1147–1151.

GIBLETT, E. R. 1969. *Genetic Markers in Human Blood.* Oxford: Blackwell Scientific Publ., pp. 267–345.

KLEIN, J. 1979. The major histocompatibility complex of the mouse. *Science* 203: 516–521.

LANDSTEINER, K., AND WIENER, A. S. 1940. An agglutinable factor in human blood recognized by immune sera for rhesus blood. *Proc. Soc. Exp. Biol.* 43: 223.
LEVINE, P., ROBINSON, E., CELANO, M., BRIGGS, O., AND FALKINBURG, L. 1955. Gene interaction resulting in suppression of blood group substance B. *Blood* 10: 1100–1108.
LEVINE, P., AND STETSON, R. E. 1939. An unusual case of intragroup agglutination. *J. Am. Med. Assoc.* 113: 126–127.
MC MICHAEL, A., AND MC DEVITT, H. 1977. The association between the HLA system and disease. *Prog. Med. Genet.* n.s. 2: 39–100.
MEO, T., ATKINSON, J. P., BERNOCO, M., BERNOCO, D., AND CEPPELLINI, R. 1977. Structural heterogeneity of C2 complement protein and its genetic variants in man: A new polymorphism of the HLA region. *Proc. Natl. Acad. Sci.* (U.S.A.) 74: 1672–1675.
MILLER, L. H., MASON, S. J., CLYDE, D. F., AND MC GINNISS, M. H. 1976. The resistance factor to *Plasmodium vivax* in blacks. *N. Engl. J. Med.* 295: 302–304.
MORGAN, W. T. J., AND WATKINS, W. M. 1969. Genetic and biochemical aspects of human blood-group A-, B-, H-, Le^a-, and Le^b-specificity. *Br. Med. Bull.* 25: 30–34.
MOURANT, A. E., KOPEĆ, A. C., AND DOMANIEWSKA-SOBCZAK, K. 1976. *The Distribution of the Human Blood Groups,* 2nd Ed. London: Oxford Univ. Press, 1055 pp.
NAGAI, M., DAVÈ, V., KAPLAN, B. E., AND YOSHIDA, A. 1978*a*. Human blood group glycosyltransferases. I. Purification of *N*-acetylgalactosaminyltransferase. *J. Biol. Chem.* 253: 377–379.
NAGAI, M., DAVÈ, V., MUENSCH, H., AND YOSHIDA, A. 1978*b*. Human blood group glycosyltransferase. II. Purification of galactosyltransferase. *J. Biol. Chem.* 253: 380–381.
RACE, R. R., AND SANGER, R. 1975. *Blood Groups in Man,* 6th Ed. Oxford: Blackwell Scientific Publ., 659 pp.
SACHS, J. A., AND BREWERTON, D. A. 1978. HLA, ankylosing spondylitis and rheumatoid arthritis. *Br. Med. Bull.* 34: 275–278.
SNELL, G. D., DAUSSET, J., AND NATHENSON, S. 1976. *Histocompatibility.* New York: Academic Press, 401 pp.
TERASAKI, P. I., GJERTSON, D., BERNOCO, D., PERDUE, S., MICKEY, M. R., AND BOND, J. 1978. Twins with two different fathers identified by HLA. *N. Engl. J. Med.* 299: 590–592.
WHO Scientific Group. 1971. Prevention of Rh sensitization. *World Health Organization Technical Report Series,* no. 468.

REVIEW QUESTIONS

1. Define:

incomplete antibodies
Coombs test
complement
maternal-fetal incompatibility
private antigen
histocompatibility

monospecific antisera	allograft
lectins	isograft
Bombay type	xenograft
erythroblastosis fetalis	mixed lymphocyte culture test

2. How would you set up a procedure for ABO typing, including A_1 versus A_2 subtyping, if you were unable to purchase commercial antisera and standard cells? Assume that you know yourself to be type A_2. How would your strategy change if you were type O?

3. In marriages of Rh− (*rr*) women with husbands of Rh types R_1r, R_1R_2, R_0R_0, rr″, and rr, which marriages would have a risk of children with hemolytic disease of the newborn? Which would have the highest risks?

4. You need a kidney transplant. Tissue typing is not available. Arrange the following potential donors in the order in which they are most likely to provide a successful transplant: (a) father, (b) maternal uncle, (c) identical twin, (d) friend of same age and sex, (e) brother, (f) fraternal twin.

5. Fill in the following table with predicted (+) or (−) reactions. AB represents I^AI^B. *H*, *Se*, and *Le* represent the dominant alleles, in either homozygous or heterozygous combination.

	Antigens on red cells					Antigens in secretions				
Genotype	A	B	H	Le^a	Le^b	A	B	H	Le^a	Le^b
AB, H, Se, Le										
AB, H, sese, Le										
AB, H, Se, lele										
AB, H, sese, lele										
AB, hh, Se, Le										
AB, hh, Se, lele										

6. What Rh genotypes correspond to the following reactions? Give a most probable genotype if two or more are possible.

Anti-	C	D	E	c	e	Genotype
a.	−	−	−	+	+	
b.	+	+	+	+	+	
c.	+	+	−	+	+	
d.	−	+	−	+	+	
e.	+	−	−	+	+	
f.	+	+	−	−	+	

CHAPTER FOURTEEN

SOMATIC CELL GENETICS

The ability to culture mammalian cells for extended periods outside the body has opened new possibilities for studying gene action under carefully controlled conditions and has provided new techniques for answering such varied questions as the location of genes on specific chromosomes and the clonal versus multiple origin of tumors. A few human cell lines, such as that designated HeLa, have been in culture for several decades. These were isolated from malignant tissue, however, and are adapted for growth under conditions that will not support growth of normal tissues. The chromosome constitutions of the various strains of HeLa now available have diverged not only from the original, presumably normal karyotype but also from each other.

The search for conditions that will support long-term growth of normal cells has been spurred by the hope that the ability to manipulate cells in much the same way that bacteria can be studied would permit very much greater insight into cell metabolism and gene expression and interaction. The substantial successes in this direction and the development of cell fusion techniques have justified the hopes and provided opportunities for prenatal diagnosis and linkage studies not originally foreseen.

METHODS OF CELL CULTURE

If a bit of fresh tissue is placed in nutrient media, only a few of the cells appear to grow and divide, moving away from the tissue onto the supporting surface. These are the fibroblasts. Depending on the tissue, other cell types may also grow but generally at a much slower rate. Since fibroblasts are, in some ways, a more generalized cell than many highly differentiated tissues from which they can be isolated, it was thought at one time that "dedifferentiation" accompanied the ability of the differentiated cells to grow in culture. It is now appreciated that the

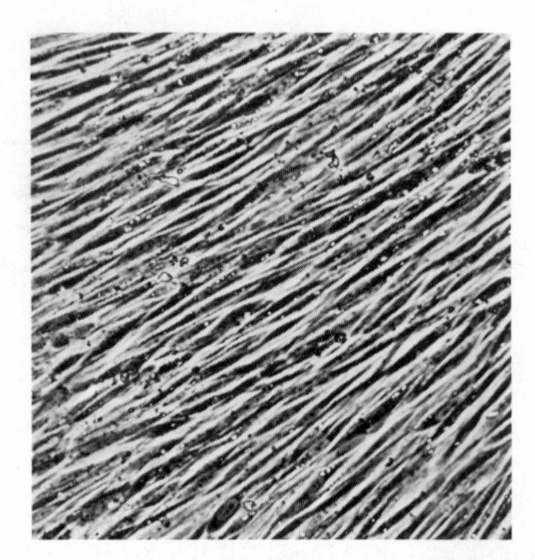

Figure 14-1 Monolayer of human fibroblast cells grown to confluence on a glass surface. (Furnished by T. R. Chen.)

very widely distributed fibroblasts simply outgrow the more differentiated cells under culture conditions but that the differentiated cells do retain their characteristics in culture. Many differentiated cells will grow and exhibit the expected biochemical properties of that tissue.

Fibroblast Cultures. By far the most extensively used cell cultures involve fibroblasts. They can be rather easily started by placing a bit of tissue, usually skin, in medium and allowing it to incubate at 37°C. Fibroblasts require a surface to which they can attach, and they will grow and spread to form a confluent monolayer of cells (Figure 14-1). When the surface is covered, growth ceases due to a phenomenon known as *contact inhibition.* Occasionally a cell may lose the property of contact inhibition, sometimes associated with infection by a virus, at other times for unknown reasons. In such instances, cell growth continues, leading to a clone of cells that behaves in some ways as if it were malignant. Such a change is referred to as *transformation.*

Newly established fibroblast cultures from normal tissue grow well initially. However, growth ceases after about 50 generations. The basis of this senescence has not been established. During the period of growth, the original karyotype is usually preserved. With transformation, a variety of karyotypic changes may occur. Cells that have not transformed and have retained their original karyotype are sometimes said to be *homonuclear.* Cells that have an altered karyotype are *heteronuclear* or *heteroploid.*

A key technique in the development of bacteriology was the ability to grow pure cultures from isolated single cells. This has proved more difficult in the case of mammalian cells, especially untransformed cells such as fibroblasts. The difficulty is also certainly attributable to lack of knowledge of trace nutritional requirements. Nevertheless, the ratio of volume of medium to inoculum size must not be excessive if a culture is to grow. An important advance in cell culture methods was achieved by Puck when he developed a method of plating cells over a confluent layer of feeder cells that had been treated with ultraviolet radiation. The radiation destroys the ability to replicate further but permits metabolic activity. These cells can then condition the milieu into which the isolated test cells are introduced, and the test cells in turn can grow into a colony, each originating from a single cell (Figure 14-2). The system is thus comparable to isolation of bacterial colonies on an agar plate. The plating efficiency (the proportion of introduced cells that give rise to clones) is substantially less than 100%.

Permanent Cell Lines. If transformation occurs, a permanent cell *line* may result. Cell lines do not show senescence, and they can grow in suspension under certain conditions. While some have origi-

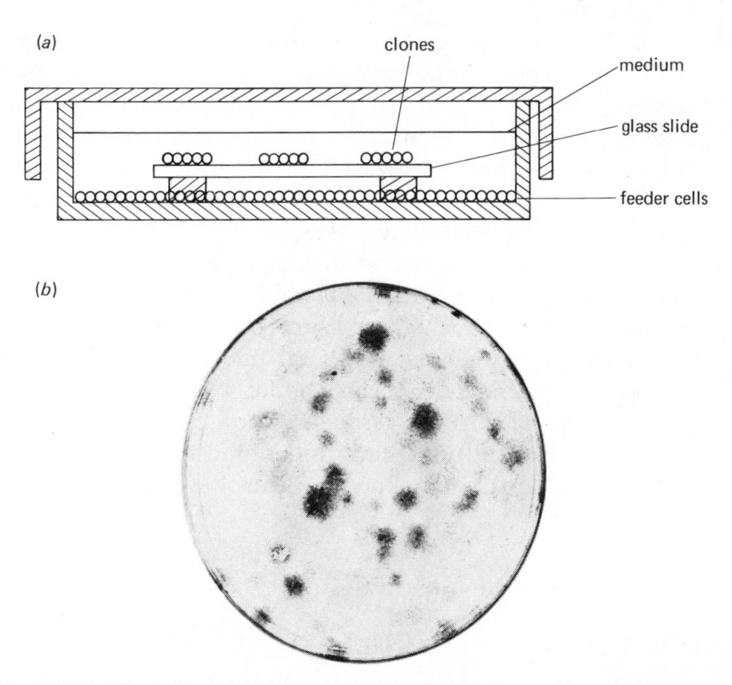

Figure 14-2 Growth of clones of cells using a feeder layer of inactivated cells. (*a*) Diagram of the growth chamber showing a layer of inactivated HeLa cells on the bottom of the Petri dish. A dilute suspension of cells to be cloned is added to the glass slide suspended above the inactivated feeder cells. The size of the cells and clones is exaggerated in this diagram. (*b*) Photograph of a slide showing clones of human cells. (*Figure 14-2a redrawn from Puck and Marcus, 1955; Figure 14-2b furnished by T.R. Chen.*)

nated from cultures of normal cells, most in common use were first isolated from malignant tumors. Such cell lines continue to exhibit the features of the tissue from which they were derived, secreting plasma proteins in the case of liver cells or manufacturing specific hormone receptors. Most established cell lines, such as HeLa, are heteroploid, which limits their use for some purposes. An advantage of cell lines is the ease with which they can be cloned. Usually single cells in soft agar will grow to form a colony that can be used to generate clones of identical cells. This is especially useful in isolating mutants.

Lymphoblastoid Cultures. Lymphocytes appear to grow primarily as a response to antigenic stimulation. They are grown suspended in culture media rather than on glass surfaces. An important ingredient of the medium is phytohemagglutinin (PHA), a plant extract originally added because of its ability to agglutinate unwanted red blood cells, but subsequently discovered to have growth-stimulating activity (mitogenic activity) as well, apparently due to antigenic stimulation. The importance of antigenic stimulation is illustrated by mixed lymphocyte cultures (Chapter 13).

A number of permanent cell lines derived from lymphocytes have been established. These cell lines, designated *lymphoblastoid,* promise to be useful in much the same way that bacterial cultures have proved useful (Glade and Beratis, 1976). They are chromosomally stable for long periods and they grow well in suspension. They are difficult to isolate *de novo,* but success is greater if lymphocytes are taken from persons with certain viral diseases, such as infectious mononucleosis. Some investigators have had success by incorporating cell-free medium taken from previous lymphoblastoid cultures. For these reasons, it is thought that transformation from lymphocyte to lymphoblastoid cultures may be associated with viral infection of the cultures, and E-B virus (Epstein-Barr virus) can regularly be detected in lymphoblastoid lines. Since this virus is associated with certain tumors, it may be that lymphoblastoid cultures have undergone what is in effect a malignant transformation. Lymphoblastoid lines have rather a low cloning efficiency.

METABOLIC STUDIES IN CULTURED CELLS

Although fibroblasts may seem not to be representative of all somatic cells, they have a remarkably diverse metabolism, expressing a very large number of genes concerned with intermediary metabolism. Many genes are not expressed, such as those for hemoglobin and for certain nerve and liver cell functions. Nevertheless, fibroblast cultures have been an exceedingly rewarding approach to the study of normal metabolism and of metabolic disorders. Not only are many genes expressed in fibroblast cultures, but also these cultures make possible experiments that could never be carried out with human subjects. The environment of cultured cells can be rigidly controlled, and the cells can be exposed to alien surroundings that may be highly enlightening to the scientist but incompatible with continued survival or normal function of experimental mammals, much less man. Further, an experiment with fibroblasts can be repeated many times, and careful controls can be set up. With human subjects, it may be virtually impossible to repeat a study if a disease is very rare, and the deliberate exposure of normal control subjects to abnormal levels of metabolites or drugs raises ethical as well as scientific questions.

A final advantage offered by fibroblasts for investigation of metabolic disorders is the ability to work with fibroblasts from a particular disorder whether or not the donor patient himself is available. Many laboratories maintain stocks of fibroblasts from a variety of patients, long after the patients are dead or have become otherwise inaccessible. The stocks can be frozen in liquid nitrogen indefinitely and returned to normal growth when desired and convenient. Cells can be shipped by mail to all parts of the world, so that several laboratories may be working simultaneously on a single patient. An investigator who wishes to study a rare disorder may be able to do so by ordering cell cultures from a central source. These include not only inherited metabolic disorders but cells with chromosome aberrations as well. Finally, it is possible to make crosses between human and nonhuman cells, such as mouse cells, in a parasexual cycle that would never be possible with whole animals!

Identification of Genetic Defects

Many metabolic defects, especially blocks, have been identified by investigators working with intact patients. In other instances, the nature of the defect may be more subtle than a complete block, and the relevant metabolic reactions may be too responsive to environmental variables. In these instances, cell culture methods sometimes permit observations that are subject to much more rigorous interpretation. Not all metabolic defects can be so studied. Some, as in the case of phenylketonuria, result from defects in enzymes found only in liver. At present it is not possible to culture normal liver cells. Others involve transport defects, and these may not be expressed in single cells. But there are good examples of defects for which real understanding had to await development of cell culture methods.

Hypercholesterolemia is an example of a metabolic defect that would have been nearly impossible to work out in intact persons (Chapter 11). The exposure of cells to artificially high or low levels of LDL (low density lipoprotein), necessary to relate cholesterol synthesis to plasma levels of LDL, could not be achieved except *in vitro*. Fortunately, fibroblasts express the defect in binding of LDL and therefore can serve as an excellent model for cholesterol regulation.

An additional example is the binding of dihydrotestosterone (DHT). Cultured cells from both males and females normally bind DHT to surface receptors. Fibroblasts from some persons with testicular feminization syndrome lack such binding, as evidenced by the inability to bind radioactively labeled DHT (Chapter 7). However, this particular defect does not occur in all persons with the syndrome (Amrhein et al., 1976). Additional studies of DHT binding by fibroblasts from normal persons show that binding is greater for fibroblasts obtained from genital areas than from nongenital areas, even though the fibroblasts have gone through a number of replicative cycles (Kaufman et al., 1977).

In both examples, the defect is in a membrane function. These would be much more difficult to study in tissue homogenates, where the cell architecture is destroyed and the chain of biochemical events disrupted. But even in simple enzyme blocks, the *in vitro* studies with cultured cells greatly expand the range of possible experiments. The need to vary the milieu of the somatic cells in defined ways cannot always be met in a whole organism. All somatic cells carry a full gene complement, although all genes may not be expressed. On the other hand, many are expressed *in vivo* in tissues in which they are thought not to play an important role and also *in vitro* in cell cultures.

Complementation Studies

One of the standard techniques of experimental genetics is the complementation tests. As ordinarily carried out, cells or an organism with two mutant recessive genes in the trans configuration are tested for presence or absence of mutant phenotypes. If absent, it is concluded that the organism is heterozygous at two functionally different sites, and the two mutations are said to be complementary to each other. On the other hand, if the trans double heterozygote is of mutant phenotype, it is concluded that the same function must be defective on both chromosomes.

Complementation can also be carried out using heterokaryons, that is, cells with two or more separate nuclei. This technique has been especially used in *Neurospora*. An additional type of complementation test is when different cells are able to compensate for different genetic defects. If one cell line is defective in function A and the other in function B, then sometimes when the two cells are grown together, the two lines are able to supply the missing function for each other, provided the two gene products are able to diffuse outside the cell.

Reference was made in Chapter 11 to the distinction between

Hurler's syndrome and Hunter's syndrome on the basis of complementation of fibroblast cultures from these patients. Prior to identification of the different specific enzyme defects, it was demonstrated that metachromatic granules, which form in cultures from these patients, do not form if the two cell lines are grown together. The existence of several complementation groups in xeroderma pigmentosum patients, demonstrated in mixed cell cultures, tells us that the disorder can result from several primary genetic defects, presumably reflecting a similar number of gene loci.

The use of complementation in this manner is primarily of interest when two diseases show very similar, possibly identical, phenotypes. If the gene products are released into the surrounding medium, there is at least the possibility that nonidentity of genetic function can be established this way. This is probably more likely to be the case with degradative metabolic pathways than with biosynthetic pathways, but the possibility should not be overlooked in any case.

Prenatal Diagnosis

References have already been made to the possibility of genetic diagnosis form cells in amniotic fluid. When an amniotic tap is made, one may do direct chemical examination of the fluid, one may do direct studies on the few cells in the amniotic fluid, or one may culture fibroblasts from these cells and do studies on the cultures. Each has its advantages for particular disorders.

Examination of cultured cells is especially useful for enzymes that are not secreted into the amniotic fluid. Under these circumstances, the number of cells that can be isolated from the amniotic fluid is apt to be too low for reliable assays of enzyme activities. Therefore, it is common to culture such cells for a period of two to four weeks, increasing manyfold the amount of material available for study.

Some dozens of metabolic diseases are now diagnosable in cultured amniotic cells (Milunsky, 1976). Many others are potentially diagnosable. The length of the list illustrates the fact that many genes are active in cultured cells. It is the hope that ways can be found to turn additional genes on. Cell hybridization, discussed later, offers some possibilities for this.

MUTATIONS IN CULTURED CELLS

Although most of the cell lines that have been investigated have been derived from patients with specific hereditary defects, it has proved possible to isolate newly arising mutants by the use of various selective

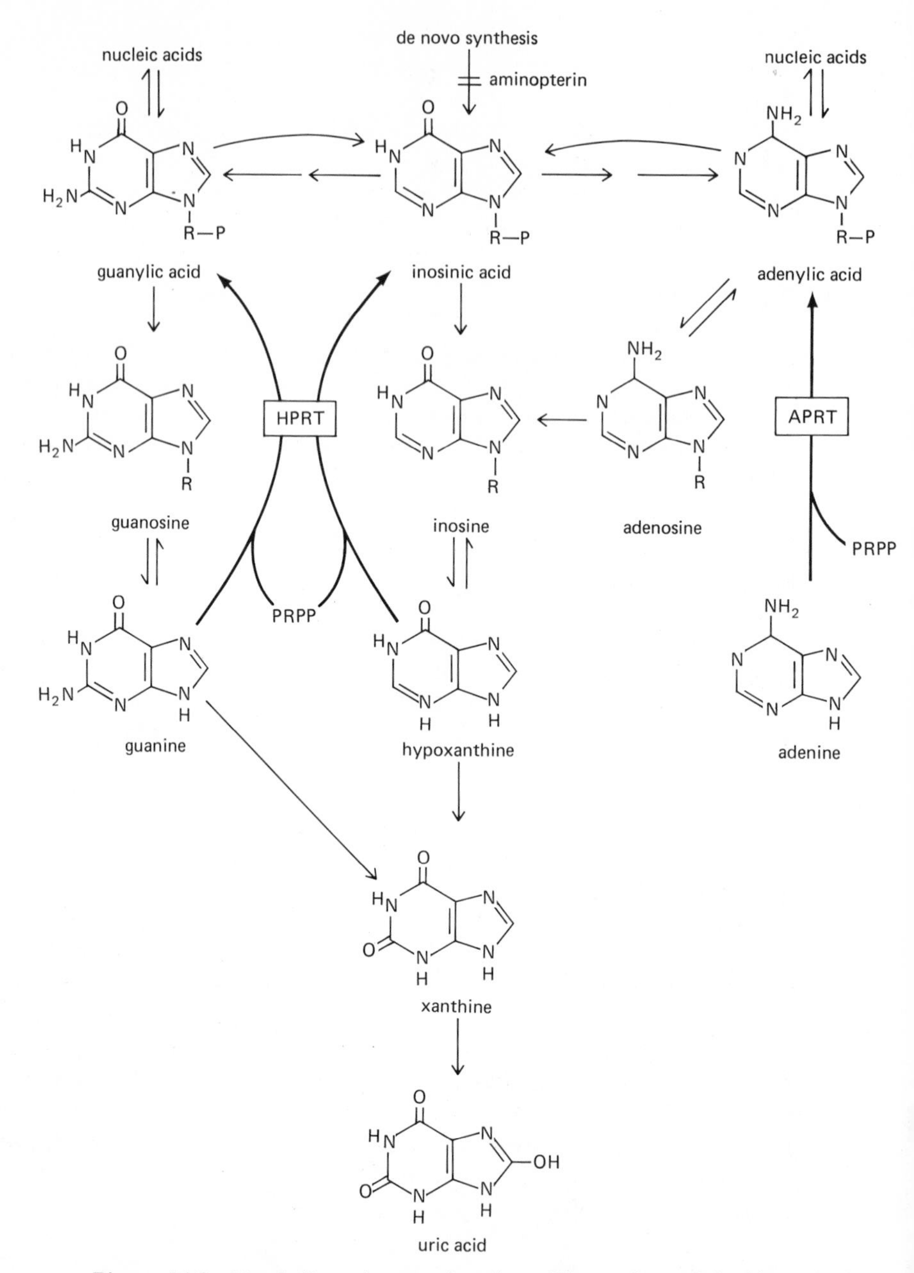

Figure 14-3 Metabolic pathways of purines. The product of the biosynthetic pathway is inosinic acid, which is then coverted into guanylic acid (guanidine) or adenylic acid. The primary breakdown products are guanine and hypoxanthine, respectively, which may then either be further catabolyzed to xanthine and uric acid, or they may be "salvaged" by conversion back to the respective nucleotides. The salvage pathways are shown in heavy arrows, the principal pathways being catalyzed by HPRT. Free adenine, which is not a major breakdown product of purine metabolism, also is salvaged by APRT. If *de novo* synthesis is blocked in mamma-

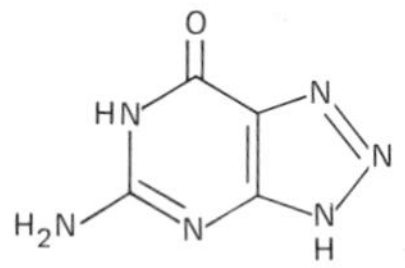

Figure 14-4 8-Azaguanine.

media. The best known example is the use of selective media to isolate cells deficient in hypoxanthine phosphoribosyltransferase (HPRT), the same deficiency responsible for Lesch–Nyhan syndrome. This enzyme function is shown in Figure 14-3. It is responsible for what is sometimes known as the salvage pathway for purines. Mammalian cells can manufacture purines *de novo,* but they also have the enzymes necessary to recover purines from breakdown of nucleic acids and from dietary sources rather than break them down completely and resynthesize them from nonpurine starting materials. Cells will utilize exogenously supplied purines in preference to making their own purines, but probably in most cells both pathways operate to some extent. In HPRT− cells, the salvage pathway is blocked and cells must manufacture their purines from nonpurine material.

If one adds 8-azaguanine (Figure 14-4) to cultures of normal cells, the structural similarity of this antimetabolite to guanine permits it to enter into some of the normal reactions of guanine. Azaguanine appears to be incorporated into DNA, where it interferes with normal growth and kills the cells. Since this incorporation is dependent on the salvage pathway, those cells that lack HPRT activity are unable to utilize azaguanine and therefore can grow in its presence. This constitutes a selective medium that permits isolation of particular kinds of mutants, so-called azaguanine-resistant (AG^r) mutants. Using this technique, several investigators have isolated new HPRT− mutants from cultures of normal cells and have been able to do a variety of interesting studies on these mutants (Albertini and De Mars, 1970; De Mars and Held, 1972).

A selective system that operates on the same enzyme but in the opposite direction is known as HAT medium. HAT medium contains hypoxanthine, aminopterin, and thymidine. Aminopterin is an antimetabolite that prevents *de novo* synthesis of purines and methylation of deoxyuridylate to form thymidylate. Cells growing in the presence of aminopterin are dependent on exogenous sources of purines and of thymidine. Hypoxanthine can serve as an exogenous purine source pro-

---

lian cells, hypoxanthine can serve as the sole source of purines, provided the HPRT pathway is active.

Abbreviations: HPRT, hypoxanthine-guanine phosphoribosyltransferase; APRT, adenine phosphoribosyltransferase; PRPP, phosphoribosylpyrophosphate; R, ribose; P, phosphate.

vided the HPRT is available to convert it to inosinic acid (Figure 14-3). A medium containing hypoxanthine, aminopterin, and thymidine will therefore support growth of cells with active HPRT but not cells that are HPRT−. Therefore, one can select for HPRT+ mutants from cultures of cells that are HPRT−.

Another system that has been used with human cells as well as other mammalian cells is selection for thymidine kinase negative (TK−) mutants in the presence of TK+ cells. This can be accomplished by adding bromodeoxyuridine (BrdU) to the medium, since BrdU enters into the same metabolic reactions as thymidine, including incorporation into DNA (Figure 14-5). The phosphorylation of thymidine is necessary only if exogenous thymidine is the obligatory source. So long as the cell can manufacture its own thymine nucleotides, TK− mutants are able to grow. If BrdU is present, TK+ cells incorporate it into DNA, which ultimately leads to cell death. Therefore, in the presence of BrdU, all nonmutant TK+ cells are killed, while TK− cells survive. TK+ cells also can be selected for in HAT medium, which contains both aminopterin and thymidine, making the cells dependent on the conversion of thymidine for DNA synthesis. HAT medium thus re-

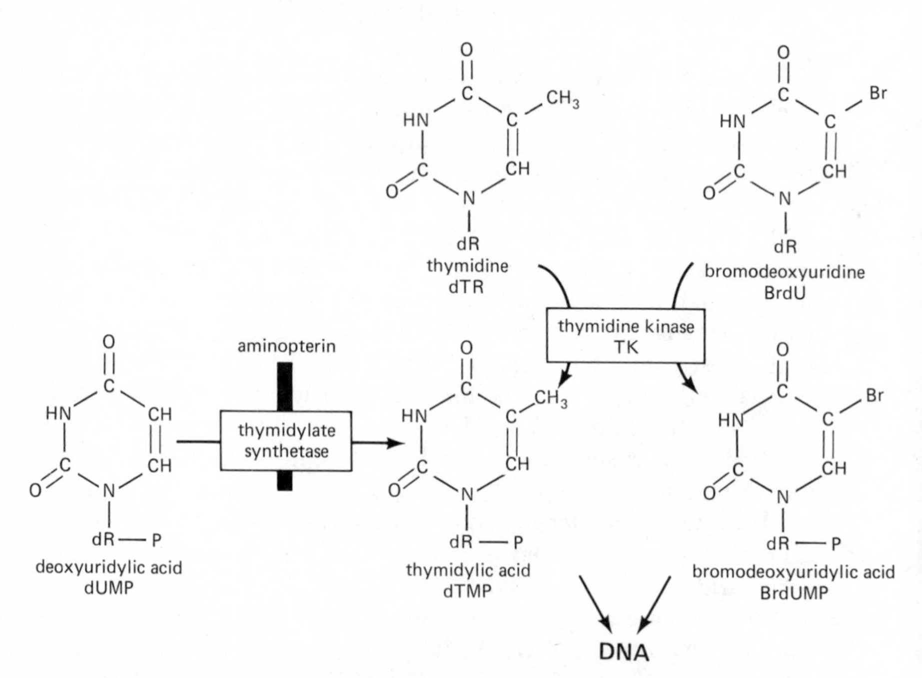

Figure 14-5 Production of thymidylic acid from biosynthetic pathway (dUMP) or from salvage pathway (dTR). The biosynthesis is blocked by the folic acid antagonist aminopterin, but dTMP can be produced from thymidine if thymidine kinase (TK) is present. BrdU also is phosphorylated by TK and incorporated into DNA, where it causes lethal mutations. TK− cells cannot use thymidine or BrdU and are dependent on the biosynthetic pathway. If that is blocked, they cannot grow.

quires both HPRT+ and TK+ alleles for growth. A number of other selective systems have been developed for specific purposes (Chu and Powell, 1976).

The ability to isolate mutants *in vitro* has expanded opportunities for study of mutation rates and for identification of new mutants not encountered in natural populations. The biochemical geneticist is no longer dependent only on those mutants that are found in human beings.

HYBRID FORMATION BY MAMMALIAN CELLS

In 1960, Barski, Sorieul, and Cornefert were able to demonstrate for the first time the fusion of mammalian cells, leading to a mononucleate hybrid capable of continued proliferation. The early studies were done with mouse cells with distinguishable heteroploid karyotypes. The frequency of fusion is very low, however, so that isolation of hybrid cells was difficult.

Littlefield (1964) devised the first effective selective medium, a medium that still is used as the primary means of isolating hybrid cells. This HAT medium was described in the previous section. If two parental lines are used, one being HPRT−, the other being TK−, neither can grow in HAT medium. The HPRT− cells cannot use hypoxanthine as a source of purines, and the TK− cells cannot phosphorylate thymidine to form the essential thymidylate. The parental stocks would be HPRT−/TK+ × HPRT+/TK−. Hybrid cells would have both parental complements and would be able to grow in HAT medium (Figure 14-6). This medium proved effective, not only with mouse–mouse hybrids but also with hybrids between other species so long as one parental strain is HPRT− and the other TK−.

The frequency of cell fusion is very low, on the order of one per mil-

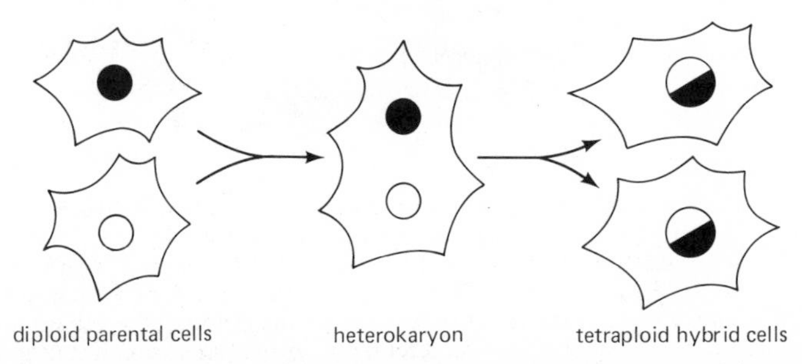

Figure 14-6 Diagram showing fusion of two parental cell lines to form a heterokaryon. When the heterokaryon undergoes mitosis, the parental chromosomes fuse in the daughter cells to yield single tetraploid nuclei.

lion cells in culture. A major advance was the recognition that the rate of fusion could be enormously increased if cells are exposed to inactivated Sendai virus (Okada, 1962). The virus can be inactivated by exposure to ultraviolet irradiation, destroying its ability to replicate. However, it still binds to cells and causes them to fuse. Under optimum conditions, some 90% of cells will be fused. This technique has been especially useful when human and mouse fibroblasts are fused. Cells also will fuse with high frequency if exposed to polyethylene glycol, which has become a standard technique in cell hybridization.

The utility of hybrid cells depends in part on their expression of parental genes in the hybrid combination. This appears generally to be the case. A number of enzyme systems have been analyzed in hybrids, and the parental enzymes can regularly be detected in the hybrids, so long as the chromosomes on which their genes are located are present. In the case of some enzymes, new combinations of protein monomers occur, illustrating convincingly the activity of the parental genes and the functional homology of their products (Figure 14-7).

The utility of hybrid cells also stems in part from the fact that such tetraploid cells tend to lose chromosomes. In the case of mouse-human hybrid cells, human chromosomes are preferentially lost for reasons that are not understood. This permits association of certain cell traits with the presence of certain human chromosomes, a technique that has proved very useful in chromosome mapping (Ruddle, 1973). To illustrate, since the hybrid cells depend on the presence of the human TK+

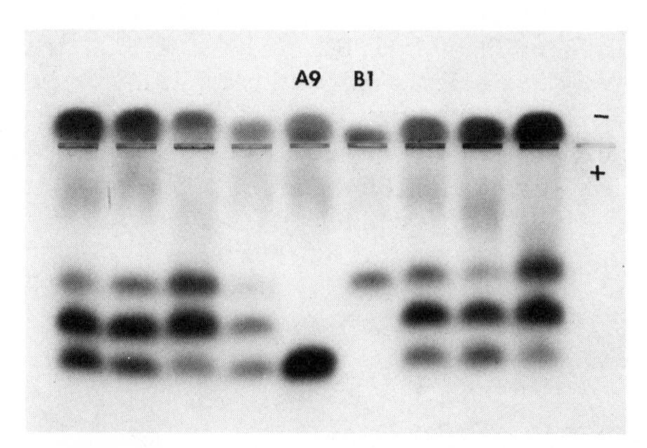

Figure 14-7 Electrophoretic patterns of malate dehydrogenase (MDH) from mouse (A9) and hamster (B1) cells and from their hybrids. The supernatant MDH migrates toward the anode. Each unlabeled slot is from a different clone of hybrid cells. MDH is a dimer. Therefore, in hybrids, the dimeric molecule might be represented as M_2, MH, or H_2, depending on the origin of the subunits. The MH form is a new combination not represented in either parental cell line and migrates with a mobility intermediate between pure M and pure H dimers. (*From B. R. Migeon and B. Childs, 1970. Hybridization of mammalian somatic cells.* Prog. Med. Genet. 7:1, by permission.)

gene in order to grow in HAT medium, any cell that lost the chromosome bearing the TK+ gene would not survive. The loss of chromosomes is random, so that if a number of such cultures are observed, the combinations of human chromosomes still remaining would be a random sample of human chromosomes with the single exception that the TK+ gene must always be present. In analyzing such a series, it became apparent that only human chromosome 17 was essential to continued survival of the cells. Therefore, the TK+ gene could be assigned to chromosome 17. Other examples will be given in Chapter 19 along with extensions of this technique that have been developed especially in conjunction with the study of linkage

The fact that hybrid cells tend to lose chromosomes to return to a quasi diploid state provides a parasexual cycle like that known in certain fungi such as *Aspergillus*. It is not known to what extent crossing over may occur in somatic cells. If it does occur, it might be possible to develop a system of formal genetic analysis using laboratory "matings" of cultured cells rather than depending on matings of whole organisms. Many more gene and chromosome markers are needed that are expressed in cell culture. When these become available, there should be no limit to the kinds of matings, within species or among mammalian species, that can be carried out in the laboratory.

An additional approach to introduction of specific genes and chromosomes into cultured cells is the direct incorporation of isolated chromosomes of DNA into cell lines. Mouse fibroblast or L-cell lines immersed in a suspension of Chinese hamster or human chromosomes will acquire the ability to produce Chinese hamster or human hypoxanthine phosphoribosyl transferase (McBride and Ozer, 1973; Willecke and Ruddle, 1975; Wullems et al., 1975). Such transfer apparently does not involve incorporation of the human genes into the mouse genome, although that possibility is not ruled out as an occasional result.

Cell hybrids are especially useful in studying interaction of different mutant genes in what is in effect a *cis-trans* test. In one interesting application, rat cells that were HPRT− were fused with human cells, with resulting production of rat HPRT (Croce et al., 1973). It is concluded from such a cross that the rat mutation involved a control gene rather than the structural gene, which appears to be intact. In a similar cross, a mouse hepatoma line that produces mouse albumin was fused with human leukocytes that do not. The hybrid produced both mouse and human albumin (Darlington et al., 1974). This is indication that the genes that normally function in liver cells can be made to function in other cells, given the appropriate signals. In other words, differentiation, in this case, is reversible.

The possibility of activating genes by cell fusion was demonstrated by Rankin and Darlington (1979), who fused mouse hepatoma cells with cells from human amniotic fluid. These human cells ordinarily do not produce liver proteins, but the hybrids produced the human serum proteins albumin, transferrin, α_1-antitrypsin, and ceruloplasmin, which normally are manufactured in the liver. This opens up the possibility that selective fusion of amniotic cells with differentiated tumor cells may make it possible to diagnose defects in liver function prenatally.

The various hybrids do not follow a single pattern of gene repression or activation. Rather, some structural genes are extinguished in a hybrid with another cell type. Others are activated. Cell hybrids will continue to be useful in understanding gene action and regulation and will undoubtedly be most important in unraveling the complexities of differentiation.

One additional technical accomplishment deserves special note. That is the potential incorporation of human genes into mice. The teratocarcinoma mouse tumor cell line is a totipotent line that can be injected into mouse blastocysts and then differentiate with the recipient cells to form a chimeric mouse (Dewey et al., 1977). The *allophenic* mice are quite normal, even though they have four parents. By treating the teratocarcinoma cells with a mutagen and selecting for HPRT− cells before blastocyst injection, it is possible to create mice chimeric for HPRT+ and HPRT− tissue. Thus the treatment and selection occurred *in vitro*, but the resulting cells have been transferred to *in vivo* systems. Since gonadal tissue may include the teratocarcinoma-derived cells, the offspring of such mice occasionally would carry the induced mutation.

Rather than using mutagenesis to generate variant teratocarcinoma cells, Illmensee et al. (1978) fused the teratocarcinoma cells (which are TK−) with human fibrosarcoma cells (which were HPRT−). The hybrids were grown in HAT medium, which would select for human TK+/mouse HPRT+ cells. The hybrid rapidly lost human chromosomes, except for 17, which carries the TK gene. These cells could then be injected into blastocysts, producing chimeric mice with at least one human chromosome in some tissues. Such an approach offers the opportunity of transferring selected human genes to inbred strains of mice for types of experiments that could never be carried out on human beings.

SUMMARY

1. Fibroblasts are the human cells most extensively used for cultures. Normally they are grown on a glass or other surface, forming a confluent layer of single cells. They grow relatively well for about 50 cell divisions. Rarely the contact inhibition that characterizes normal fibroblasts is lost, and the cells are said to be transformed into permanent cell lines. Cell lines are also obtained from tumors.

2. Lymphocytes grow in culture, although ordinarily only for a few divisions. Long-term "lymphoblastoid" cultures have been obtained, probably as a result of virus infection and transformation. Isolation of clones grown from single cells, either fibroblast or lymphoblastoid, is difficult but can be accomplished by using a "feeder layer" of killed cells. Isolation of clones from permanent cell lines is generally easier.

3. Cultured cells have several important uses. They permit the study of many metabolic disorders under laboratory conditions without directly involving the patient. This allows more rigorous control of the cell milieu and also allows study of metabolic potential under conditions that cannot be duplicated *in vivo.* For example, fibroblasts from patients with the dominant form of hypercholesterolemia express the metabolic disorder *in vitro.* It has been shown that the primary defect is the binding of low density lipoprotein to the cell membrane, a defect that would be very difficult to establish if study were limited to patients. Similarly, the binding of dihydrotestosterone by fibroblasts has been shown to be defective in testicular feminization, a finding also difficult to establish directly in human beings.

4. Cell cultures can be used to study gene complementation if the gene products are diffusable. This was demonstrated in Hurler and Hunter syndromes, which are phenotypically similar, although one is an autosomal recessive trait and the other X-linked recessive. When grown separately, cells from these patients form metachromatic granules. When grown together in mixed culture, they appear normal.

5. Cell cultures have been especially useful in prenatal diagnosis of inherited defects. Cells that will grow can be obtained from amniotic fluid, and controlled tests of enzyme function can be carried out on such cultures without risk to the fetus. It is possible to recognize the presence of a large number of metabolic and chromosomal defects in such cultures.

6. Mutant cells and clones can be isolated from cells grown in culture. For example, many mutants for the enzyme hypoxanthine phosphoribosyl transferase (HPRT) have been isolated from normal cells. Forward mutations (HPRT−) are selected by their ability to grow in 8-azaguanine, which kills HPRT+ cells. Reverse HPRT+ mutants are selected by their growth in HAT (hypoxanthine-aminopterin-thymidine) medium. Thymidine kinase (TK−) mutants can be selected by their resistance to bromodeoxyuridine. Reverse mutants (TK−) are also selected by HAT medium. Many other special selection systems have also been developed.

7. One of the striking properties of cells in culture is their ability to fuse, especially in the presence of agents such as inactivated Sendai virus or polyethylene glycol. When grown in selective media, this permits isolation of cells bearing two parental complements of chromosomes. Of special interest is the ability to fuse cells from different species, such as human and mouse. The human chromosomes are preferentially lost from such hybrids, a fact that permits assignment of genes to particular human chromosomes by virtue of the association between the trait and a single human chromosome.

8. Cell hybridization has permitted the study of effects of different mutations in the same cell. For example, fusion of HPRT− rat cells with human HPRT− cells caused production of rat HPRT but not human. Fusion of mouse hepatoma cells with human amniotic cells causes expression of human liver function in the hybrid, possibly useful in prenatal diagnosis.

9. Introduction of genes into cultured cells can also be accomplished by immersing the cells in chromosome preparations. By fusing human cells with mouse teratocarcinoma cells, followed by mixing the fused cells with mouse blastocyst cells to form allophenic mice, it is possible to introduce human genes into mice.

REFERENCES AND SUGGESTED READING

ALBERTINI, R., AND DE MARS, R. 1970. Diploid azaguanine-resistant mutants of cultured human fibroblasts. *Science* 169: 482–485.

AMRHEIN, J. A., MEYER, W. J. III, JONES, H. W. Jr, AND MIGEON, C. J. 1976. Androgen insensitivity in man: Evidence for genetic heterogeneity. *Proc. Natl. Acad. Sci.* (U.S.A.) 73: 891–894.

BARSKI, G., SORIEUL, S., AND CORNEFERT, F. 1960. Production dans des cultures in vitro de deux souches cellulaires en association, de cellules de charactère "hybrid." *C. R. Acad. Sci.* 251: 1825–1827.

BOONE, C. M., AND RUDDLE, F. H. 1969. Interspecific hybridization between human and mouse somatic cells: enzyme and linkage studies. *Biochem. Genet.* 3: 119–136.

CHU, E. H. Y., AND POWELL, S. S. 1976. Selective systems in somatic cell genetics. *Adv. Hum. Genet.* 7: 189–258.

CROCE, C. M., BAKAY, B., NYHAN, W. L., AND KOPROWSKI, H. 1973. Reexpression of the rat hypoxanthine phosphoribosyltransferase gene in rat-human hybrids. *Proc. Natl. Acad. Sci.* (U.S.A.) 70: 2590–2594.

DARLINGTON, G. J., BERNHARD, H. P., AND RUDDLE, F. H. 1974. Human serum albumin phenotype activation in mouse hepatoma-human leukocyte cell hybrids. *Science* 185: 859–862.

DEMARS, R., AND HELD, K.R. 1972. The spontaneous azaguanine resistant mutants of diploid human fibroblasts. *Humangenetik* 16: 87–110.

DEWEY, M. J., MARTIN, D. W. JR., MARTIN, G. R., AND MINTZ, B. 1977. Mosaic mice with teratocarcinoma-derived mutant cells deficient in hypoxanthine phosphoribosyltransferase. *Proc. Natl. Acad. Sci.* (U.S.A) 74: 5564–5568.

EPHRUSSI, B., AND WEISS, M. C. 1965. Interspecific hybridization of somatic cells. *Proc. Natl. Acad. Sci.* (U.S.A.) 53: 1040–1042.

GLADE, P. R., AND BERATIS, N. G. 1976. Long-term lymphoid cell lines in the study of human genetics. *Prog. Med. Genet.* n. s. 1: 1–48.

HARRIS, H. 1974. *Nucleus and Cytoplasm,* 3rd Ed. Oxford: Clarendon Press. 186 pp.

ILLMENSEE, K., HOPPE, P. C., AND CROCE, C. M. 1978. Chimeric mice derived from human-mouse hybrid cells. *Proc. Natl. Acad. Sci.* (U.S.A.) 75: 1914–1918.

KAUFMAN, M., STRAISFELD, C., AND PINSKY, L. 1977. Expression of androgen-responsive properties in human skin fibroblast strains of genital and nongenital origin. *Som. Cell Genet.* 3: 17–25.

LITTLEFIELD, J. W. 1964. Selection of hybrids from matings of fibroblasts in vitro and their presumed recombinants. *Science* 145: 709–710.

MC BRIDE, O. W., AND OZER, H. L. 1973. Transfer of genetic information by purified metaphase chromosomes. *Proc. Natl. Acad. Sci.* (U.S.A.) 70: 1258–1262.

MIGEON, B. R., AND CHILDS, B. 1970. Hybridization of mammalian somatic cells. *Prog. Med. Genet.* 7: 1–28.

MILUNSKY, A. 1976. Prenatal diagnosis of genetic disorders. *N. Engl. J. Med.* 295: 377–380.

MINTZ, B., CRONMILLER, C., AND CUSTER, R. P. 1978. Somatic cell origin of teratocarcinomas. *Proc. Natl. Acad. Sci.* (U.S.A.) 75: 2834–2838.

OKADA, Y. 1962. Analysis of giant polynuclear cell formation caused by HVJ virus from Ehrlich's ascites tumor cells. *Expt. Cell Res.* 26: 98–107.

POLLACK, R. (Ed.). 1973. *Readings in Mammaliam Cell Culture.* Cold Spring Harbor Laboratory, Cold Spring Harbor, N. Y. 735 pp.

PUCK, T. T., AND MARCUS, P. I. 1955. A rapid method for viable cell titration and clone production with HeLa cells in tissue: the use of x-irradiated cells to supply conditioning factors. *Proc. Natl. Acad. Sci.* (U.S.A.) 41: 432–437.

RANKIN, J. K., AND DARLINGTON, G. J. 1979. Expression of human hepatic genes in mouse hepatoma-human amniocyte hybrids. *Somat. Cell Genet.* 5: 1–10.

RUDDLE, F. H. 1973. Linkage analysis in man by somatic cell genetics. *Nature* 242: 165–169.

WATANABE, T., DEWEY, M. J., AND MINTZ, B. 1978. Teratocarcinoma cells as vehicles for introducing specific mutant mitochondrial genes into mice. *Proc. Natl. Acad. Sci.* (U.S.A.) 75: 5113–5117.

WIGLER, M., PELLICER, A., SILVERSTEIN, S., AXEL, R., URLAUB, G., AND CHASIN, L. 1979. DNA-mediated transfer of the adenine phosphoribosyltransferase locus into mammalian cells. *Proc. Natl. Acad. Sci.* (U.S.A.) 76: 1373–1376.

WILLECKE, K., AND RUDDLE, F. H. 1975. Transfer of the human gene for hypoxanthine-guanine phosphoribosyltransferase via isolated human

metaphase chromosomes into murine L-cells. *Proc. Natl. Acad. Sci.* (U.S.A) 72: 1792–1796.
WULLEMS, G. J., VAN DER HORST, J., AND BOOTSMA, D. 1975. Incorporation of isolated chromosomes and induction of hypoxanthine phosphoribosyltransferase in Chinese hamster cells. *Somat. Cell Genet.* 1: 137–152.

REVIEW QUESTIONS

1. Define:

fibroblast	lymphoblastoid	selective medium
contact inhibition	cell line	HAT medium
transformation	mitogen	hepatoma
homonuclear	complementation	allophenic
heteronuclear	heterokaryon	blastocyst
heteroploid		

2. How are fibroblast clones obtained from human cell cultures?

3. What is the difference between a fibroblast and a lymphoblastoid culture?

4. What are some of the advantages that cell cultures have in studies of gene function? What are the disadvantages?

5. How can viral etiology of a tumor be reconciled with monoclonal origin of the tumor?

6. What is a parasexual cycle?

7. Although human-mouse hybrids in which the human line supplies the essential TK+ gene virtually always retain human chromosome 17 in HAT medium, there are rare exceptions in which chromosome 17 is lost, even though the culture is viable. How might this be explained?

8. Galactosemia is a disorder due to absence of the enzyme galactose-1-phosphate uridyl transferase. H. L. Nadler, C. M. Chacko, and M. Rachmeler made hybrid cells from two patients with this disorder and found that the hybrid cells were able to make the enzyme (*Proc. Nat. Acad. Sci.* 67: 976, 1970). How might this be explained?

9. A family with one child affected with Tay–Sachs disease came in for prenatal diagnosis of the next pregnancy. Amniocentesis and cultivation of the amniotic cells showed hexosaminidase A activity characteristic of a normal fetus. The pregnancy was allowed to continue, giving rise to a child with Tay–Sachs disease. What is the most likely source of the error? How might it have been avoided?

CHAPTER FIFTEEN

PATTERNS OF INHERITANCE: SEGREGATION ANALYSIS

When a trait or disease is thought to be inherited, confirmation may be obtained by showing that transmission follows a definite predictable pattern. The first step in deciding the mode of inheritance is the formulation of a genetic hypothesis—usually that the trait is transmitted as a simple Mendelian dominant or recessive. Such a hypothesis entails predictions, which may be shown to be incorrect, proving the hypothesis to be incorrect. If the hypothesis cannot be proved incorrect, it may be accepted until additional tests can be made. Even if a trait is clearly inherited as a dominant or recessive, it is sometimes useful to know whether the true segregation ratio is exactly as predicted by Mendelian theory or whether the ratio may be altered by such factors as differential survival.

The nature of scientific proof should be kept clearly in mind. It is never possible to prove that a hypothesis is correct. Rather, one can only prove that the alternatives are incorrect or unlikely. If such "proof" is to carry weight, all reasonable hypotheses must be tested for compatibility with the observations. As knowledge increases, previously "unreasonable" hypotheses sometimes become reasonable, necessitating reexamination of older proofs. Failure to disprove a hypothesis may happen because the hypothesis is correct or because experimental observations are inadequate.

In many experimental organisms, genetic hypotheses are relatively easy to test. The usual procedure is to cross two pure strains to give an F_1 hybrid, which in turn is allowed to self-fertilize to give the F_2 or which may be backcrossed to the apparently recessive parental strain. If the variation under study is influenced by a single pair of alleles, then the F_2 generation yields $AA:Aa:aa$ in a 1:2:1 ratio ideally. The backcross yields Aa and aa in equal numbers. Deviation from either ratio is sufficient evidence that the trait does not follow simple Mendelian rules. It would then be necessary to consider the influence of other genetic loci and the environment.

Human traits are often more difficult to test, due in large part to the need to locate families in which the loci of interest are segregating. For some traits, especially rare autosomal dominant or X-linked traits, pedigree analysis is the most efficient means of establishing the transmission pattern (Chapter 2). For rare recessive traits, pedigree analysis is not useful, since only one generation is ordinarily affected. In those cases, as well as in autosomal dominant and X-linked traits, segregation ratios can be estimated, even in human populations. It is necessary to know how the affected persons were *ascertained,* so that corrections for ascertainment biases can be made.

In the following sections, some of the frequently used tests for segregation of autosomally inherited human traits are considered. The availability of computers has permitted the development of much more complex methods, but these will not be considered.

TESTS FOR DOMINANT INHERITANCE

Dominant inheritance can usually be recognized by inspection of pedigrees, since every affected child should have an affected parent and a pair of unaffected parents should not have an affected child. Exceptions to these rules may occur when persons with the dominant genotype fail to express the related phenotype. This may happen because other genetic or environmental factors influence the gene expression. Some genes are not expressed until a person is an adult or perhaps even in advanced age, an obvious example being baldness. In such cases, a person may die before he expresses his true phenotype, leading to apparent inconsistencies with simple genetic interpretation. Such factors as mutation and illegitimacy also may give rise to deviation from these rules.

Dominant inheritance is the simplest type of inheritance to test. If the dominant gene is rare, the likelihood of homozygous dominant individuals is very small. Every person with the trait may thus be considered heterozygous. Since such persons will nearly always marry homozygous recessive partners, the marriages are comparable to a backcross of F_1 to homozygous recessive parental type. The ratio of children with the dominant gene to children who lack it is 1 : 1. Since these sibships can be identified through a parent who exhibits the trait, all the sibships can be pooled to give a good estimate of the genetic ratio. If the ratio of affected to normal offspring is 1 : 1 among marriages of affected to normal, the trait may be considered to result from a dominant gene.

For most rare dominant traits, there is little error in assessing the genetic ratio, provided adequate attention is given to the manner in which the data are collected. In actual practice, each individual or family bearing the trait of interest must be ascertained (identified as possessing the trait). Serious bias may be introduced by the method of ascertainment. In the case of simple dominant inheritance, bias can be avoided if only the descendants of propositi are considered. (The *propositus,* also called *proband* or *index case,* is the person bearing the trait under investigation through whom a family is located.) Inclusion of the sibships of the propositi leads to error because of the greater likelihood of locating sibships that by chance have more than the usual number of affected persons. A correction can be introduced, as in the following section on recessive inheritance, enabling use of index sibships. However, the usual practice is to avoid use of index sibships.

As an example of testing for simple dominant inheritance, consider the sibships given in Figure 15-1. Sibships of two and three persons are given with distributions of dominant and homozygous recessive phenotypes in the combinations predicted by the binomial theorem (see Appendix A). In each case, the parent was the propositus, so that sibships with no dominant phenotypes are observed, and sibships with different numbers of dominant phenotypes are found in the same relative frequency in which they occur in the population. There are 32 offspring total, 16 with the dominant phenotype and 16 homozygous recessive. This conforms to the expected 1 : 1 distribution in a backcross. But in

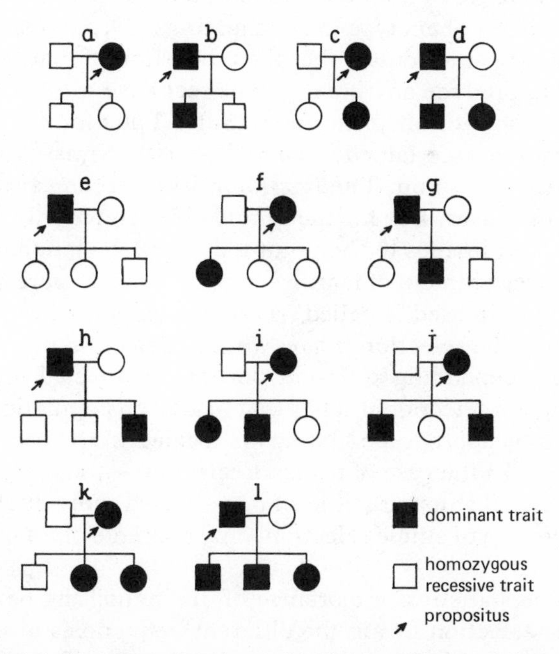

Figure 15-1 Dominant inheritance in sibships of two and three persons.

some sibships, all members carry the dominant gene, and in other sibships none carries it.

While this test for dominant inheritance appears to be simple to apply, it is sometimes rather difficult with the available data. This is because pedigrees of very rare dominant genes tend to be noted because of an unusual concentration in the pedigree of persons having the dominant phenotype. An unsophisticated observer is apt to be more impressed by family *l* of Figure 15-1 with four members of dominant phenotype than by family *e* with only one, although the latter is just as important in genetic analysis as the former. Thus most published pedigrees suggesting dominant inheritance are useless for testing segregation ratios. The basic assumption that all families will come to the attention of the investigator in the same proportion that they occur in the general population is not met in these cases. At times, it is possible to correct for this assumption, but in most cases the correction cannot be applied with confidence.

TESTS FOR RECESSIVE INHERITANCE

Recessive traits are more difficult to test than dominant because matings of two heterozygous persons are usually ascertained only through the existence of homozygous recessive offspring. But the probability of a dominant phenotype in such matings is $\frac{3}{4}$, and using the binomial theorem we can calculate that the proportion of matings which by chance fails to produce any homozygous recessive offspring is $(\frac{3}{4})^n$, where n is the number of offspring in a family. Thus for $n = 1$, $\frac{3}{4}$ of the matings will not be ascertained; for $n = 2$, $\frac{9}{16}$ will be missed; for $n = 3$, $\frac{27}{64}$ will be missed, and so on. The fraction missed becomes smaller as the family size increases. But for the family sizes commonly found, this fraction never diminishes to the point that it can be ignored.

Ascertainment of families in which one or more combinations of offspring is missed is called *truncate* selection. If every type of family is located, the selection is *nontruncate.* Selection may also be complete or single, depending on what portion of affected sibships are located. Complete selection implies that in a given population every sibship produced by appropriate matings is located in the case of nontruncate selection. In the case of truncate selection—the more common situation—every sibship with at least one affected individual should be located. In the case of single selection, only a sample of total possible sibships is located.

The statistical importance of distinguishing between *complete* and *single* selection lies in the different frequencies of various types of sibships under the two situations. In the case of complete selection, sibships are found in the same proportion in which they occur in the population. In single selection, sibships with two affected persons are twice as likely to be ascertained as those with one; those with three affected are three times as likely; and so on. The differences between complete and single truncate selection are illustrated in Figure 15-2 for sibships of size three.

Since the binomial theorem enables us to predict the proportion of families missed, it is possible to test for Mendelian ratios, even in recessive inheritance. The corrections consist either in modifying the expected segregation ratios so that the ascertainment bias is compensated (a priori methods) or in introducing the compensation in the observations so that they conform to simple Mendelian ratios (a posteriori

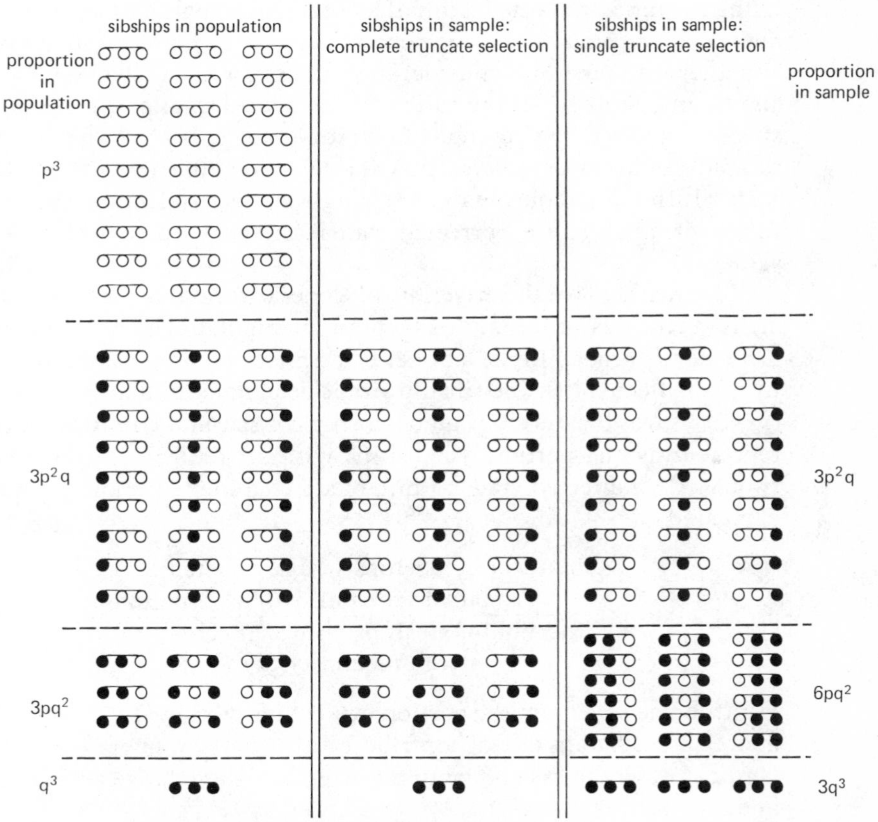

Figure 15-2 Sibships of size three in the proportions in which they occur: (1) in the general population, (2) in complete truncate selection, and (3) in single truncate selection. Only sibships that are offspring of two heterozygous parents are considered. Thus, p = probability of nonaffected offspring = $\frac{3}{4}$; q = probability of affected offspring = $\frac{1}{4}$. Solid circles indicate affected and open circles nonaffected offspring.

methods). The methods to be discussed are appropriate for matings of two heterozygotes; matings in which one parent also shows the proposed recessive phenotype should not be considered. For most recessive conditions, particularly those resulting in abnormalities, the genes are rare enough so that only matings of two heterozygotes lead to homozygous recessive offspring.

Genetic ratios can also be tested in the case of recessive traits which are sufficiently common so that matings of homozygous recessive with heterozygous person are likely to occur. Treatment of this situation requires first a consideration of population genetics (Chapter 20).

The a Priori Method (Method of Apert)

One procedure for testing for ratios in recessive sibships involves calculating the ratio of affected to unaffected in the actual sample. In order to make such a calculation, it is necessary to assume a ratio, such as 3:1, and alter it to take into consideration the manner of selection. For example, in Figure 15-2, the ratio of nonaffected to affected in the total sibships produced by matings of heterozygotes is 144:48 or 3:1. In the sample found in complete truncate selection, the ratio is 63:48 or 1.313:1. In the sample obtained in single truncate selection, the ratio is 72:72 or 1:1. These corrected ratios can be used to test observed values.

The mathematical derivation of general formulas for the a priori method requires only simple algebraic manipulations of the binomial. Let p be the probability of a nonaffected offspring ($\frac{3}{4}$) and q the probability of an affected ($\frac{1}{4}$). The sibship size is designated n. The expansion of $(p + q)^n$ gives the distribution of sibships of size n, the individual terms representing the portion of sibships with zero affected, one affected, two, . . . , n affected. The portion of affected in the sample, q', may be expressed

$$q' = \frac{\Sigma(\text{portion of sibships with } x \text{ affected})(x)\,(\text{relative probability of ascertainment})}{\Sigma(\text{portion of sibships with } x \text{ affected})(n)\,(\text{relative probability of ascertainment})}.$$

For truncate selection, the portion of sibships with $x=0$ affected is zero; hence such sibships do not contribute either to the numerator or the denominator. In the case of complete selection, the relative probability of finding sibships with 1, 2, . . . , n affected is 1, since they are found in exactly the proportions in which they occur in the population. In single selection, the relative probability of ascertainment is x, the number of affected per sibship.

As an example of derivation of the general formula for q' for complete selection, consider the special case where $n = 3$. The binomial distribution is

$$p^3 + 3p^2q + 3pq^2 + q^3.$$

The portion of affected in the total group of sibships is

$$q = \frac{(0)(p^3) + (1)(3p^2q) + (2)(3pq^2) + (3)(q^3)}{(3)(p^3) + (3)(3p^2q) + (3)(3pq^2) + (3)(q^3)}.$$

Since the first term in the denominator represents the sibships with no affected persons, this term would not appear in the calculation of q', and

$$q' = \frac{3p^2q + 6pq^2 + 3q^3}{9p^2q + 9pq^2 + 3q^3}.$$

Dividing by 3 and remembering that $3p^2q + 3pq^2 + q^3 = 1 - p^3$,

$$q' = \frac{q(p^2 + 2pq + q^2)}{1 - p^3}$$

$$= \frac{q(p + q)^2}{1 - p^3}$$

$$= \frac{q}{1 - p^3}.$$

It can be shown that for any value of n,

$$q' = \frac{q}{1 - p^n}.$$

For $n = 3$, $q' = (\frac{1}{4})/[1 - (\frac{3}{4})^3] = \frac{64}{148}$, which is the same as the 48:63 ratio obtained earlier.

For single ascertainment, and again considering the case $n = 3$,

$$q' = \frac{(1)(3p^2q) + (2)(6pq^2) + (3)(3q^3)}{3[3p^2q + (2)(3pq^2) + (3)(q^3)]}.$$

Dividing numerator and denominator by $3q$,

$$q' = \frac{p^2 + 4pq + 3q^2}{3}$$

$$= \frac{p^2 + 2pq + q^2 + 2pq + 2q^2}{3}$$

$$= \frac{(p + q)^2 + 2q(p + q)}{3}$$

$$= \frac{1 + 2q}{3}.$$

For any value of n,

$$q' = \frac{1 + (n - 1)q}{n}.$$

For $n = 3$, $q' = [1 + 2(\frac{1}{4})]/3 = \frac{1}{2}$. The ratio obtained from Figure 15-2 was 1:1, showing agreement.

Since the values of q' are dependent solely on q and n, they need be calculated only once. Table 15-1 gives the values of q' for the two methods of ascertainment. This method is not used for single ascertainment, since the simple sib method (below) is easier to apply and gives equivalent answers. The following example illustrates the use of the a priori method for complete ascertainment.

Stevenson and Cheeseman located all persons with hereditary deaf mutism in Northern Ireland. Many of these had one or both parents also deaf. However, there were 309 sibships with at least one deaf member, both parents having normal hearing. The data on these sibships are given in Table 15-2. There are 1151 normal and 453 deaf persons. Is this compatible with a corrected 3:1 ratio? The expected number of affected persons is calculated as indicated. Sibships consisting of one affected person only are ignored, since these can only have a ratio of 1 affected:0 unaffected and therefore are uninformative.

The large χ^2 is incompatible with simple recessive inheritance as the explanation for all cases of deaf mutism. Inspection of Table 15-2 indicates a deficiency of affected persons for sibships of all sizes greater than three. More extensive analysis of these and related data indicates that, although a large portion of congenital deafness is due to homozygosity for recessive genes, there are many cases whose etiology is differ-

Table 15-1 **CORRECTED EXPECTATION OF q (q') FOR COMPLETE AND SINGLE TRUNCATE ASCERTAINMENT FOR SIBSHIPS OF VARIOUS SIZES n. THE PRODUCT nq' ALSO IS GIVEN FOR CONVENIENCE**

	Complete ascertainment		Single ascertainment	
n	q'	nq'	q'	nq'
1	1.0000	1.0000	1.0000	1.0000
2	.5714	1.1428	.6250	1.2500
3	.4324	1.2972	.5000	1.5000
4	.3657	1.4628	.4375	1.7500
5	.3278	1.6390	.4000	2.0000
6	.3041	1.8246	.3750	2.2500
7	.2885	2.0195	.3571	2.5000
8	.2778	2.2224	.3438	2.7500
9	.2703	2.4327	.3333	3.0000
10	.2649	2.6490	.3250	3.2500
11	.2610	2.8710	.3182	3.5000
12	.2582	3.0984	.3125	3.7500
13	.2561	3.3293	.3077	4.0000

Table 15-2 **DISTRIBUTION OF DEAF AND NORMAL OFFSPRING IN FAMILIES IN WHICH BOTH PARENTS ARE NORMAL***

Sibship size n	Number of sibships x	Total offspring nx	Observed affected	Observed unaffected	Expected affected $nq'x$
1	21	21	21	0	21.00
2	35	70	40	30	40.00
3	39	117	50	67	50.59
4	34	136	41	95	49.74
5	35	175	48	127	57.37
6	49	294	68	226	89.41
7	34	238	57	181	68.66
8	33	264	69	195	73.34
9	15	135	32	103	36.49
10	6	60	10	50	15.89
11	3	33	6	27	8.61
12	4	48	10	38	12.39
13	1	13	1	12	3.33
	309	1604	453	1151	526.82

Observed affected (less $n = 1$ sibships) = 432.
Observed unaffected = 1151.
Expected affected (less $n = 1$ sibships) = 505.82.
Expected unaffected: 1583 − 505.82 = 1077.18.
$\chi^2 = 15.83$, 1 df, $P < 0.01$.
* Data are from Stevenson and Cheeseman (1956).

ent. These cases might be due to more complex inheritance or to environmental factors. The data of Table 15-2 have an excess of families with only one affected person. Presumably some of these are not due to simple recessive inheritance.

Simple Sib Method

Where the a priori method corrects the expected 3:1 ratio to match a particular set of observations, the *simple sib method* corrects the observed values so that they can be tested against the theoretical 3:1 ratio. The rationale behind the sib method is found in the relationships of the terms in the expansion $(p + q)^n$ to those in $(p + q)^{n-1}$. Let us consider the specific case of $n = 5$, giving on expansion

$$p^5 + 5p^4q + 10p^3q^2 + 10p^2q^3 + 5pq^4 + q^5.$$

Multiplying each term by the number of affected gives

$$(0)p^5 + (1)5p^4q + (2)10p^3q^2 + (3)10p^2q^3 + (4)5pq^4 + (5)q^5.$$

Dividing each term by the common factor $5q$ does not change the *relative* magnitude of the terms and gives

$$p^4 + 4p^3q + 6p^2q^2 + 4pq^3 + q^4.$$

But this is the expansion of $(p + q)^4$.

The utility of this relationship arises from the fact that truncate selection provides a series of terms complete except for the first. These terms are present in a definite proportion. By adjusting the size of the terms by multiplying by a factor equal to the number of affected individuals in each sibship, a new series is obtained that has all the terms corresponding to sibships of one less person. If we then consider each term as representing the combination of individuals for the $n - 1$ series, a 3:1 ratio of dominant and recessive phenotypes should result. In this procedure, sibships of five persons with one affected, represented in the original expansion by $5p^4q$, would be *counted* as sibships of four persons with no affected.

An example drawn from Figure 15-2 will illustrate the procedure. In the population, the ratio of one affected, two affected, and three affected among sibships of three is 27:9:1. This is also the ratio in complete selection. Multiplying each number by the affected persons gives (1)27:(2)9:(3)1 or 27:18:3 (9:6:1). The sibships of Figure 15-2 would then be counted as 27 with *two unaffected sibs,* 18 with *one unaffected* and *one affected,* and 3 with *two affected.* This gives a total of 72 unaffected and 24 affected persons—a 3:1 ratio.

A sample drawn by single selection is already altered by the influence of the number of affected persons on the likelihood of ascertaining a sibship. As demonstrated earlier, the effect is to increase each term of the expansion by a factor equal to the number of affected persons in the sibship. This is the same operation just indicated as the simple sib correction; hence in single ascertainment the sibships already occur in the corrected proportions. It is merely necessary to subtract one affected person from each sibship before counting. Thus in Figure 15-2 the sample resulting from single truncate ascertainment has 27 sibships with one affected, 18 with two affected, and 3 with three affected. These are to be counted as 27 with two unaffected, 18 with one affected and one unaffected, and 3 with two affected, giving 72 unaffected and 24 affected, as in the earlier example.

Once values are obtained for the corrected number of unaffected and affected persons, they can be tested for conformity to a 3:1 ratio by the usual χ^2 test.

The Method of Discarding the Singles

Li and Mantel (1968) have proposed a simple method of estimating the segregation ratio for complete truncate ascertainment. The method, known as "discarding the singles," consists in omitting the affected person from sibships with only one affected. Expressed symbolically,

$$q' = \frac{R - J}{T - J},$$

where q' is the estimate of the segregation ratio, R is the total number of affected persons in a sample of sibships, T is the total number of persons in the sample, and J is the number of sibships with only one affected.

For example, in Figure 15-2, there are 27 sibships with only one affected, 9 sibships with two affected, and 1 sibship with three affected. For such observations, $q' = (48 - 27)/(111 - 27) = 21/84 = 0.25$. One can show this to be the expected result for sibships of any size by solving the appropriate algebraic identity. For sibships of size 3,

$$q' = \frac{(2)3pq^2 + (3)q^3}{(2)3p^2q + (3)3pq^2 + (3)q^3},$$

where the numbers in parentheses in the numerator are the numbers of affected persons counted in sibships with two and three affected, respectively, and the numbers in parentheses in the denominator are the total numbers of persons counted in sibships with one, two, and three affected, respectively.

Li and Mantel (1968) applied the method to data collected on sibships with the Ellis-van Creveld syndrome, a rare form of dwarfism that occurs with high frequency in certain Amish populations. Their results are shown in Table 15-3. The reliability (variance) of the calculated value of $q' = 0.215$ can be estimated with the help of tables provided by Li and Mantel.

DIFFICULTIES IN TESTING GENETIC HYPOTHESES

It is relatively simple to develop corrections for truncate ascertainment and for complete versus single ascertainment. Collecting data for which these corrections are strictly applicable is quite another matter. Complete ascertainment means that no homozygous recessive person should escape detection. With the improved medical diagnosis and recording of the present time, complete ascertainment can sometimes be approached if the homozygous recessive state results in a readily characterized condition. In general, ascertainment is possible because a person has sought medical aid; hence only those conditions which require medical aid are candidates for this type of investigation.

Single ascertainment assumes that sibships are detected in proportion to the number of homozygous recessives per sibship. A variety of

Table 15-3 **ESTIMATE OF THE SEGREGATION RATIO IN SIBSHIPS WITH ELLIS-VAN CREVELD SYNDROME***

Family no.	Conditions of children (+, affected; −, normal)	No. of children	No. of affected	No. of "singles"
14	+−	2	1	1
23	?−+	2	1	1
1	+−+	3	2	—
10	+−−	3	1	1
16	+++	3	3	—
2	−−−+	4	1	1
5	−+−−	4	1	1
9	−−+−	4	1	1
15	++−−	4	2	—
20	+−−−	4	1	1
21	−−+−	4	1	1
3	−+++−−	6	3	—
6	−−−−−+	6	1	1
8	−−+−−−	6	1	1
27	+−−−−−	6	1	1
7	−+−−−++	7	3	—
17	−−−−+++	7	3	—
13	−−−−+++−	8	3	—
19	+−−−−−−−	8	1	1
29	+−+−−−−−	8	2	—
30	−−+−−−−−	8	1	1
4	−−−++−−+−	9	3	—
24	−−−+−+++−	9	4	—
28	−−−+−−−−−+	10	2	—
25	−−+−−−−−+−−	11	2	—
12	−−+−−+−−−−−−	12	2	—
11	−−−−−−−−+−−−−−	14	1	1
Total		$T = 172$	$R = 48$	$J = 14$

* From Li and Mantel (1968), based on data collected by V. A. McKusick, J. A. Egeland, R. Eldridge, and D. E. Krusen.

factors make this assumption incorrect. Not all segments of society are equally likely to seek medical attention. In some cases, a person is more likely to visit a doctor if a relative with the same apparent condition has benefited from medical attention. Families may more readily recognize the need for medical attention if they have already gained experience with a particular disease. This would lead to an excess of multiply affected sibships compared to singly affected sibships.

A variety of corrections have been developed to compensate for deviations from the ideal situations treated above. These corrections are somewhat complicated and are beyond the scope of this text. Further discussion can be found in the references listed at the end of the chapter. In practice, many investigators attempt as nearly as possible to collect data that conform to the assumptions outlined above, with the hope that uncorrected deviations will not jeopardize their conclusions. This is generally possible in a *prospective* study. Sometimes the data are collected for other purposes and are made available to the geneticist to use in the best way possible. In this case, the data are less likely to match one of the simple models discussed in this chapter, and more complex models must be used.

Penetrance, Sex Limitation, and Expressivity

Penetrance is a commonly used word that is frequently misunderstood. It may be defined as the probability of detecting a particular combination of genes when they are present. Most often, one makes use of a particular means of observation. As the means of observation changes, the level of penetrance may change also. Thus it is not only a function of the gene in question. Genes at other loci may also influence penetrance. For example, detection of gene A may be possible only when it is in combination with M. When A occurs in combination with m it would be considered nonpenetrant. Unless we recognize the variation at the M locus, it may be very difficult to decide why A sometimes is expressed and other times is not. To say that a gene has less than 100% penetrance is purely a probability statement regarding the likelihood of detecting the presence of the gene. As such, it is a useful term for predicting the likelihood of showing genetic effects given a certain genotype.

Sex-limited inheritance refers to genes that can be expressed only in one sex, either males or females. Inherited variations in primary and secondary sex characteristics would be obvious examples. It is very difficult to study patterns of beard growth in females, although females could very well transmit genes influencing beard growth.

Snyder and Yingling (1935) provided an interesting example of sex-influenced inheritance in their study of pattern baldness. This form of baldness is characterized by extensive loss of hair, usually about age 25. It was found to be due to an autosomal gene that is dominant in males but recessive in females. Two copies of the gene are necessary to produce baldness in females, but only one copy is sufficient in males.

The inheritance of baldness may in fact be more complicated than that, but the relationship of baldness to hormonal status is a well-founded observation. Genes may be sex-influenced in their penetrance and expression without being strictly limited to one sex. The hormonal and other chemical differences between males and females may alter the threshold required for a trait to occur. For this reason, care must be exercised in interpreting pedigrees in which one sex is more often affected. Inheritance by sex chromosomes is not necessarily indicated.

Expressivity refers to the variety of ways in which a particular gene manifests its presence. A classical example of variation in expression is found in the condition osteogenesis imperfecta. The three principal features of this disease are blue sclerae, very fragile bones, and deafness. The condition is inherited as a dominant with nearly 100% penetrance. However, the expressivity varies greatly. A person with the gene may have any one or any combination of these three traits. The bone fragility may be a minor defect or it may be a major problem, the patient having dozens of fractures during his lifetime. Given that a person has the gene for osteogenesis imperfecta, there is no way to predict what combination of traits he will have and how severe they will be. As with penetrance, expressivity undoubtedly depends upon interactions with many environmental and genetic factors. In medical genetics, it is important to recognize that a given gene may vary widely in its expression.

The terms penetrance and expressivity should not be confused. Penetrance refers to the presence or absence of any effect of the gene. Expressivity refers to the type of manifestation.

Phenocopies. Occasionally in experimental organisms, manipulation of the environment leads to production of a phenotype that mimics a particular genetic trait, even though the gene ordinarily responsible for the trait is known not to be present. The nongenetic form of the trait is said to be a *phenocopy*. Although phenocopies may cause little difficulty in the usual genetic analysis of human traits, they may be troublesome in the analysis of very rare events, such as mutation. This is particularly true if the trait cannot be investigated at the chemical level. Good examples of human phenocopies are uncommon, but some of the congenital malformations may prove to be genetic in some cases and environmental in others, the latter therefore qualifying for designation as phenocopies (Lenz, 1971). Methemoglobinemia is most often due to heredity—either an abnormal hemoglobin or a defect in the methemoglobin reductase system. It also can be due to ingestion of high levels of nitrate. Laboratory examination of the blood could of course distinguish readily among these causes.

Gene Interaction and Epistasis

Most of the examples of genetic variation discussed have been traits related in a simple manner to the variant genes. Since observations of gene action are many steps removed from the primary gene action, it is to be expected that a given trait will sometimes reflect variation at several loci.

Several levels of gene interaction may occur. Interaction at the chromosome level has already been discussed (Chapter 4). There is also interaction at the level of protein-synthesizing mechanisms, although examples of genetic variation in these mechanisms are not known. More commonly, interaction occurs at the metabolic level. The effects of metabolic interaction vary with the systems of interest and may be very complex. It will not be possible to consider all modes of interaction, but apparent departures from simple inheritance frequently are the result of such interaction.

Certain designations have arisen to describe types of gene interaction. A locus is said to be *epistatic* to another if variation at the first obscures variation at the second. An example of epistasis in mice is the masking of genes for variations in hair color by the gene for albinism. Mice that are homozygous for the recessive albino allele may have genes for a variety of coat patterns at other loci, but only the albino condition is expressed. The albino allele is thus epistatic to the other loci.

An example of epistasis in man occurs in the Bombay type of the ABO blood group system (Chapter 13). The action of the normal Bombay allele *H* seems to be to direct synthesis of H substance, which in turn serves as a precursor for A and B substances. Homozygous *hh* persons cannot synthesize H substance. The *h* allele is therefore epistatic to the alleles at the *ABO* locus.

SUMMARY

1. The mode of transmission of a trait can be tested by the agreement of the segregation ratios in offspring of various mating types with ratios predicted by Mendelian laws. If parental genotypes can be identified, the offspring phenotypes in human sibships can be pooled from similar matings to give results that can be tested against Mendelian expectation. In human subjects, this is most often possible for rare dominant traits, in which the families can be ascertained through affected parents only.

2. For rare recessive traits, families are most often ascertained through affected offspring of matings between heterozygotes. Because of the small family sizes, many such matings produce by chance only normal offspring and therefore cannot be identified (truncate selection). The loss of these normal offspring from the data leads to observed values of recessive phenotypes greater than one-fourth.

3. Another factor biasing observations is the completeness with which affected sibships are identified. If all sibships with at least one affected person are identified (complete selection), the relative frequencies with which sibships with one, two, three, or more affected persons appear in the sample are not distorted. However, if only a small sample of sibships is collected, those sibships with more affected are more likely to appear in the sample.

4. The a priori method (Apert) permits compensation for the truncation bias by providing a corrected theoretical segregation ratio. It is most often used in the case of complete selection.

5. Segregation ratios can be estimated by the simple sib method in the case of single truncate selection. This consists of omitting the index case from the data, thereby reducing the number of persons in each sibship by one. Results then can be compared with the theoretical one-fourth for a recessive trait.

6. The method of discarding singles (Li and Mantel) is useful for complete truncate selection. This procedure involves omitting the affected person from all sibships in which there is a single affected person. The resulting ratio is compared to the theoretical one-fourth.

7. Deviations from simple Mendelian ratios or inheritance may occur for many reasons. An allele that is present but not expressed is said to lack penetrance. Traits may be limited to one sex, or their expression may be influenced by the sex of the person. If an allele shows phenotypic variation among different persons, it is said to have variable expressivity. A trait that is usually inherited but that may be induced by nongenetic causes is said to be a phenocopy if environmentally induced.

8. Alleles of one locus may influence the penetrance and expression of alleles at another. If variation at one locus obscures variation at a second, the first locus is said to be epistatic to the second.

REFERENCES AND SUGGESTED READING

BAILEY, N. T. J. 1951. A classification of methods of ascertainment and analysis in estimating the frequencies of recessives in man. *Ann. Eugen.* 16: 223–225.

FISHER, R. A. 1934. The effect of methods of ascertainment upon the estimation of frequencies. *Ann. Eugen.* 6: 13–25.

HALDANE, J. B. S. 1932. A method for investigating recessive characters in man. *J. Genet.* 25: 251–255.

HOGBEN, L. 1931. The genetic analysis of familial traits. I. Single gene substitutions. *J. Genet.* 25: 97–112.

LENZ, W. 1971. Phenocopies. *J. Med. Genet.* 10: 34–49.

LI, C. C., AND MANTEL, N. 1968. A simple method of estimating the segregation ratio under complete ascertainment. *Am. J. Hum. Genet.* 20: 61–81.

MORTON, N. E. 1959. Genetic tests under incomplete ascertainment. *Am. J. Hum. Genet.* 11: 1–16.

SNYDER, L. H., AND YINGLING, H. C. 1935. Studies in human inheritance. XII. The application of the gene-frequency methods of analysis of sex-influenced factors, with especial reference to baldness. *Hum. Biol.* 7: 608–615.

STEVENSON, A. C., AND CHEESEMAN, E. A. 1956. Hereditary deaf mutism, with particular reference to Northern Ireland. *Ann. Hum. Genet.* 20: 177–231.

WEINBERG, W. 1912. Weitere Beiträge zur Theorie der Vererbung. 4. Über Methode und Fehlerquellen der Untersuchung auf Mendelsche Zahlen beim Mencshen. *Arch. Rass. u. Ges. Biol.* 9: 165–174.

REVIEW QUESTIONS

1. Define:

propositus	complete selection	expressivity
ascertainment	proband	epistasis
single selection	penetrance	phenocopy
truncate selection		

2. Huntington's chorea is a rare autosomal dominant trait in man. Among 200 offspring of affected individuals, 85 are found also to be affected. Is this a significant departure from theoretical expectation?

3. In many pedigrees, brown eyes appear to be inherited as a dominant trait. Matings of brown-eyed × blue-eyed parents, where the brown-eyed parent had a blue-eyed parent, were ascertained. These matings subsequently produced 45 brown-eyed and 55 blue-eyed offspring. Is this ratio consistent with simple dominant inheritance?

4. In a certain population, there are a total of 110 sibships of the variety of muscular dystrophy thought to be inherited as an autosomal recessive. Of these, 25 are single-child sibships, 40 have two children, 35 have three, and 10 have four. A total of 147 affected individuals were observed in this group. Is this ratio compatible with simple recessive inheritance?

5. Suppose that you wish to test the theory that deafness is inherited as a simple recessive trait. A random sample of 100 sibships, each with one or more deaf persons, was found to consist of the following: 17 with a single affected individual; 15 with two affected; 43 with one affected and one unaffected; 5 with three affected; 5 with two affected and one unaffected; and 15 with one affected and two unaffected. Are these results compatible with a simple recessive theory?

6. Cystic fibrosis is thought to be caused by homozygosity for an autosomal recessive gene. A small sample of 100 sibships, each with one or more affected persons, yielded 30 two-child sibships, 40 three-child sibships, and 30 four-child sibships. These contained 165 affected individuals. Is this number compatible with simple autosomal recessive inheritance?

7. For any of the above examples that deviate from expectation, what explanations might be offered in addition to incorrect genetic theory?

8. Prove the identity of $q' = [(2)3pq^2 + (3)q^3]/[(2)3p^2q + (3)3pq^2 + (3)q^3] = 0.25$ given in the section on the method of discarding the singles.

9. All cases of congenital deafness born in a 30-year period in a small Midwestern city were ascertained. Subsequent study of their families yielded the following sibships:

Sibship size	No. deaf in sibship	No. sibships observed
1	1	4
2	1	9
2	2	3
3	1	16
3	2	4
3	3	1
4	2	1

Using the method of discarding the singles, estimate the segregation ratio of deafness on the hypothesis of a simple recessive trait.

10. Assume that the result in the previous question is significantly different from one-fourth. What might be some reasons for this deviation?

CHAPTER SIXTEEN

COMPLEX PATTERNS OF INHERITANCE

The previous chapters have dealt with traits that vary as a result of single gene differences. Many human traits, especially diseases or traits that are observed at the biochemical level, can be described as single gene variations, or simple Mendelian traits, as they are frequently called. On the other hand, many very important traits, such as intelligence, special aptitudes, or susceptibility to heart disease, often cannot be described in such simple terms. Nevertheless, it is legitimate to ask whether the obvious variations in these more complex traits can be ascribed in part or entirely to genetic variation.

It will be useful to distinguish between two kinds of complexity. One is heterogeneity of genotypes. We may not be able to distinguish persons who have a particular phenotype associated with a variety of genotypes. For example, gout may be due to many different genetic defects. One rare kind is due to very limited activity of the enzyme hypoxanthine phosphoribosyl transferase (HPRT). Complete deficiency causes Lesch–Nyhan syndrome. The structural gene for HPRT is on the X chromosome, and this form of gout is X-linked. But most gout clearly is not X-linked, and HPRT is normal. Thus a variety of genotypes—and environmental factors—may be associated with a single phenotype.

Complexity may also result from the requirement for specific allelic combinations at several loci and possibly the requirement for selected environmental experiences as well. Thus all the persons with a particular phenotype would have the same combinations of alleles, but these would involve many loci. The appearance of the trait in families would not resemble simple Mendelian inheritance, even though its appearance is entirely due to genetic variation.

CONTINUOUS VARIATION

A trait that is measured quantitatively and that may assume any value on a scale (or limited portion of a scale) is said to be continuous. An example is height. Between the shortest person and the tallest person, there are no values that cannot correspond to someone's height. If the distribution of heights in a population is plotted, the values fall into a curve similar to the normal distribution of Figure 16-1. It is apparent that the techniques used to detect the effects of heredity on discontinuous traits or attributes cannot be applied without modification to continuous traits.

Traits that show continuous variation are difficult to study but present no difficulties so far as genetic theory is concerned. The genes involved in continuous traits are not unusual. The only special feature is that variation in several loci often influences one trait. Such traits are said to be *polygenic, multigenic, or multifactorial.* (Some use multifactorial for traits that are strongly influenced by environmental factors as well as genetic.) If there exist only the alternatives A and a to influence some trait—for example, skin color—persons will be relatively simple to classify with respect to the A locus. But if, at the same time, variation at the B locus is important, the task is more difficult, since there are nine genotypes to consider: $AABB$, $AABb$, $AAbb$, $AaBB$, $AaBb$, $Aabb$, $aaBB$, $aaBb$, and $aabb$. Dominance might reduce the number of phenotypic distinctions. Even if A is completely dominant to a and B to b, there remain four distinctive combinations. Addition of a third and fourth locus rapidly leads to a system so complicated that analysis would be difficult.

In the case of skin color, it would be necessary to recognize very subtle differences in order to attempt an analysis based even on four loci. Environmental influences and errors of measurement must be consid-

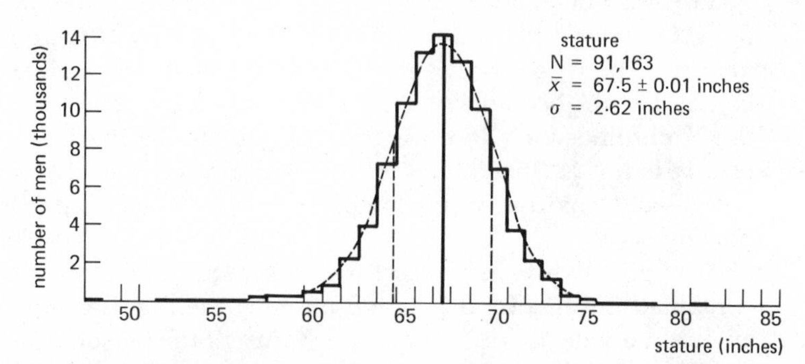

Figure 16-1 Distribution of stature in young Englishmen called up for military service in 1939. The position of the mean ($\bar{x}$) is shown by a continuous line and of the standard deviation (σ) as broken lines. A Gaussian curve, calculated from the mean and standard deviation of the data, is also shown. (*From G. A. Harrison, J. M. Tanner, N. A. Barnicot, 1964.* Human Biology. *New York: Oxford University Press.*)

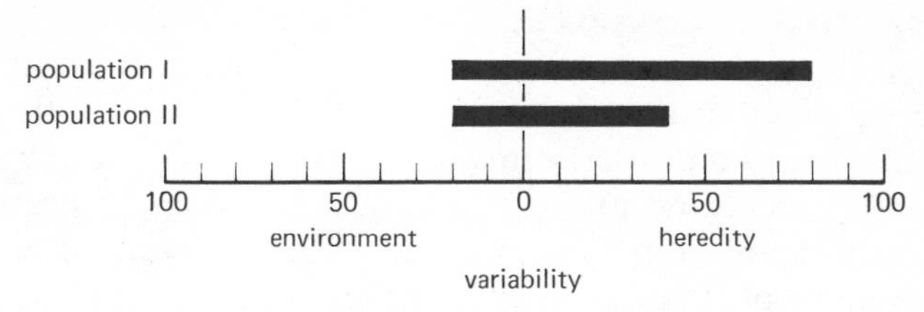

Figure 16-2 Diagram showing relative contributions of heredity and environment for two populations with equal environmental variability but different total variability. The scale is in arbitrary units of variability.

ered also. A person's skin may be lighter or darker, depending on exposure to sun. And every measure has its limit of accuracy that imposes a limit on the recognizable number of gene combinations.

In spite of these difficulties, it is frequently useful to have an estimate, however crude, of the contribution of heredity to a given variation. A variety of techniques appropriate to different experimental organisms has been developed to ascertain genetic components of variation. Plant and animal breeders in particular have been interested in knowing to what extent a certain variation is hereditary in order to decide whether the environment or selective breeding would be more likely to influence the trait.

It is important to understand the nature of estimates of heritability. As in all areas of genetics, *variation* is the object of investigation. A statement that height is 80% inherited means that 80% of the variation in height can be attributed to variation of genes. But such a statement is meaningless without reference to a specific population and environment.

Consider two populations that are under investigation for heritability of skin color and assume that environmental variables are equal for the two populations. On an arbitrary scale of variability, population I might be represented as in Figure 16-2. Suppose, however, that population II is homozygous at several of the loci that influence pigment and for which population I has two or more alleles. The genetic variability of population II will be less than that of population I, and the total variability also will be less. Heritability estimates in the two populations would be 80% for I and 67% for II. The corresponding contributions of environment would be 20% and 33%, even though the absolute magnitude of the environmental component is the same. Similarly, if the environmental variation increases, the portion of the total variation attributable to genetic variation will decrease. For example, skin pigment might show very little genetic variation within an African population or within a Nordic population. A hybrid population derived from these two stocks would show a large genetic variability.

The complexity of genetic and environmental interactions is also illustrated by certain skills. Skill at a game such as tennis is generally considered to depend largely on good instruction and practice. Estimation of the genetic component of skill of the total population would very likely yield an answer near zero. Many persons will have played very little and would perform poorly; those who do play would perform largely according to their amount of training. Environment would appear to be all-important because the total variation in skill is so large and most of it is due to environment.

On the other hand, an outstanding player must have excellent neuromuscular coordination, and persons vary in the limits to which their coordination can be trained. Among a cross section of the population exposed to intensive instruction and practice in tennis, some would excel and others would do poorly. At least part of the reason for difference in performance would lie in inherent—presumably genetic—differences in neuromuscular make-up. By a reduction in the environmental variation through training, the genetic components may have an opportunity to be expressed.

Still other results might be obtained if observations were limited to champion tennis players. Is the achievement of such persons limited primarily by inherited anatomical and physiological variations, or is practice all-important? Our democratic ideology emphasizes environment and individual opportunity. These are the variables that can be manipulated readily. We should be aware, however, of the limitations and opportunities attributable to genetic variation.

It is thus very difficult to arrive at estimates of genetic or environmental contributions that have general significance. Any such estimate is valid at best for a particular population in a particular environment. A statement that the variation in yield of corn from a specific strain in a certain environment is 50% heritable is useful, since it suggests that selective breeding might increase the average yield. Conversely, if the variation in yield is 50% environmental, then control of the environment should also increase the yield. In the case of human populations, estimates of heritability (and nonheritability) are useful as indicators of the traits that are most likely to respond to environmental manipulation. But it must not be forgotten that persons at one end of the scale may show a responsiveness to environmental changes different from those in the middle or at the other end of the scale. Not everyone can be trained to solve differential equations, no matter how intensive the effort.

FAMILIAL CORRELATIONS WITH COMPLEX TRAITS

We recognize many traits to involve resemblance among members of a family. Some are intuitively attributed to heredity and others to environment, depending on one's judgment and biases. Galton attempted to express such resemblances in a more objective way and invented the correlation coefficient to do so. Briefly, the correlation coefficient, r, is a measure of association between two variables. If the association is perfect, that is, if the values of one variable tell precisely the values of the other variable, the correlation coefficient is $+1.0$ (if both increase or decrease together) or -1.0 (if one increases when the other decreases). If the two values are completely independent, the correlation is zero. Values between -1.0 and $+1.0$ indicate the degree of association. For example, if $r = 0.5$, one variable is useful in predicting the other, but the prediction is a statistical statement subject to error.

Correlation coefficients may concern a single measure made on pairs of individuals (for example, I.Q. of husbands versus that of wives), or they may concern two different measures made on single persons (for example, stature and weight). We will consider primarily the former. A more detailed discussion of the calculation of correlation coefficients is given in Appendix A.

Mendelian Inheritance and Intrafamilial Correlations

Galton had no knowledge of Mendelian inheritance but assumed that the contribution from each parent is equal. In a classical paper, Fisher (1918) arrived at the expected correlation coefficients on the basis of Mendelian inheritance. To begin, let us consider traits that are completely hereditary, that is, whose variation is due only to genetic variation. Further, let us consider matings that are random with respect to these particular genetic traits. In these families, the correlation coefficient r between husband and wife should be zero. Each child will receive exactly half his (autosomal) genes from each parent. Therefore, he should resemble both parents. For genes without dominance, his expected correlation will be 0.5 with each parent, but for any one trait he may be like one or the other parent, he may be intermediate, or he may be outside the range of variation of the parents. If dominant genes affect the trait measured, the expected correlation is reduced.

Although each child in a family receives half his genes from each parent, it is not the same half for consecutive children of a mating. Theoretically two sibs might by chance receive identical complements of genes, or they might have no genes that are identical through descent. The likelihood of either of these two extremes is infinitesimal. Rather, the most probable situation is for sibs to share half their genes on the average. This is a statistical expectation, in contrast to the exact relation between parent and child. But in view of the large number of chromosomes and the recombination through crossing over that occurs, this expectation ordinarily is met if the number of loci under consideration is large.

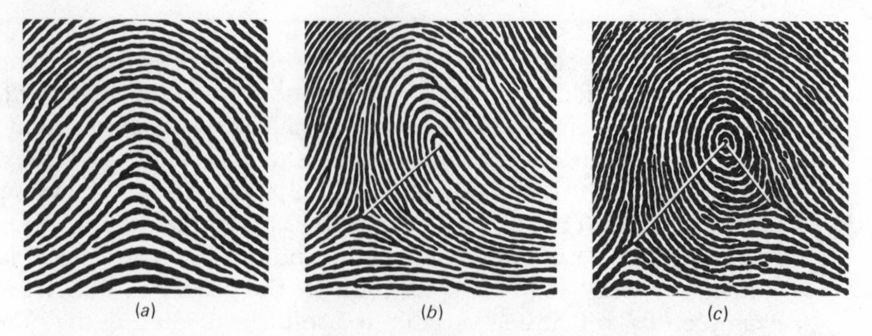

Figure 16-3 Examples of the three basic fingerprint patterns. (*a*) Arch (no triradius). There is no line of count, and the score is 0. (*b*) Loop (one triradius). The triradius is on the left at the junction of three ridge systems. The straight line joining the point of the triradius to the center of the loop illustrates the method of ridge counting. The number of ridges cutting the line is 13. (*c*) Whorl (two triradii). The straight lines are the two lines of count, one from each triradius to the center of the whorl. The ridge count on the left of the pattern is 17, that on the right is 8. The higher count is used. (*From Holt, 1961a.*)

Fingerprint Ridge Counts. The individuality of fingerprints has been recognized at least since the work of Galton, and many studies have been carried out on inheritance of ridge patterns. Since the dermal ridges develop during the embryonic period and remain constant throughout the life of the individual, any environmental influences must be exerted in early embryogenesis. The increased frequency of certain patterns in chromosomal disorders has stimulated much recent interest in dermatoglyphics.

A good example of multifactorial inheritance is provided by counts of fingerprint ridges. Counts are made as shown in Figure 16-3. Many comparisons can be made with such data, but especially interesting results have been obtained by S. B. Holt (1961*a*, 1961*b*) using the total ridge count for all ten fingers. The mean count for 825 British males was 144.98, with standard deviation of 51.08. For 825 females it was 127.23, with standard deviation of 52.51. The range in both sexes was 0 to 300.

Calculation of the correlations between relatives is given in Table 16-1. There is impressive agreement with the theoretical values on the assumption of complete genetic control by genes with additive effect. The nature of these genes is unknown. The ridge count is reduced in Down's syndrome, indicating that perhaps some genes on chromosome 21 influence ridge count. The effect of the sex chromosomes is evident in the difference between males and females, but the effect of this difference has been removed in the calculation of the correlation coefficients given in Table 16-1. But in spite of the inability to talk about individual genes, one can conclude with confidence that the variability in ridge count is virtually all genetic.

Table 16-1 **CORRELATIONS BETWEEN RELATIVES FOR TOTAL DERMAL RIDGE COUNT***

Relationship	No. of pairs	Observed correlation coefficient	Theoretical correlation coefficient†
Parent–child	810	0.48	0.50
Mother–child	405	0.48 ± 0.04	0.50
Father–child	405	0.49 ± 0.04	0.50
Father–mother	200	0.05 ± 0.07	0.00
Midparent–child	405	0.66 ± 0.03	0.71
Sib–sib	642	0.50 ± 0.04	0.50
Monozygotic twins	80	0.95 ± 0.01	1.00
Dizygotic twins	92	0.49 ± 0.08	0.50

* From Holt (1961a).
† Assuming variation is due only to additive genes.

Cholesterol Levels. Although calculation of correlation coefficients is a useful technique, comparison of mean values of a trait among relatives of index cases with mean values for a control group may also be used to demonstrate the importance of genetic variation. As an example, let us consider the intrafamily resemblances of blood levels of cholesterol. Hypercholesterolemia, which often leads to coronary disease, occurs as a family trait, at least in some families. Indeed, as noted in Chapter 11, some families appear to have a dominant gene for hypercholesterolemia, but even in such families the classification of persons as having high or normal blood levels of cholesterol is somewhat arbitrary, since the levels are continuously distributed. Goldstein and coworkers (1973) measured blood levels of cholesterol in the families of 500 survivors of myocardial infarction and obtained the results shown in Table 16-2. In the case of a trait such as cholesterol level it is necessary to adjust the observations so that the effects of sex and age are removed. Males typically have higher levels of cholesterol, and the levels increase in both sexes with age. The effects of such variables can be readily removed statistically, leaving, in the case of hypercholesterolemia, a residual increased level of cholesterol in first-degree relatives (parents, sibs, offspring) of hypercholesterolemic patients but not of patients with normal cholesterol values.

To what extent do such observations result from genetic as opposed to environmental factors? It should be noted first that significant correlations among appropriate family members do not rule out environmental contributions to the variation. If the environmental contribution is large, one would expect the correlation to be diminished in some instances. However, families share environments as well as genes, and

Table 16-2 **BLOOD CHOLESTEROL AND TRIGLYCERIDE LEVELS IN FIRST-DEGREE RELATIVES OF PATIENTS SURVIVING MYOCARDIAL INFARCTION***

Subjects	Number	Cholesterol mg/100ml ± SD	Triglyceride mg/100ml ± SD
A. Controls	950	218 ± 41	93 ± 48
B. Relatives of hyperlipidemic† survivors	645	235 ± 53	126 ± 174
C. Relatives of hypercholesterolemic survivors	379	247 ± 56	
D. Relatives of normolipidemic survivors	113	215 ± 39	94 ± 54

* From data of Goldstein and co-workers (1973).
† Increased levels of cholesterol and/or triglycerides.

sibs may resemble each other because they share a common diet. Husbands and wives also share a common diet, and if the correlation between them is zero in spite of diet and other shared elements of the environment, one may reasonably attribute the parent-offspring and sib-sib correlations to heredity. It might be noted that the husband-wife correlation for cholesterol level is —0.023, a value not significantly different from zero (Goldstein and co-workers, 1973).

Assortative Mating

A somewhat different line of reasoning may be required for certain externally expressed traits, such as stature and intelligence. For both traits, extensive studies have shown strong intrafamilial correlations. The problem here is that the correlation between husbands and wives is positive. While environmental factors are known to influence growth rate in children, adult height is essentially constant, and marriage partners do not adjust their heights to match each other more closely. Rather the effect is attributed to *assortative mating,* that is, to tall persons selecting tall mates and short persons short mates. There are many exceptions, but the effect of assortative mating is strong enough to be readily detected statistically (Table 16-3). In spite of this husband-wife correlation, we still attribute the parent-child and sib-sib correlation largely to heredity because we are unable to identify environmental variables adequate to explain the results. Care should be exercised with such conclusions, however. Malnutrition of several types may decrease the growth rate of children. If the population under investigation in-

Table 16-3 **EXAMPLES OF CORRELATIONS BETWEEN HUSBANDS AND WIVES FOR SELECTED PHYSICAL MEASUREMENTS***

Trait	*r*	Trait	*r*
weight	0.08†	foot length	0.11†
stature	0.29‡	sitting height	0.18
forearm length	0.43‡	head length	0.07†
hand length	0.18	head breadth	0.20‡
middle finger length	0.61‡	total face height	−0.001†
bi-iliac breadth	0.29‡	nose height	0.05†
minimum neck circumference	0.20‡	nose breadth	0.09†
minimum waist circumference	0.38‡	interpupillary breadth	0.20‡
maximum hip circumference	0.22‡	ear length	0.40‡

* The sample studied was a U.S. white population living in Ann Arbor, Michigan. From a study by J. N. Spuhler, reported in Spuhler, 1968.
† Significant at the 5% level.
‡ Significant at the 1% level.

cludes a substantial number of undernourished families, this may diminish the importance of genetic variation but still lead to significant intrafamily correlations.

DISCONTINUOUS VARIATION

Many discontinuous traits may show complex patterns of inheritance, even though expression of the trait is limited to the simple alternatives of presence or absence. The classical work in this field is that of Sewall Wright on number of toes in guinea pigs. Wright showed that although most animals have three toes on each paw, rare animals have four. By selective breeding of these animals, the probability of an animal having four toes could be greatly increased. Any one locus made only a small contribution to the tendency for four toes, and only rarely would a combination of alleles occur naturally that led to four toes. By selective breeding, the frequency of these alleles is enriched.

This and other studies support the idea of a threshold model for appearance of certain discontinuous traits. One may imagine some physiological variable whose values are continuously distributed as in Figure 16-4. In the general population, a certain proportion will exceed the threshold. Various subpopulations differing in genotype may have relatively more or fewer persons exceeding the threshold. The threshold need not be a sharp dividing line as shown in the figure. In practice, it is

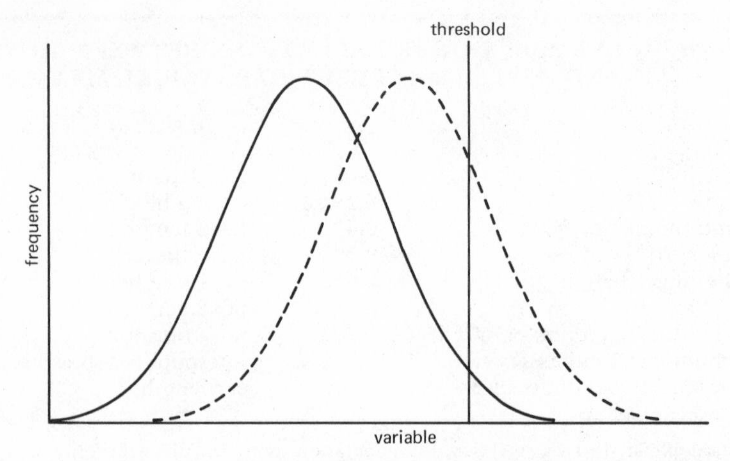

Figure 16-4 Hypothetical distribution of some physiological variables, showing a small portion of individuals exceeding a threshold associated with a discontinuous trait that depends on this variable. Solid line: general population; broken line: persons in the population with a particular genotype. Many more of the latter exceed the threshold.

probably more often a sharp increase in risk, the realization of which may depend on undefined factors (including environmental factors).

Especially important examples in man include the late onset diseases, such as cancer and coronary disease. In both instances, rare families are known showing simple Mendelian inheritance. Familial hypercholesterolemia was discussed in Chapter 11, with homozygotes having a 100% probability of coronary disease and heterozygotes having a high risk. Rare forms of cancer, such as retinoblastoma and multiple polyposis of the colon, are transmitted as simple dominants, heterozygotes having a probability near unity of developing malignant tumors. However, in most cases of cancer and heart disease, genetic factors are not so clearly seen. One may nevertheless question whether the risk is due in part to genetic factors, either multigenic factors or single loci with small effects.

Examination of risks in relatives of affected persons may be used to test the hypothesis of genetic factors in the predisposition to disease. This, of course, is exactly the procedure in testing for simple Mendelian inheritance, except that in Mendelian inheritance a definite predictive model is used. With polygenic inheritance, the hypothesis is that relatives have more genes in common with an index case than do unrelated persons and that they should therefore have a greater risk of suffering from the same diseases that are present in the index case.

Breast Cancer. This form of cancer provides an example of a discontinuous trait that has been studied extensively for the involvement of genetic factors. One such study is summarized in Table 16-4.

Table 16-4 **FREQUENCY OF BREAST CANCER IN DECEASED RELATIVES OF WOMEN WITH BREAST CANCER, WITH OTHER MALIGNANCIES, AND WITHOUT KNOWN MALIGNANCY***

Index cases with:		Mothers	Sisters
Breast cancer	Number dead	213	132
	Number breast cancers	11	14
Other malignancies	Number dead	203	162
	Number breast cancers	4	6
Control	Number dead	158	74
	Number breast cancers	3	5

* From Macklin (1959).

The risk of having breast cancer in first-degree relatives is increased some threefold. In a study by D. E. Anderson (1974), the risk for a sister of a breast cancer patient whose mother also had breast cancer was approximately 30%, a risk greatly elevated over that in sisters of control patients with other malignancies. Other comparisons indicated that this risk is transmitted by fathers as well as by mothers. These studies also show that the risk in breast cancer is heterogeneous; that is, in some families in which breast cancer occurs, the risk is higher than in other families in which breast cancer occurs.

The interpretation of such increased risk as due to genetic factors is subject to reservation. Environmental factors, including exposure to oncogenic viruses, might lead to similar results. Nevertheless, such results are anticipated if genetic factors are involved, and they encourage a search for more clearly defined single gene factors.

Diabetes Mellitus. There are many genetic studies of diabetes, with generally uniform if not very satisfying results (Simpson, 1962; Barrai and Cann, 1965; Neel et al., 1965; Degnbol and Green, 1978). Diabetes clearly is concentrated within certain families, suggesting a genetic etiology. At least two forms are recognized on the basis of familial risk: juvenile and late onset. If a young person is affected with diabetes, there is a risk that other young persons in the family will be affected but no large risk that adults will become diabetic in older age groups. Conversely, the presence of late onset diabetes in a family does not carry increased risk of juvenile diabetes. Supporting this difference is the observation that juvenile diabetics are typically "insulin-dependent," whereas late onset diabetes can be managed by diet and drugs.

Almost 100% of monozygotic twins with late onset diabetes are concordant. If one has it, so does the other. By contrast, monozygotic twins with juvenile diabetes are only about 50% concordant. This latter observation clearly is inconsistent with a purely genetic explanation. Rubenstein et al. (1977) made the interesting observation that the risk of juvenile diabetes is associated with HLA inheritance, and they presented evidence that susceptibility is a result of homozygosity for a recessive gene closely linked to HLA. Only 50% of homozygotes actually develop the disease however. The pathogenesis of juvenile diabetes has since been shown probably to result from infection of susceptible persons with a Coxsackievirus, which destroys the insulin-producing beta cells of the pancreas (Yoon et al., 1979). The genetic component of juvenile diabetes therefore is a result of homozygosity for a gene in the MHC region of chromosome 6 that reduces resistence to infection with the virus. But actual disease requires exposure to and development of the infection. Late onset diabetes has a different pathogenesis and shows no such dependence on nongenetic events.

TWINS IN GENETIC RESEARCH

One experiment of nature useful in the study of complex traits is the occurrence of multiple births—twins, triplets, and so on. Because of their greater prevalence, twin births provide most of the opportunities for observations. The utility of twin births arises from the existence of two kinds of twins, the so-called identical (monozygotic) and fraternal (dizygotic) twins. Monozygotic (MZ) twins arise from a single zygote that divides into two separate embryos very early in gestation. Since only one zygote is involved, the twins are genetically identical. Dizygotic (DZ) twins arise from two zygotes that are produced by fertilization of two separate ova. Thus they have the same genetic relationship as ordinary sibs. They may be either like-sexed (two boys or two girls) or unlike-sexed (a boy and a girl). Monozygotic twins must be like-sexed.

Persons who are genetically identical provide a means of estimating the effects of environment, since any differences between co-twins must have a nongenetic origin. Environment, as defined by this test, comprises all nongenetic effects, including the internal environment of the body and unequal distribution of cytoplasmic constituents in cell division. Dizygotic twins may differ from each other both for genetic as well as for environmental reasons.

Twins do not reveal *how* genes operate, but they may tell us *whether* genetic variation is present and approximately how much of the total variation is genetic. The question to be answered is whether monozygotic twins vary less than dizygotic twins. An affirmative answer supports the hypothesis of a genetic component of variation. Ideally, the most effective use of twins would be a comparison of monozygotic twins reared apart with pairs of unrelated persons of the same age, socioeconomic level, and so on. In practice, it is difficult to assemble such a group of twins. However, the use of dizygotic twins as the comparison group for monozygotic twins reared together permits conclusions as to the contribution of genetic variation for a particular trait or measure.

Diagnosis of Twin Zygosity

The usefulness of twins in genetic research depends on correctly assessing the zygosity. The criteria of zygosity determination vary somewhat with the investigator but ultimately depend on whether or not a *genetic* difference can be demonstrated between members of a twin pair. If a difference is present, the twins are classified as dizygotic. If no reliable difference can be detected, they are classified as monozygotic. The thoroughness with which one investigates zygosity varies with the situation. For a comparison of measurements on 100 MZ twins with those on 100 DZ twins, the presence of a few DZ twins among the MZ sets is not likely to influence conclusions. However, for successful transplantation of an organ from one twin to the other, zygosity is of great importance.

The first step in establishing zygosity is to classify subjects for simple genetic traits. The simplest is sex; as indicated earlier, twins of different sex arise from different zygotes. Other traits include the blood groups, plasma protein types, and salivary secretion of antigens, all of which follow simple Mendelian patterns of inheritance. A difference in any one system is adequate evidence for classifying a pair of twins as DZ, although, purely by chance, DZ twins may agree for all of the traits tested. More complex traits may also be considered: finger and palm prints, hair color, eye color, and morphology of skeletal and other tissues, such as teeth, ears, and nostrils. Reliable judgments of complex traits can be made by experienced observers.

Zygosity of twins has often been judged erroneously by the nature of the fetal membranes. Both MZ and DZ twins may have two separate placentas with separate amnions and chorions. Conversely, both types of twins may have a single fused placenta. DZ twins always have two chorions, but they may be so fused that they cannot be distinguished without special examination. A high portion of MZ twins have separate amnions with a single chorion. The birth membranes may sometimes help establish monozygosity, but more than casual examination is required.

The most rigorous test of zygosity is skin grafting. Because of the surgical procedure involved, transplanting of skin is done only if it is important to establishing zygosity with certainty. Permanent acceptance of a skin graft ordinarily is evidence of genetic identity (Chapter 13).

Concordance in Twins

The simplest use of twins is in testing for concordance of monozygotic twins for a trait suspected to be hereditary. If a trait is entirely genetic, all monozygotic twin pairs should be concordant. All exceptions are evidence against hereditary factors as sole causes. Increased concordance of MZ twins as compared to DZ twins for a trait suggests a genetic predisposition for the trait, even though other factors may be important in whether the trait is expressed. This use of twins for studying discontinuous traits has long been used in attempts to understand the causes of complex diseases, such as diabetes and schizophrenia.

Twin Studies for Continuous Traits

Many complex traits show continuous variation, and twins cannot be classified as concordant or discordant. The differences between members of a twin pair are measured on a continuous scale and can be readily analyzed statistically to detect greater similarity of MZ twin pairs as compared to DZ pairs.

If one considers pairs of unrelated persons drawn at random from a population, the differences between members of the pairs provide a measure of the total variability, both genetic and environmental, of the population. Expressed symbolically,

$$V_U = V_G + V_E,$$

where V_U is the variability among unrelated persons, V_G is the portion attributable to gene differences, and V_E is the portion attributable to environmental differences. Some pairs by chance will have similar environments; others will have very different environments. But the average difference in environment within pairs should be similar.

Since $V_G = 0$ for MZ twins,

$$V_M = V_E,$$

where V_M is the variability within monozygotic twin pairs, and

$$V_G = V_U - V_M.$$

This formula is straightforward, but requires a class of persons—MZ twins reared apart—that is too small and difficult to locate to be of practical use. The alternative is to use MZ twins reared together. In this case, the twins are exposed to similar environments, and it is necessary to compare them to pairs of related persons living together so that V_E will be constant. Since age and sex are important in determining environment, DZ twins of like sex are used. The assumption is that the similarity of environment for MZ twins is equal to that for DZ twins. For many traits this probably is correct; for others, such as personality traits, it is not. The experience of being an identical twin is apt to be different from the experience of being a nonidentical twin. The relationships among twins may be symbolized as before:

$$V_{DZ} = V_G + V_E;$$
$$V_{MZ} = V_E;$$
$$V_G = V_{DZ} - V_{MZ}.$$

But the meanings have changed slightly. For twins reared together, V_E is certainly smaller than the same quantity for persons reared apart. Furthermore, V_G for DZ twins is much smaller than the genetic variability of unrelated persons, since DZ twins share many genes in common by virtue of having the same parents. For these reasons, the results obtained from comparisons of MZ and DZ twins are not true measures of the size of the genetic and environmental components. But if V_G can be shown to be greater than zero, then genetic factors are considered to contribute to the variability of the trait in question.

A variety of statistical techniques has been used for the analysis of twin data. If the distribution of twin -pair differences is Gaussian, then V may be measured as the variance and the ratio V_{DZ}/V_{MZ} used to test for the contribution of genetic variation. The calculation of variance ratios is discussed in most textbooks of statistics. When appropriate, this is the most satisfactory method of testing for genetic effects. The answer obtained—when genetic effects can be detected—is usually only a crude estimate of the magnitude of the effects. If a very large number of twin pairs is examined, sufficiently good estimates of V_{DZ} and V_{MZ} might be obtained so that V_G could be computed. With the number usually available, it is not meaningful to attempt to calculate a value for V_G. In an experimental study, it is also necessary to account for the error variance, that is, the variation associated with test reproducibility.

Less elegant statistical techniques may be used with sacrifice of some of the information in the data. An example of the distribution of differences between twin pairs for height is shown in Figure 16-5. As expected, the most frequently encountered values are very low, since DZ twins share many genes in common as well as a common environment. However, the large values that may occur in DZ twins are very rare in MZ twins. From inspection of such a distribution, it may be surmised that the variability of DZ twins is larger than MZ twins, and hence genetic factors must be operating.

One technique that may be used is to divide the distribution of values arbitrarily so that a χ^2 test of heterogeneity may be applied. This is done by selecting some value, such as the median for the combined data, to classify the twins as having "large" or "small" differences. Fifty percent of the twins will have small differences by this definition, and 50% will have large differences. If there is no association of mag-

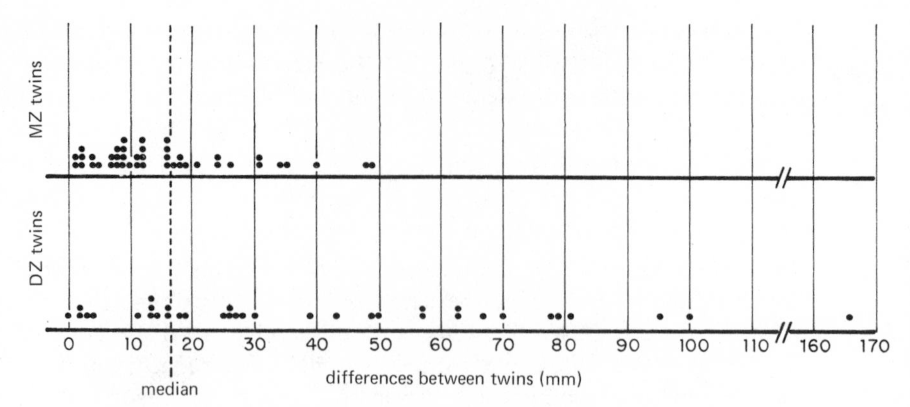

Figure 16-5 Distribution for differences in height between MZ twins and DZ twins.

nitude of the difference with zygosity, then 50% of the MZ twins should fall into each category as should 50% of the DZ twins. On the other hand, if MZ twin pairs contribute relatively more of the small values and DZ pairs more of the large, then a χ^2 test will show heterogeneity in the combined distribution.

In the example of Figure 16-5, division at the median gives 28 MZ twins with low values and 15 with high. There are 12 DZ twins with low and 24 with high. A 2 × 2 test for heterogeneity gives a χ^2 of 7.92 with one degree of freedom, a value significant at the 1% level. Since the two types of twins therefore do not represent one population, the conclusion is that they are different with respect to height, and it is thought that the two populations are different because DZ twins have genetic variation as well as environmental, but MZ twins have only environmental variation.

For many traits, the differences within pairs of MZ twins are sufficiently small compared to the differences within DZ twins pairs, so that demonstration of a genetic component is relatively simple. Quantitative measures of the genetic component are more difficult.

SUMMARY

1. Mendelian heredity is most easily observed in discontinuous traits. However, many traits under genetic control are continuous, and special approaches are used to show that these reflect genetic variation. Often the variation is subject to the influence of several loci, in which case the trait is said to be polygenic, multigenic, or multifactorial. The relative contributions of genetic and environmental variation to such complex traits depend on the frequencies of the alleles involved as well as on the extent of environmental variation. Thus the heritable component of variation may increase or decrease as a result of environmental changes only.

2. For complex traits, genetic variability is detected by the resemblance among relatives of certain degrees of relationship. This resemblance is often expressed by means of the correlation coefficient, r. For first-degree relatives, the expected r is 0.50. Counts of fingerprint dermal ridges are an example of a complex trait whose correlations among family members are almost exactly as predicted on the hypothesis of multifactorial genetic control. Comparison of blood cholesterol levels among relatives of patients with high cholesterol shows close relatives (but not spouses) also to have high levels, presumably because of genes in common.

3. For traits such as stature and intelligence, analysis is complicated somewhat by assortative mating. Marriage partners tend to select each other on the basis of perceived similarities, leading to positive correlations between them. Nevertheless, evidence supports genetic variation as the primary source of variation in these traits.

4. Multifactorial inheritance may also be involved in complex discontinuous traits. Coronary disease is often associated with high levels of blood cholesterol. While the distribution of cholesterol values is continuous, the risk of coronary disease increases sharply if cholesterol levels exceed certain thresholds. Breast cancer is another trait that occurs more often in relatives of patients with breast cancer. The observations agree with a strong genetic component in the risk.

5. Diabetes is another example of a discontinuous trait that is determined by multifactorial inheritance. The two major types of diabetes—juvenile and late onset—are genetically independent. Juvenile diabetes appears to be due to homozygosity for a gene in the MHC complex that confers susceptibility to Coxsackievirus. Infection by the virus of some susceptible persons destroys the insulin-producing beta cells of the pancreas.

6. Twins are very useful in studying genetic variation. Monozygotic (MZ) twins, being genetically identical, can differ only as a result of nongenetic factors, whereas dizygotic (DZ) twins arise from entirely separate zygotes. For discontinuous traits, the most convenient comparison is for concordance.

7. Twins can be used to study continuous variation also. The differences between co-twins for a measure should be smaller if the twins are MZ as compared to DZ, on the hypothesis that the trait is influenced by genetic variation. Results from a study of stature are given as an example. The close resemblance of MZ co-twins as compared to the sometimes large divergence between DZ co-twins argues strongly for heredity as a major determiner of stature.

REFERENCES AND SUGGESTED READING

ALLEN, G. 1965. Twin research: Problems and prospects. *Prog. Med. Genet.* 4: 242–269.

ANDERSON, D. E. 1974. Genetic study of breast cancer: Identification of a high risk group. *Cancer* 34: 1090–1097.

ANDERSON, V. E., GOODMAN, H. O., AND REED, S. C. 1958. *Variables Related to Human Breast Cancer.* Minneapolis: Univ. of Minnesota Press, 172 pp.

BARRAI, I., AND CANN, H. M. 1965. Segregation analysis of juvenile diabetes mellitus. *J. Med. Genet.* 2: 8–11.

BULMER, M. G. 1970. *The Biology of Twinning in Man.* London: Oxford Univ. Press, 205 pp.

DAHLBERG, G. 1926. *Twin Births and Twins from a Hereditary Point of View.* Bokförlags-A. B. Tidens Tryckeri, Stockholm.

DEGNBOL, B., AND GREEN, A. 1978. Diabetes mellitus among first- and second-degree relatives of early onset diabetes. *Ann. Hum. Genet.* 42: 25–47.

FISHER, R. A. 1918. The correlation between relatives on the supposition of Mendelian inheritance. *Trans. Roy. Soc. Edinburgh* 52: 399–433.

GALTON, F. 1876. The history of twins as a criterion of the relative powers of nature and nurture. *J. Anthropol. Inst.* (Lond.) 5: 391.

GOLDSTEIN, J. L., SCHROTT, H. G., HAZZARD, W. R., BIERMAN, E. L., AND MOTULSKY, A. G. 1973. Hyperlipidemia in coronary heart disease. II. Genetic analysis of lipid levels in 176 families and delineation of a new inherited disorder, combined hyperlipidemia. *J. Clin. Inv.* 52: 1544–1568.

HARRISON, G. A., WEINER, J. S., TANNER, J. M., AND BARNICOT, N. A. 1964. *Human Biology.* New York and Oxford: Oxford Univ. Press, 536 pp.

HOLT, S. B. 1961*a*. Quantitative genetics of fingerprint patterns. *Br. Med. Bull.* 17: 247–250.

HOLT, S. B. 1961*b*. Inheritance of dermal ridge patterns. In *Recent Advances in Human Genetics* (L. S. Penrose, Ed.). London: J. and A. Churchill, pp. 101–119.

HRUBEC, Z. 1973. The effect of diagnostic ascertainment in twins on the assessment of the genetic factor in disease etiology. *Am. J. Hum. Genet.* 25: 15–28.

MACKLIN M. T. 1959. Genetic considerations in human breast and gastric cancer. In *Genetics and Cancer.* Austin: Univ. of Texas Press, pp. 408–425.

NEEL, J. V., FAJANS, S. S., CONN, J. W., AND DAVIDSON, R. T. 1965. Diabetes mellitus. In *Genetics and the Epidemiology of Chronic Diseases* (J. V. Neel, M. W. Shaw, W. J. Schull, Eds.). Public Health Service Publ. No. 1163. U.S. Government Printing Office, Washington, D.C. pp. 105–132.

NEWMAN, H. H., FREEMAN, F. N., AND HOLZINGER, J. K. 1937. *Twins, a Study of Heredity and Environment.* Chicago: Univ. of Chicago Press, 369 pp.

ROBERTS, J. A. F. 1964. Multifactorial inheritance and human disease. *Prog. Med. Genet.* 3: 178–216.
RUBINSTEIN, P., SUCIU-FOCA, N., AND NICHOLSON, J. F. 1977. Genetics of juvenile diabetes mellitus. *N. Engl. J. Med.* 297: 1036–1040.
SIMPSON, N. E. 1962. The genetics of diabetes: A study of 233 families of juvenile diabetics. *Ann. Hum. Genet.* 26: 1–21.
SPUHLER, J. N. 1968. Assortative mating with respect to physical characteristics. *Eugenics Quart.* 15: 128–140.
YOON, J.-W., AUSTIN, M., ONODERA, T., AND NOTKINS, A. L. 1979. Virus-induced diabetes mellitus. *N. Engl. J. Med.* 300: 1173–1179.

REVIEW QUESTIONS

1. Define:

continuous variation
discontinuous variation
polygenic
multigenic
multifactorial inheritance
correlation coefficient
assortative mating
monozygotic twins
dizygotic twins
placenta
amnion
chorion
Gaussian distribution
variance

2. How are polygenes different from the traditional Mendelian genes?

3. Outstanding athletic ability is often found in several members of a family. Devise a study to determine to what extent athletic ability is inherited.

4. A number of studies have shown variations in stature to be almost entirely due to heredity. Yet average height has increased substantially since the Middle Ages, and the increase in height of children of immigrants to the United States, as compared with height of the immigrants themselves, is especially noteworthy. How can these observations be reconciled?

5. It has been said the best way to assure a long life is to pick the right parents. How would you design a study to see whether longevity is inherited?

6. Would you expect uniformly good nutrition to increase or decrease the heritability of height?

7. How would you distinguish between multifactorial inheritance and monofactorial inheritance with low penetrance for an uncommon trait? Would your answer change if the trait were common?

CHAPTER SEVENTEEN

GENETICS OF DEVELOPMENT

Much of the previous discussion has dealt with human cells and tissues in the same terms that would be used for bacteria. This may be very useful up to a point. The insights into molecular processes of hemoglobin synthesis do not require understanding the events that lead to differentiation of hemoglobin-forming tissue from other tissue. But such differentiation does occur as a regular part of normal development. Many genes are expressed only in one tissue or at most in a limited number of tissues. During embryogenesis, many genes become active or inactive at regular stages in development. What are the mechanisms that control genes during development? Are they the same as or different from the regulation of genes in a fully developed organism?

At a more complex level, during embryogenesis there are many instances in which tissues must interact for normal development to occur. This interaction may involve direct contact or it may be by means of humoral agents. At what levels is this interaction programmed by genes, and to what extent is variation in the interaction due to genetic variation? At a practical level, to what extent are congenital malformations attributed to heredity?

Answers to these questions are tentative in some instances and nonexistent in others. Much of the information on development is derived from organisms other than man, since the study of early human embryos is technically very difficult. The examples in the following sections have been selected to illustrate those areas of developmental genetics in which some progress has been made and to define insofar as possible some of the major problems remaining. As has been true in other areas of biology, the study of mutant genes should be a powerful tool in elucidating the normal steps of development.

GENE ACTION DURING DEVELOPMENT AND DIFFERENTIATION

There is a wealth of information on the levels of enzymes during development in a variety of experimental organisms. Such information is much more scarce in early human embryonic development. It is quite clear though that specific enzymes appear at characteristic times during embryogenesis and that their appearance must signal the activation of previously inactive genes, presumably by initiating transcription. This is nowhere better illustrated than in the case of human hemoglobins (Figure 10-4). The ζ chain has been identified only in early embryos and the ϵ, γ, δ, and β chains appear in sequence during development, the last two appearing together as the adult pattern. The mutation that causes hereditary persistence of fetal hemoglobin (HPFH) prevents the normal switchover from γ to β and δ chain synthesis and thus can be classed as a developmental mutant.

Many enzymes are known to increase near the time of birth, and indeed these may be used as a measure of fetal maturity (Kretchmer and co-workers, 1963). Delays in appearance of certain enzymes occur fairly often. Premature infants often have elevated blood levels of phenylalanine and tyrosine as a result of low levels of tyrosine transaminase and *p*-hydroxyphenylpyruvic acid oxidase. Assays of the tyrosine oxidizing systems in livers of fetal rats, rabbits, and human beings show negligible activity. In the case of rats, adrenalectomy immediately after birth prevents the normal increase in tyrosine transaminase that occurs two hours after birth, indicating dependence on adrenal hormones for gene activation (Sereni and co-workers, 1959).

Elevations in the tyrosine metabolites in premature human infants are transient, disappearing as the enzymes become manufactured in adequate quantities. With some test systems, the increased phenylalanine has led to confusion with phenylketonuria, but in phenylketonuria the tyrosine is not elevated. There is no evidence that the delay in appearance of these enzymes has a genetic base. This system does represent a clear instance in which structural genes become active (or at least their translational products appear) late in development.

Another example of differential gene action is found in the lactate dehydrogenase (LDH) system. Mammalian LDH is a tetrameric enzyme, consisting of two types of monomeric subunits, A and B. These may combine in any combination of four to give active enzyme, that is, A_4, A_3B, A_2B_2, AB_3, and B_4. Since A and B have different electrical charges, the five combinations can be separated by electrophoresis (Figure 17-1). The A and B subunits are determined by separate structural genes located, in human beings, on different chromosomes. These two genes may well have originated from a common ancestral gene that duplicated and diverged, both in structure and location. Of special interest is the fact that the two genes have different activities in different tissues (reviewed in Vesell, 1965). This is shown in Figure 17-1.

The physiological significance of these different forms is not known, but it is assumed that some functional differences must exist, the different combination of *isozymes* being adapted to the various tissue functions. It would be interesting to know the mechanisms by which differential activity of the two loci occurs, especially since they seemingly can have various degrees of activity. Although structural mutations are known for both loci in man as well as in some other mammals, no instance of absent enzyme is known nor have persons been observed with the "wrong" pattern for a particular tissue. This may be due to lack of opportunity to observe such errors, in view of the need to do biopsies in order to carry out an assay.

These examples have focused on one locus at a time (or two closely linked loci in the case of the β and δ genes of hemoglobin). It may be that all differentiation and development occurs this way. However, since many genes become active or inactive at approximately the same time, it is tempting to invoke processes by which clusters of genes are regulated together. This could occur through some mechanism yet to be defined, perhaps involving hormones or related activators of gene transcription.

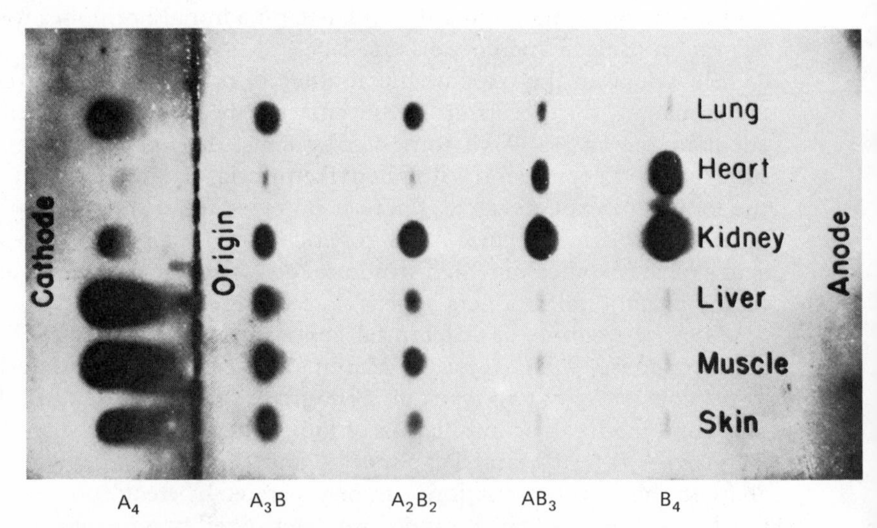

Figure 17-1 Gel electrophoresis of lactate dehydrogenase (LDH) isozymes from various normal human tissues. The five LDH bands are designated by their subunit composition. (*From Vesell and co-workers, 1962.*)

GENETIC DEFECTS IN DEVELOPMENT

The designation of an inherited defect as developmental or nondevelopmental is somewhat arbitrary since, in many instances, a single gene whose primary action is well understood may have consequences for seemingly remote morphological traits. For example, sickle cell anemia is associated with many sequelae, including alterations in the skull shape owing to increased activity of bone marrow. Yet it seems of limited usefulness to label sickle cell anemia a developmental disorder. In this discussion, consideration will be limited to those alterations that affect tissue differentiation or that influence tissue and organ interactions during embryonic and fetal development.

Congenital malformations constitute a major problem in human health. In one study carried out by the World Health Organization in 24 centers in 16 countries, a frequency of 12.7 major malformations per 1000 single births was observed (Stevenson and co-workers, 1966). A number of other studies more limited in scope have given frequencies on the order of 10 children with malformations per 1000 births. Thus some 1% of newborns have a readily recognizable defect in development.

Certain populations are noted for high frequencies of a particular malformation, for example, spina bifida and anencephaly among the Irish, cleft lip and palate in Japanese, and congenital dislocation of the hip in some American Indians. Table 17-1 compares the frequencies of six major malformations in Japanese and Caucasians.

The extent to which the major malformations are inherited has been

Table 17-1 **COMPARISON OF THE FREQUENCIES OF OCCURRENCE OF SIX READILY DIAGNOSED MALFORMATIONS IN JAPANESE AND WHITES***

	Japanese	White
Total number of births	113,441	120,101
Number affected per 1000 births:		
Anencephaly	0.64	0.90
Spina bifida	0.21	1.29
Anophthalmos/microphthalmos	0.23	0.08
Atresia ani	0.26	0.37
Harelip/cleft palate	2.34	1.52
Polydactyly	1.02	0.62
Total	4.71	4.79

* The results of several studies were pooled by Neel (1958) to yield these data. The table here is taken from Table 2 of his publication.

the object of many studies. Clearly many newborns with chromosomal defects have multiple abnormalities. These are caused by errors in the genetic apparatus but are transmitted only in the case of balanced translocations. Many malformations do cluster in families and populations in a manner consistent with genetic origin, but such clustering must be distinguished from environmentally induced clustering.

Congenital Malformations Due to Single Genes

Many Mendelian traits can be classified as errors in development, since the variant genes are expressed during embryogenesis and influence the formation of organs and tissues. Brachydactyly, the first Mendelian trait reported in man, is noted especially for shortened middle phalanges, but other minor skeletal defects also occur, including short stature. Knowledge of developmental mutations in man is highly biased by the difficulty in studying conditions that lead to embryonic or fetal death. Even dominantly inherited malformations that are consistent with postnatal survival are not readily recognized as hereditary if affected persons cannot reproduce. Therefore, we know much more about the inheritance of mild than severe malformations. Of course, from one point of view, the mildly affected person who requires medical attention is a greater problem than the early embryonic abortion that is noted only as a missed menstrual period.

Defects in Sexual Differentiation. A variety of inherited defects in human sexual differentiation are known, partly because such errors are not life-threatening and the affected persons survive to be seen. Many of the defects have been reviewed in Chapter 7. Here we mention only one group, the male pseudohermaphrodites, including testicular feminization. Their development is of special interest because it involves failure of tissues to respond to a normal signal.

Briefly, the development of a penis and other masculine characteristics is in response to androgens secreted by interstitial cells of the testes. In testicular feminization, androgen secretion is normal, but tissues are unresponsive. This is thought to occur through absence or inactivity of the protein that binds and transports testosterone into the cells. Once inside the cell or perhaps the cell nucleus, the androgens activate certain genes that lead ultimately to a male phenotype. In certain of the male pseudohermaphrodites, the process is only partially defective, but whether the defect always involves the same mechanism as in testicular feminization is not known.

Ohno (1974) has summarized studies indicating that the testicular feminization locus (*Tfm*) in mice normally produces a receptor protein for androgens and that the receptor-androgen combination acts as an inducer of a number of other loci, presumably by binding to specific chromosome receptors. Whatever the precise mechanism, this clearly is a mutation in one important developmental system that may well serve as a model for other developmental systems.

Achondroplasia. This disorder has been recognized as an entity for over a century. It is a simple dominant trait, occurring with a frequency of approximately 20 per million births. Affected persons are characterized by very short long bones, an enlarged head, and depressed nasal bridge (Figure 17-2). The trunk is normal in size. Additional skeletal defects are noted on x-ray examination. Adults average about four feet in height. The primary defect has not been identified, although it apparently concerns bone formation. There are several other forms of dwarfism that resemble achondroplasia but that have somewhat different clinical features and in some instances recessive rather than dominant heredity.

Achondroplasia is of genetic interest as one of the few loci for which the mutation rate has been estimated (Chapter 9). Although there is

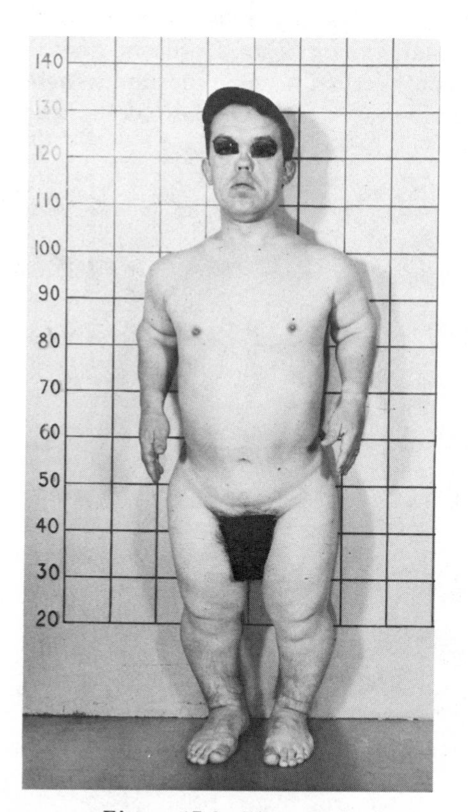

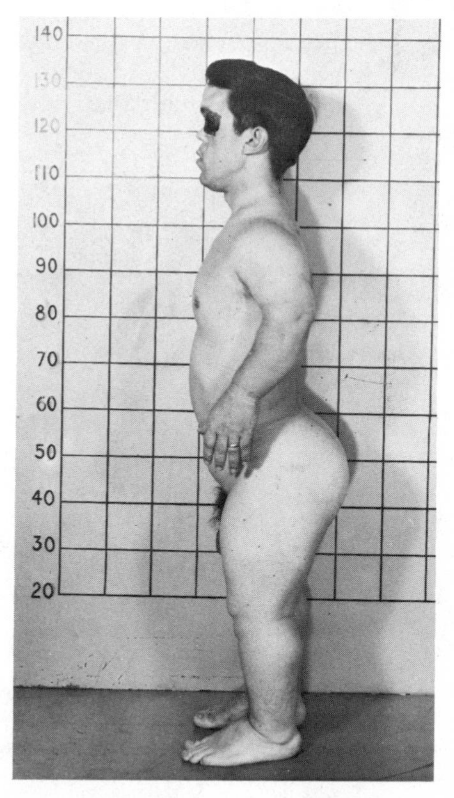

Figure 17-2 Photograph of a patient with typical features of achondroplasia. (*From Hall and co-workers, 1969.*)

variability in expression of the gene, essentially all affected persons can be recognized. Affected persons are fertile, but reproduction is low. Some 80% of cases result from new mutations. Most rare dominant traits are known only in the heterozygous combination. Two families have been reported in which matings between two achondroplastics gave rise to two severely affected children thought to be homozygous (Hall and co-workers, 1969).

Nail-Patella Syndrome. This autosomal dominant disorder is characterized by defects in mesodermal and ectodermal derivatives, giving rise to deformed hypoplastic fingernails and sometimes toenails, small or absent patella, dysplasia of the elbow, and frequent "iliac horns" (Figure 17-3). Many affected persons have a form of nephritis, occasionally leading to death.

A number of pedigrees are known in which the gene for nail-patella syndrome is segregating. Study of these families has shown the *Np* locus to be only 10 map units away from the ABO blood-type locus. This was one of the first linkage groups to be established in man (Renwick and Lawler, 1955). Renwick (1956) has presented evidence that several normal alleles *(isoalleles)* may exist at the *Np* locus, distinguishable only by the degree of defect that results when they are in heterozygous combination with the allele that causes the nail-patella syn-

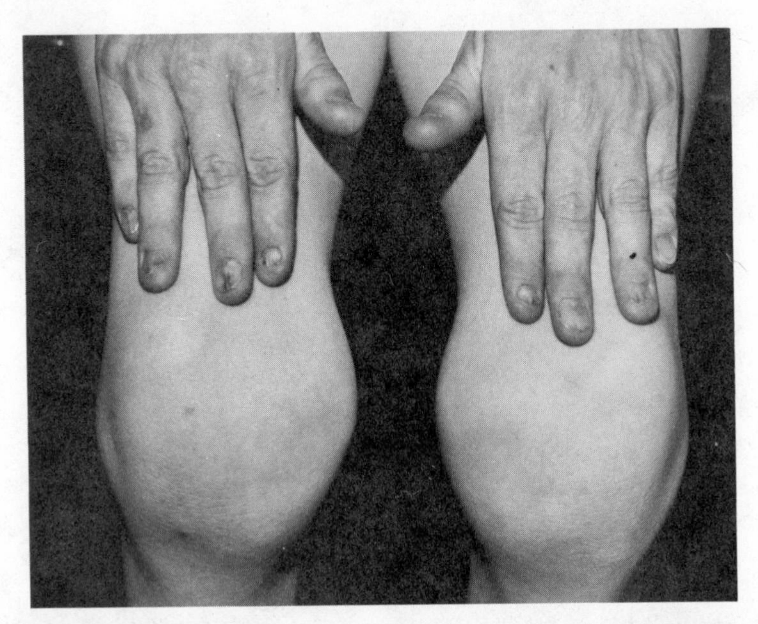

Figure 17-3 Hands and knees of a person affected with nail-patella syndrome. Often the nails are nearly absent. (*From G. L. Lucas and J. M. Opitz, 1966.* J. Pediatr. *68: 273.*)

drome. Thus there was high correlation in severity of defect between sibs, but zero correlation between parents and their offspring.

Congenital Malformations of Complex Etiology

Among the more common malformations there are no examples that are the result of simple Mendelian inheritance. One can nevertheless assess the contribution of genetic variation by a variety of techniques (discussed more fully in Chapter 16). For example, monozygotic twins are genetically identical and should always be concordant for a trait due entirely to heredity. In the early studies of Down's syndrome, the nearly complete concordance in monozygotic twins was an important clue to the genetic basis of the disorder. The very low concordance in dizygotic twins was inconsistent with a simple recessive trait, however.

Another approach sometimes useful is to compare the frequency of affected persons among first-degree relatives (parents, sibs, or offspring) or more distant relatives with the frequency in the general population. The lack of ability to reproduce or to survive associated with some malformations, such an anencephaly, may limit such comparisons to sibs or to second- or third-degree relatives. Since families share environments as well as genes, some care must be exercised in interpreting the basis for an increased risk among relatives. Genetic theory predicts similar increased risks among relatives of the same degree, for example, sibs and offspring.

Although the major malformations do not follow simple patterns of inheritance, the results can be explained by assuming more complex inheritance. For example, multifactorial inheritance involving several loci is consistent with the observed recurrence risks for several. Unfortunately, the data are equally consistent with models involving environmental influences on a single locus or with still other models.

Cleft Lip and Cleft Palate. Perhaps the most widely studied malformations from a genetic viewpoint are cleft lip and cleft palate. Clefts occur when the embryonic palate shelves fail to fuse, usually because they do not move into position at the critical time when fusion can occur. Beginning with the large study in Denmark by Fogh-Andersen (1942), it has been recognized that the oral cleft disorders are etiologically diverse. There are many rare inherited conditions that may sometimes have cleft lip and/or cleft palate as a feature. In addition, cleft lip and cleft palate occur as isolated malformations with frequencies of some 2 cases per 1000 births (Table 17-1). The many studies of cleft lip and cleft palate exemplify the methodological approaches used in the study of malformations.

On the basis of type of defect recurring in families, there appear to be two distinct disorders: (1) cleft palate and (2) cleft lip with or without

Table 17-2 **CONCORDANCE RATES IN CO-TWINS OF PROBANDS WITH CLEFT PALATE, CP, OR WITH CLEFT LIP WITH OR WITHOUT CLEFT PALATE, CL(P)***

	CP		CL(P)	
Twin zygosity	Number of pairs	Concordance %	Number of pairs	Concordance %
MZ	17	23.5	53	37.7
DZ	20	10.0	86	8.1

* From R. J. Gorlin. Cited in Fraser (1970).

cleft palate. Relatives of persons with cleft palate only are at risk for cleft palate but not for cleft lip. On the other hand, relatives of persons with cleft lip only are at risk for cleft lip and may also have cleft palate. The two conditions are often abbreviated (1) CP and (2) CL(P).

Genetic components in the risk of CP and CL(P) are indicated in twin data summarized in Table 17-2. While the concordance is less than the 100% expected for a trait completely determined by heredity, nevertheless the risk for an MZ co-twin of an affected twin is much greater than for a DZ co-twin. In the case of CL(P), conclusions are borne out in studies of relatives of other degrees of relationship (Table 17-3). There is a consistent increase in risk with closeness of rela-

Table 17-3 **PERCENTAGE OF AFFECTED FIRST-, SECOND-, AND THIRD-DEGREE RELATIVES OF INDEX PATIENTS WITH CLEFT LIP WITH OR WITHOUT CLEFT PALATE, AS REPORTED IN THREE SEPARATE STUDIES***

	Study		
Relationship	Copenhagen	Utah	London
First-degree relatives			
Sibs	4.9 ± 1.0†	4.6 ± 0.6	3.2 ± 0.6
Offspring	—	4.3 ± 1.6	3.0 ± 0.7
Second-degree relatives			
Aunts and uncles	0.8 ± 0.1	0.7 ± 0.1	0.6 ± 0.2
Nieces and nephews	—	0.8 ± 0.3	0.7 ± 0.3
Third-degree relatives			
First cousins	0.3 ± 0.1	0.4 ± 0.1	0.2 ± 0.1

* Tabulated by C. O. Carter. 1965. The inheritance of common congenital malformations. *Prog. Med. Genet.* 4: 59. Reprinted by permission.
† Values are percentages ± standard error.

tionship, even though relatives of the same degree are in different generations. If environmental factors were predominant, one would expect sibs to have a higher risk than offspring, since sibs are exposed in general to prenatal environments much more like that of the index case than are offspring of the index case. This argument would apply with even more force to nieces and nephews as compared to aunts and uncles, since these are more distantly related to each other, although they have the same degree of relationship to the index cases. Similar data have not been collected for CP, although the risk in sibs of index cases is clearly elevated, approximately 3.5%, with some evidence of greater risk in female sibs (Fraser, 1970).

The most reasonable explanation for inheritance of a trait such as CL(P) or CP involves variation at two or more loci, with possible environmental factors also of importance. The situation can be visualized as in Figure 17-4. The horizontal axis might be a variable such as time in development at which the palatal shelves reach a position in which fusion is possible. A person affected with CL(P) has some minimum combination of alleles necessary for expression of the trait. His first-degree relatives will have a high frequency of such alleles, perhaps sufficient for delayed closure but not necessarily. The time of shelf movement is viewed as distributed about some mean, the mean in first-degree

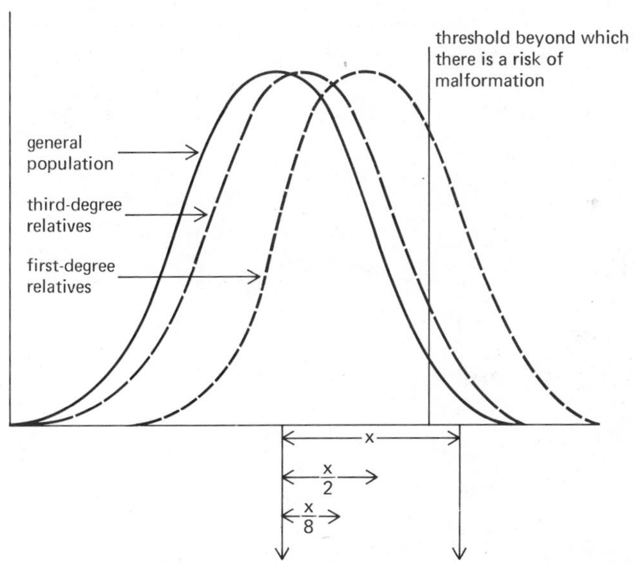

Figure 17-4 Model for inheritance of cleft lip with or without cleft palate assuming contribution of several loci to deviation from the mean. Affected persons are those who exceed the threshold. Their close relatives would also be more likely to exceed the threshold. (*From Carter, 1969.*)

relatives being shifted with respect to the general population but not so extreme as in affected persons. More distant relatives have distributions that are still closer to the general population, but there remains a slightly increased risk that the combination of alleles leading to malformation will occur. There is evidence that facial morphology of sibs of CL(P) patients deviates somewhat from the general population, although the differences observed were too small to be of diagnostic value (Erickson, 1974).

Other Common Malformations. Carter (1969) has summarized familial risks in some other congenital malformations as shown in Table 17-4. The least genetic component in this group is associated with spina bifida and anencephaly. These two malformations result from errors in neural tube closure. Neither has a substantial genetic component, but data are difficult to obtain on relatives, since both are often lethal prenatally and result in abortion. Nevertheless, extensive studies fail to show marked concentration within families. One can construct genetic models using a number of loci with high-frequency alleles that are consistent with the low familial risks in spina bifida and anencephaly, but one would also search for non-genetic factors that might be important in the etiology of these malformations.

Table 17-4 **RISKS IN RELATIVES OF PERSONS WITH SOME COMMON CONGENITAL MALFORMATIONS AS COMPARED WITH RISKS IN GENERAL POPULATION***

Congenital malformation	Frequency in general population	Mono-zygotic twins	Risk compared with risk in general population in: 1st-degree relatives	2d-degree relatives	3d-degree relatives
Cleft lip ± cleft palate	0.001	×400	×40	×7	×3
Talipes equinovarus	0.001	×300	×25	×5	×2
Congenital dislocation of hip (males only)	0.002	×200	×25	×3	×2
Pyloric stenosis (females only)	0.005	×80	×10	×5	×1.5
Spina bifida and anencephaly	0.008	—	×7	—	—

* From Carter (1969).

In the case of simply inherited malformations, it is reasonable to expect ultimately to identify a biochemical lesion associated with a single locus. Of course, the effect on the gene product may be quantitative rather than qualitative, and such quantitative effects are notoriously difficult to interpret in terms of gene action. More complex malformations are correspondingly more difficult to study at the biochemical level, since the malformation may require simultaneously quantitative defects at several loci and there may be several genotypes (and biochemical patterns) associated with a particular malformation.

Heterogeneity is a common feature of many malformations. The basis for such heterogeneity is in part genetic. Evidence for this comes from the expression of defect in index cases grouped on the basis of phenotypic similarities. For example, in cleft lip, the risk among relatives of index cases with severe defect is greater than if the index cases are less severely affected. If the index case has bilateral cleft lip with cleft palate, the risk for sibs is 5.6% (Fraser, 1970). If the index case has right or left cleft lip with cleft palate, the risk is 4.1%. If there is only cleft lip, the risk drops to 2.6%. This could be due to the greater number of contributing loci in severe defect, to the presence of more "defective" alleles in patients with severe defect, or to different combinations of loci (and alleles) in different patients.

An additional example is found in the case of pyloric stenosis. Males are about five times as often affected as females. Among the sons of male index patients, 5.5% are also affected. Among the sons of female index patients, 19.4% are also affected (Carter, 1969). The rates for daughters are 2.4% and 7.3%, respectively. Whatever the basis for the sex difference, the gene combinations that cause defect in females are more likely to be expressed in males.

ENVIRONMENTAL INTERACTIONS IN DEVELOPMENTAL DISORDERS

Small changes in timing of events in embryogenesis can have profound effects on development. It is thus hardly surprising that the embryo is considered the most sensitive stage of the human life cycle. Indeed, the embryo appears to be much more sensitive to environmental insult than are the preimplantation stages, when there appears to be a greater capacity to recover from disruption. There is a very large literature on teratogenic effects of environmental agents, such as drugs or other chemicals. For the most part, such studies are carried out with inbred strains of animals selected for their ability to give a specific response. Other strains and species may be avoided because they do not respond to a particular teratogen. The difference in response may be attributed to genetic differences, but only rarely does the genotype become the object of study in these systems. There are occasional opportunities to carry out epidemiological studies on a human group, but these studies typically cannot provide information on differences in genetic susceptibility (Smithells, 1976).

Species Differences in Response to Teratogens. A single example will illustrate the importance of species differences with respect to teratogenesis. The tranquilizer thalidomide was administered to large numbers of persons in Germany, the Netherlands, and the United Kingdom during the period 1959 to 1961. The drug had been tested in several experimental animals and failed to show evidence of teratogenicity. When administered to pregnant women, it proved to be a powerful teratogen, producing a condition known as phocomelia, involving major malformation of the arms and often the ears. In several cases, the long bones of the arms are absent. The sensitive period for the embryo is between the 34th and 50th day following the last menstrual period. Lenz (1964) has estimated that 5400 children were affected in West Germany before the cause was established and thalidomide was removed from the market.

Additional studies of thalidomide in a wider range of experimental animals have disclosed some species and strains that do give a positive response. In the case of man, it is not known whether some persons are more sensitive to the effects of thalidomide because of their genotype. Fortunately, the "experiment" was terminated before familial risks could be assessed.

Thalidomide is one of the very few examples of a teratogenic agent that has been directly related to congenital malformations in man. Therefore, it is necessary to look to experimental animals for controlled studies of malformations as related to genotype and treatment.

Inherited Susceptibility to Teratogens. Information on genetic variation in susceptibility to teratogens comes entirely from experimental organisms. For the most part, the evidence consists in differences between strains rather than differences attributable to single genes. However, there has been little effort to resolve strain differences into the component single gene differences.

One of the more thoroughly investigated teratogenic responses is the production of cleft palate in mice. As in man, the palatine shelves must come together and fuse at a critical period in embryogenesis. Prior to this period, the tongue lies between the shelves. In order for the shelves to approach each other they must force the tongue out of the way. Any factor that delays movement of the palatine shelves can cause cleft palate since the shelves will no longer be able to touch if the head has grown. Decreased force by the shelves or increased resistance by the

Table 17-5 **FREQUENCY OF CLEFT PALATE IN MICE OF VARIOUS GENOTYPES, PRODUCED BY INJECTING THE MOTHER WITH CORTISONE***

Cross	Mother	Father	Number of offspring	Percent with cleft palate
1	A	A	36	100
2	C57BL	C57BL	75	19
3	A	C57BL	46	43
4	C57BL	A	82	4
5	[A × C57BL]	A	116	22
6	[C57BL × A]	A	71	25

* The purebred stains are A and C57BL. In crosses 5 and 6, the mothers are F_1 hybrids, the maternal line being listed on the left. From Fraser and co-workers (1957), based on data from Kalter (1954).

tongue can cause cleft palate, as can a skull that is too broad to permit close approach of the shelves.

F. C. Fraser and his colleagues demonstrated some 25 years ago that injection of cortisone at a critical time in embryogenesis can cause a high frequency of cleft palate in certain mice. Illustrative results are presented in Table 17-5. Strain A is especially susceptible to cleft palate as compared with the C57BL strain. The F_1 embryos differ from the parental strains, being intermediate if the mother is A but lower than either if the mother is C57BL. Apparently the genotypes of the mother and the embryo both are important in susceptibility. This is further borne out by the crosses in which the mothers are hybrid.

Nutrition also has been demonstrated to influence the appearance of many malformations in experimental animals. Goldstein and co-workers (1963) produced cleft palate in mice using 6-aminonicotinamide, a metabolic antagonist to the B vitamin nicotinamide. They found strain differences similar to those observed with cortisone. In a related study, Warburton and co-workers (1962) found that with certain dietary conditions, the C57BL strain produced more cleft palates than the A strain. Kalter and Warkany (1957) made similar observations using diets containing galactoflavin, an antagonist of the B vitamin riboflavin. Under the conditions of their study, 2.9% of A strain mice had clefts, 13% of C57BL mice had clefts, and 41% of a third strain (DBA) had clefts. This reversal of sensitivity with diet compared to the results in earlier cortisone studies indicates a nutrition-genotype interaction rather than a general instability of palate development. High levels of vitamin A act as an effective inducer of cleft palates in several strains of mice (Kalter and Warkany, 1961).

The significance of these studies for man is one of principle rather than assessment of specific risks. The great variability in the results from different animal species and from different strains makes it impossible to draw firm conclusions concerning human risks from teratogenic agents. Unfortunately, the observations must be made on man himself. The studies do suggest that many environmental agents may have adverse effects, given the right human genotype (both maternal and embryonic), the right nutrition, and the right timing. Efforts to relate the common human malformations, such as cleft lip, to environmental agents have been unsuccessful.

The studies also emphasize that certain persons in the population may be genetically much more sensitive to a specific agent than are other persons. One person may absorb and metabolize a particular nutrient or drug at a different rate than another person. Embryos that they may carry could be exposed to very different levels of the agent, and the embryos themselves may differ in their abilities to respond. With so many possibilities for variation, it is small wonder that so little has been learned about the etiology of the major malformations. However, the high frequency of malformations makes this an important area for further study.

SUMMARY

1. Genes are activated at specific times in development and in specific tissues. The ζ, ϵ, γ, δ, and β chains of hemoglobin are produced by genes that are active only at certain developmental stages and in limited tissues. Lactate dehydrogenase occurs widely in tissues, but the two loci involved differ in relative activity according to the tissue. Some enzymes, such as tyrosine transaminase, appear at time of birth but may be delayed on occasion.

2. Many Mendelian traits in man can be classified as developmental mutants. Examples include brachydactyly, the first Mendelian trait recognized in man, and many of the disorders in sexual development, such as testicular feminization. Achondroplasia is a dominant trait with defective formation of long bones. It has been especially useful in estimating mutation rates. The nail-patella syndrome involves defective formation of fingernails, toenails, patella, and elbows. The locus for this dominant trait is closely linked to the ABO blood group locus.

3. Approximately 1% of newborns have a congenital malformation, usually not attributable to single genes. The familial concentration of cases suggests that genetic variation contributes to the cause of malformations such as cleft lip (with or without cleft palate). The risk that close relatives of index patients also will be affected is substantially greater than for unrelated persons. In the case of other malformations, such as anencephaly, inherited variation may play only a minor role in the etiology.

4. There is evidence of genetic heterogeneity for some malformations. For example, in severe cases of cleft lip with cleft palate, affected relatives are also likely to be more severely affected. In pyloric stenosis, males are more often affected than females. The risk of being affected is greater among relatives of female probands than male probands.

5. Studies of teratogenesis in experimental animals have demonstrated the importance of species and strain differences in response to environmental agents. The induction of cleft palate by cortisone occurs in only 19% of offspring in the C57BL strain of mice under conditions that lead to 100% affected offspring in the A strain. Malnutrition—for example, deficiencies of riboflavin and nicotinamide and excess of vitamin A—causes cleft palate in mice, the magnitude of the effect varying with the strain. These studies emphasize the need to make observations directly on man rather than on other species for reliable information on teratogenesis.

REFERENCES AND SUGGESTED READING

CARTER, C. O. 1965. The inheritance of common congenital malformations. *Prog. Med. Genet.* 4: 59–84.

CARTER, C. O. 1969. Genetics of common disorders. *Br. Med. Bull.* 25: 52–57.

CARTER, C. O. 1976. Genetics of common single malformations. *Br. Med. Bull.* 32: 21–26.

CHUNG, C. S., CHING, H. S., AND MORTON, N. E. 1974. A genetic study of cleft lip and palate in Hawaii. II. Complex segregation analysis and genetic risks. *Am. J. Hum. Genet.* 26: 177–188.

ERICKSON, J. D. 1974. Facial and oral form in sibs of children with cleft lip with or without cleft palate. *Ann. Hum. Genet.* 38: 77–88.

FOGH-ANDERSEN, P. 1942. *Inheritance of Harelip and Cleft Palate.* Copenhagen: Ejnar Munksgaard.

FRASER, F. C. 1970. The genetics of cleft lip and cleft palate. *Am. J. Hum. Genet.* 22: 336–352.

FRASER, F. C., WALKER, B. E., AND TRASLER, D. G. 1957. Experimental production of congenital cleft palate: genetic and environmental factors. *Pediatrics* 19: 782–787.

GOLDSTEIN, M., PINSKY, M. F., AND FRASER, F. C. 1963. Genetically determined organ specific responses to the teratogenic action of 6-aminonicotinamide in the mouse. *Genet. Res.* 4: 258–265.

GRÜNEBERG, H. 1963. *The Pathology of Development: A Study of Inherited*

Skeletal Disorders in Animals. Oxford: Blackwell Scientific Publ., 309 pp.

HALL, J. G., DORST, J. P., TAYBI, H., SCOTT, C. I., LANGER, L. O. JR., AND MC KUSICK, V. A. 1969. Two probable cases of homozygosity for the achondroplasia gene. *Birth Defects: Original Article Series* Vol. 5, No. 4. pp. 24–34.

KALTER, H. 1954. The inheritance of susceptibility to the teratogenic action of cortisone in mice. *Genetics* 39: 185–196.

KALTER, H., AND WARKANY, J. 1957. Congenital malformations in inbred strains of mice induced by riboflavin-deficient, galactoflavin-containing diets. *J. Exp. Zool.* 136: 531–565.

KALTER, H., AND WARKANY, J. 1961. Experimental production of congenital malformations in strains of inbred mice by maternal treatment with hypervitaminosis A. *Am. J. Pathol.* 38: 1–22.

KRETCHMER, N., GREENBERG, R. E., AND SERENI, F. 1963. Biochemical basis of immaturity. *Annu. Rev. Med.* 14: 407–426.

LENZ, W. 1964. Chemicals and malformations in man. In *Proceedings of the Second International Conference on Congenital Malformations.* The International Medical Congress, New York. pp. 263–276.

LUCAS, G. L., AND OPITZ, J. M. 1966. The nail-patella syndrome. *J. Pediatr.* 68: 273–288.

MURDOCH, J. L., WALKER, B. A., HALL, J. G., ABBEY, H., SMITH, K. K., AND MC KUSICK, V. A. 1970. Achondroplasia—a genetic and statistical survey. *Ann. Hum. Genet.* 33: 227–244.

NEEL, J. V. 1958. A study of major congenital defects in Japanese infants. *Am. J. Human Genet.* 10: 398–445.

OHNO, S. 1974. Regulatory genetics of sex differentiation. In *Birth Defects* (A.G. Motulsky and W. Lenz, Eds.). Amsterdam: Excerpta Medica, pp. 148–154.

RENWICK, J. H. 1956. Nail-patella syndrome: evidence for modification by alleles at the main locus. *Ann. Hum. Genet.* 21: 159–169.

RENWICK, J. H. AND LAWLER, S. D. 1955. Genetical linkage between the ABO and nail-patella loci. *Ann. Hum. Genet.* 19: 312–331.

SERENI, F., KENNEY, F. T., AND KRETCHMER, N. 1959. Factors influencing the development of tyrosine-α-ketoglutarate transaminase in rat liver. *J. Biol. Chem.* 234: 609–612.

SMITHELLS, R. W. 1976. Environmental teratogens of man. *Br. Med. Bull.* 32: 27–33.

STEVENSON, A. C., JOHNSTON, H. A., STEWART, M. I. P., AND GOLDING, D. R. 1966. Congenital malformations: a report of a study of series of consecutive births in 24 centres. *Bull. WHO* 34 (Suppl.). 127 pp.

VESELL, E. S. 1965. Genetic control of isozyme patterns in human tissues. *Prog. Med. Genet.* 4: 128–175.

VESELL, E. S., OSTERLAND, K. C., BEARN, A. G., AND KUNKEL, H. G. 1962. Isozymes of lactic dehydrogenase; their alterations in arthritic synovial fluid and sera. *J. Clin. Inv.* 41: 2012–2019.

WARBURTON, D., TRASLER, D. G., NAYLOR, A., MILLER, J. R., AND FRASER, F. C. 1962. Pitfalls in tests for teratogenicity. *Lancet* 2: 1116–1117.

REVIEW QUESTIONS

1. Define:

achondroplasia	dizygotic twins	pyloric stenosis
anencephaly	first-degree relatives	spina bifida
anophthalmos	isoalleles	talipes equinovarus
atresia ani	monozygotic twins	teratogen
congenital	polydactyly	

2. Why would a positive correlation between sibs but not between parents and offspring for expression of a dominant trait such as nail-patella syndrome be evidence for isoalleles?

3. What information would you seek to rule out anencephaly as a simply inherited trait?

4. A rare defect lethal in the fetal period is suspected of being an autosomal dominant trait. What evidence would support this hypothesis?

5. What evidence would you seek to test whether achondroplasia of different degrees of severity is due to different alleles? How would you test to see whether achondroplasia is due to more than one locus?

6. Even in the case of a teratogen such as thalidomide it has been proposed that some persons may be genetically more susceptible. How would you test this hypothesis?

CHAPTER EIGHTEEN

BEHAVIOR GENETICS

Behavior, like all other physiological traits, is determined by the genotype acting within a particular enviroment. That this should be so is perhaps not quite so obvious as for other traits. When we speak of culture as inherited, we do not ordinarily think of the biological potential for culture but rather the particular variety that may be of interest. Similarly, the patterns of behavior that interest us are apt to be those that occur as normal or near normal variations. These variations are likely to be very complex in their origin, involving the interaction of many inherited potentials with a great variety of environmental experiences.

We will define behavior broadly, including intelligence, aptitudes, interpersonal relationships, etc. First, those variations that can be associated with clear cut genetic causes, either chromosomal changes or single Mendelian genes, will be examined. We will then consider more complex situations to learn, if possible, to what extent genetic variation may influence these traits also.

BEHAVIORAL CONSEQUENCES OF ANEUPLOIDY

Descriptions of various aneuploid states were given in Chapters 5–7. Autosomal trisomies, at least those that survive beyond the neonatal period, as well as many other unbalanced chromosome complements, are associated with mental retardation. In most trisomies, retardation is severe, which, coupled with the greatly diminished life expectancy, has discouraged research into other behavioral aspects. Down's syndrome involves less severe mental retardation and good viability. The personalities of patients with Down's syndrome are very characteristic and are not typical of other patients with equal retardation. They tend to be very friendly and congenial and respond warmly to attention. Whatever the nature of the metabolic imbalances caused by trisomy 21, the effects on personality are as characteristic as are the effects on anatomical development.

Aneuploidy of the sex chromosomes is less predictable. XXY Klinefelter's syndrome patients have a slightly increased risk of diminished intellect and of emotional instability, but most XXY males are normal for these characteristics. XXXY and XXXXY males are consistently impaired intellectually.

Females with Turner's syndrome (XO) have been especially interesting. They do not appear to have increased risk of mental retardation, and most have quite normal intelligence. But they often show a deficit in tests involving visual memory and related special abilities (Waber, 1979). In view of the fact that males also have only one X chromosome and often perform better than females in tests of spatial imagery, the results on Turner's patients are unexpected. They cannot be attributed to hemizygosity for X-linked genes.

The problems of XYY males have been the most publicized. The first report, that XYY is associated with criminal behavior, stimulated great interest in verifying or disproving this association. As noted in Chapter 5, it seems certain that at least some XYY males have a higher risk of criminal behavior. It is not certain whether all XYY males have the higher risk but only some actually engage in such behavior or whether only some XYY males have the increased risk. Of those who do develop social problems, the expression is one of increased impulsiveness with little perception of future consequences (Money et al., 1974).

There thus appears to be a general small increase in risk of behavioral problems associated with abnormal chromosome complements. Surveys of psychiatric hospitals usually show a small excess of patients with a variety of aneuploid karyotypes. Individually, the risk seems small, and it is often difficult to determine whether the problem involves a direct physiological influence on behavior or whether associated physical abnormalities cause the person to be maladjusted in society.

SINGLE GENE EFFECTS ON BEHAVIOR

Many simple Mendelian genes are associated with variations in behavior. Most conspicuous are those, such as phenylketonuria, that are associated with severe mental retardation. There are many examples of mental retardation associated with metabolic diseases or other known genes. In addition, there are many examples of inherited mental retardation whose biochemical basis is obscure.

An example of X-linked mental retardation is shown in Figure 18-1. Although X-linked retardation is uncommon, the pattern of transmis-

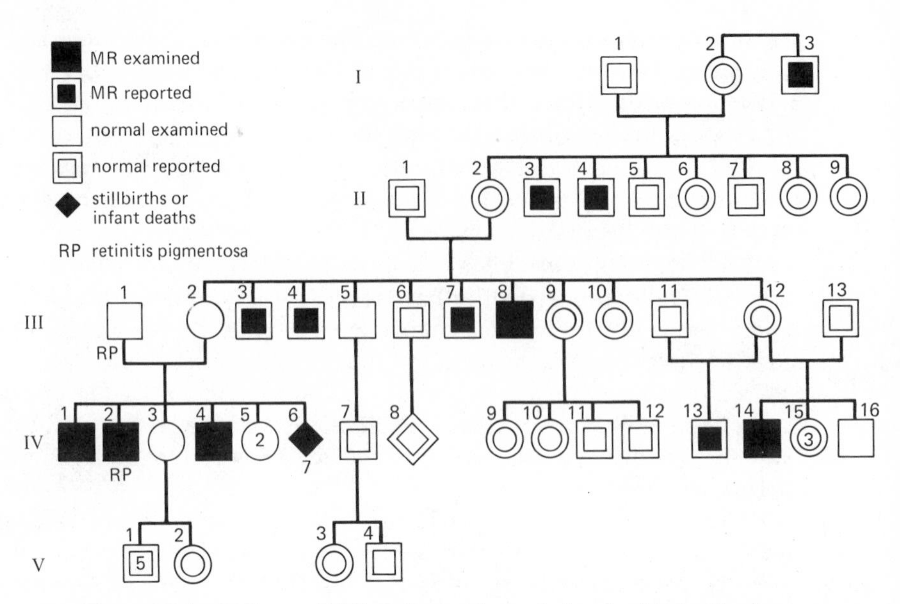

Figure 18-1 Pedigree of X-linked mental retardation. In this family, the autosomal dominant trait retinitis pigmentosa (RP) was introduced by III-1, who came from a large family with many other cases of retinitis pigmentosa but no reported cases of mental retardation. (*Based on a pedigree published by Allan and Herndon, 1944.*)

sion makes it somewhat easy to identify if there is sufficient information on family members. Autosomal recessive mental retardation is more difficult to recognize because it occurs in isolated sibships. If two or more within a sibship are affected, one would suspect recessive inheritance. However, since sibs share similar environments, familial occurrence of cases does not in itself prove heredity. Autosomal dominant inheritance of severe mental retardation is rarely observed, because severely retarded persons do not ordinarily reproduce.

One characteristic of simply inherited retardation is the marked difference that often exists between affected persons and other members of their families. Heterozygotes often show no effects of the gene and therefore are normal. By contrast, borderline retardates are likely to be so because of multifactorial inheritance and are more likely to resemble their parents in this regard. Therefore one does not see the sharp distinction between "affected" and "nonaffected" in families with borderline intelligence.

Lesch–Nyhan Syndrome. The X-linked deficiency of hypoxanthine phosphoribosyl transferase (HPRT) provides one of the more bizarre examples of aberrant behavior associated with a single gene. Persons who inherit alleles of this locus that produce intermediate levels of enzyme activity have high levels of blood uric acid and gout (Chapter 11), but they are of normal intelligence and other behavior. Persons with alleles that cause complete deficiency of HPRT have Lesch–Nyhan syndrome. Since this is a rare X-linked trait, only males are affected. It was originally thought that they are severely retarded, but some doubt has been cast on this. The most notable feature is the compulsion for self-mutilation. In extreme cases, such patients bite off their fingers and lips. This is not due to absence of pain receptors. It is not characteristic of mentally retarded children in general. The events that lead to the aberrant behavior are unknown even though the primary defect is.

Diseases Involving Neurological Degeneration. Many inherited diseases are known that involve neurological degeneration. The classical example is Huntington's disease (Huntington's chorea), an autosomal dominant trait. Persons are normal until adulthood, although a few have onset of symptoms earlier, and some have onset of symptoms as late as the seventh decade (Chapter 24). The disease is characterized by degeneration of the central nervous system, with development of choreiform movements and mental deterioration. Often the first symptom is a departure in behavior from previous patterns, with increased sexual promiscuity and breakup of family relationships. The biochemical basis of Huntington's disease is unknown. It is a clear example, however, of one disorder in which variability in behavior has a simple genetic basis.

Other inherited degenerative disorders, such as Tay-Sachs disease, also may be viewed from the point of view of their effects on behavior. Although normal at birth and in early development, affected children lose their abilities to carry out previously learned activities, such as speech and motor activities. These disorders are often not considered in the context of behavior genetics, since they involve such obvious pathology. However, were it not for the progressive nature of the diseases, the early stages might well correspond to behavioral variants in the normal population.

Affective Illness. There has been substantial debate on the extent to which affective illness (manic–depressive psychosis) is inherited and how it is inherited (Gershon et al., 1977). Affective illness is found in some one percent of the population. A distinction is made between persons who have experienced both manic and depressive episodes (bipolar illness) and those who have experienced only depressive episodes (unipolar illness). Patients with bipolar illness are generally more responsive to lithium treatment suggesting a biological basis for this distinction. Bipolar illness tends to cluster in families, although relatives of bipolar patients also have a higher risk of unipolar depression. On the other hand, many families with unipolar illness do not seem to be at risk for bipolar illness. Studies of manic-depressive adopted children show that the biological parents are more likely also to have affective disorder than are adoptive parents (Mendlewicz and Rainer, 1977).

Familial clustering may be due either to hereditary or environmental variations (or both), and one would especially suspect environmental factors in such a behavioral trait. Examination of families has shown that bipolar illness often appears to be transmitted as an X-linked dominant trait (Mendlewicz and Rainer, 1974). Such a conclusion is clouded first by the traditional problems of diagnosis of psychiatric disorders. Further, not everyone with the potential to develop bipolar illness may have done so. The disorder may be heterogeneous, with some families having the X-linked gene, others having other causes. That this must be true in part is indicated by families with apparent father-to-son transmission, not possible with an X-linked trait.

The true explanation of all cases of affective illness must await further research. It seems likely though that such research will show a strong genetic component in many, perhaps most cases of bipolar illness.

Mendelian Traits with Occasional Behavioral Effects. Simply inherited diseases not often considered to be behavioral may nevertheless influence behavior on occasion. We will exclude from discussion those situations involving primarily psychological reaction to an inherited medical illness and consider those in which the disease process influences behavior directly.

One instructive example is found in Rh incompatibility (Chapter 13). Prior to the development of effective means of preventing hemolytic disease of the newborn, children born with this problem often became very jaundiced, with permanent damage to the central nervous system due to the high levels of plasma bilirubin. As a result, mental retardation is a risk, very small to be sure, associated with the Rh-positive alleles. The fact that we do not often associate the Rh problem with behavior genetics derives from our knowledge that it is first an immunological problem. But if we didn't know the physiological basis for the behavioral deficit, we could very well think of it as an example of inherited mental defect, one that shows familial clustering but that is complex in its inheritance. Complicating the issue would be other causes of jaundice, some inherited, some not.

Many other inherited diseases have central nervous system manifestations as a frequent or occasional expression. Sometimes these may be the first expression. For example, Wilson's disease (hepatolenticular degeneration), an autosomal recessive trait, is a defect in copper metabolism associated with very low levels of the copper-transporting protein ceruloplasmin in the blood. The deposition of copper in tissues, including brain, causes irreversible damage that in turn may cause psychosis. In at least one case, the psychosis preceded other obvious evidence of the disease, and the patient was hospitalized for psychosis. One wonders in such a case whether this particular patient had other genetic predisposition that caused his central nervous system to be especially sensitive to the toxic effects of liver damage. If so, might such sensitivity have been expressed in the presence of other milder forms of liver damage?

COMPLEX TRAITS

Just as behavior is very complex, it is not surprising that the genetic basis for variations in behavior is also often complex. The single gene effects that we know about tend to be marked deviations from normal. But normal variation and many major deviations as well show the influence of many loci. The evidence that heredity is involved at all depends on comparisons of relatives, with results that are difficult to explain except by genetic variation (Chapter 16). The following examples are some of the traits that have been particularly well investigated, although the complete answers are still somewhat elusive.

Intelligence

Perhaps no trait has been as extensively studied, and with as much controversy, as has intelligence. Intelligence is one of man's most valued attributes, and suggestions that the mental abilities of persons are limited by their genes are not apt to be well received by the persons concerned. In addition, certain political theories emphasize the importance of environmental opportunities and find genetic theories inconvenient. An improved understanding of genetics will often diminish the significance of such conflicts.

Just what is measured by IQ tests is heavily debated, but the measure correlates with ability to perform in school and in other situations. IQ tests have been criticized because most are obviously biased by culture. This is not a large problem so long as the analysis is restricted to persons from the same background, but it becomes a problem in cross-cultural comparisons.

There are conflicting theories as to whether intelligence is one thing—a general factor—or whether it is a composite of special abilities. If the latter is the case, one would expect the various special abilities, such as number ability or verbal ability, to show more clear-cut patterns of inheritance than if they are averaged together into a single score

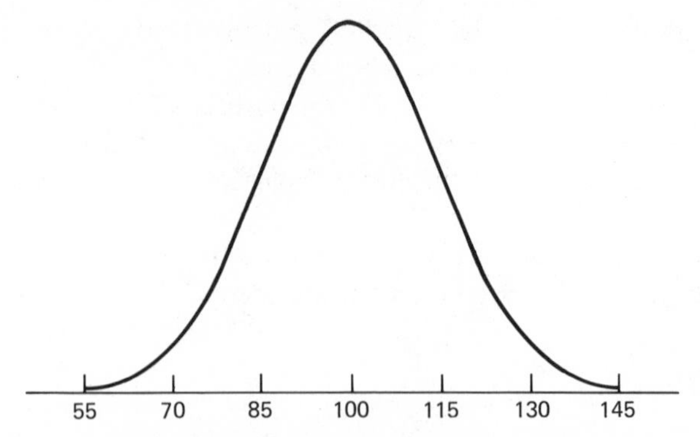

Figure 18-2 Distribution of IQ scores in a U.S. white population. The scores are normally distributed with a mean of 100 and a standard deviation of 15. Thus approximately two-thirds of the population falls in the range 85–115, and 95% are in the range 70–130. A score of 70 is often considered the lower limit of normal, although there are many examples of persons with measured IQ's below 70 who are able to function outside an institution. Several studies have shown an excess of persons with very low and very high scores as compared to the predictions of the normal distribution.

of performance. However, there is no obvious way to design tests that will correspond to whatever unit biological aptitudes may exist. Hence, the so-called tests of special abilities are likely also to be averages of several more primary abilities. The data in fact are not very different for general tests of intelligence and for tests of special abilities. Because the general tests have been used much more extensively, we will consider only those results. The distribution of IQ values in a test population is shown in Figure 18-2.

The evidence for genetic variation in intelligence comes from sib comparisons, including twins, and parent/offspring comparisons on intelligence tests, especially some form of IQ test. Although there are claims of changes in IQ in persons, these changes are small with respect to the full range of values for IQ, and some may represent only increased sophistication in being tested.

The results of a number of studies of intrafamily comparisons of IQ are summarized in Figure 18-3. In spite of the deviations from theory obtained in some studies, there is overall agreement with a high genetic contribution to intelligence.

Efforts to quantify how much of the variation in intelligence is hereditary have faced difficulties not yet fully resolved. Such estimates are statistical in nature and will not be discussed in detail here. The simplest models assume random mating with respect to intelligence and random placement of adopted children with foster parents. As Figure 18-3 illustrates, the foster parent-child correlations and those of unrelated persons reared together (generally foster sibs) are uniformly

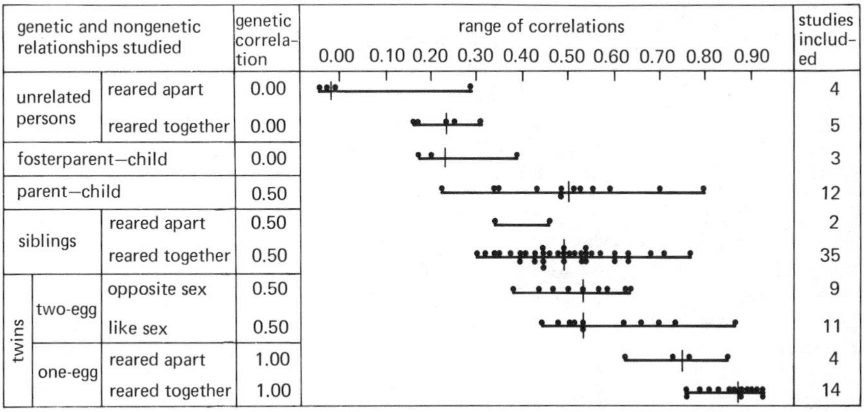

genetic and nongenetic relationships studied			genetic correlation	studies included
unrelated persons		reared apart	0.00	4
		reared together	0.00	5
fosterparent–child			0.00	3
parent–child			0.50	12
siblings		reared apart	0.50	2
		reared together	0.50	35
twins	two-egg	opposite sex	0.50	9
		like sex	0.50	11
	one-egg	reared apart	1.00	4
		reared together	1.00	14

range of correlations: 0.00 0.10 0.20 0.30 0.40 0.50 0.60 0.70 0.80 0.90

Figure 18-3 A summary of correlations from 52 studies of intelligence test scores. Over two-thirds of the tests were IQ; the remainder were special tests. Correlation coefficients are represented by dark circles. The median values are shown as vertical lines. (*From L. Erlenmeyer-Kimling and L. Jarvik, 1963.* Science *142: 1477–1479. © American Association for Advancement of Science. The genetic correlations were added by Lerner, 1968, p. 159.*)

greater than zero. Does this indicate the effects of environment or of errors in the assumptions of randomness? It is a well known fact that adoption agencies have tried to match children with prospective foster parents for many traits. The most obvious is race. Even within a racial group, there is usually a conscious effort to match conspicuous physical traits, such as hair color and complexion, of the biological and foster parents, so that the adopted child will look like a member of the family. And educational level of the biological and foster parents has almost certainly been used as a guide to placement. To the extent that educational level (generally college vs. noncollege) is associated with IQ, one would expect a positive correlation between adopted children and foster parents *if* intelligence is inherited.

The assumption of random mating with respect to intelligence is clearly not warranted. Intelligence is one of the traits for which there is strong assortative mating (Table 18-1). While these data were obtained from already married couples and are therefore subject to the formal criticism that the marriage partners may have become more alike after marriage, a large study by Reed and Reed (1965) using premarriage IQ tests also gave correlation coefficients of 0.33 and 0.46, depending on how the data were grouped.

Brain damage and malnutrition during early childhood may permanently diminish intelligence (Loehlin et al., 1975). Good nutrition does not appear to increase intelligence beyond the level for which we are genetically programmed.

Table 18-1 **ASSORTATIVE MATING FOR TEST SCORES FOR VARIOUS INTELLIGENCE TESTS***

Test	Number of couples	Correlation coefficient $r \pm$ S.D.	Reference†
Stanford–Binet	174	0.47 ± 0.04	Burks (1928)
Otis	150	0.49 ± 0.04	Freeman and co-workers (1928)
Army Alpha	105	0.60 ± 0.04	Jones (1928)
Progressive matrices	324	0.76	Halperin (1946)
Progressive matrices	180	0.40	Spuhler (1962)
Vocabulary	108	0.21 ± 0.06	Carter (1932)
Chicago verbal			
–total right	151	0.31	Spuhler (1962)
–proportion right	148	0.73	Spuhler (1962)
Arithmetic	108	0.03 ± 0.06	Carter (1932)
Mental grade	100	0.44	Penrose (1933)
Various tests	433	0.19 ± 0.03	Smith (1941)

* From Spuhler (1972).
† References to the original reports may be found in Spuhler (1972).

Dyslexia. Some 5 to 10% of the population has specific reading disability (*dyslexia*) that is in sharp contrast to normal development of other abilities in these same persons. It has long been suspected that dyslexia is inherited. This hypothesis is based on several observations. (1) Dyslexia frequently occurs in persons who perform well on all intelligence tests except those that involve reading. (2) There is some clustering in families. (3) Within those families, there is often sharp discontinuity between those who read well and those who read with great difficulty. (4) Monozygotic twins are concordant for dyslexia, whereas dizygotic twins often are not. (5) Within families with both affected and nonaffected children, there is no obvious environmental factor that might explain the difference.

Dyslexia is not a simple Mendelian trait, however. It is much more prevalent in males than in females, but X-linked recessive inheritance can be ruled out readily with pedigrees. One difficulty in studying dyslexia is that many affected persons do learn to read or to compensate for their difficulty in reading. These persons must be identified for any study that includes adults. In one such study, 45% of the first-degree relatives of parents of dyslexic children were also affected, with the expected preponderance of males (Finucci et al., 1976). This very high frequency is difficult to reconcile with any simple genetic mechanism, and the authors suggested that dyslexia is perhaps heterogeneous. Whatever the final explanation, it remains one of the very interesting unsolved problems of behavior (Herschel, 1978).

Schizophrenia. The relative importance of genetic and non-genetic factors in the etiology of mental disorders has been especially debated in the case of schizophrenia. It has been estimated that 0.8% of the population is affected with schizophrenia. Many studies have shown also that close relatives of schizophrenic patients have an increased risk of being affected. This can be attributed as readily to a common "schizophrenogenic" environment as to common genes, and there are many advocates of an environmental cause for schizophrenia.

In order to avoid the problem of learned abnormal behavior, geneticists have especially made use of adoptive studies. In these cases, offspring of schizophrenic parents are located that have been placed in foster homes early in infancy. A summary of two such studies is given in Table 18-2. It is seen that even though these offspring were not exposed to the schizophrenic parent or to the environment of the biological family, many nevertheless became schizophrenic. In contrast, none of the control children of nonschizophrenic parents, also adopted, developed schizophrenia. These results support strongly the hypothesis that genetic factors predispose to schizophrenia.

Carrying this approach somewhat further, Kety and co-workers (1976) compared paternal half-sibs, one of whom had been adopted into an unrelated foster family at birth. If the adopted member of the sib-pair was subsequently diagnosed as schizophrenic, the half-sib also had increased risk of schizophrenia as compared to half-sibs of adopted persons without the diagnosis (Table 18-3). Since the only common bond between the half-sibs is the father's sperm, the common risk should be attributed to genetic factors.

The comparison of parents with their children reared in foster homes or of sibs reared in separate foster homes is a powerful approach to the detection of genetic factors in the etiology of complex traits. It is

Table 18-2 **SCHIZOPHRENIA IN ADOPTED OFFSPRING OF SCHIZOPHRENIC AND NONSCHIZOPHRENIC PARENTS***

	Offspring			
Parents	Schizophrenic	Borderline or schizoid	Not affected	Total
Heston (1966):				
Schizophrenic parents	5	8	34	47
Nonschizophrenic parents	0	0	50	50
Rosenthal and co-workers (1971):				
Schizophrenic parents	3	19	47	69
Nonschizophrenic parents	0	12	55	67

* These data are summarized by Wender (1972), from which this table was derived.

Table 18-3 **SCHIZOPHRENIA IN PATERNAL HALF SIBS OF ADOPTED CHILDREN WHO HAVE BECOME SCHIZOPHRENIC***

	Paternal half-sibs of	
	Schizophrenic probands	Controls
Total	63	64
Definite schizophrenia	8	1
Definite + uncertain schizophrenia	14	2

* From Kety et al. (1976).

exceedingly difficult to organize such a study, since adoption practices often do not permit identification of biologically related persons.

A number of studies on concordance of twins for schizophrenia have been reported (Allen et al., 1972), and several of the larger studies are summarized in Table 18-4. The studies are consistent in showing much higher concordance rates for MZ as compared to DZ twins. The concordance rates differ among the different studies as a result of different methods of ascertainment and different diagnostic criteria. For example, the low concordance rates of the study by Allen and co-workers are very likely due to the fact that the subjects were all veterans of World War II and would have to have been reasonably functional for both to have been inducted into service. The important comparison is the MZ/DZ ratio within a study.

Table 18-4 **COMPARISON OF CONCORDANCE FOR SCHIZOPHRENIA IN MZ AND DZ TWINS***

	Investigator				
	Kallman (1946)	Slater (1961)	Inouye (1961)	Kringlen (1967)	Allen and co-workers (1972)
MZ Twins					
No. of pairs	174	41	55	55	95
Concordance %†	69–86	68–76	60–76	25	27
DZ Twins					
No. of pairs	517	115	17	172	125
Concordance %†	10–15	11–14	12–22	4	5
MZ/DZ ratio	5.7	5.4	3.5	3.8	5.7

* A fuller summary of twin studies is given in Allen and co-workers (1972), from which information in the above table was taken.
† The second figure is corrected for age.

Pioneers in this field, such as Franz Kallman, interpreted the twin studies as evidence of the genetic basis for schizophrenia, but others have questioned whether having an identical twin might constitute such a different experience from having a fraternal twin that the twin method would be inappropriate. The fact that studies of adopted children supported the results of the twin studies suggests that the twin method may be valid after all for this disorder.

An interesting extension of the twin method as applied to schizophrenia has been reported by Wyatt and co-workers (1973). They had noted previously that the activity of the enzyme monoamine oxidase is lower in blood platelets of schizophrenics than in normal controls. In 23 pairs of MZ twins discordant for schizophrenia, the monoamine oxidase levels were low in the normal twins as well as in the affected. The authors suggested that low monoamine oxidase reflects the genetic predisposition of the normal twins to schizophrenia. Through such studies, it may be possible to identify not only the genetic risk factors but also the environmental factors.

SUMMARY

1. Aneuploid chromosome complements often have effects on behavior. All viable autosomal trisomies as well as many unbalanced chromosome complements cause mental retardation. In the case of Down's syndrome, the retardation is less severe, and Down's patients are characteristically very friendly and warm. Sex chromosome aneuploidies are less consistent. XYY males appear to have a higher risk of antisocial behavior, and XO females typically have trouble with spatial imagery.

2. Single Mendelian genes are sometimes associated with variant behavior. X-linked mental retardation is well known and relatively easy to identify in large pedigrees. Autosomal recessive retardation also occurs, often with a known biochemical basis, as in phenylketonuria. Mental retardation due to simple genetic mechanisms may involve marked differences from sibs or parents, whereas borderline retardation is more likely due to complex mechanisms.

3. Examples of Mendelian traits with variant behavior include Lesch–Nyhan syndrome, in which there is absence of hypoxanthine phosphoribosyl transferase (HPRT) activity. Persons affected with this X-linked recessive trait have a compulsion for self-mutilation that includes biting off their lips, fingers, etc. Huntington's disease is an autosomal dominant trait involving central nervous system degeneration. Difficulty in relationships with other persons is an early symptom.

4. Affective illness (manic–depressive psychosis) has been suggested as an inherited trait, possibly X-linked dominant. The evidence is stronger for bipolar illness (with both manic and depressive episodes) than for unipolar illness (depression only). The disease is probably heterogeneous, with some pedigrees not in agreement with X linkage.

5. Some Mendelian traits are not generally classified as behavioral but nevertheless sometimes affect behavior. An example is Rh incompatibility, in which Rh-positive alleles in a newborn may lead to hemolytic disease, with jaundice and possible brain damage.

6. Many behavioral traits are inherited in a complex manner. Intelligence is strongly influenced by genotype. However, it is very difficult to say exactly how much of the variation in intelligence, as measured by IQ tests, is due to genetic variation.

7. Dyslexia, or specific reading disability, clusters in certain families and is thought by many to be inherited. Analysis of pedigrees shows that relatives of affected persons have a very high risk of being affected also, especially males. But no simple Mendelian pattern occurs. Possibly dyslexia is heterogeneous in its etiology.

8. Schizophrenia shows strong evidence of clustering within families and has been suggested as inherited. Children of schizophrenics adopted into other families have an increased risk of developing schizophrenia, and MZ twins resemble each other in risk more than do DZ twins. The genetic predisposition to schizophrenia appears to be real, but the biochemical basis is not known, nor are the environmental factors that cause some to express the trait.

REFERENCES AND SUGGESTED READING

ALLAN, W., AND HERNDON, C. N. 1944. Retinitis pigmentosa and apparently sex-linked idiocy in a single sibship. *J. Heredity* 35: 40–43.

ALLEN, M. G., COHEN, S., AND POLLIN, W. 1972. Schizophrenia in veteran twins: a diagnostic review. *Am. J. Psychiat.* 128: 939–945.

CANCRO, R. (Ed.). 1971. *Intelligence: Genetic and Environmental Influences.* New York: Grune and Stratton, 312 pp.

CHILDS, B., FINUCCI, J. M., PRESTON, M. S., AND PULVER, A. E. 1976. Human behavior genetics. *Adv. Hum. Genet.* 7: 57–97.

CHRISTENSEN, K. R., AND NIELSEN, J. 1974. Incidence of chromosome aberrations in a child psychiatric hospital. *Clin. Genet.* 5: 205–210.

EHRMAN, L., OMENN, G. S., AND CASPARI, E. (Eds.). 1972. *Genetics, Environment, and Behavior.* New York: Academic Press.

EHRMAN, L., AND PARSONS, P. A. 1976. *The Genetics of Behavior.* Sinauer Associates, Sunderland, Mass. 390 pp.

ERLENMEYER-KIMLING, L., AND JARVIK, L. F. 1963. Genetics and intelligence: a review. *Science* 142: 1477–1479.

FIEVE, R. R., ROSENTHAL, D., AND BRILL, H. (Eds.). 1975. *Genetic Research in Psychiatry.* Baltimore: Johns Hopkins Press, 301 pp.

FINUCCI, J. M., GUTHRIE, J. T., CHILDS, A. L., ABBEY, H., CHILDS, B. 1976. The genetics of specific reading disability. *Ann. Hum. Genet.* 40: 1–23.

FULLER, J. L., AND THOMPSON, W. R. 1978. *Foundations of Behavior Genetics.* St. Louis. C. V. Mosby, 533 pp.

GERSHON, E. S., TARGUM, S. D., KESSLER, L. R., MAZURE, C. M., AND BUNNEY, W. E. Jr. 1977. Genetic studies and biologic strategies in the affective disorders. *Prog. Med. Genet.* n.s. 2: 101–164.

GOTTESMAN, I. I., AND SHIELDS, J. 1972. *Schizophrenia and Genetics: A Twin Study Vantage Point.* New York: Academic Press, 443 pp.

HERSCHEL, M. 1978. Dyslexia revisited. *Hum. Genet.* 40: 115–134.

HESTON, L. L. 1966. Psychiatric disorders in foster home reared children of schizophrenic mothers. *Br. J. Psychiat.* 112: 819–825.

HORN, J. M., LOEHLIN. J. C., AND WILLERMAN, L. 1979. Intellectual resemblance among adoptive and biological relatives: the Texas Adoption Project. *Behav. Genet.* 9: 177–207.

KETY, S. S., ROSENTHAL, D., WENDER, P. H., SCHULSINGER, F., AND JACOBSEN, B. 1976. Mental illness in the biological and adoptive families of adopted individuals who have become schizophrenic. *Behav. Genet.* 6: 219–225.

LERNER, I. M. 1968. *Heredity, Evolution, and Society.* San Francisco: W. H. Freeman, 307 pp.

LOEHLIN, J. C., LINDZEY, G., AND SPUHLER, J. N. 1975. *Race Differences in Intelligence.* San Francisco: W. H. Freeman, 380 pp.

LOEHLIN, J. C., AND NICHOLS, R. C. 1976. *Heredity, Environment, and Personality.* Austin: Univ. of Texas Press, 202 pp.

MENDLEWICZ, J., AND RAINER, J. D. 1974. Morbidity risk and genetic transmission in manic-depressive illness. *Am. J. Hum. Genet.* 26: 692–701.

MENDLEWICZ, J., AND RAINER, J. D. 1977. Adoption study supporting genetic transmission in manic-depressive illness. *Nature* 268: 327–329.

MONEY, J., ANNECILLO, C., VAN ORMAN, B., AND BORGAONKAR, D. S. 1974. Cytogenetics, hormones and behavior disability: Comparison of XYY and XXY syndromes. *Clin. Genet.* 6: 370–382.

REED, E. W., AND REED, S. C. 1965. *Mental Retardation: A Family Study.* Philadelphia: W. B. Saunders, 719 pp.

ROSENTHAL, D., WENDER, P. H., KETY, S. S., WELNER, J., AND SCHULSINGER, F. 1971. The adopted-away offspring of schizophrenics. *Am. J. Psychiat.* 128: 307–311.

SLATER, E., AND COWIE, V. 1971. *The Genetics of Mental Disorder.* London: Oxford Univ. Press., 413 pp.

SPUHLER, J. N. 1972. Behavior and mating patterns in human populations. In *The Structure of Human Populations* (G. A. Harrison and A. J. Boyce, Eds.). Oxford: The Clarendon Press, pp. 165–191.

TSUANG, M. T. 1978. Genetic counseling for psychiatric patients and their families. *Am. J. Psychiat.* 135: 1465–1475.

VANDENBERG, S. G. (Ed.). 1965. *Methods and Goals in Human Behavior Genetics.* New York: Academic Press, 351 pp.

WABER, D. P. 1979. Neuropsychological aspects of Turner's syndrome. *Develop. Med. Child Neurol.* 21: 58–70.

WENDER, P. H. 1972. Adopted children and their families in the evaluation of nature-nurture interactions in the schizophrenic disorders. *Annu. Rev. Med.* 23: 355–372.

WYATT, R. J., MURPHY, D. L., BELMAKER, R., COHEN S., DONNELLY, C. H., AND POLLIN, W. 1973. Reduced monoamine oxidase activity in platelets: a possible genetic marker for vulnerability to schizophrenia. *Science* 179: 916–918.

REVIEW QUESTIONS

1. Define:

dyslexia	Huntington's disease
schizophrenia	affective illness
intelligence	jaundice
psychosis	random mating
Lesch–Nyhan syndrome	assortative mating

2. It has sometimes been said that even though severe mental deficiency may be inherited, normal intelligence is not. What is wrong with such a statement?

3. Some persons describe themselves as "morning" types, that is, they wake up alert and do their best work in the morning. Others are "night" types, who are most alert in the evening. How would you determine whether there is a genetic basis for such "types?"

4. If schizophrenia is found to be inherited, how does this influence your thinking on therapeutic measures? In particular, does this mean that psychotherapy would not be useful?

CHAPTER NINETEEN

LINKAGE

Mendel's second law states that different loci assort into gametes independently of one another. The genetic markers that he used did assort independently, as do most pairs of markers chosen at random in any organism. Mendel's work was prior to the recognition of chromosomes as the physical structures on which genes are located. Had he been aware of this, he would have realized that there are more genes than chromosomes and therefore some genes must be together on the same chromosome. Since chromosomes retain their integrity during meiosis and cell division, not all the genes on a particular chromosome can assort independently.

Linkage and Crossing Over. Formal proof of the lack of independent assortment of genes on the same chromosome was provided by Sturtevant in 1913. Using genes on the X chromosome of *Drosophila,* he found that the loci do not reassort freely; that is, the particular allelic combinations in the parents tend to be preserved. He attributed the failure to recombine freely to the close position of these two loci on the X chromosome. Although some recombination does occur, it may happen rarely if two loci are very close.

Subsequent studies showed that recombination in eukaryotes occurs by physical exchange of chromosome segments at homologous points on a pair of chromosomes. Such an exchange is known as crossing over. Crossing over can be observed physically in the form of chiasmata. During meiosis, homologous chromosomes synapse. As they begin to separate, it can be seen that the attachment at certain points is very strong, forming "bridges" (chiasmata), which might serve as points of exchange of chromatid segments.

Chiasmata are seen to form at positions all along the chromosome. Presumably, crossing over can occur anywhere, although the likelihood need not be uniform. The greater the physical separation of two genes on a chromosome, the more likely that crossing over will occur between them. Thus it should be possible to measure the relative position of genes along a chromosome by the relative frequencies of crossing over between pairs of loci, provided, of course, that genes are arranged in a linear sequence along the chromosomes. This hypothesis has been amply confirmed. The distance between two loci is measured as the percent of recombinant (crossover) gametes, expressed as centimorgans (cM) or map units. (One percent recombination = 1 cM = 1 map unit.) When three loci are studied, the distance between the outer pair, measured directly, is the sum of the distances between the middle and outer loci. Stated symbolically, in the sequence *ABC*, $AC = AB + BC$.

Experimentally, one must have a system in which recombinant and nonrecombinant gametes can be distinguished. Sturtevant used F_1 females from the cross ♀ *MMpp* × ♂ *mP*, where *M* (normal wings) is dominant to *m* (rudimentary wings) and *P* (normal red eyes) is dominant to *p* (vermilion eyes). (See Figure 19-1.) The F_1 females were double heterozygotes, *MmPp*. Furthermore, *M* and *p* had to be on one chromosome, and *m* and *P* on the other, because of the parental genotypes. When these *Mp*/*mP* females were mated with the F_1 males (*Mp*/*Y*), the recombinants and nonrecombinants could be distinguished in F_2 males. *Mp* and *mP* chromosomes were present in the F_1 females, but *MP* and *mp* chromosomes could arise only by crossing over between the two loci. Of 405 F_2 males tested, 109 were recombinant, giving a crossover frequency of 26.9% (26.9 crossover units).

For autosomal loci, an analogous method of testing is used, the F_1 double heterozygote being backcrossed to a double homozygous recessive. For example, the parental cross *AB*/*AB* × *ab*/*ab* gives F_1 type *AB*/*ab*. If this is crossed to *ab*/*ab*, nonrecombinant offspring are of types *AB*/*ab* and *ab*/*ab*. Recombinants are *Ab*/*ab* and *aB*/*ab*. The arrangement of the alleles in the F_1 generation is unimportant so long as it is known. The F_1 could have been *Ab*/*aB*, the nonrecombinants then being *Ab*/*ab* and *aB*/*ab*.

Alleles that are on the same chromosome are said to be *cis* with respect to each other. If they are on opposite chromosomes, they are in the *trans* position. The terms *coupling* and *repulsion*, equivalent to *cis* and *trans* respectively, were introduced earlier by Bateson and are still encountered.

Crossover Frequencies and Map Distances. Unlinked genes recombine with a frequency of 50%. If the *A* and *B* loci are on different chromosomes, the double heterozygote *AaBb* will form four types of gametes, *AB*, *Ab*, *aB*, and *ab*, in a 1 : 1 : 1 : 1 ratio, regardless of the parental origin of the alleles. Fifty percent is thus the maximum recombination frequency that can be directly observed. At map distances approaching 50%, the frequency of double crossovers becomes important but cannot be estimated with only two loci. A double crossover between two marker genes preserves the original combination of alleles and is

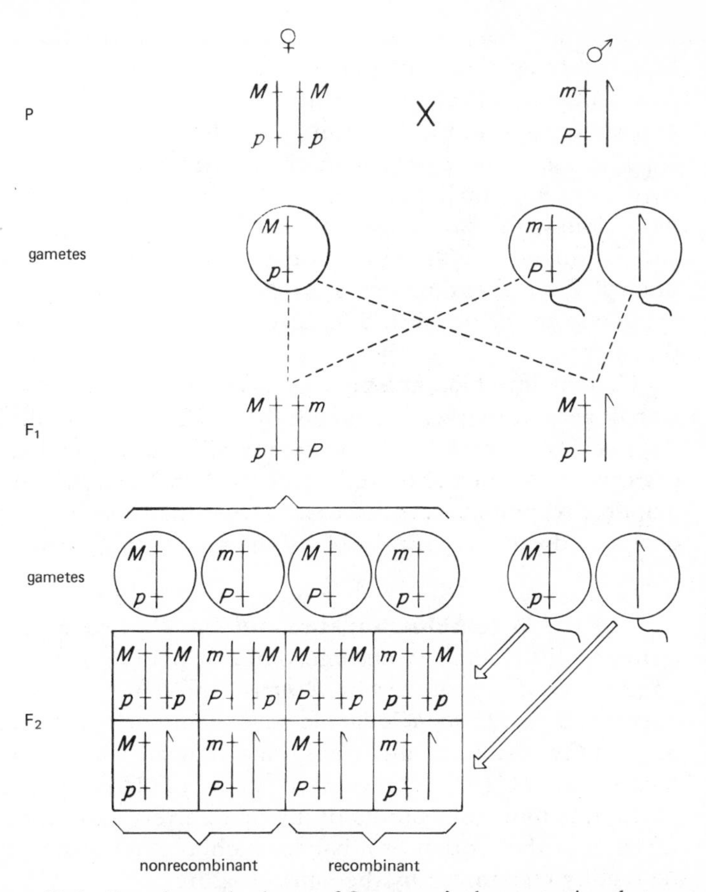

Figure 19-1 Experimental scheme of Sturtevant's demonstration that genes on the same chromosome do not reassort freely. *M* (normal wings) is dominant to *m* (rudimentary wings), and *P* (normal red eyes) is dominant to *p* (vermilion eyes). Both loci are on the X chromosome. Independent assortment would predict equal numbers of recombinant and nonrecombinant offspring. In fact, there were fewer recombinant offspring as compared to nonrecombinant.

counted as a nonrecombinant, although it should be counted as two. For this reason, measurements made directly on widely separated loci are underestimates and will be smaller than the sum of component map distances using intermediate loci. The best figures for long distances therefore are obtained by summing component short distances.

Many chromosomes are so long that multiple crossing over is likely to occur in every synapsis. Summing the map distances between close genetic markers gives values greater than 100 map units. Genes located at opposite ends of these chromosomes assort at random in meiosis. They belong to the same linkage group—that is, they are *syntenic*—but they fail to show linkage in a direct test. Failure to find linkage between a pair of loci therefore does not constitute evidence that they are on different chromosomes. In practice, it is usually not feasible to demonstrate linkage between loci greater than 40 map units apart, and in some experimental studies, the limit is still lower.

The use of three genetic markers permits detection of double crossovers. For intermediate distances, the two crossovers occur more or less at random with respect to each other. If the three markers are very close, the likelihood of a double crossover is less than would be expected by chance. This phenomenon is called *interference* and is visualized as resulting from the physical difficulty in forming two unions in so short a distance.

Certain lower organisms (bacteria, viruses) also show recombination of genetic markers, although the mechanism may differ from crossing over. It is possible to study extremely short map distances in these organisms. At such short distances (0.01 to 0.1 units), the likelihood of a double recombinant is increased rather than decreased. This has been designated *negative interference*. The mechanism is not understood.

The Recombination Map and The Physical Map. In most organisms, little detail in chromosome morphology can be observed. Exceptions are found in certain *Diptera*, including *Drosophila*. The salivary glands of *Drosophila* larvae have giant polytene chromosomes that are readily visible at interphase and that are rich in deeply staining bands (Painter, 1934). The chromosomes of old larvae undergo somatic pairing, homologous points lining up in a very precise manner. Any alteration in the sequence of bands, such as deletions or inversions, can be readily recognized by the pairing figures.

Because of the individuality of bands, short deletions of chromosomal material can be readily recognized. The deleted genes sometimes can be identified by standard genetic procedures, and the genes can be mapped in the manner outlined earlier in the chapter. It is then possible to compare the recombination map, based on crossover frequencies, with the cytological map.

Painter's initial studies showed that genes map in the same sequence in which they occur on the chromosome. However, units of recombination are not equal along the chromosome. In some areas, crossing over is frequent; in others, it is infrequent. The demonstration that the recombination map has physical meaning was an important milestone in genetics.

Experimental organisms have generally shown recombination to be higher in females than in males. Opportunities to make such observations in human beings have been limited, but for several pairs of loci there also seems to be a higher frequency of recombination in females (Renwick and Schulze, 1965; Fenger and Sørensen, 1975; Robson et al., 1977). The differences is not large, however, and is compounded by a possible effect of age on recombination rate in males (Elston et al., 1976).

THE DETECTION AND MEASUREMENT OF LINKAGE

In theory, the techniques for measuring linkage in experimental eukaryote organisms can also be applied to man. In practice, the opportunities to do so are very limited. The basic requirement is a double heterozygote of known phase (that is, which alleles are coupled) crossed to a person whose genotype permits recognition of the allelic combinations in the offspring. Such matings do occur, but, for rare dominant traits, additional information is usually available in a pedigree, and more effective ways have been developed to extract the additional information.

It is primarily with X-linked traits that one encounters in human pedigrees a procedure that is derived with little modification from linkage studies in experimental organisms. Genes that are on the X chromosome clearly are syntenic, whether or not they are sufficiently close to show linkage. In the ideal situation, a woman who is a double heterozygote for two X-linked genes is descended from a father whose genotype also is known. This establishes the coupling in the woman, since crossing over cannot occur with the single X chromosome of her father (Figure 19-2). Her sons can be classified as recombinant or nonrecombinant by simple enumeration, and a relatively few such pedigrees can

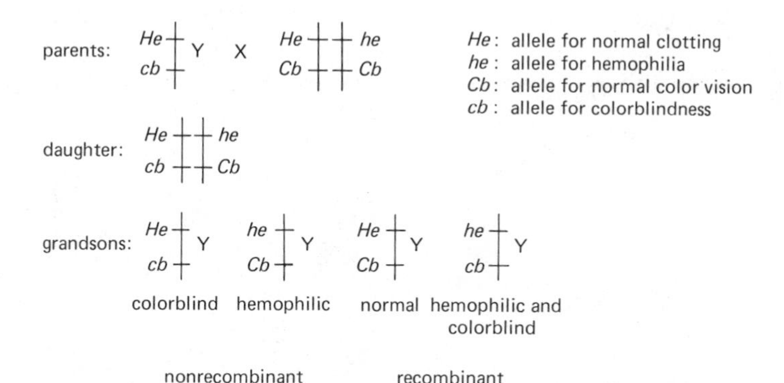

Figure 19-2 Recombination of X-linked genes. The daughter is a double heterozygote for hemophilia and colorblindness, and the coupling is established by the presence of only colorblindness in her father. The distance between these two loci is therefore the proportion of recombinant sons among all her sons. Because of the small size of human families, a number of such families would be necessary to establish the distance between the two loci with confidence.

establish the recombination distance between two loci. Since the sons' father contributes a Y chromosome, his genotype is unimportant.

Several variations of the above have been introduced in attempts to exploit the more typical human pedigrees involving autosomal traits. For the most part these attempts are of historical interest only, since they were not very useful and have been displaced. The following method has proved highly efficient and has led to the detection of a number of cases of linkage.

The Lod Score Method

The lod (log odds) score method, attributed especially to Morton, makes use of recombination between simultaneously segregating markers in pedigrees. It is therefore an extension of the conventional approach of experimental genetics. The questions are phrased somewhat differently, however, in order to take advantage of all the information in a pedigree. Basically, one asks for a particular pedigree, "What is the likelihood of obtaining this pedigree on the assumption that the recombination frequency θ between two markers has a specific value—say, 0.10—as compared with the likelihood if $\theta = 0.50$?" If the ratio of these two expectations is very large, one may accept linkage ($\theta < 0.50$) rather than nonlinkage ($\theta = 0.50$) as an explanation for the nonrecombination of markers in the pedigree.

As an example, consider the pedigree in Figure 19-3. One does not know the coupling phase of the *Rh* locus and the *El* (elliptocytosis) locus in the doubly heterozygous father. If R^1 and *El* were received from the same parent, then R^1 and *El* would be coupled in the man in generation I (as would R^2 and *el*), and four of the five children in II would be nonrecombinant. On the other hand, if R^1 were coupled with *el*, four of the five children in II would be recombinant. A priori we have no reason to assume one coupling over the other, so equal probability is

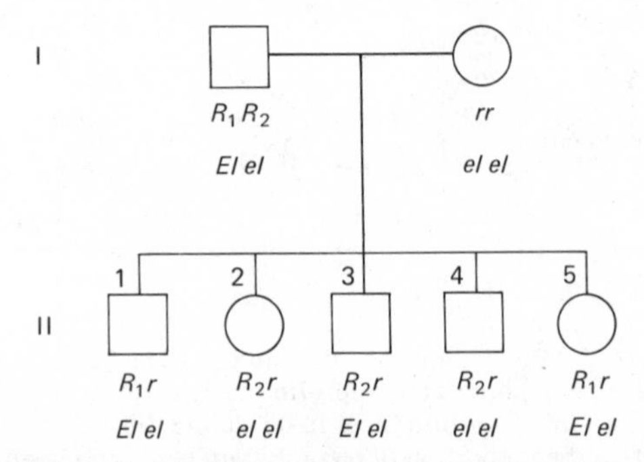

Figure 19-3 Pedigree showing simultaneous segregation of *Rh* and elliptocytosis (*El*). Either II-3 is recombinant and the remainder of generation II are nonrecombinant, or II-3 is nonrecombinant and the remainder are recombinant.

Table 19-1 **PROBABILITIES OF OBTAINING THE PEDIGREE IN FIGURE 19-3 FOR VARIOUS VALUES OF THE RECOMBINATION FRACTION θ**

θ	$P(F_1\|\theta)$	$\frac{P(F_1\|\theta)}{P(F_1\|\theta = 0.5)}$	z
0	0	0	—
.05	.0204	.652	−.186
.10	.0328	1.048	+.020
.15	.0394	1.259	.100
.20	.0416	1.329	.124
.25	.0410	1.310	.117
.30	.0389	1.243	.094
.35	.0361	1.153	.062
.40	.0336	1.073	.031
.45	.0319	1.019	.008
.50	.0313	1.000	.000

assigned. The probability of obtaining the pedigree for any value of θ is then

$$P(F_1|\theta) = \tfrac{1}{2}[\theta(1 - \theta)^4 + \theta^4(1 - \theta)],$$

where the left-hand side of the equation is read "The probability of obtaining F_1, given a value of θ." Various values of θ from 0 to 0.5 are substituted in the equation and tabulated as in Table 19-1. The greatest value corresponds to $\theta = 0.20$, and calculation of smaller increments of θ shows the maximum value actually to correspond to $\theta = 0.21$. This is the recombination fraction most likely to have given this pedigree. Since the pedigree is small, values of this probability do not vary sharply with different values of θ.

The values of $P(F_1|\theta)$ are conveniently expressed as the ratio to $P(F_1|\theta = 0.5)$. This in turn is usually expressed as a lod score (z), where

$$z = \log \frac{P(F_1|\theta)}{P(F_1|\theta = 0.5)}.$$

Values of z greater than zero favor linkage; those less than zero are against linkage. An advantage of z scores is the ease with which observations from different pedigrees can be combined. The total probability for a series of pedigrees is the product of the individual probabilities. However, z scores may be combined by adding them, since they are logarithms. One may therefore accumulate data sequentially until the total z achieves a significant magnitude, usually taken as $z \geq 3.0$, for some value of θ less than 0.5, or until it clearly is maximum at $\theta = 0.5$.

Large pedigrees are much more informative than the small pedigree of the present example. The calculation of z scores is correspondingly

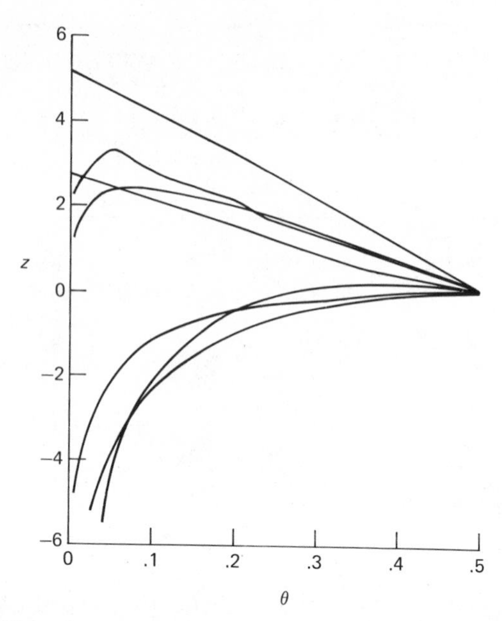

Figure 19-4 Plot of z scores for various values of θ for seven large families in which elliptocytosis and rhesus blood types were segregating. Four of the families show maximum values at or near $\theta = 0.0$, indicating close linkage. The remaining three have maximum lod scores at $\theta = 0.5$, indicating independent assortment. (*From Morton, 1956.*)

more complicated. Tables for calculation of lod scores have been prepared by Maynard Smith et al. (1961; see also Smith, 1968, 1969; Race and Sanger, 1975; Emery, 1976). These simplify the calculations required for two- and three-generation pedigrees.

An early example of the power of this approach was given by Morton (1956), who obtained the results shown in Figure 19-4 when he analyzed seven families in which the dominant gene for elliptocytosis (elliptical-shaped red blood cells) was segregating. Among the other markers tested was the *Rh* locus. Four of the families showed close linkage between elliptocytosis and Rh, the z scores maximizing near zero. The other three families showed no linkage between elliptocytosis and Rh. This was the first suggestion that two different gene mutations can cause elliptocytosis, one locus, designated El_1, being closely linked to the *Rh* locus, the other, El_2, not being linked.

Studies of Somatic Cell Hybrids

The principles of working with somatic cells, including formation of hybrids between human cells and cells from other species, are given in Chapter 14. In such hybrids—for example, in mouse-human hybrids—

human chromosomes tend to be preferentially lost, returning the hybrid to a quasidiploid state. The order in which the human chromosomes are lost is not regular, so that a series of subclones from a particular hybrid will show different combinations of human chromosomes still present.

Comparison of the persistence of specific human characteristics with the persistence of human chromosomes has permitted assignment of many such characteristics to specific chromosomes. In the first such example, hybrid cells exposed the HAT selective medium could survive only if they had retained the human TK^+ (thymidine kinase) gene, since the mouse genome was TK^-. Analysis of such clones revealed that only human chromosome 17 was always retained. It was therefore concluded that the *TK* locus must be on chromosome 17.

A selective medium is not necessary for assignment of genes to chromosomes if one can recognize the human trait. A major technique for accomplishing this is gel electrophoresis. Very often the mouse and human forms of an enzyme differ in electrophoretic mobility (Figure 14-7), as a result of diverging structures during evolution. Analysis of a large number of clones may show a 1 : 1 relationship between presence of the human enzyme and a single chromosome, permitting the conclusion that the human gene is located on that chromosome. An example is given in Table 19-2.

Additional information can sometimes be achieved by using as the human parental cell a culture from a person with a chromosomal translocation. For example, a particular enzyme may have been shown to be associated with chromosome 10. If a translocation involving chromosome 10 is available, the human enzyme may be shown to be associated

Table 19-2 **SEGREGATION OF LOCUS FOR HUMAN α-GLUCOSIDASE (αGLU) AND MARKERS FOR 5 HUMAN CHROMOSOMES IN SOMATIC CELL HYBRIDS PREPARED FROM HUMAN FIBROBLASTS AND A MOUSE CELL LINE**

Chromosome:		5		11		17		18		21	
Marker:		HEX B		LDH A		GALK		PEP A		SOD A	
		+	−	+	−	+	−	+	−	+	−
αGLU	+	29	3	26	4	19	0	30	0	17	3
	−	2	0	1	3	0	4	2	2	3	1

The numbers are the hybrid clones that are positive or negative for αGLU *and* a gene product whose locus is known to be on the chromosome indicated. Each chromosome marker shows discordant clones (αGLU+, marker− or αGLU−, marker+) except GALK on chromosome 17, which shows only concordant clones (both + or both−). Thus αGLU must be on the same chromosome as GALK, chromosome 17. Markers for the remaining chromosomes were also studied; all were discordant in segregation. (Based on data of Solomon et al., 1979.)

Abbreviations: HEX B, hexosaminidase B; LDH A, lactate dehydrogenase A; GALK, galactokinase; PEP A, peptidase A; SOD A, superoxide dismutase A.

with only the translocated portion of chromosome 10, or it may be associated with only the complementary segment of chromosome 10. In this way, it is possible to localize genes to regions of chromosomes. The discrimination is limited by the ability to resolve different regions of the chromosome by current banding techniques and by the availability of cell lines with different translocations.

The study of recombination in families permits measurement of distances between loci and the establishment of linkage groups, but it is not useful for assigning loci to chromosomes (unless one of the loci happens to be a morphological variant of a chromosome). In contrast, the cell hybridization technique permits assignment of genes to chromosomes and the establishment therefore of syntenic groups, but it does not yield information on the recombination distances between loci. (Assignment to different ends of a chromosome obviously will be associated with large recombination distance.) The two methods therefore are complementary. Given sufficient time and the variety of families and translocations, either might be used to arrive at a detailed map of the human genome. Having the two approaches available simultaneously has permitted much greater progress than would have been possible with either alone.

One potential method for measuring linkage of synthetic loci is provided by the incorporation of isolated chromosomes into cultured cells (Chapter 14). If cultured mouse cells are suspended with human chromosomes, incorporation occurs as a rare event. If the human chromosome carries an essential locus, such as $HPRT^+$, and the mouse line is HPRT^-, clones can be selected in HAT medium that necessarily carry the human *HPRT* locus. The entire X chromosome is not incorporated. Rather, the incorporation seems to involve much smaller fragments. But closely linked loci may be cotransferred with the *HPRT* locus (McBride et al., 1978; Miller and Ruddle, 1978). This is reminiscent of the cotransfer of bacterial loci used to map bacterial genes, but the mechanisms are different. As yet, the method has not been used to establish any new close linkages.

A related technique that may prove useful involves the incorporation of human DNA segments into bacteria using recombinant DNA techniques. Again, this is likely to prove useful only with closely linked loci. An example is provided by the hemoglobin β and δ loci, which have been incorporated together into a plasmid (Chapter 10).

Deletion Mapping

One of the early uses of banding in *Drosophila* salivary chromosomes was to localize small deletions. Many mutations are in fact small deletions, and the high resolution achievable with polytene chromosomes makes it possible to detect very small deletions. By comparison, the detection of deletions in human chromosomes is very crude. Nevertheless, persons do survive with visible deletions, and it should be possible occasionally to recognize hemizygosity for one or more loci in such persons. Conversely, heterozygosity for a locus in such a person clearly places it elsewhere than in the deleted region.

One of the earliest examples of deletion mapping was provided by Ferguson-Smith and co-workers (1973), who were able to assign the locus for acid phosphatase to the distal portion of the short arm of chromosome 2 (Figure 19-5).

A further example of deletion mapping was provided by Marsh and co-workers (1974). These investigators observed that nucleated hematopoietic precursor cells of a man suffering from myelofibrosis consisted of two types: Rh-positive cells with normal karotype and Rh-negative cells with a complex chromosomal rearrangement, including deletion of the terminal region of the short arm of chromosome 1 (1pter → 1p32). Family studies indicated the man should have been heterozygous R^1r. Since the *Rh* locus had already been assigned to chromosome 1, it was concluded that the Rh-positive allele (R^1) was lost in the deletion, leaving the abnormal cells of the patient hemizygous for *r*.

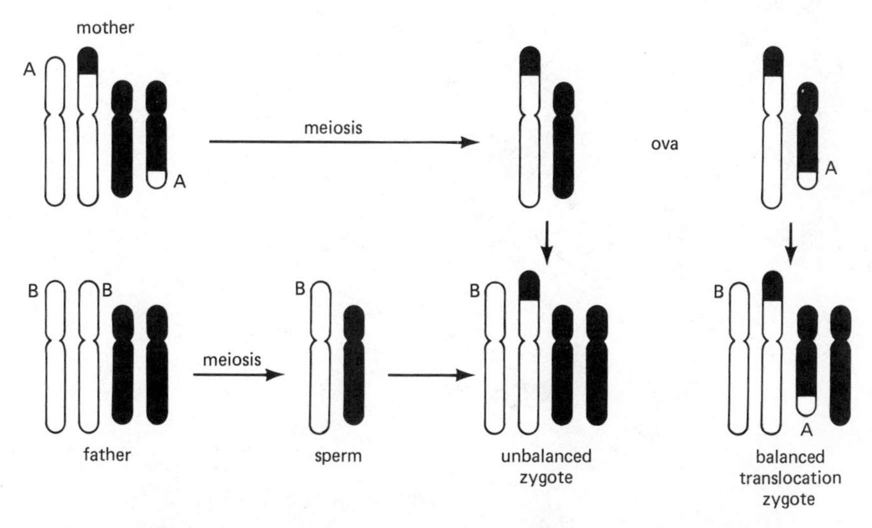

Figure 19-5 Diagram showing loss of acid phosphatase locus (*ACP*-1) in unbalanced gamete from a mother who has a balanced translocation between chromosomes 2 (open) and 5 (shaded). In the unbalanced ovum, the terminal region of 2p is missing and the terminal region of 5q is present in duplicate. The resulting zygote is therefore hemizygous for the terminal region of 2p. The child who developed from such a zygote expressed only the ACP^B allele received from his father and gave no evidence of having received an *ACP* allele from his mother, who was homozygous for ACP^A. It was concluded that the missing terminal portion of 2p must contain the *ACP*-1 locus. (*From Ferguson-Smith et al., 1973.*)

Nucleic Acid Hybridization

When DNA is heated, the hydrogen bonds holding the strands together are disrupted, permitting the strands to separate. Incubation at lower temperatures allows the complementary strands to reanneal. If an excess of RNA complementary to DNA is present during the reannealing process, the RNA may enter into the double helix formation with DNA, but only in chromosomal regions where the RNA and DNA have complementary sequences. These would be the regions where the RNA originally was transcribed from the DNA.

RNA can be labeled radioactively to serve as a cytological marker for complementary DNA sequences. Such *in situ* hybridization is carried out on a standard chromosome preparation on a slide, and the binding of label to chromosomes is recognized by exposure to film. Sources of pure RNA, that is, RNA of a single transcriptional type, are difficult to obtain. However, a few classes of RNA, such as ribosomal RNA, appear to represent transcription from a limited number of DNA sites. Using this technique with ribosomal RNA, Henderson, Warburton, and Atwood (1972) were able to show that the tritium label is bound only to the satellite regions of chromosomes 13, 14, 15, 21, and 22 (Figure 19-6).

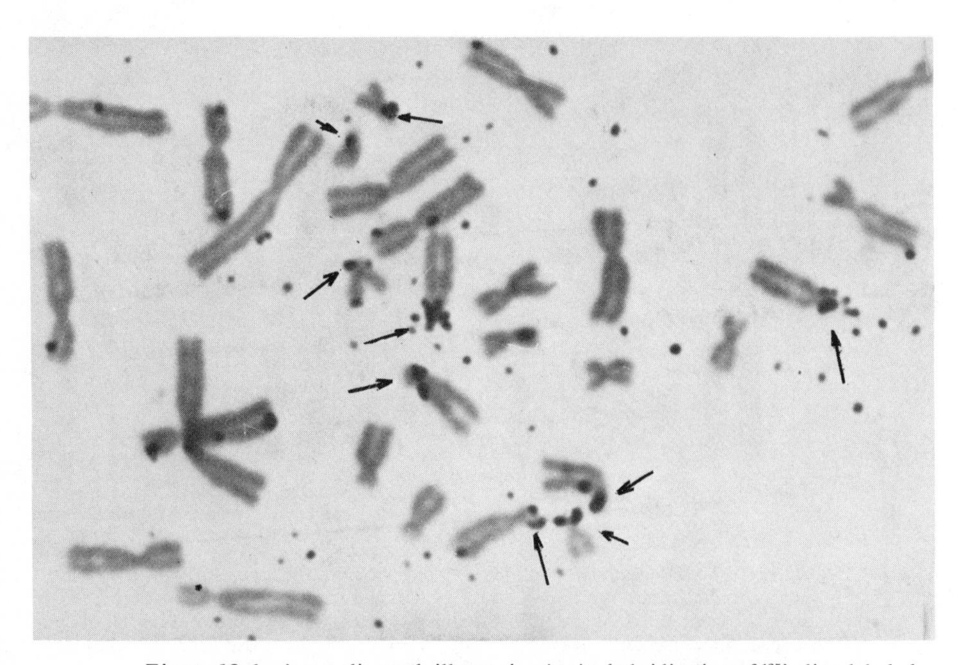

Figure 19-6 Autoradiograph illustrating *in situ* hybridization of ^{125}iodine-labeled ribosomal RNA to human chromosomes. Specific labeling occurs over the satellites of chromosomes 13, 14, 15, 21, and 22, as reported earlier by Henderson, Warburton, and Atwood (1972). Arrows indicate D and G group chromosomes. (*Furnished by A. S. Henderson, D. Warburton, and K. C. Atwood.*)

As yet this technique has not been fully exploited, owing in part to the lack of quantities of pure messenger RNA and to the intense label required to recognize binding by a unique sequence of DNA, as opposed to the multiple copies of the ribosomal genes and other repeated sequences. If these problems can be overcome, the *in situ* hybridization technique should become a powerful means of localizing structural genes on chromosomes.

ESTABLISHED LINKAGE GROUPS

The assignment of human genes to chromosomes and to linkage groups has been especially rapid since the development of the cell hybridization techniques. A summary of loci assigned to specific autosomes is given in Appendix B. Some of these assignments are tentative, being based on limited observations. Only in a few instances has it been possible to assign loci to specific regions of chromosomes.

Some 90 loci have been assigned to the X chromosome. While all are obviously syntenic, the location of most on the X chromosome is unknown, as is the proximity of the loci to each other. There are two very useful marker genes on the X chromosome. These are red-green colorblindness and the red blood cell antigen Xg^a. Approximately 8% of males are colorblind, and 17% of females are heterozygous for colorblindness. Normal daughters of colorblind men are heterozygous for the gene for color vision. The Xg^a red cell type has been recognized since 1962. Red cells of persons who have the antigen agglutinate with immune antibodies formed against Xg(a+) cells. Approximately 60% of males are Xg(a+), a favorable frequency for linkage studies. A third marker is available in some populations. The gene that results in an abnormal type of the enzyme glucose-6-phosphate dehydrogenase (G6PD) is on the X chromosome. Most populations do not have the G6PD deficiency gene. It has a high frequency in some other populations, however, such as Sardinians, Oriental Jews, and African blacks. In addition to the allele that results in deficiency, blacks have an allele that leads to enzyme of normal activity but with altered electrophoretic mobility.

In some instances, it has been possible to study recombination among X-linked genes, especially with the polymorphic loci. Two clusters of linked genes have been detected. One includes the deutan and protan colorblindness loci, hemophilia A, glucose-6-phosphate dehydrogenase, phosphoglycerate kinase, α-galactosidase (Fabry's disease), hypoxanthine phosphoribosyltransferase, phosphoribosyl pyrophosphate synthetase, Duchenne muscular dystrophy, the Xm serum protein type, and Hunter's syndrome. The second cluster includes the Xg blood type, X-linked ichthyosis, retinoschisis, and ocular albinism. Many other loci are known not to be linked to either of these clusters. The total length of the X chromosome has been estimated to be on the order of 200 cM.

The only example of a double crossover in man involves genes on the X chromosome (Graham and co-workers, 1962). The hemophilia B locus (He_B) is known to be outside the segment between *cb* and *Xg*, although the direction is not known. In the pedigree of Figure 19-7, the mother is heterozygous for Xg^a (Xg^a/Xg), hemophilia B ($He_B{}^+/he_B$), and colorblindness (Cb^+/cb). Since her parents could not be examined, the coupling of the genes on her X chromosomes cannot be established. There are two possible sequences of genes and four possible coupling options. Whichever is correct, at least one of the sons resulted from a double crossover.

Linkage Groups in Other Mammals. A comparison of human linkage groups with those of other mammals has been possible primarily for the X chromosome, where homology can be established because of the role in sex determination. In addition, the homology of autosomes of the higher primates is now largely established through banding techniques. One may then inquire whether the arrangement of loci has been conserved during evolution.

The earlier possibility of assigning loci to the X chromosome relatively easily led Ohno (1967) to propose that the X chromosome has been very conservative in vertebrate evolution. More recent studies have supported this idea, although the cytological features are due largely to heterochromatic regions, this would not contradict conservation of the euchromatic regions.

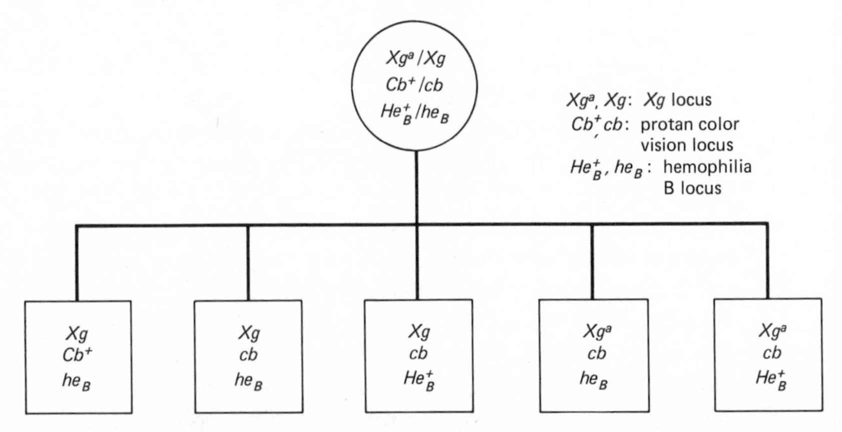

Figure 19-7 Pedigree showing recombination of loci on X chromosome, with at least one son being a double recombinant. (*From Graham and co-workers, 1962.*)

In the case of autosomes, cytological homology is evident among man and the higher primates, and cell hybridization techniques can be used to assign loci in other primates as in man. A number of such studies have shown that the location of genes has remained stable in recent evolution (summarized in Grouchy et al., 1978). This in turn helps in the identification of genetically homologous but cytologically divergent chromosomes in more distant species. Through such approaches, the broad outlines of chromosome evolution in primates is beginning to emerge.

SUMMARY

1. Genes close together on a chromosome do not assort independently in meiosis. However, the farther apart they are, the more likely crossing over is to occur between them, leading to new chromosomal combinations of alleles. The distance between two loci can be measured by the fraction of gametes that are recombinant. This distance is often expressed as centimorgans (cM) or map units, where 1 cM = 1 map unit = 1% recombination.

2. In practice, linkage between loci that are far apart on a chromosome cannot be detected directly, since random assortment yields 50% recombinant gametes. When linkage to intermediate loci is shown, recombination maps greater than 50 cM can be constructed, and chromosomes with map distances greater than 100 cM are common. Loci that are on the same chromosome are said to be *syntenic*.

3. The distance between loci is measured by the frequency of recombinant gametes produced by a person heterozygous at both loci. This requires that the allelic combinations in offspring be distinguishable. Large pedigrees are especially informative. X-linked traits can be studied readily in human families, since the hemizygous sons of a doubly heterozygous mother express X-linked alleles, regardless of whether they are dominant or recessive.

4. The lod score method is most often used to measure linkage in human pedigrees. This method can be applied to complex families, which often are very informative. It consists in comparing the likelihood of obtaining the pedigree on the assumption of various recombination values between two loci as compared with a recombination value of 0.50. Data from different pedigrees can be conveniently combined by adding together the logs of the probabilities (log odds or lod) calculated for each pedigree. The recombination value at which the lod score is maximal is the most probable value.

5. Studies of hybrids between cultured human and mouse cells have been especially helpful in assigning loci to specific chromosomes. Such hybrids tend to lose human chromosomes preferentially. By comparing the persistence of a particular human gene expressed in cell culture with the human chromosome complements remaining in a series of subclones of the hybrid, it is often possible to recognize a one-to-one correspondence between a particular human chromosome and the marker gene. Syntenic groups, that is, groups of loci on the same chromosome, can be identified in this way, but the proximity of the loci to each other is not indicated. By using human cell lines with translocations, it often is possible to localize the human gene to a particular region of the chromosome.

6. Human genes also may be localized to a particular chromosome region by deletion mapping. This was demonstrated in the case of acid phosphatase by a family with a deletion of the distal portion of the short arm of chromosome 2. This approach may be used in somatic mosaics as well, as illustrated by a person who had a clone of hematopoietic precursor cells that had lost both the terminal region of the short arm of chromosome 1 and the *Rh* gene, previously known to be on that chromosome.

7. Some genes can be localized by hybridization with radioactive RNA. For example, labeled ribosomal RNA can be isolated in relatively pure form. If a chromosome preparation is denatured, causing the DNA strands to separate, and then allowed to reanneal in the presence of the RNA, the RNA will combine with complementary regions of the DNA. These regions presumably contain the structural genes for the ribosomal RNA. This method should be useful for any messenger RNA that can be isolated in pure form and labeled sufficiently.

8. A summary of established human linkage groups is given in Appendix B. Where it has been possible to study the same genetic markers in other primates, the markers are found on the same chromosomes as in man. Thus the arrangement of loci on chromosomes has been conserved during evolution. This conservation in turn helps identify homologous chromosomes in more distant species where the cytological appearance of the chromosomes differs from the pattern of the higher primates.

REFERENCES AND SUGGESTED READING

ELSTON, R. C., LANGE, K., AND NAMBOODIRI, K. K. 1976. Age trends in human chiasma frequencies and recombination fractions. II. Method for analyzing recombination fractions and applications to the ABO:Nail-patella linkage. *Am. J. Hum. Genet.* 28: 69–76.

EMERY, A. E. H. 1976. *Methodology in Medical Genetics.* Edinburgh: Churchill Livingstone. 157 pp.

FENGER, K., AND SØRENSEN, S. A. 1975. Evaluation of a possible sex difference in recombination for the ABO-AK linkage. *Am. J. Hum. Genet.* 27: 784–788.

FERGUSON-SMITH, M. A., NEWMAN, B. F., ELLIS, P. M., THOMSON, D. M. G., AND RILEY, I. D. 1973. Assignment by deletion of human red cell acid phosphatase gene locus to the short arm of chromosome 2. *Nature New Biol.* 243: 271–274.

GRAHAM, J. B., TARLETON, H. L., RACE, R. R., AND SANGER, R. 1962. A human double cross-over. *Nature* 195: 834.

GROUCHY, J. DE, TURLEAU, C., AND FINAZ, C. 1978. Chromosome phylogeny of the primates. *Annu. Rev. Genet.* 12: 289–328.

HENDERSON, A. S., WARBURTON, D., AND ATWOOD, K. C. 1972. Location of ribosomal DNA in the human chromosome complement. *Proc. Natl. Acad. Sci.* (U.S.A.) 69: 3394–3398.

KUCHERLAPATI, R. S., AND RUDDLE, F. H. 1976. Advances in human gene mapping by parasexual procedures. *Prog. Med. Genet.* n. s. 1: 121–144.

MARSH, W. L., CHAGANTI, R. S. K., GARDNER, F. H., MAYER, K., NOWELL, P. C., AND GERMAN, J. 1974. Mapping human autosomes: evidence supporting assignment of rhesus to the short arm of chromosome no. 1. *Science* 183: 966–968.

MAYNARD SMITH, S., PENROSE, L. S., AND SMITH, C. A. B. 1961. *Mathematical Tables for Research Workers in Human Genetics.* J. and A. Churchill, London. 74 pp.

MC BRIDE, O. W., BURCH, J. W., AND RUDDLE, F. H. 1978. Cotransfer of thymidine kinase and galactokinase genes by chromosome-mediated gene transfer. *Proc. Natl. Acad. Sci.* (U.S.A.) 75: 914–918.

MC KUSICK, V. A., AND RUDDLE, F. H. 1977. The status of the gene map of the human chromosomes. *Science* 196: 390–405.

MILLER, C. L., AND RUDDLE, F. H. 1978. Co-transfer of human X-linked markers into murine somatic cells via isolated metaphase chromosomes. *Proc. Natl. Acad. Sci.* (U.S.A.) 75: 3346–3350.

MOHR, J. 1963. The Lutheran-secretor linkage: estimation from combined available data. *Acta. Genet. Statist. Med.* 13: 334–342.

MORTON, N. E. 1955. Sequential tests for the detection of linkage. *Am. J. Hum. Genet.* 7: 277–318.

MORTON, N. E. 1956. The detection and estimation of linkage between the genes for elliptocytosis and the Rh blood type. *Am. J. Hum. Genet.* 8: 80–96.

MORTON, N. E. 1957. Further scoring types in sequential linkage tests, with a critical review of autosomal and partial sex linkage in man. *Am. J. Hum. Genet.* 9: 55–75.

OHNO, S. 1967. *Sex Chromosomes and Sex-linked Genes.* Berlin-Heidelberg-New York: Springer-Verlag, 192 pp.

PAINTER, T. S. 1934. Salivary chromosomes and the attack on the gene. *J. Heredity* 25: 465–476.

PORTER, I. H., SCHULZE, J., AND MC KUSICK, V. A. 1962. Genetical linkage between the loci for glucose-6-phosphate dehydrogenase deficiency and colour-blindness in American Negroes. *Ann. Hum. Genet.* 26: 107–122.

RACE, R. R., AND SANGER, R. 1975. *Blood Groups in Man,* 6th Ed. Oxford: Blackwell Scientific Publ., 659 pp.

RENWICK, J. H. 1971. The mapping of human chromosomes. *Annu. Rev. Genet.* 5: 81–120.

RENWICK, J. H., AND SCHULZE, J. 1965. Male and female recombination fractions for the nail-patella: ABO linkage in man. *Ann. Hum. Genet.* 28: 379–392.

RINALDI, A., VELIVASAKIS, M., LATTE, B., AND FILIPPI, G. 1978. Triplo-X constitution of mother explains apparent occurrence of two recombinants in sibship segregating at two closely X-linked loci (G6PD and deutan). *Am. J. Hum. Genet.* 30: 339–345.

ROBSON, E. B., COOK, P. J. L., AND BUCKTON, K. E. 1977. Family studies with the chromosome 9 markers *ABO*, AK_1, $ACON_S$, and *9qh*. *Ann. Hum. Genet.* 41: 53–60.

RUDDLE, F. H. 1972. Linkage analysis using somatic cell hybrids. *Adv. Hum. Genet.* 3: 173–235.

STURTEVANT, A. H. 1913. The linear arrangement of six sex-linked factors in Drosophila, as shown by their mode of association. *J. Exp. Zool.* 14: 43–59.

SMITH, C. A. B. 1968. Linkage scores and corrections in simple two- and three-generation families. *Ann. Hum. Genet.* 32: 127–150.

SMITH, C. A. B. 1969. Further linkage scores and corrections in two- and three-generation families. *Ann. Hum. Genet.* 33: 207–223.

SOLOMON, E., SWALLOW, D., BURGESS, S., AND EVANS, L. 1979. Assignment of the human acid α-glucosidase gene (*αGLU*) to chromosome 17 using somatic cell hybrids. *Ann. Hum. Genet.* 42: 273–281.

REVIEW QUESTIONS

1. Define:

recombination	*cis*	interference
crossing over	*trans*	synteny
chiasma	coupling	morgan
linkage group	repulsion	centimorgan

2. A strain of mice homozygous for dominant genes *AABB* is crossed to a strain recessive at both loci, *aabb*. The F_1 is backcrossed to the recessive parent, producing 115 *AaBb*, 88 *Aabb*, 86 *aaBb*, and 111 *aabb* offspring. Are these loci linked?

3. In mice, albinism may result from homozygosity for recessive genes at either of two loci, *A* and *B*. When the F_1 from a cross of *AABB* × *aabb* was backcrossed to the double recessive *aabb*, 288 offspring were albinos and 112 were pigmented. Are the two loci linked?

4. A man with nail-patella syndrome and type B blood married a normal woman with type A blood. The man's father also had the syndrome and type O blood. The woman's father was normal with type O blood.

With respect to these two loci, what types of offspring would be expected and in what proportions? Assume these loci to be 13 cM apart.

5. In the following pedigree, which of the offspring in generation III are nonrecombinant, which are recombinant, and which are uncertain?

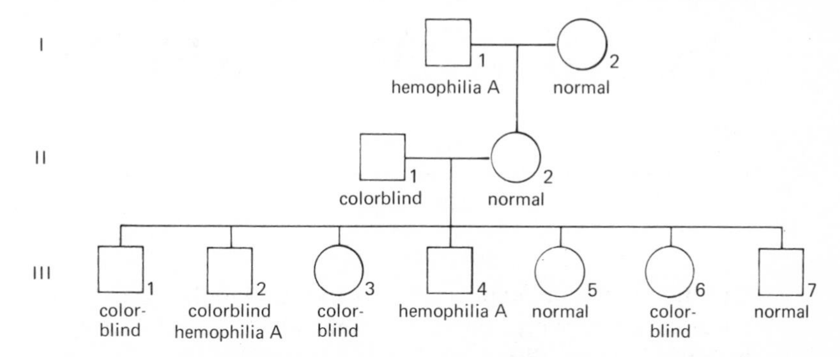

6. The loci for G6PD deficiency and colorblindness are approximately 0.05 cM apart. If a man with G6PD deficiency and colorblindness marries a woman who is homozygous normal, what will be the expected genetic status of his grandchildren? Give all possible combinations and the probability for each. Assume that only normal genes are introduced through marriages.

7. A woman with type A blood and nail-patella syndrome is married to a normal man with type B blood. Her father also had nail-patella syndrome and type O blood. She is pregnant, and a few red cells obtained from the fetus are type O. What is the likelihood that the fetus has the nail-patella gene?

CHAPTER TWENTY

POPULATION GENETICS

Although genes are expressed only in individuals, many problems of genetics concern groups of persons and can be solved only by considering the entire group. This aspect of genetics is known as population genetics. There are major health issues that rely heavily on concepts of population genetics for formulation and interpretation.

Population genetics is a quantitative science. Mathematical models are used extensively, and the assumptions and properties of the models must be understood in order to appreciate their limitations in biological applications.

THE HARDY–WEINBERG LAW

One of the foundations of population genetics is the Hardy–Weinberg law. Although this law is only a special case of the binomial theorem for $n = 2$, its applicability to genetics was not fully appreciated until 1908 when G. H. Hardy, a British mathematician, and W. Weinberg, a German physician, independently pointed out its utility. The short paper by Hardy is particularly interesting as an example of a concise statement of a very important principle.

The basis of the Hardy–Weinberg law is the randomness with which two alleles are combined in persons. The frequency of an allele in a population may be expressed without regard to the genotypic combinations in which it actually occurs. If there are two alleles, A and a, the frequency of A may vary from zero to one, with the frequency of a being the difference between 1.0 and the frequency of A. It is customary to let p and q stand for the frequencies of two alleles, in this case A and a, respectively. If there are only two alleles in the population, $p + q = 1$, and $p = 1 - q$.

The basic form of the binomial is $(p + q)^n$, where n is the number of events taken at a time. For any one autosomal locus, a person may be

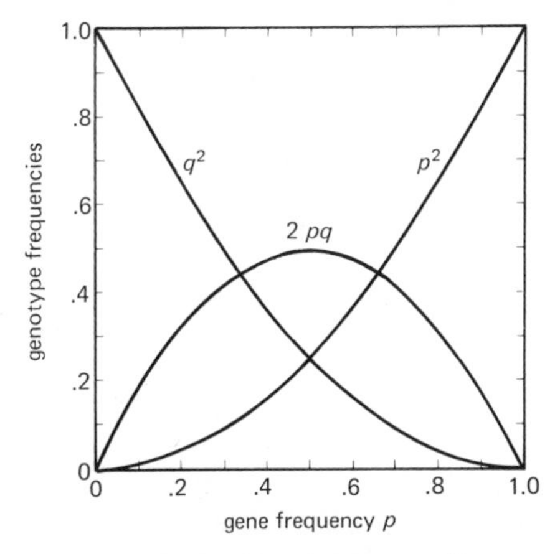

Figure 20-1 Relative frequencies of homozygous and heterozygous persons plotted against gene frequency p, where $p + q = 1$.

considered a random sample of alleles taken two at a time; therefore, $n = 2$. Expanding the binomial gives the familiar

$$(p + q)^2 = p^2 + 2pq + q^2,$$

where p^2 is the frequency of persons both of whose alleles are A, $2pq$ is the frequency of heterozygous Aa persons, and q^2 is the frequency of aa persons. If 40% of the alleles at a particular locus are A and 60% are a, then $p = 0.4$ and $q = 0.6$. Substituting into the formula gives 16% AA, 48% Aa, and 36% aa persons.

The frequencies of the three genotypes can be calculated readily from the frequency of either allele. The change in genotype frequencies with gene frequency is shown in Figure 20-1. The frequency of heterozygotes never exceeds 0.5, a value it attains when $p = q = 0.5$. On the other hand, the heterozygotes are never the least frequent class. They increase rapidly as p or q moves away from zero, and for values of p or q between 0.33 and 0.67, the heterozygotes are the prevalent class.

Multiple alleles can be handled in the same way, using the formula $(p + q + r + \cdots + s)^2$ to calculate the expected frequencies of each genotype. If two alleles are indistinguishable with a particular testing system, such as the A_1 and A_2 blood groups, their combined frequencies and the expected frequencies of various genotypic combinations are represented by the sum of the individual gene frequencies, $p = p_1 + p_2$.

The Assumptions of the Hardy–Weinberg Law

The Hardy–Weinberg law can be expected to apply only if the assumptions upon which it is based are met in the populations to which it is applied. The most important requirement is random mating or *pan-*

Table 20-1 **FREQUENCIES OF PARENTAL AND OFFSPRING GENOTYPES FOR A TWO-ALLELE SYSTEM**

Mating type	Frequency	Probability of offspring			Mating frequency × probability of offspring		
		AA	Aa	aa	AA	Aa	aa
$AA \times AA$	p^4	1	0	0	p^4	0	0
$AA \times Aa$	$4p^3q$	$\frac{1}{2}$	$\frac{1}{2}$	0	$2p^3q$	$2p^3q$	0
$AA \times aa$	$2p^2q^2$	0	1	0	0	$2p^2q^2$	0
$Aa \times Aa$	$4p^2q^2$	$\frac{1}{4}$	$\frac{1}{2}$	$\frac{1}{4}$	p^2q^2	$2p^2q^2$	p^2q^2
$Aa \times aa$	$4pq^3$	0	$\frac{1}{2}$	$\frac{1}{2}$	0	$2pq^3$	$2pq^3$
$aa \times aa$	q^4	0	0	1	0	0	q^4

Total AA offspring $= p^4 + 2p^3q + p^2q^2 = p^2(p^2 + 2pq + q^2) = p^2$.
Total Aa offspring $= 2p^3q + 4p^2q^2 + 2pq^3 = 2pq(p^2 + 2pq + q^2) = 2pq$.
Total aa offspring $= p^2q^2 + 2pq^3 + q^4 = q^2(p^2 + 2pq + q^2) = q^2$.

mixis. When mates are chosen, the various genotypes must have a chance of being chosen exactly proportional to their frequencies. The frequencies of the mating types can be expressed by expansion of $(p^2 + 2pq + q^2)^2$ to give the results in Table 20-1. As is seen from the offspring totals, the frequencies of genotypes among the offspring generation are also random and are the same as among their randomly mated parents.

A second assumption is equal viability of the different genotypes. If the genotype aa does not survive as well as AA or Aa, then the frequency of aa will be less than q^2. In an extreme case, it may approach zero, and many such examples are known. Differential viability of genotypes is known as selection and will be discussed in Chapter 21.

ESTIMATION OF GENE FREQUENCIES

Autosomal Genes

Since genes cannot be observed directly, their presence must be inferred from phenotypes. In some systems, this is easy; in others, it is difficult. In the MN blood types, each genotype corresponds to a distinguishable phenotype. The numbers of L^M and L^N alleles can be counted accurately by observing the frequencies of M, MN, and N persons. The frequency of L^M is the number of L^M alleles divided by the $L^M + L^N$ alleles. Such an estimate is frequently called a direct gene count, even though it is based on phenotypes only.

In the MN system, the alleles are codominant. Had one been recessive, the heterozygote would not have been distinguishable from the homozygous dominant and an allele count would not be possible. The

Rh blood groups, considering only the distinction Rh+ and Rh−, behave as a two-allele system with the *Rh* + allele dominant to the *Rh* −. The *Rh* + allele can be subdivided into several alleles (Chapter 13), but these can be grouped under the allelic designation *R*. The allele that, when homozygous, gives Rh− phenotype is designated *r*. On the basis of these two alleles, there are three genotypes, *RR*, *Rr*, and *rr*. *RR* and *Rr* produce red cells that react with anti-Rh_0 serum; *rr* cells do not react and are therefore Rh−.

The frequencies of the alleles cannot be counted directly, since the proportion of Rh positives who are *RR* versus those who are *Rr* cannot be ascertained. However, if the population is assumed to be in Hardy–Weinberg equilibrium, then the frequency of *rr* should be q^2, where q is the frequency of the r allele. If 16% of the population is Rh−, the frequency of r should be $\sqrt{q^2} = \sqrt{0.16} = 0.4$. The frequency of R is $1 - 0.4 = 0.6$. The frequency of *RR* ($p^2 = 0.36$) and of *Rr* ($2pq = 0.48$) in the total population can then be calculated.

If more than two alleles are present at a locus, the preferred method for estimating frequencies will depend on the exact dominance relationships among the alleles. In theory, the frequency of a recessive allele can always be estimated as the square root of the phenotype frequency. In practice, this estimate may be inaccurate if the frequency of the recessive phenotype is low.

The ABO system will serve as an example of a more complex system. Consider the three alleles I^A, I^B, and I^O, with gene frequencies p, q, and r, respectively. Recalling from Chapter 13 (Table 13-2) the dominance relationships of the various allelic combinations, the frequencies of the types would be A, $p^2 + 2pr$; B, $q^2 + 2qr$; AB, $2pq$; and O, r^2. If observed genotype frequencies matched exactly the expected, it would be a simple matter to calculate p, q, and r from the observed phenotypes:

$$\hat{r} = \sqrt{\mathrm{O}},$$
$$\hat{p} = \sqrt{\mathrm{A} + \mathrm{O}} - r,$$
$$\hat{q} = 1 - (p + r),$$

where A and O are the proportions of the population with these phenotypes. In practice, the problems of sampling in finite populations introduces error into these estimates, especially since estimates $\hat{p}$ and $\hat{q}$ are based on an estimate $\hat{r}$.

Several more accurate methods for estimating ABO frequencies have been developed, of which the one by Bernstein (1930) is given. Preliminary estimates of $\hat{p}$, $\hat{q}$, and $\hat{r}$ are as follows:

$$\hat{r} = \sqrt{\mathrm{O}},$$
$$\hat{p} = 1 - \sqrt{\mathrm{B} + \mathrm{O}},$$
$$\hat{q} = 1 - \sqrt{\mathrm{A} + \mathrm{O}}.$$

With real observations, these estimated values do not add up to one. The deviation $D = 1 - (\hat{p} + \hat{q} + \hat{r})$ can be used to correct the estimates with the following formulas:

$$p^\star = \hat{p}\left(1 + \frac{D}{2}\right),$$

$$q^\star = \hat{q}\left(1 + \frac{D}{2}\right),$$

$$r^\star = \left(\hat{r} + \frac{D}{2}\right)\left(1 + \frac{D}{2}\right).$$

The corrected estimates should not differ greatly from the preliminary estimates but are more efficient statistically.

If several phenotypic distinctions are possible in a multiallele system, the most accurate estimates would come from simultaneous consideration of all the phenotypes. This can be done by the *maximum likelihood method.* Information on this method should be sought in more advanced texts. Briefly, it consists in finding the gene frequency values that, when substituted into the equation describing the distribution of phenotypes, are the most likely to have produced the observations.

Sex-linked Genes

The fact that males have only one X chromosome makes estimates of genes on the X chromosome particularly easy. The phenotypes of X-linked genes ordinarily are expressed in males, so that the presence of colorblindness or normal vision, hemophilia or normal clotting time, glucose-6-phosphate dehydrogenase deficiency or normal enzyme are reliable expressions of the male genotype. The frequency of these genes is thus the frequency of phenotypes among males. In a population in which 8% of the males are colorblind, the gene frequency of colorblindness is 0.08.

Sex-linked genes in females follow the same rules in general as autosomal genes. It should be possible therefore to estimate gene frequencies from females or at least to include data from females in a general formula for distribution of phenotypes. Difficulties arise because of the large variability of expression of X-linked genes in females, due to X-chromosome inactivation. Females heterozygous for X-linked genes may show primarily the effects of only one allele. Females heterozygous for glucose-6-phosphate dehydrogenase deficiency may have normal, intermediate, or deficient amounts of enzyme. The error in classification is very large as a consequence. For other sex-linked genes, such as colorblindness, the error in classifying females is much smaller.

However, the rarity of females homozygous for many recessive traits requires examination of very large numbers of females in order to obtain reliable estimates of phenotype frequencies. If the frequency of colorblindness in males (q) is 0.08, the expected frequency in females (q^2) would be 0.0064. Examination of 1000 males would yield approximately 80 affected persons; in 1000 females, only 6 would be expected, with a proportionally larger chance of sampling error. Furthermore, a few heterozygous females have sufficient color vision defect to make classification difficult.

Accuracy of Gene Frequency Estimates

All measures of gene frequency are made from observations of a finite sample that is considered to be representative of a larger population. If a very large sample is observed, the gene frequency estimate may be quite accurate. If only 20 persons are observed, the estimate may be unreliable.

The usual means of expressing confidence in an estimate is by calculation of the *standard error of the mean.* This quantity, symbolized as SE, expresses the chances that the real frequency in the population, as opposed to the frequency found in a sample of the population, lies within a certain range of the observed value. The statement that $p = 0.50$ and SE $= 0.10$, often written $p = 0.50 \pm 0.10$, means that the observed value of p is 0.50 and that in 68% of samples the real value of p will lie between 0.40 and 0.60. Two standard errors, ± 2 SE, includes 95% of the real values, and ± 3 SE includes 99.7%. (The derivation of these quantities may be found in elementary statistics textbooks.) A standard error of 0.02 indicates a more reliable observation, at least in terms of sampling, than a standard error of 0.10.

Calculation of the standard error for direct allele counts is simple. The general formula for proportions is

$$\text{SE} = \sqrt{\frac{pq}{n}},$$

where n is the total number of alleles in a sample, or twice the number of persons in the case of autosomal loci. (Compare with the formula for standard deviation in Appendix A.) If, among 50 persons, there are 10 of blood type M, 23 of type MN, and 17 of type N, the frequency p of L^M would be 0.43 and the frequency q of L^N would be 0.57. The standard error would be 0.0495. Thus 68% of the time, the frequency of L^M in the parent population from which the sample of 50 persons was drawn will be between 0.38 and 0.48. Ninety-five percent of the time the correct value will lie between 0.33 and 0.53.

When the gene frequency must be calculated from the frequency of homozygous recessive phenotypes, the formula for the standard error is

$$\text{SE} = \sqrt{\frac{1 - q^2}{2n}}.$$

For large values of q, the standard error is relatively small for a given sample size. For small values, it is quite large.

DEVIATIONS FROM HARDY–WEINBERG EQUILIBRIUM

In a random mating population, each genotype should be present in a frequency that is a function only of the frequencies of various alleles in that population. If it can be shown that the genotypes are not distributed randomly, it is then necessary to examine the population or the genetic theory, or both, to see in what way these fail to satisfy the Hardy–Weinberg law.

Test for Equilibrium

The usual procedure for testing for equilibrium is to calculate the expected genotypes on the basis of gene frequencies and to test for deviation from the observed frequencies. In most applications, the gene frequencies have to be ascertained from the sample population. For direct allele counts, the gene frequency estimates are not based on assumptions of equilibrium. Consequently, these can be used to test for equilibrium. In the example used earlier of 50 persons consisting of 10 M, 23 MN, and 17 N, the frequency of L^M was 0.43 and of L^N was 0.57. If these alleles were distributed at random among the 50 persons, the distribution of genotypes would be $50\ (p^2 + 2pq + q^2) = 50[0.43^2 + 2(0.43)(0.57) + 0.57^2] = 9.25$ M + 24.51 MN + 16.25 N. These figures closely match the 10, 23, and 17 observed, and therefore this sample would indicate the population to be at equilibrium.

The observed and expected values do not always agree so closely. In that event, it is necessary to use a statistical test to assess the chances that the deviation is significant. A χ^2 can be calculated as described in Appendix A. For the examples given, the procedure would be

$$\begin{aligned}\chi^2 &= \sum \frac{(O - E)^2}{E} \\ &= \frac{(10 - 9.25)^2}{9.25} + \frac{(23 - 24.51)^2}{24.51} + \frac{(17 - 16.25)^2}{16.25} \\ &= 0.188.\end{aligned}$$

Although there are three terms in this χ^2, there is only one degree of freedom, since the expected values were derived from the observed data. For problems of this particular type, the number of degrees of freedom is equal to the number of phenotypes minus the number of alleles. For this value of χ^2 and 1 df, $.70 > P > .50$. Thus the observed deviation could happen by chance more than half the time. Had the observed deviation been so large that it would happen by chance rarely, it would be appropriate to conclude that the population from which the sample was drawn probably was not in equilibrium. While the test described here is widely used to test for departures from equilibrium, it should be noted that small deviations may not be detected unless the sample size is very large.

In many genetic systems, gene frequencies are computed from the square root of the recessive phenotype(s). This procedure *assumes* the population to be in equilibrium; hence it is not possible subsequently to test for equilibrium with the same data.

Reasons for Departure from Equilibrium

The demonstration that a population is not in equilibrium is evidence that the assumptions necessary for equilibrium have not been met. Two primary assumptions were noted earlier—random mating and equal likelihood of survival of different genotypes. Most departures from equilibrium are due to nonrandom mating. In genetic terms, nonrandom mating means that mates are selected because of their genotype. It is convenient to distinguish several causes of nonrandom mating, although the genetic effects are equivalent.

Stratification. When a population consists of two or more subpopulations with different gene frequencies, it is said to be stratified. The most common example in the United States is the black and white populations. These two groups differ in the frequencies of a number of genes, and failure to recognize the presence both of blacks and whites in a population may lead to erroneous gene frequency estimates. There are many other subpopulations that may also be important. The distribution of genes varies within and among the countries of Europe, and populations that are closely related may yet differ in the frequencies of some alleles. A typical American city, with populations originally from Ireland, Germany, Poland, Italy, West Africa, and so on, has many groups that must be recognized for genetic studies. America may be the melting pot, but the mixture hasn't completed melting yet.

For many purposes, it may be possible to group together related populations with similar gene frequencies. The decision will depend on the use to be made of the information. Deviation from expected equilibrium values is frequently small when two populations are mixed, even though the gene frequencies may differ. In a pooled sample, there are more homozygotes and fewer heterozygotes than would be expected from the allele frequencies. The test for equilibrium makes detection of pooling of two similar populations unlikely unless the data are based on large numbers of observations.

Assortative Mating. Persons who choose mates because of particular traits are said to mate assortatively. If mating partners resemble each other for a trait, the assortative mating is positive; if partners are chosen because they are different, assortative mating is negative. Every mating is assortative with respect to some traits and random with respect to others. People tend to marry into their own ethnic group, thereby maintaining the population strata. This may not be a question of ethnic preference so much as the opportunity for social contacts. Persons often are encouraged to marry within the religious denomination to which they belong. In many countries, certain denominations are drawn primarily from one ethnic group, and marriages within the denomination therefore are likely to be between persons of similar ethnic and genetic background. Numerous other social institutions might be cited that tend to perpetuate ethnic differences.

These examples of assortative mating are basically the same as stratification, since the preferred mate is selected from the same subpopulation. In addition, there is assortative mating within a population if mates select each other for similar characteristics. Tall people tend to marry tall; short people marry short; intelligent people marry intelligent; and so on. This type of assortative mating is of genetic interest only to the extent that the traits selected are under genetic control. Only traits apparent to the marriage partners can be selected. Generally, blood types are not known, and even if known, they are not likely to play an important role in mate selection. Most of the traits important in assortative mating are complex genetically.

Unequal Viability of Genotypes. Even though the mating structure of a population may be random, persons of different genotypes do not necessarily have the same chances of survival. An extreme case would be complete lack of viability of one genotype. This occurs with zygotes homozygous for certain genes. A genetically equivalent situation would be failure of persons born with a certain genotype to survive to an age when they would be included in the survey of genotypes. Ideally, surveys should be carried out on newborn infants. A more common procedure is to work entirely with adults.

There are many examples of unequal viability of genotypes. For most recessive traits where viability of the homozygous recessive is greatly reduced, the frequency of the recessive gene has so diminished that surveys of random groups of people are unrewarding. The frequency can be more efficiently, if less accurately estimated from ascertainment of affected persons only, provided the size of the population from which they are drawn is known.

An example of a common deleterious trait is sickle cell anemia. In many populations, there is a moderately high frequency of the gene for abnormal hemoglobin, Hb_{β}^{S}, which in the homozygous state causes sickle cell anemia. Persons who are homozygous rarely survive to adulthood, but in certain parts of the world heterozygotes seem to have a slight advantage over persons homozygous for the normal allele, Hb_{β}^{A}, because of increased resistance to malaria. A survey of the frequency of these alleles among infants would show the three genotypes to be in approximately the proportions predicted by the Hardy–Weinberg law. If the survey were conducted among adults, there would be only a few persons found homozygous for Hb_{β}^{S}.

Unequal Ascertainment of Genotypes. Genotypes may be equally viable but nevertheless may not be present in a particular sample of the population in the same proportions in which they occur at birth. This would be the case if the persons actually examined were not representative of the parent population because of associations between genotype and likelihood of being selected for study.

In practice, a random sample of a population is virtually impossible to assemble. Institutionalized persons are almost never included as part of the population from which they are drawn. Genotypes that are likely to lead to institutionalization will be underrepresented in the noninstitutional population. The same is true of hospitalized patients or persons whose health does not permit them to assume an active role. In some instances, surveys are limited to a narrow segment of the population—for example, workers in a particular factory, admissions to a maternity hospital, or students in a school. These particular examples might lead to overascertainment of genes associated with physical robustness, fertility, or intelligence, respectively. To the extent that a gene under investigation is associated with the above traits, the genotypes may be disproportionately represented in the sample.

It will be easier to avoid confusion concerning ascertainment if the concepts of *incidence* and *prevalence* are clearly separated. Incidence is defined as the number of persons who acquire a characteristic in a defined population. Prevalence is the frequency of a characteristic in the population at a particular time or in a particular time period. Two diseases might have equal incidence in that they affect an equal portion of the population. But if one disease lasts twice as long, there would be twice as many persons affected at a given time (if the risk were constant with time). Incidence of genetic traits often refers to the frequency at birth of a given genotype among all genotypes. Prevalence refers to the frequency of the genotype at a given time after natural selection for and against various gene combinations has occurred.

Incorrect Genetic Theory. In a two-allele system, the maximum frequency of heterozygotes is 0.50 (Fig. 20-1). Occasionally, the heterozygote frequency is higher. This may be caused by sampling error, but χ^2 analysis should show the distribution of genotypes not to be significantly different from that predicted by the Hardy–Weinberg law. Rarely, the frequency of heterozygotes may be much higher. This is more likely to be observed if genetic theory has been formulated on one racial group and sampling is then done on another. For example, when Gm types were first detected, the three types were Gm (a+b−), Gm (a+b+), and Gm (a−b+), corresponding to genotypes Gm^a/Gm^a, Gm^a/Gm^b, and Gm^b/Gm^b (Chapter 12). In a sample of American Negroes, Steinberg, Stauffer, and Boyer (1960) found the frequency of Gm (a+b+) to be 0.95, a figure greatly in excess of that predicted by the Hardy–Weinberg law. They proposed and subsequently established that in Negroes there exists an allele Gm^{ab} that leads to the reactions of both Gm^a and Gm^b and that is present in very high frequency. The frequency of the Gm^{ab} allele is very high in Negroes but very low in other groups tested. It was the high frequency of apparent heterozygotes that led to recognition of the new allele. Such possibilities must be kept in mind when testing new populations for genetic traits.

Chance. Finally, one should never forget that chance deviations do occur as a problem in sampling. A statistical test that permits one to conclude that an observed deviation would occur by chance only 5% of the time also requires one to question whether a particular observation may belong to the 5%.

GENE FREQUENCIES IN MIXED POPULATIONS

There are numerous examples in the history of man in which two or more populations, each with its characteristic gene frequencies, have united to form a single new panmictic population. The frequency of inherited traits in the new population is a function of the frequencies in the parent populations. In order to arrive at the new frequencies, one must keep in mind that the frequency of various allele combinations is expressed in terms of the frequency of the individual alleles. Calculation of the allele frequencies in the new population is therefore the first step in predicting the frequency of allelic combinations.

The general formula for averaging frequencies is

$$p_t = ap_a + bp_b + cp_c + \cdots + np_n$$

where p_t is the frequency of a specific allele in the new population; a, $b, \ldots, n$ are the relative contributions of the parent populations; and $p_a, p_b, \ldots, p_n$ are the corresponding allele frequencies in the parent populations. The various $p_a, p_b, \ldots, p_n$ may have any value from zero to one, but $a + b + \cdots + n = 1$.

Consider the example of Rh− persons who are homozygous for the r allele. Among North American whites, the frequency of Rh− persons is approximately 15%. Among the Bantu of Africa, it is about 5%. On the basis of other genetic systems, American blacks are known to have received about 78% of their genes from African ancestors and about 22% from whites. To arrive at the predicted frequency of Rh− persons among American blacks, it is necessary first to calculate the frequency of the r allele in the present population. Thus the frequency in Bantus q_B is $\sqrt{0.05} = 0.22$, and the frequency in whites q_C is $\sqrt{0.15} = 0.39$. The gene frequency in American blacks is then

$$q_N = (0.78)(0.22) + (0.22)(0.39) = 0.26.$$

The frequency of Rh− persons among American blacks should be $(0.26)^2$ or 0.068. This corresponds to an observed frequency of approximately 8%.

Often it is desirable to calculate the relative contributions of two parent populations to a new population, knowing the phenotype frequencies in all three populations. In this case, p_t, p_a, and p_b are known. Substitution into the general equation, remembering that $a + b = 1$, permits calculation of the values of a and b.

When two distinct populations mix and undergo one generation of random mating, the genotypic combinations taken one locus at a time follow Hardy–Weinberg expectations. If two loci at a time are considered, associations representing the parental allelic combinations persist over several generations before the alleles become random with respect to each other. This is most easily illustrated if population I is homozygous for alleles A and B at the two loci, while population II is homozygous for A' and B'. Each can produce only one type of gamete, AB and $A'B'$, respectively. If these unite at random, those homozygous for A will also be homozygous for B, and similarly for the $A'B'$ homozygotes. In each successive generation, this group of parental homozygotes will constitute a smaller proportion of the total, and as heterozygous combinations appear, reassortment becomes possible. It should be remembered that this association is not based on linkage but rather on association of unlinked genes. Linkage would tend to delay the reassortment process still farther.

TESTS OF GENETIC HYPOTHESES FOR POLYMORPHIC TRAITS

In addition to the analysis of segregation and of pedigrees covered in earlier chapters, L. H. Snyder pointed out that ratios of offspring can also be anticipated from the Hardy–Weinberg law and used to validate the pattern of transmission, even though the parental genotypes are not always known. The two situations that occur are marriages in which both partners have the putative dominant phenotype and those in which one is dominant and the other recessive. Marriages between two persons, both of whom have the apparently recessive phenotype should produce only children with the same phenotype.

Dominant × Dominant. These matings are three types, $AA \times AA$, $AA \times Aa$, and $Aa \times Aa$. Their relative frequencies are p^4, $4p^3q$, and $4p^2q^2$, respectively, where p and q are the gene frequencies of A and a. Only the last can produce recessive offspring, the expectation being one-fourth. The fraction of recessives among all such matings should then be

$$
\begin{aligned}
P(aa) &= \frac{(4p^2q^2)/4}{p^4 + 4p^3q + 4p^2q^2}, \\
&= \frac{q^2}{p^2 + 4pq + 4q^2}, \\
&= \left(\frac{q}{p + 2q}\right)^2, \\
&= \left(\frac{q}{1 + q}\right)^2.
\end{aligned}
$$

Thus for a trait thought to be recessive, the fraction of recessive offspring from dominant × dominant matings should conform to the above formula and can be tested statistically.

Dominant × Recessive. There are two mating types that correspond to these phenotypes: $AA \times aa$ and $Aa \times aa$. By similar reasoning, the expected proportion of recessive offspring from such matings is:

$$
\begin{aligned}
P(aa) &= \frac{(4pq^3)/2}{2p^2q^2 + 4pq^3}, \\
&= \frac{q}{1 + q}.
\end{aligned}
$$

SUMMARY

1. According to the Hardy–Weinberg law, populations are composed of persons who may be considered as random samples of the gene pool. This can be represented by the binomial expansion $(p + q)^2$, where p and q are the frequencies of two alleles. Random mating (panmixis) and equal viability of genotypes are assumed.

2. Gene frequencies can most readily be estimated by direct counting if there is a one-to-one correspondence between genotype and phenotype. However, if one of the alleles is recessive, the frequency of the recessive allele is the square root of the recessive phenotype. The frequencies of sex-linked genes are most readily obtained by the frequency of the phenotypes in hemizygous males.

3. Populations can be tested for conformity to the Hardy–Weinberg law if allele frequencies are obtained by direct count. The expected numbers of each genotype are calculated by the binomial expansion, and these expected numbers can then be compared with the observed. A population may differ from the theoretical distribution for several reasons: It may actually be a mixture of two or more populations with different gene frequencies; there may be assortive mating; there may be differential survival of genotypes or unequal ascertainment of genotypes; and the genetic theory may be incorrect.

4. When two populations with different allele frequencies intermix, the new phenotype frequencies of the hybrid population can be calculated by averaging the allele frequencies of the parent populations weighted according to the contribution of each parental population to the hybrid.

5. In matings where one or both parents have a dominant phenotype, the Hardy–Weinberg law predicts the frequencies of dominant and recessive offspring as a function of gene frequency. This is especially useful in verifying Mendelian inheritance for polymorphic traits.

REFERENCES AND SUGGESTED READING

BERNSTEIN, F. 1925. Zusammenfassende Betrachtungen über die erblichen Blutstrukturen des Menschen. *Zeitschr. Abstgs. u. Vererbgsl.* 37: 237–270.

CAVALLI-SFORZA, L. L., AND BODMER, W. F. 1971. *The Genetics of Human Populations.* San Francisco: W. H. Freeman, 965 pp.

CEPPELLINI, R., SINISCALCO, M., AND SMITH, C. A. B. 1955. Estimation of gene frequencies in a random-mating population. *Ann. Hum. Genet.* 20: 97–115.

EMERY, A. E. H. 1976. *Methodology in Medical Genetics.* Edinburgh: Churchill Livingstone, 157 pp.

HARDY, G. H. 1908. Mendelian proportions in a mixed population. *Science* 28: 49–50.

HARRISON, G. A., AND BOYCE, A. J. (Eds.). 1972. *The Structure of Human Populations.* Oxford: Clarendon Press, 447 pp.

LI, C. C. 1977. *First Course in Population Genetics.* Pacific Grove, Calif.: Boxwood Press, 631 pp.

SMITH, C. A. B. 1967. Notes on gene frequency estimation with multiple alleles. *Ann. Hum. Genet.* 31: 99–107.

SNYDER, L. H. 1932. Studies in human inheritance. IX. The inheritance of taste deficiency in man. *Ohio J. Sci.* 32: 436–440.

STEINBERG, A. G., STAUFFER, R., AND BOYER, S. H. 1960. Evidence for a Gm^{ab} allele in the Gm system of American Negroes. *Nature* 188: 169–170.

REVIEW QUESTIONS

1. Define:

panmixis
mean
standard error of mean
stratification
assortative mating
incidence
prevalence

2. Blue eyes has been proposed as a simple recessive trait. In a certain panmictic population, 16% of the persons have blue eyes.

a. What portion of the brown-eyes persons would be heterozygous for blue eyes?

b. What would be the predicted frequency of blue-eyed children among offspring of parents, both of whom are brown-eyed?

3. Phenylketonuria has been reported to occur once in approximately 10,000 births. Pedigree studies indicate the disease to be due to homozygosity for a recessive gene.

a. What is the frequency of the gene?

b. What percentage of the population is heterozygous for the gene?

c. What will be the frequency of marriages involving two heterozygotes?

4. In a certain population, color blindness occurs 25 times more frequently in males than in females. What is the frequency of the gene for color blindness? What is the frequency of heterozygous females?

5. Cystic fibrosis of the pancreas is due to an autosomal recessive gene. The frequency of the gene is approximately 0.02 in white populations.

a. What is the frequency of affected individuals
b. What is the frequency of carriers?
c. What is the frequency of marriages of carriers with homozygous normals?
d. What is the frequency of marriages of carriers with carriers?

6. The observed frequencies of blood groups at the MN locus in a particular population are: Type M, 110; type MN, 180; type N, 110. What are the gene frequencies of L^M and L^N? Is the population at equilibrium? If this population undergoes random mating with one twice as large in which the frequency of $L^M = 0.7$, what will be the expected frequencies of phenotypes in the new population?

7. Cystic fibrosis is an autosomal recessive trait that occurs in 1 in 2500 births of Western Europeans. A couple, neither of whom is affected, has a child with cystic fibrosis. The mother subsequently marries a man unrelated to the first husband. What is the likelihood that their first child will be affected? How would your answer change if the second husband were the brother of the first husband?

CHAPTER TWENTY-ONE

SELECTION

REPRODUCTIVE FITNESS

Selection operates when there is inequality in fitness of genotypes. Fitness is to be understood in a broad sense, being measured ultimately in terms of the number of offspring produced. A person may be fit functionally in that he is healthy, and he may contribute to survival of the population, but, if he fails to leave offspring, he is scored genetically as "dead." By the same token, long survival beyond the reproductive age does not make a person more fit than one who leaves as many offspring but dies at a young age. Everyone in the population may produce the same number of offspring, but those who do so sooner are more fit biologically than those who reproduce at later ages.

Fitness is a property of a whole organism in a particular environment. Therefore the factors that contribute to fitness are many and complex. It is possible, nevertheless, to measure the relative fitness of two groups of phenotypes that differ on the average by a single particular gene.

It is common to speak of *positive* or *negative* selective advantage. The terms are somewhat arbitrary, since there is no absolute scale with unit advantage. But an allele can be said to have positive selective advantage if it leads to greater fitness than the alternate alleles under consideration. It has negative selective advantage if it leads to lesser fitness.

Sets of alleles at a particular locus are sometimes said to be selectively neutral if the various genotypic combinations appear to have equal fitness. It is possible that true selective neutrality never exists. Every difference, no matter how slight, could lead to slight but significant differences in fitness. However, alleles whose only difference is a substitution of one amino acid for another very similar amino acid in some protein may have very similar fitness values, and many geneticists accept the idea that truly neutral alleles are common. The estimates of relative fitness must be based on observation rather than biochemical reasoning, since very small changes, as in hemoglobins S, C, and A, may cause significant differences in fitness.

Selection occurs at all stages of the life cycle. Most familiar is the selection against certain genotypes subsequent to birth. Persons homozygous for hemoglobin S rarely survive to reproduce. Until the availability of insulin, persons with juvenile diabetes were at a great disadvantage compared to nondiabetics. Resistance to infectious diseases, although poorly understood from the genetic point of view, was an important feature of selection prior to the advent of antibiotics.

It has been suggested that modern medical practices have largely eliminated natural selection. It is true that persons with diseases have a much better chance of survival than formerly. But survival of persons is only the most obvious area in which selection operates. If the entire life cycle is considered—formation of gametes, survival of gametes, formation of zygote, development of fetus, survival of newborn through reproductive period—there are many points at which selection can operate without leading to visible elimination of certain genotypes and without medical intervention.

An example of selection in gametogenesis is found in *Drosophila.* The *segregation-distorter* (*SD*) locus has two known alleles. Normal segregation of chromosomes occurs with the wild-type allele. Males heterozygous for the mutant allele, under certain conditions of environment, show among progeny a marked departure from the expected 1:1 ratio in favor of the *SD* allele. This interaction of homologous genes during meiosis is called *meiotic drive.* The extent to which it may occur in man is unknown. Loss of gametes is experimentally difficult to establish as are small deviations from Mendelian ratios.

Even though mature gametes are formed, not all are equally likely to give rise to a zygote. This is most clearly seen in *gametic* or *prezygotic* selection involving sperm. The cervical secretions of many women contain antibodies against certain of the ABO blood groups. Sperm contain these blood group substances. If a man has antigens not present in his wife, she may well form antibodies that act against his sperm, thereby reducing the fertility of this particular genetic mating (Gershowitz and co-workers, 1958).

There is little information on the genetic constitution of aborted fetuses apart from chromosome studies. Spontaneous abortions, many too early to be recognized as such, are frequent. Many abortions involve malformed fetuses; in other cases, a defect is not obvious, although many metabolic defects would be undetected. Judging from studies on laboratory animals, a large portion of abortions are genetic in origin. Such abortions represent natural selection in operation.

Finally, in spite of the great progress with many human diseases, there are still many that have not yielded to therapy. Some of these are inherited and subject to selection.

SELECTION AGAINST DELETERIOUS TRAITS

The simplest situations to consider quantitatively are those in which a mutant gene, either dominant or recessive, leads to reduced fitness compared to the nonmutant "wild-type" phenotype.

Selection against Dominant Genes

If an allele is lethal in heterozygous combination, resulting either in death or in sterility, selection is rapid and effective. A fully penetrant gene with zero fitness would be virtually eliminated in one generation, since it would not be transmitted to offspring. Mutation would produce new copies of the mutant gene, but, since the mutation rate is characteristically very low, few persons would carry the dominant gene because of a new mutation.

Examples of deleterious genes meeting these requirements are not known. There are many rare conditions in which affected persons are unable to reproduce. In order to prove that a trait is genetic, it must be transmitted. Dominants that cannot be transmitted therefore cannot be distinguished usually from environmentally induced traits, although one may make the reasonable assumption that a trait is inherited if it is associated with an altered amino acid sequence, for example.

Although not a single-gene mutation, trisomy 21 syndrome illustrates some of the difficulties in interpretation of defects that result in sterility. (The few cases of fertility now known among trisomy 21 patients were mostly found after the chromosomal basis of the disease was established.) The disease occurs too often for each case to represent a new mutation of a conventional gene. However, the only association with environmental factors was with maternal age. The absence of opportunity for transmission to offspring of affected persons prevented recognition of the genetic basis of the disease. It was as easily explained by unidentified environmental factors.

It seems likely that all genes undergo mutation. Many such mutations, in heterozygous combination, almost certainly lead to nonviability. This could result in a stillbirth or a spontaneous abortion, the latter possibly occurring so early in gestation as to be unrecognized. Or the mutation may be compatible with viability in the somatic sense but not in the genetic; that is, the person is sterile. The frequency of a condition of this type due to mutation at a specific locus would be very rare, approximately twice the mutation rate. Based on an average mutation rate of 1×10^{-6} per locus per gamete, zygotes heterozygous for specific new mutations would occur with an incidence of about 1/500,000. Thus some of the very rare abnormalities may result from new mutations that cannot be demonstrated by conventional genetic means.

Many dominant genes, although deleterious, are still compatible with partial fertility. The efficiency of selection is less than in the previous case and is proportional to the loss of fitness of heterozygous persons. The loss of fitness of a genotype can be expressed by the *coefficient of selection, s.* The fitness of a genotype is symbolized by $w = 1 - s$. If the frequency, p_0, of a dominant gene A is very low in the parent or zeroth generation, so that AA persons are rare, the effects of selection for one generation can be expressed by

$$p_1 = \frac{2p_0q_0w}{2(2p_0q_0w + q_0^2)} = \frac{p_0w}{2p_0w + q_0}.$$

As p becomes very small compared to q, the denominator approaches 1. The equation is then

$$p_1 \cong p_0w.$$

For additional generations

$$p_n = p_0w^n, \tag{1}$$

where n is the number of generations of selection.

For any value of w less than 1, the frequency p will approach zero as n becomes very large. Theoretically, it will not become zero, but as the number of persons with the gene becomes small, chance may cause fluctuations from the predicted frequency. Eventually, since the deleterious genes are also being formed through mutation, the rate of loss through selection will equal the rate of gain through mutation (Chapter 9).

Selection against Recessive Genes

Deleterious recessive genes are removed from the population just as are dominants. However, the process is much slower because only a portion of the genes are in the homozygous combination and therefore subject to removal. If the appropriate terms of the binomial expansion are multiplied by the number of recessive alleles, the fraction of recessive genes in homozygotes is given by the expression

$$\frac{2q^2}{2pq + 2q^2}, \tag{2}$$

which reduces to q. For rare genes, only a small portion would be subject to removal by selection; even for common genes, selection is slow.

For complete selection against the recessive phenotype corresponding to genotype aa, the rate of change of gene frequency, q, can be readily calculated. Consider a population consisting of the usual three gene combinations involving two alleles. The distribution would be

$$p_0^2 + 2p_0q_0 + q_0^2.$$

If the aa persons represented by the term q_0^2 do not reproduce, then the frequency of a in the next generation will be the same as in the $AA + Aa$ persons of the original generation; thus

$$q_1 = \frac{2p_o q_o}{2(p_o^2 + 2p_o q_0)}. \tag{3}$$

It is necessary to multiply the denominator by 2 since each person has two alleles, whatever the genotype. Dividing by $2p_o$ and remembering that $p_o = 1 - q_o$ yields

$$q_1 = \frac{q_o}{1 + q_o}. \tag{4}$$

The frequency of recessive genotypes would be $q_1^2 = q_0^2/(1 + q_0)^2$. The frequency of a after two generations of selection can be obtained by substituting for q_1 in the formula $q_2 = q_1/(1 + q_1)$. This yields

$$q_2 = \frac{q_o}{1 + 2q_o}. \tag{5}$$

Similarly, it can be shown that for any number of generations n,

$$q_n = \frac{q_o}{1 + nq_o}. \tag{6}$$

The change of q versus n is shown in Figure 21-1. Although q approaches zero, it does so very slowly. This is because an increasing portion of the recessive genes are in heterozygous combination and not selected against as q becomes small. For any particular value of q, the portion of alleles lost is q. Therefore, even the most rigorous selection against aa accomplishes little once the frequency of a is low.

In spite of the inefficiency of selection against recessive genes, the frequency does decrease steadily. The number of generations required to accomplish a given decrease in gene frequency can be calculated by rearranging formula (6) in the form

$$n = \frac{1}{q_n} - \frac{1}{q_o}. \tag{7}$$

In order to reduce the frequency of a gene from 0.10 to 0.05, ten generations of complete selection would be necessary. A reduction from 0.05 to 0.025 would require another twenty generations.

Incomplete selection against aa may also occur and can be readily treated mathematically by adding a q_0^2w term to both numerator and denominator of equation (3), where again w is the genetic fitness of aa. Thus

$$q_1 = \frac{2p_o q_o + 2q_o^2 w}{2(p_o^2 + 2p_o q_o + q_o^2 w)}$$

$$= \frac{q_o(1 - sq_o)}{1 - sq_o^2}, \tag{8}$$

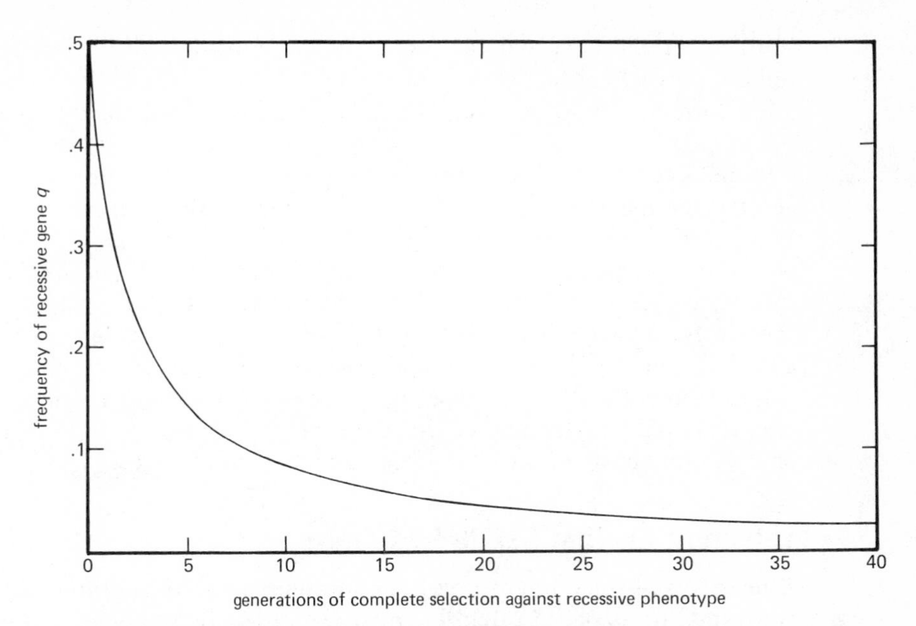

Figure 21-1 Change in gene frequency q of a recessive gene as a result of complete selection against homozygous recessives. The gene frequency approaches zero but theoretically never attains it. Eventually, the number of genes lost through selection will be so few that they may be replaced by mutation. At that point, the population would be in equilibrium.

where $s = 1 - w$. As s, the coefficient of selection, approaches 1, (8) approaches (4). Iteration of this formula will give the new gene frequency for subsequent generations of selection. For $s < 1$, selection is less effective than total selection against the *aa* genotype. The frequency of *a* will decrease nevertheless to the equilibrium value at which removal of *a* by selection and addition of *a* by mutation are exactly balanced, that is,

$$q_e = \sqrt{\mu},$$

where μ is the mutation rate (Chapter 9).

As was noted earlier, the proportion of recessive genes in homozygous combination is equal to the gene frequency q and is therefore low for a deleterious gene. A corollary is that even minor expression of the gene in the heterozygote may overwhelm the selection operating against the homozygote. For example, if $q = 0.01$, then 99% of the alleles would be in heterozygotes. If there were some expression of the recessive allele in heterozygotes so that such persons had only 98% fitness, than as many genes would be lost through heterozygotes as through homozygotes. As q becomes smaller, the situation corresponds to selection against a mildly deleterious rare dominant allele. The detection of very small decrements of fitness is virtually impossible in human populations, and no such systems are known. Similar arguments apply when the heterozygote has greater fitness than the homozygous normal, as will be discussed in the section on balanced polymorphisms.

The situations of selection against completely dominant or recessive genotypes are special cases. A more general statement of selection permits any fitness to be associated with any genotype. For example, when two alleles at a locus are considered, the relative fitness of the three genotypes AA, Aa, and aa may be designated w_1, w_2, and w_3, respectively. Where more than two alleles exist, additional terms can be added as needed. It is theoretically possible for the w's to have any value. In practice the genotype that is most fit is assigned a fitness of one. It should be remembered that the fitness of a genotype is a function of environment and may change as environment changes.

Selection against Sex-linked Traits

The effects of selection on sex-linked traits are different from those on autosomal traits only in that we must recognize the special patterns of transmission resulting from hemizygosity of the X chromosome in males. For traits with a high fitness, one-third of the genes are in males, two-thirds in females. But for recessive traits with a low fitness, a larger portion of genes are in males. This is because affected males do not transmit their X chromosomes, although heterozygous females transmit an X chromosome bearing an abnormal gene equally to sons and daughters.

For rare recessive traits, selection occurs only in males, since homozygous recessive females are virtually nonexistent. Mathematical treatment assumes that heterozygous females are normal. If the fitness of affected males is zero, half of the genes are eliminated each generation because of affected males; the other half are transmitted to females and survive. This is analogous to the situation of an autosomal dominant gene with $w = \frac{1}{2}$. The gene frequency of q after n generations of selection is

$$q_n = q_0(\tfrac{1}{2})^n \tag{9}$$

Selection is not as efficient as for an autosomal dominant trait of zero fitness, but it is efficient nevertheless. The treatment for $w > 0$ is complicated by the fact that the proportion of genes carried by males is itself a function of w.

No dominant sex-linked traits that are highly deleterious are known. The frequency of such a trait would be very rare in any event, since all persons with the gene would be subject to selection. The discussion of the difficulty in recognizing autosomal dominant genes when $w = 0$ also applies to sex-linked dominant genes. The only clue to the genetic nature of such a trait and to its location on the X chromosome would be the 2 : 1 ratio of affected females and males. On the other hand, nongenetic factors could also produce such a ratio by chance.

POLYMORPHIC TRAITS

The word *polymorphic* means existing in multiple forms. As applied to biology, it means that a trait under consideration occurs in two or more forms in a species. Geneticists further restrict the meaning of the word by requiring that the different forms be present in frequencies greater than can be maintained by mutation. In the previous section, we saw that the effects of selection were to reduce the frequency of an allele to a value so low that the allele is maintained in the population only because mutation continually replaces alleles lost through selection. Since mutation is infrequent, traits maintained solely by this mechanism are necessarily infrequent also.

The Prevalence of Genetic Polymorphism

With the development of techniques for electrophoretic separation of enzymes and other proteins, it became apparent that the number of genetic polymorphisms was much greater than previously thought. Studies in natural populations of *Drosophila* indicated a third of randomly chosen loci to be polymorphic for electrophoretic variants (Hubby and Lewontin, 1966; Lewontin and Hubby, 1966). Since nonelectrophoretic variants would have been missed, the actual extent of polymorphism should be higher.

A similar study in man has been reported by Harris and his associates (Harris, 1966; 1969). Using gel electrophoresis, they examined a randomly chosen set of enzymes, representing 71 structural loci, in European populations. Twenty of the loci were polymorphic (Table 21-1). Again, only electrophoretic polymorphisms were detected, so that the true extent of heterozygosity at polymorphic loci should be about three times greater.

The Biological Significance of Polymorphism

Polymorphisms may be classified as transient, neutral, or balanced. A *transient* polymorphism is one in which one allele is in the process of displacing another allele. It is a temporary situation, although if the two alleles are very nearly equal in fitness, a very long time might be required for the new allele to reach fixation. We will include under transient polymorphisms those instances in which a deleterious gene

Table 21-1 **A LIST OF THE 20 ENZYME LOCI FOUND BY ELECTROPHORESIS TO BE POLYMORPHIC IN EUROPEANS***

Enzyme locus	Average heterozygosity
1. Acid phosphatase (red cell)	0.52
2. Phosphoglucomutase: PGM_1	0.36
3. Phosphoglucomutase: PGM_3	0.38
4. Adenylate kinase	0.09
5. Peptidase A	0.37
6. Peptidase C	0.02
7. Peptidase D	0.02
8. Adenosine deaminase	0.11
9. 6-Phosphogluconate dehydrogenase	0.05
10. Placental alkaline phosphatase	0.53
11. Amylase (pancreatic)	0.09
12. Glutamate-pyruvate transaminase (soluble)	0.50
13. Glutamate-oxaloacetate transaminase (mitochondrial)	0.03
14. Galactose-1-phosphate uridyl transferase	0.11
15. Alcohol dehydrogenase: ADH_2	0.07
16. Alcohol dehydrogenase: ADH_3	0.48
17. Pepsinogen	0.47
18. Acetylcholinesterase	0.23
19. Malic enzyme (NADP), mitochondrial	0.30
20. Hexokinase (white cell)	0.05

* A total of 71 randomly selected structural loci not previously studied for electrophoretic polymorphism were investigated. The remaining 51 were not polymorphic. The average heterozygosity is the portion of the population that is heterozygous for alleles at that locus. (From Harris and Hopkinson, 1972.)

achieves polymorphic frequencies through genetic drift or other random processes associated with small populations. A *neutral* polymorphism is one in which the two or more alleles are exactly equal in fitness. The allele frequencies are dependent solely on random events such as genetic drift. A *balanced* polymorphism is one in which two or more alleles are maintained by selection. This may happen in several patterns, all characterized by reduced fitness for certain genotypes at some stage of the life cycle.

It is often very difficult to distinguish among these mechanisms for a particular polymorphism in a real population. One would need to study the change in allele frequencies over many generations. This has not been possible in human generations, and there would be no assurance in any event that fitness would remain equal over long periods of time with changing environment.

In spite of these uncertainties, the study of polymorphisms has attracted much attention from geneticists, both for theoretical and practical reason. From the theoretical side, it is interesting to know to what extent selection operates in maintaining population heterozygosity. On the practical side, if polymorphisms are maintained by selection, to what extent does this contribute to the ill health of the population?

Neutral Polymorphisms

The impetus for believing that neutral polymorphisms may be widespread is based primarily on the difficulty in accepting the operation of selection, i.e. reduced fitness, in maintaining balanced polymorphisms at one-third or more of all the loci. Even if the reduction in fitness per locus is small, when multipled by the large number of loci, the overall fitness would be very low. This problem would be avoided if most polymorphisms resulted from chance increases in a few of the very large number of possible neutral mutations at a structural locus. Rarely, such a neutral mutant might become the prevalent allele and might go to fixation (increase in frequency to become the wild type).

The likelihood of this happening is a function of population size, very large populations being resistant to change by chance phenomena. However, during most of his evolution, man has lived in small tribes, so that such chance fluctuations could have occurred.

Balanced Polymorphisms

A polymorphism is said to be balanced if there exists a stable equilibrium maintained by selection. The simplest situation would be when the heterozygote is more fit than either homozygote; that is, $w_2 > w_1$ and w_3. Under such circumstances there exists a value of q such that any chance deviations in q are counterbalanced by increased numbers of homozygotes. The effect is to restore q to its equilibrium value.

The equilibrium values of p and q can be calculated purely as a function of the fitness of the various genotypes. For example, starting with a genotype frequency distribution p^2, $2pq$, and q^2, after selection the respective frequencies are w_1p^2, $2w_2pq$, w_3q^2. The allele frequencies in the next generation will be

$$p' = \frac{2w_1p^2 + 2w_2pq}{2(w_1p^2 + 2w_2pq + w_3q^2)};$$

$$q' = \frac{2w_2pq + 2w_3q^2}{2(w_1p^2 + 2w_2pq + w_3q^2)}.$$

The ratio of the allele frequencies is

$$\frac{p'}{q'} = \frac{p(w_1p + w_2q)}{q(w_2p + w_3q)}.$$

At equilibrium, there will be no further change in p and q. Therefore $p' = p$ and $q' = q$. After rearrangement, the equation becomes

$$w_1p + w_2q = w_2p + w_3q,$$

from whence, remembering that $q = 1 - p$, the equlibrium value is obtained:

$$p_e = \frac{(w_2 - w_3)}{(w_2 - w_3) + (w_2 - w_1)}.$$

If w_2 is assigned a value of 1, the equation can be more simply stated in terms of selection coefficients, where $s = 1 - w$:

$$p_e = \frac{s_3}{s_1 + s_3}.$$

Similarly,

$$q_e = \frac{s_1}{s_1 + s_3},$$

and

$$\frac{p}{q} = \frac{s_3}{s_1}.$$

If the two homozygotes are equal in fitness ($s_1 = s_3$), then $p_e = q_e = 0.5$, whatever the actual values of s_1 and s_3. If $s_3 = 2s_1$, then $p_e = 0.67$ and $q_e = 0.33$. Any chance deviation from these values will be counterbalanced by selection to restore the gene frequencies to the equilibrium values.

Perhaps the best explored examples of balanced polymorphism are in the abnormal hemoglobins, particularly sickle cell hemoglobin. Persons homozygous for Hb S have very low fitness, few surviving to adulthood ($s_3 \cong 1$). On the other hand, heterozygotes seem to have some protection from malaria. In some populations of Africa, the frequency of the Hb S gene is such that, if the populations have in fact achieved a balanced equilibrium, persons homozygous for normal Hb A are only 75% as fit as those heterozygous for Hb S ($s_1 = 0.25$).

Unstable Polymorphisms

In the special situation in which the heterozygote is less fit than either homozygote, equilibrium frequencies may occur, but any chance deviation leads to selection in favor of one allele. The best known example is Rh incompatibility. For purposes of illustration, the rather complex Rh genetics may be simplified to a two-allele system, *R* and *r*, the genotypes *RR* and *Rr* being Rh+, and *rr* being Rh−. Mothers who are Rh− sometimes form antibodies against Rh+ fetuses, giving rise to hemolytic disease of the newborn. Such a baby can arise from two types of matings: ♂ *RR* × ♀ *rr* and ♂ *Rr* × ♀ *rr*. In either case, in order for the child to be "incompatible" with the mother, it must have an *R* allele and thus the genotype *Rr*. Failure to survive would lead to loss of equal numbers of *R* and *r* alleles. If the population frequencies of *R* and *r* were 0.5, then loss of *Rr* children would not alter the frequency of either allele. If the frequencies were other than 0.5, loss of *Rr* persons would cause a greater relative loss of the less frequent allele. Eventually, the less frequent allele should become rare. A system is said to be in equilibrium when selection does not alter the gene frequencies. This is an unstable equilibrium, however, since any deviation will lead to loss of one allele. The exact point of equilibrium will depend on the relative advantages of the two homozygous genotypes. The *Rh* locus is characterized by very high and very low frequencies of *r* in different populations of the world. It has been suggested that these populations were in the process of eliminating *R* or *r* alleles, but modern migrations of populations have changed the picture.

The Rh system also illustrates another possibility in selection. The fitness of the heterozygote varies with the gene frequencies. The portion of *Rr* offspring originating from *rr* mothers decreases as the frequency of *r* decreases. With very high frequencies of *r*, most *Rr* children would come from *RR* × *rr* matings, in half of which the mother would be the Rh− parent. These cases would have the potential of producing erythroblastotic babies. If *r* is rare, most *Rr* persons would come from *RR* × *Rr* matings, which do not produce babies with hemolytic disease of the newborn.

SELECTION IN HUMAN POPULATIONS

The operation of selection in many inherited traits is obvious. A large portion of the long list of inherited diseases is associated with decreased viability or fertility. The frequencies of many of these diseases are very low, suggesting that selection against them is effective and that they are maintained by mutation pressure alone. Intervention by modern medicine has done little to change the overall picture. In other instances, selection appears to be operating strongly against certain genotypes, but the gene frequencies are much too high to be maintained by mutation pressure. The question arises in these instances as to the possibility of balanced selection. Also, many of the polymorphisms, such as those observed with blood groups and serum proteins, occur worldwide. These appear to be neutral polymorphisms, but might some of them be maintained as balanced polymorphisms with low selection coefficients?

Some of the complexities of defining fitness were pointed out at the beginning of the chapter. The measurement of fitness is equally difficult. Ideally one might start with panels of newborns of the genotypes to be compared and follow them to see how many offspring each produced. Few investigators are so patient as to commit themselves to such a long-term study. Rarely it may be possible to construct such groups from past records, but these usually are less than perfect. In spite of the problems, some of the puzzling inherited traits have been successfully studied, and it will be instructive to consider the approaches used in several.

Huntington's Disease. This late-onset dominant disease stimulated interest in terms of natural selection for several reasons. Even though it is relatively rare, affecting approximately 1 in 10,000 persons in the United States, an early erroneous report suggested that *all* cases were descended from two brothers and hence the frequency of the gene must be increasing. Many persons do not express the gene until after the usual reproductive period, and one of the frequent expressions is sexual licentiousness. Finally, a study in Minnesota had reported that affected persons have more offspring than their nonaffected sibs.

Reed and Neel (1959) did an exhaustive study of Huntington's disease in the state of Michigan, identifying 231 such persons. For studies of reproductive fitness, it is common to use sibs as controls, thus correcting for socioeconomic level, genetic background, and so on. Reed and Neel also used as a comparison group the general population of Michigan. They found that the relative fitness of affected heterozygotes compared with that of their normal sibs was 1.01. The fertility of affected males was less than that of affected females. In comparison with fitness in the general population, however, the relative fitness was only 0.81. Thus within families the affected persons are not at a disadvantage. However, the presence of the disease in a family influenced both those with and those without the gene.

Such a study represents the findings in a particular culture at a particular time. As ideals of family size change and as contraception and abortion become more widely used, the fitness of the Huntington's disease gene might well change also. Were an early test for Huntington's disease to become available, this would undoubtedly reduce the relative fitness substantially.

Cystic Fibrosis. This disease has been an enigma because of its low fitness and its high frequency in European populations. The reproductive fitness of homozygotes is virtually zero, but in the United States white population, the incidence of affected persons is 1 in 2500 births, a rate thought to be too high for a recessive lethal unless heterozygotes have some undetected advantage. The frequency is much lower in non-European populations. To maintain a gene frequency for cystic fibrosis of $q = \sqrt{.0004} = 0.02$ would require a selection coefficient $s_1 = qs_3/p = (.02)(1.0)/(.98) = 0.02$. In other words, a 2% advantage of heterozygotes as compared to homozygous normal persons would account for the high frequency of cystic fibrosis.

Two groups (Danks and co-workers, 1965; Knudson and co-workers, 1967) have examined this problem in families in which cystic fibrosis had occurred compared with control families. The sizes of the sibships of the parents were compared, since these would have resulted for the most part from grandparental matings of the type $AA \times Aa$, where a is the allele for cystic fibrosis. Furthermore, since cystic fibrosis would not have occurred yet, any differences between matings that subsequently proved to have the a allele and those that did not could not be attributed to knowledge of the disease.

The results of one study are shown in Table 21-2. On the average, matings of type $AA \times Aa$ have nearly one additional child compared

Table 21-2 **COMPARISON OF NUMBERS OF LIVE OFFSPRING OF GRANDPARENTS***

Number of offspring per family	CF families		Control families	
	Number	Total sibs	Number	Total sibs
1	9	9	15	15
2	20	40	37	74
3	33	99	23	69
4	15	60	15	60
5	21	105	11	55
6	10	60	7	42
7	10	70	2	14
8	6	48	2	16
9	—	—	3	27
10	3	30	2	20
11	1	11	—	—
12	2	24	—	—
13	1	13	—	—
14	—	—	1	14
Total	131	569	118	406
Mean number of offspring		4.34		3.43
Variance		6.07		5.12
Variance of mean		0.0463		0.0434
Difference of means			0.91	
Sum of variances of means			0.0897	
Standard error of difference of means			0.30	
$\frac{\text{Difference of means}}{\text{Standard error}}$			3.00 ($P < 0.01$)	

* From Knudson and co-workers, 1967.

with $AA \times AA$ matings. This corresponds to a selection against homozygous normals of 0.21. While this may by chance be high, the fact that both studies gave evidence of greater reproductive fitness strongly supports the hypothesis that cystic fibrosis is a balanced polymorphism. If the selection coefficient s_1 is as high as 0.10 the corresponding gene frequency $q = s_1/(s_1 + s_3) = 0.10/(0.10 + 1.0) = 0.09$, and the frequency of the disease should be $(0.09)^2 = 0.0081$, or twenty times higher than at present. This has led to the suggestion that the heterozygote advantage may be of fairly recent origin in terms of generations and the equilibrium has yet to be attained.

Tay-Sachs Disease (TSD). This disease presents much the same problem for Ashkenazi Jewish populations that cystic fibrosis does for Europeans. In a study by Myrianthopoulos and Aronson (1966), evidence was obtained that heterozygotes for TSD may have greater reproductive fitness than homozygous normal controls. It is especially interesting that only one of the three major Jewish groups that dispersed in the first century A.D. has a high frequency of the TSD gene. One possible explanation is that the Ashkenazi Jews, who have lived nearly 2000 years in urban ghettos, have been exposed to different selective pressures than the other Jewish groups or than the surrounding gentiles. Such pressures might include crowding and infectious disease, but many other differences also must have existed. Supporting the selection hypothesis, Myrianthopoulos and Aronson note that Gaucher's disease and Niemann–Pick disease also have a high incidence in Ashkenazi Jews and physiologically have many similarities to TSD, although they are genetically distinct.

Blood Groups and Disease. The blood group polymorphisms have been the subject of much speculation concerning selective factors that might operate to maintain them. Maternal-fetal incompatibility, which occurs to some extent with all of them, operates against heterozygotes. Since the biological function of blood groups is not understood, it is difficult to devise selective mechanisms that would account for present distributions of ABO groups, MN groups, and so on.

One approach that has been used to search for evidence of selection is association with diseases. The studies have been done in various ways, but usually the frequencies of ABO types, for example, in persons with a particular disease are compared to the frequencies in their normal sibs or in a population matched as closely as possible for race, socioeconomic level, and so on. The first such association was reported in 1953 between ABO groups and cancer of the stomach, persons of type A having a greater likelihood of having the cancer. Subsequently, an association between type O and peptic ulcer was reported, also an association between type A and pernicious anemia. These studies have been replicated in many parts of the world, with generally consistent results. Clarke (1961) has summarized many of the studies and quotes a relative risk of duodenal ulcer in type O persons of 1.38 as compared with a risk of 1.0 in other blood types. For gastric ulcer the risk is 1.20 in type O persons. An association with secretor type also has been noted, and the risk of duodenal ulcer in group O nonsecretors is 2.49 times that in A, B, and AB secretors.

Still other associations have been reported but with more meager data. To what extent these associations have anything to do with selection is unclear, but one would expect small effect in general. The diseases are "adult" diseases, occurring later in life and interfering very little with reproductive fitness. Thus it is difficult to see how these particular disease associations could have much influence on gene frequencies. Of course, it may be that what is expressed as differential risk for late onset diseases also has other early effects that do influence fertility.

A somewhat different approach to blood groups and disease has been taken by F. Vogel. There was evidence that the smallpox virus cross reacts immunologically with blood group A. He reasoned that persons of blood group A would therefore manufacture antibodies against smallpox less effectively than would persons lacking the A antigen because of the phenomenon of immune tolerance. Several studies have been carried out by Vogel and his associates (for example, see Vogel and Chakravartti, 1966), primarily in India where smallpox existed as an endemic disease. In general the hypothesis has been confirmed. If correct, smallpox could have been a powerful selective agent against the *A* allele. This in itself would not explain the polymorphism as such, but it could be one part of the complex of selective factors. Until the development of antibiotics, infectious disease was a major cause of early death and probably a major selective agent. The epidemics of smallpox and plague could easily have altered the gene frequencies of loci producing differential susceptibility.

Malaria and Selection. One of the major selective agents in some populations has been malaria. There are four main types of malaria. All occur in the tropical regions, with the predominant malarial parasite being *Plasmodium vivax*, the only one that also occurs in temperate zones. A more severe form of malaria is produced by *P. falciparum*, which occurs only in tropical and subtropical regions. Falciparum malaria is often fatal. All forms of malaria depend on the mosquito for an obligatory part of the life cycle; hence, malaria occurs only where mosquitos can breed. Part of the life cycle also includes invasion of red cells. The red cells are then ingested by other mosquitos to continue the life cycle.

Several genetic systems appear to have mutant alleles that make red

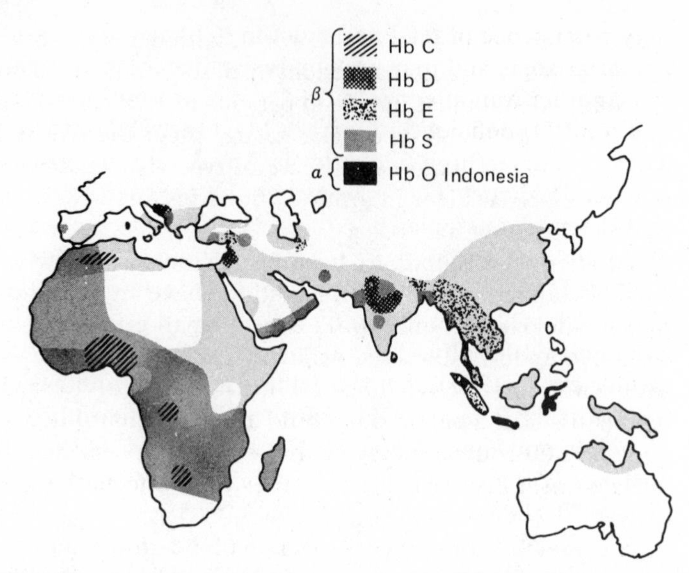

Figure 21-2 Map showing the distribution of the major hemoglobin variants that are polymorphic. The light gray areas indicate presence of falciparum malaria. (*From Salthe, 1972.*)

cells more resistant to malarial parasites. This causes the alleles to have a selective advantage and to increase in frequency. The best known is sickle cell anemia (Chapter 10). The red cells of persons heterozygous for Hb S are more resistant to malarial parasites than are cells of persons homozygous for Hb A. Under conditions of low oxygen supply, such as may exist in venous blood, the *P. falciparum* parasites are killed in Hb AS cells but in Hb A cells (Friedman, 1978). Sickle cell carriers therefore suffer fewer ill effects as compared to homozygous A persons. This seems to be especially important for falciparum malaria, and the distribution of Hb S is similar to the distribution of falciparum malaria in Africa (Figure 21-2).

Hb C also appears to confer resistance to malaria. This allele has very restricted distribution, with a maximum frequency in Upper Volta. Persons homozygous for Hb C are anemic, but they do not die in childhood as do persons homozygous for Hb S. The Hb C gene therefore has some of the advantages of the Hb S gene but not all the disadvantages. Judging from its distribution, it is now in the process of displacing Hb S in those populations in which Hb S offers an advantage. This hypothesis presumes that the Hb C mutation occurred only once, perhaps in the region now known as Upper Volta, and has spread from that point.

The Hb E allele is especially prevalent in Southeast Asia, as is Hb Constant Spring. It is thought that malaria is responsible for the high frequencies of these alleles also. The thalassemias, including hereditary persistence of fetal hemoglobin (Chapter 10), are common only in malarial areas and may provide resistance for heterozygotes.

Another trait affecting red cells is glucose-6-phosphate dehydrogenase (G6PD) deficiency (Chapter 11). One allele of this X-linked trait, Gd^{A-}, occurs in high frequency in Africa. Another allele involving even greater deficiency, Gd^{Med}, occurs in the Mediterranean areas. The obvious detriment associated with these alleles under some circumstances is possibly balanced by increased resistance to malaria.

The Duffy blood group has three alleles, Fy^a, Fy^b, and a null allele, *Fy*. The last allele is generally in low frequency except in Africans, who are almost 100% *Fy*. This is possibly due to the fact, demonstrated by Miller et al. (1976), that the Duffy antigenic sites on red cells are the receptors for *P. vivax*. Of five Fy(a−b−) persons deliberately exposed to *P. vivax,* none got malaria. Of 12 positive for Duffy a or b, all developed malaria. It is thus easy to understand the very low frequency of Fy^a and Fy^b in Africa and in African descendants.

Intelligence. Much concern has been expressed over the decades that differential reproductive patterns associated with intelligence might reduce the average intelligence of the population. Several studies showed that parents with higher IQs tended to have smaller families than parents with low IQs. Therefore, the high IQ persons would contribute relatively less to the gene pool of the succeeding generation.

Bajema (1963) successfully challenged this idea by following up through their reproductive years a group of children given intelligence tests in the sixth grade in 1916 and 1917. The 979 persons gave the results in Table 21-3. These results clearly are contrary to the idea of a negative association between intelligence and fertility. The discrepancy

Table 21-3 **RELATIVE REPRODUCTIVE FITNESS VERSUS INTELLIGENCE FOR PERSONS TESTED AS CHILDREN AND FOLLOWED THROUGH THE REPRODUCTIVE PERIOD***

IQ range	Number	Number of offspring per person	Relative fitness	Percent leaving no offspring
≥120	82	2.598	1.0000	13.41
105–119	282	2.238	0.8614	17.02
95–104	318	2.019	0.7771	22.01
80–94	267	2.464	0.9484	22.47
69–79	30	1.500	0.5774	30.00
Total	979	2.236		20.22

* Data from Bajema, 1963.

with other studies may be due in part to failure in the other studies to include several important groups. Many of the studies start with families, thereby eliminating persons who never married or who were childless. In a demographic sense, the person who is shipped off to an institution should remain a member of the community from which he was shipped, whether or not he has an identifiable residence there. Persons who die prior to completion of the reproductive period should also be counted. By starting with school children, Bajema was able to include these persons, although even his study missed those who were unable to make it to the sixth grade. A much higher proportion of persons with low IQ leave no offspring at all.

From this and other studies it seems clear that higher intelligence is associated with greater reproductive fitness. This need not be true under all circumstances, since higher IQ persons may be more likely to decrease the family size in response to the burgeoning population. As culture changes, additional studies will be necessary to answer this question.

Schizophrenia. While the inheritance of schizophrenia is yet to be resolved, it is agreed by most that genetic components contribute to the disease. These should be subject to natural selection. Schizophrenia is a very maladaptive disease, and persons with it are less likely to marry and are likely to be institutionalized during a part of the reproductive period. Yet some 1% of the population is estimated to be schizophrenic. This discrepancy suggests that nonaffected carriers of the schizophrenic genes may have an advantage, thus maintaining this "polymorphism."

Erlenmeyer-Kimling and Paradowski (1966) have reviewed the studies on fitness in schizophrenia and have contributed data from a very large study in New York. The study population consisted of two groups, one admitted to state hospitals in 1934–1936, the other admitted in

Table 21-4 **FERTILITY, EXPRESSED AS NUMBER OF CHILDREN PER WOMAN, FOR SCHIZOPHRENICS, THEIR SIBLINGS AND THE GENERAL POPULATION***

	1934–1936			1954–1956		
Measures	Siblings	Schizophrenics	General population	Siblings	Schizophrenics	General population
Children per woman	1.2	0.7	1.2	2.2	1.3	1.5
Reproductivity ratios of schizophrenics	0.6 (siblings vs. schizophrenics)		0.6 (schizophrenics vs. general population)	0.6 (siblings vs. schizophrenics)		0.9 (schizophrenics vs. general population)

* From Erlenmeyer-Kimling and Paradowski, 1966.

1954–1956. A total of 3337 cases were included. The results are summarized in Table 21-4. The sibs of the earlier patients reproduced at the same rate as the general population. Since the sibs would be expected to be carriers of some of the schizophrenia genes, they should have a high reproductive fitness if the hypothesis is correct. Indeed, this is true in the later series. In both, the schizophrenic patients are less fertile than their sibs or the general population. The problem is thus not clearly resolved. It may well so remain until the inheritance is more clearly understood.

SUMMARY

1. Natural selection occurs when there is inequality in fitness of genotypes, measured solely by production of offspring. Selection is a function of genotype-environment interactions and can change as either of these changes. Selection may occur in meiosis (meiotic drive), during the gamete phase, or postzygotically.

2. Selection against dominant traits is very efficient, since every person with the allele would be subject to selection. In contrast, selection against recessive traits is very inefficient at low gene frequencies, since most of the alleles are carried in heterozygotes. A very small expression of the recessive gene in heterozygotes may alter greatly the relative importance of heterozygotes and homozygotes, since so many more alleles are in heterozygous combination. Selection against sex-linked traits is efficient, since one-third or more of the alleles would be expressed in hemizygous males, irrespective of gene frequency.

3. Polymorphic traits are those in which the least frequent form occurs more frequently than can be explained by recurrent mutation. Studies in *Drosophila* and man of electrophoretic variants of proteins indicate a third of all loci to be polymorphic. The actual frequency, including nonelectrophoretic variants, is still greater. Polymorphisms may be transient, neutral, or balanced. In a transient polymorphism, one allele is in the process of displacing another associated with less fitness. In a neutral polymorphism, the two or more alleles are of equal fitness and are maintained together by chance. In a balanced polymorphism, the heterozygote is superior to either homozygote, resulting in equilibrium gene frequencies that reflect the relative fitness of the homozygotes. Sickle cell anemia has been suggested as a balanced polymorphism, the heterozygote being superior to homozygous normals in resistance to malaria. Unstable polymorphisms may exist if the heterozygote is less fit than either homozygote. In this case there may be an equilibrium gene frequency, but chance deviation from it will lead to fixation of one of the alleles.

4. Most inherited diseases occur at low frequency, suggesting that selection against them is effective. For example, Huntington's disease is an autosomal dominant affecting 1 in 10,000 persons in the United States. Although affected persons are fertile, in one large study the average fitness was only 0.81 compared with the general population. Fitness in this instance would likely change with development of early diagnostic procedures and with attitudes on family size.

5. Cystic fibrosis is a recessive disease with fitness near zero. However, the frequency of affected persons is 1 in 2500, corresponding to a gene frequency of 0.02. If this is a balanced polymorphism, it could be maintained by an advantage of only 2% on the part of heterozygotes as compared with homozygous normal persons. Tay–Sachs disease in Ashkenazi Jews may also be (or have been) maintained by heterozygote advantage.

6. Studies of associations between blood groups and diseases have shown persons of type A to have a greater risk of stomach cancer and pernicious anemia, while persons of type O more often have duodenal or gastric ulcer. There also is evidence that persons of type A are more susceptible to smallpox. The importance of natural selection in these associations has not been established.

7. Malaria has been a selective agent in Africa, the Mediterranean, and tropical and subtropical Asia. A variety of red cell traits are thought to have been increased in frequency because of associated resistance to malaria. These include Hb's S, C, E, and Constant Spring, thalassemia (including hereditary persistence of fetal hemoglobin), G6PD deficiency, and the Duffy *Fy* allele.

8. Concern has been expressed that the greater fertility of low IQ persons would result eventually in a lower average intelligence. A more thorough study shows that, in fact, persons with low IQ are less fertile than those of normal intelligence if all persons, including those who never marry, are considered.

9. The high frequency of schizophrenia, affecting some 1% of the population, has been an enigma. Schizophrenia appears to have a strong genetic component. Yet affected persons clearly do not reproduce at a rate comparable to normal persons. On the other hand, sibs of schizophrenics reproduce at the same rate as unrelated normal persons, suggesting no advantage for schizophrenia genes in normal persons.

REFERENCES AND SUGGESTED READING

ALLISON, A. C. 1954. Protection afforded by sickle-cell trait against subtertian malarial infection. *Br. Med. J.* 1: 290–294.

BAJEMA, C. J. 1963. Estimation of the direction and intensity of natural selection in relation to human intelligence by means of the intrinsic rate of natural increase. *Eugenics Quart.* 10: 175–187.

BAJEMA, C. J. (Ed.). 1971. *Natural Selection in Human Populations.* New York: John Wiley, 406 pp.

CLARKE, C. A. 1961. Blood groups and disease. *Prog. Med. Genet.* 1: 81–119.

DANKS, D. M., ALLAN, J., AND ANDERSON, C. M. 1965. A genetic study of fibrocystic disease of the pancreas. *Ann. Hum. Genet.* 28: 323–356.

ERLENMEYER-KIMLING, L., AND PARADOWSKI, W. 1966. Selection and schizophrenia. *Amer. Naturalist* 100: 651–665.

EWENS, W. J. 1977. Population genetics theory in relation to the neutralist-selectionist controversy. *Adv. Hum. Genet.* 8: 67–134.

FISHER, R. A. 1958. *The Genetical Theory of Natural Selection,* 2nd Ed. Dover, New York. 291 pp.

FRIEDMAN, M. J. 1978. Erythrocytic mechanism of sickle cell resistance to malaria. *Proc. Natl. Acad. Sci.* (U.S.A.) 75: 1994–1997.

GERSHOWITZ, H., BEHRMAN, S. J., AND NEEL, J. V. 1958. Hemagglutinins in uterine secretions. *Science* 128: 719–720.

HALDANE, J. B. S. 1961. Natural selection in man. *Prog. Med. Genet.* 1: 27–37.

HARRIS, H. 1966. Enzyme polymorphisms in man. *Proc. Roy. Soc. B.* 164: 298–310.

HARRIS, H. 1969. Genes and isozymes. *Proc. Roy. Soc. B.* 174: 1–31.

HARRIS, H. 1970. *The Principles of Human Biochemical Genetics.* Amsterdam-London: North-Holland, pp. 225–232.

HARRIS, H., AND HOPKINSON, D. A. 1972. Average heterozygosity per locus in man: an estimate based on the incidence of enzyme polymorphisms. *Ann. Hum. Genet.* 36: 9–20.

HUBBY, J. L., AND LEWONTIN, R. C. 1966. A molecular approach to the study of genic heterozygosity in natural populations. I. The number of alleles at different loci in *Drosophila pseudoobscura. Genetics* 54: 577–594.

KIMURA, M., AND OHTA, T. 1969. The average number of generations until fixation of a mutant gene in a finite population. *Genetics* 61: 763–771.

KING, J. L., AND JUKES, T. H. 1969. Non-Darwinian evolution. *Science* 164: 788–798.

KNUDSON, A. G. JR., WAYNE, L., AND HALLETT, W. Y. 1967. On the selective advantage of cystic fibrosis heterozygotes. *Amer. J. Human Genet.* 19: 388–392.

LEWONTIN, R. C., AND HUBBY, J. L. 1966. A molecular approach to the study of genic heterozygosity in natural populations. II. Amount of variation and degree of heterozygosity in natural populations of *Drosophila pseudoobscura. Genetics* 54: 595–609.

MILLER, L. H., MASON, S. J., CLYDE, D. F., AND MC GINNISS, M. H. 1976. The resistance factor to *Plasmodium vivax* in blacks. *N. Engl. J. Med.* 295: 302–304.

MYRIANTHOPOULOS, N. C., AND ARONSON, S. M. 1966. Population dynamics of Tay-Sachs disease. I. Reproductive fitness and selection. *Am. J. Hum. Genet.* 18: 313–327.

REED, T. E. 1959. The definition of relative fitness of individuals with specific genetic traits. *Am. J. Hum. Genet.* 11: 137–155.

REED, T. E., AND NEEL, J. V. 1959. Huntington's chorea in Michigan. 2. Selection and mutation. *Am. J. Hum. Genet.* 11: 107–136.

ROBERTS, D. F., AND HARRISON, G. A. (Eds.). 1959. *Natural Selection in Human Populations.* New York: Pergamon, 76 pp.

SALTHE, S. N. 1972. *Evolutionary Biology.* New York: Holt, Rinehart and Winston, 437 pp.

VOGEL, F., AND CHAKRAVARTTI, M. R. 1966. ABO blood groups and smallpox in a rural population of West Bengal and Bihar (India). *Humangenetik* 3: 166–180.

REVIEW QUESTIONS

1. Define:

fitness
selection
meiotic drive
prezygotic selection
polymorphism

2. *A* is a dominant normal gene; *a* is its recessive allele that is lethal when homozygous.

a. Starting with an initial gene frequency for *a* of 0.1, what will be the frequency after ten generations?

b. How many generations will be required to reduce the frequency to one-fifth its initial value?

3. The complete selection against individuals homozygous for hemoglobin S is counterbalanced in many areas by the relative advantage enjoyed by the heterozygote, who is more resistant to malaria than homozygous normal individuals. With the control of malaria, this advantage should no longer be present. In a population in which 20% of the adult population are heterozygotes, what will be the *gene* frequency of Hb S after one generation of malaria control? After ten generations? After 100 generations? How many generations will be required to reduce the frequency to 0.001?

4. The gene frequency of G6PD deficiency (an X-linked recessive trait) is 15% in U.S. blacks. If the fitness of males with this trait were suddenly reduced to 50% of normal, what portion of all the Gd^- alleles would be lost in one generation? What portion would be lost in the next generation?

5. In a certain balanced polymorphic system, the fitness of *AA* is 70% that of the heterozygote and the fitness of *aa* is 10%. At equilibrium, what is the frequency of *a*?

CHAPTER TWENTY-TWO

INBREEDING

Marriages of persons who are related through a common ancestor are consanguineous and their offspring are inbred. Since there are not enough ancestors for each of us to have arisen independently, most marriages are consanguineous. For example, each person has two parents, four grandparents, eight great-grandparents, and so on. The number of ancestors at any preceding generation is given by 2^n, where n is the number of generations. For $n = 32$ (ca. A.D. 1000 for ancestors of the present generation), the number of ancestors would be 4,294,967,-296. Since this figure is larger than the entire population of the world today, some persons served as ancestors through more than one line of descent.

Such remote consanguinity is of little genetic interest. So long as there is random mating and the size of the population is sufficient to maintain a large gene pool, the unavoidable remote consanguinity does little to alter the expected frequencies of gene combinations. The mating of close relatives is important, however, since it leads to a significant increase in homozygosity of offspring. To the extent that homozygosity for genes is deleterious, consanguineous marriages are deleterious. However, the deleterious effects are not shared equally by all offspring of consanguineous marriages, since the effects depend on homozygosity for specific alleles. Some matings show no deleterious effects; others show severe effects.

Rarely in history, particular families, usually of the ruling class, have practiced very close consanguinity over several generations. The best known example is found among the pharaohs of Egypt, where brother-sister matings were the rule. In spite of this apparent close inbreeding the progeny recorded seem often to have been normal. Recent more fully documented cases of brother-sister matings yielded a high proportion of defective children (Adams and Neel, 1967).

THE INCREASE IN HOMOZYGOSITY IN CONSANGUINEOUS MARRIAGES

The most important consanguineous marriages, as measured by the number of marriages and the magnitude of consanguinity effect, are marriages of first cousins. Even in cultures or special groups within cultures in which such marriages are frowned on, they do occur, sometimes by special dispensation.

The frequency of cousin marriages varies both with the population structure and the cultural concepts of incest. In a population in which mobility is low, and particularly if the population is partitioned into small social groups—a situation found in societies composed of small villages—suitable mates are often related. Urbanization and increased mobility tend to decrease the proportion of consanguineous marriages. In some cultures, such as Japan, cousin marriages traditionally have been regarded with favor. Table 22-1 is a summary of the frequency of cousin marriages in several cultures.

Consanguinity and Marriages of First Cousins

Formulas for treating the effects of consanguinity quantitatively have been developed, primarily by Dahlberg. Only autosomal genes will be considered, since genes on X chromosomes form a special situation. Males, with one X chromosome, cannot be inbred with respect to X-linked genes. Females can, and many of the relationships of autosomal genes apply also to X chromosomes of females. However, the strong selection exerted against deleterious genes on the X chromosome in hemizygous males reduces the problem of inbreeding for females.

First cousin marriages will be considered primarily in this discussion, since the effect of inbreeding is much greater in these than in

Table 22-1 **FREQUENCY OF FIRST COUSIN MARRIAGES IN SEVERAL POPULATIONS***

Population	Frequency, in percent
Japan, 3 cities	4.13
England	.61
U. S., rural	.00
Brazil, São Paulo	.66
Germany, Münster	.08
India, Marathas	10.00

* From Morton, 1961.

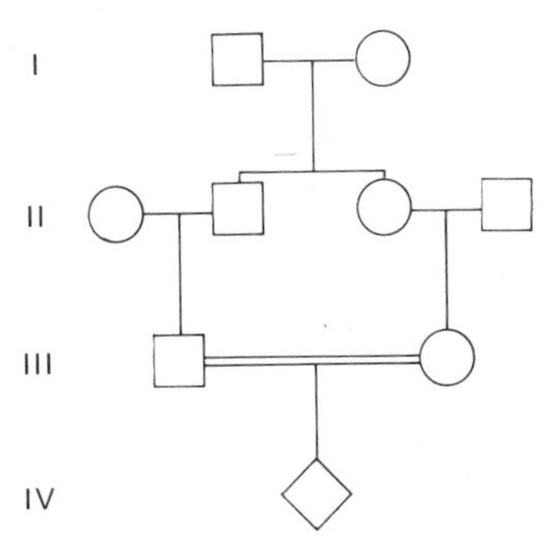

Figure 22-1 Diagram of first cousin marriage. In this pedigree, the husband is related to his wife through his father's sister. There are three other possible paths of relationship—through his father's brother, his mother's sister, and his mother's brother. For autosomal genes, these relationships are equivalent. For genes on the X chromosome, they are not. The frequencies of the four types of cousin marriages vary among different cultures.

more distant consanguineous marriages. Figure 22-1 is a diagram of a first cousin marriage. The probability that a particular autosomal allele in one parent of generation I will be transmitted to the great-grandchild of generation IV is $\frac{1}{8}$. The probability that the same allele will be transmitted through both lines of descent is $\frac{1}{8} \times \frac{1}{8} = \frac{1}{64}$. However, there are four alleles (two parents with two alleles each) at any locus in generation I, any of which might be transmitted with a probability of $\frac{1}{64}$. Therefore, the total probability that both alleles of generation IV are derived from a single allele of generation I is $4 \times \frac{1}{64} = \frac{1}{16}$. The probability that both alleles of generation IV had a different origin is $1 - \frac{1}{16} = \frac{15}{16}$.

For an allele a, the probability that the offspring in generation IV is homozygous aa because of common origin of a is $(\frac{1}{16})q$, where q is the frequency of a. The probability of homozygosity if both alleles had a different origin is $(\frac{15}{16})q^2$. Adding these together, the total probability of homozygosity, $P(aa)$ for a is

$$\begin{aligned} P(aa) &= (\tfrac{15}{16})q^2 + (\tfrac{1}{16})q \\ &= q^2 + (\tfrac{1}{16})pq. \end{aligned}$$

Through similar reasoning, it can be shown that $P(AA) = p^2 + (\frac{1}{16})pq$. However, the frequency of heterozygotes, $P(Aa)$, is decreased by an amount equal to the increase in homozygotes. Thus

$$P(Aa) = (\tfrac{15}{16})2pq = 2pq - (\tfrac{2}{16})pq.$$

Table 22-2 shows the relative frequencies of homozygous recessives among offspring of first cousin marriages and among offspring of unrelated persons for selected values of q. If q is large, there is very little difference in frequency of aa; if q is very small, the likelihood of aa is relatively much larger for children of first cousin marriages.

Table 22-2 **FREQUENCIES OF HOMOZYGOUS RECESSIVE OFFSPRING FROM MARRIAGES OF UNRELATED PERSONS AND FROM FIRST COUSIN MARRIAGES FOR VARIOUS VALUES OF THE GENE FREQUENCY**

Frequency of recessive gene q	Frequency of *aa* offspring from unrelated parents q^2	Frequency of *aa* offspring from first cousin marriages $q^2 + (\frac{1}{16})pq$	Ratio of *aa* offspring: related to unrelated
.5	.25	.2656	1.06
.2	.04	.0500	1.25
.1	.01	.0156	1.56
.05	.0025	.00547	2.19
.02	.0004	.00163	4.08
.01	.0001	.000719	7.19
.005	.000025	.000336	13.44
.002	.000004	.000129	32.25
.001	.000001	.000063	63.00

The proportion, k, of first cousin parents of *aa* offspring among all parents of *aa* offspring can be calculated for a given q, provided the frequency of first cousin marriages in the general population is known. This is given by

$$k = \frac{c(q + 15q^2)/16}{(1 - c)q^2 + c(q + 15q^2)/16},$$

$$= \frac{c(1 + 15q)}{c(1 - q) + 16q},$$

where c is the frequency of first cousin marriages. The above formula can be rearranged to

$$q = \frac{c(1 - k)}{16k - 15c - ck},$$

which permits calculation of the gene frequency q in terms of k and c only.

Coefficient of Inbreeding

A general formula for expressing the degree of inbreeding was developed by Sewall Wright. The coefficient of inbreeding, F, is the probability that both alleles in a person were derived from a single allele in some common ancestor. For nonconsanguineous marriages, $F = 0$ in the offspring. For offspring of a first cousin mating, $F = \frac{1}{16}$, as has been shown in the previous section. The formula for F is

$$F = \Sigma(\tfrac{1}{2})^N,$$

where N is the number of persons in the path of relationship, starting with one parent, counting back to the common ancestor, then continuing down the line of descent to the other parent. The component values must be calculated for each common ancestor and summed to give F. In the usual first cousin marriage, there are two common ancestors, each contributing $(\frac{1}{2})^5$ to the value of F (Figure 22-2). More complex relationships are readily handled with this formula.

If a common ancestor is himself inbred, there is increased likelihood that the alleles in an inbred descendant will be identical in origin. This can be evaluated by the more general expression,

$$F = \Sigma(\tfrac{1}{2})^N(1 + F_\alpha),$$

where F_α is the inbreeding coefficient of the ancestor.

The increase in homozygosity can be generalized in terms of F, since, in the specific case of first cousin marriages, the $\frac{1}{16}$ probability of identical origin of two alleles is merely the coefficient of inbreeding. Thus, in more general terms,

$$\begin{aligned} P(AA) &= (1 - F)p^2 + Fp, \\ P(Aa) &= (1 - F)2pq, \\ P(aa) &= (1 - F)q^2 + Fq. \end{aligned}$$

These formulas permit calculation of inbreeding effect for any degree of parental relationship.

It is sometimes useful to calculate the average inbreeding coefficient for a population. This is expressed as

$$\alpha = \Sigma p_i F_i,$$

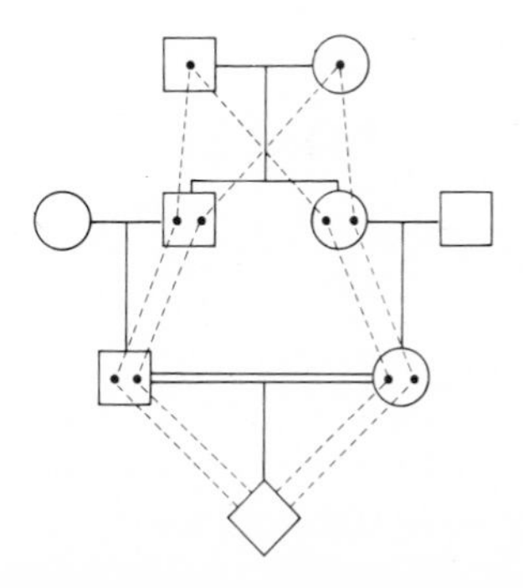

Figure 22-2 Diagram showing paths of relationship for the usual first cousin marriage. Since there are two common ancestors of the parents, there are two paths. The coefficient of inbreeding is therefore $F = (\frac{1}{2})^5 + (\frac{1}{2})^5 = \frac{1}{16}$.

where α is the average inbreeding coefficient and p_i is the frequency of persons of inbreeding coefficient F_i. Comparison of α values among populations is a rapid means of comparing the expected homozygosity through identical descent.

Consanguinity in Isolates

Many small populations are known in which inbreeding is very much higher than in the general population. This will occur if a population is isolated as a result of geographic or social factors. Some small island populations number only a few hundred persons, and matings occur almost exclusively within this group. Inevitably, over several generations, the population becomes inbred. By chance, some alleles are lost and persons become homozygous for alleles having a common origin. A similar situation has occurred among several religious isolates. Where doctrine does not permit marriage outside the group and outsiders are not taken into the group, isolation and inbreeding occur.

There are several religious isolates in the United States that have been studied genetically. Most of these isolates were founded by immigration of a small group of persons, who then set up a closed organization in this country. Some of the groups prospered and consist now of thousands of persons. As the groups have become larger, the opportunities to avoid inbreeding would seem to have increased also. This reasoning is somewhat spurious, however, since the population is restricted to the original gene pool.

An example of the degree of inbreeding sometimes encountered in isolates is given in Figure 22-3. This pedigree is from a group living in Indiana. Although marriages do not always involve close relatives, the number of consanguineous marriages that have occurred among ancestors of the most recent generation greatly increases the coefficients of inbreeding for these generations.

CONSANGUINITY AND CONCEALED DETRIMENTAL GENES

Not all genes can be identified in homozygous state, even though they may be detrimental. The effects may be on some complex attribute, such as body size, intelligence, or skeletal development. The effects may be minor or they may lead to death. Many of the defects may result from environmental as well as genetic causes or from interaction of specific genotypes and environments.

It is possible to test for the contribution of rare recessive genes to these complex traits by measuring their increase in offspring of consanguineous marriages. A number of studies have been designed to do this,

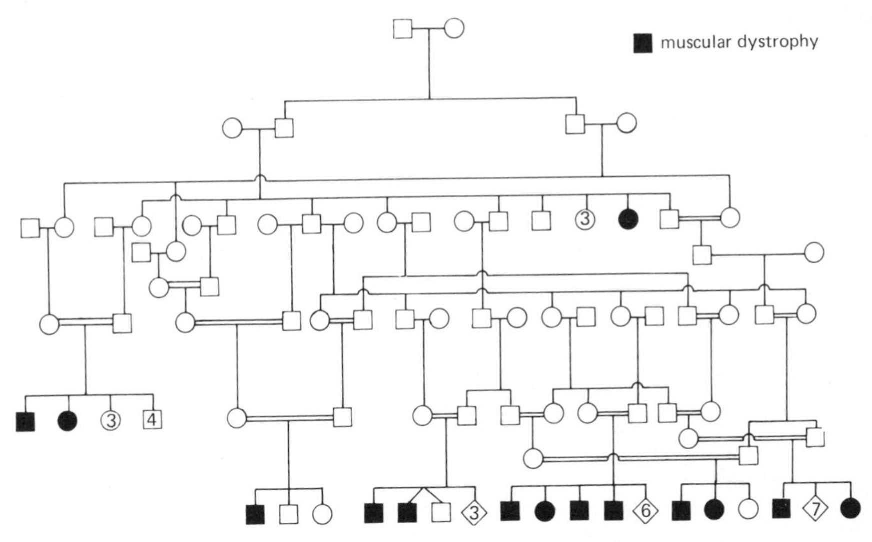

Figure 22-3 Pedigree from an isolate showing a high rate of consanguinity and the presence of a number of cases of muscular dystrophy inherited as an autosomal recessive trait. (*Redrawn from Hammond and Jackson, 1958.*)

a small portion of the results of which are presented in Table 22-3. Because of ease of assessment, the proportion of stillbirths and neonatal deaths has been studied frequently. Regardless of the population investigated, there is an increase in these events. The causes are heterogeneous. In many cases, homozygosity at a single locus may be the main factor; in other cases, simultaneous homozygosity at several loci may be important.

The most extensive study of inbreeding is that of Schull and Neel (1965) in Japan. This study was carried out in Hiroshima and Nagasaki on 3314 children of consanguineous matings and a comparable group

Table 22-3 **MORTALITY IN OFFSPRING OF FIRST COUSIN MARRIAGES COMPARED WITH THAT IN OFFSPRING OF UNRELATED PARENTS***

Trait	Unrelated	First cousins
Stillbirths and neonatal deaths	.044	.111
Infant and juvenile deaths	.089	.156
Early deaths, Hiroshima	.031	.050
Early deaths, Kure	.035	.041
Juvenile deaths	.160	.229
Postnatal deaths	.024	.081
Miscarriages	.129	.145

* From various sources, summarized by Morton, 1961.

of control children from unrelated parents. A large number of variables was studied, a small selection of which is given in Table 22-4. Nearly all variables showed some effect of inbreeding. Of particular interest are the performance tests, indicating a small but significant decrease in aptitude for most tests, even though the children were in school.

Table 22-4 **PERFORMANCE OF CHILDREN OF FIRST COUSIN MARRIAGES AS COMPARED WITH THAT OF CHILDREN OF UNRELATED PARENTS IN HIROSHIMA. ALL COMPARISONS ARE SIGNIFICANTLY DIFFERENT AT THE 1% LEVEL***

Characteristic	Sex	Average control child	Average offspring of first cousins	Inbreeding effect†	Percent change with inbreeding†
Weight (kg)	♂	26.59	26.31	2.34	.9
	♀	26.35	25.99	2.34	.9
Height (cm)	♂	129.7	129.1	.47	.4
	♀	129.8	129.1	.47	.4
Dynamometer grip					
right hand (kg)	♂	14.04	13.71	.29	2.1
	♀	12.35	12.02	.29	2.3
left hand (kg)	♂	13.23	13.01	.20	1.5
	♀	11.43	11.20	.20	1.7
Maze tests score	♂	17.92	17.50	.34	1.9
	♀	16.98	16.52	.34	2.0
W.I.S.C. verbal score	♂	58.67	55.34	2.76	4.7
	♀	57.01	53.46	2.76	4.8
W.I.S.C. performance score	♂	57.37	54.94	2.06	3.6
	♀	55.10	52.52	2.06	3.7
School performance (average grade)					
Language	♂	3.09	2.95	.10	3.2
	♀	3.28	3.10	.10	2.8
Social studies	♂	3.17	3.04	.09	2.8
	♀	3.14	2.98	.09	2.9
Mathematics	♂	3.21	3.04	.13	4.0
	♀	3.19	2.99	.13	4.1
Science	♂	3.29	3.11	.13	4.0
	♀	3.16	2.95	.13	4.1
Music	♂	2.94	2.78	.12	4.1
	♀	3.34	3.14	.12	3.6
Fine arts	♂	3.09	2.95	.10	3.2
	♀	3.40	3.23	.10	2.9
Physical	♂	3.28	3.13	.13	4.0
	♀	3.27	3.09	.13	4.0

* From Schull and Neel, 1965.
† Based on pooled results of males and females in both Hiroshima and Nagasaki.

Estimates of the average number of deleterious recessive genes per person can be made from studies of the increase in defects among offspring of consanguineous marriages. Such estimates are crude; nevertheless, they provide a basis for estimating the ability of populations to tolerate mutations. The "genetic load" of deleterious recessive genes is usually expressed in terms of lethal equivalents. A lethal equivalent is one gene that, if present in homozygous combination, would cause death, or two genes, each of which in homozygous combination would cause death half the time and so forth. A number of studies have indicated 1 to 3 lethal equivalents per genome, with the most recent average being 1.1 (Cavalli-Sforza and Bodmer, 1971). This means that on the average each gamete has 1.1 genes that, if the gamete were to combine with an identical gamete, would lead to death. In addition to genes that are lethal, there are others that are detrimental even though they may not result in death.

Although future studies may cause a revision in the estimate of the number of detrimental genes per person, it is unlikely that new data will greatly alter the magnitude. Thus, most persons carry several genes that can be detrimental in appropriate combinations with other genes and the environment. Close inbreeding will often reveal these genes. Persons who survive close inbreeding have a slightly better chance of having fewer or none of the genes.

SUMMARY

1. The offspring of consanguineous marriages are more likely to be homozygous at a locus compared with the offspring of unrelated parents. This can be expressed quantitatively by the inbreeding coefficient F, which is defined as the likelihood that the two alleles at a locus will be identical by common origin. For offspring of first cousin marriages, $F = \frac{1}{16}$.

2. Dahlberg has derived equations relating the proportion of first-cousin parents among all parents of homozygous offspring to the allele frequency and the frequency of first-cousin marriages in the population.

3. Sewall Wright developed a general formula for the inbreeding coefficient, $F = \Sigma(\frac{1}{2})^N$, where N is the number of persons in a path of relationship. The formula is especially useful for calculating inbreeding coefficients for complex pedigrees.

4. Although remote inbreeding is ordinarily of little consequence, it may be important in isolated populations in which a marriage may be consanguineous through many different ancestral paths. This may lead to unexpectedly high inbreeding coefficients associated with the appearance of rare recessive traits.

5. Studies of inbred children have been helpful in estimating the load of concealed detrimental genes. These detrimental genes include not only those causing clear-cut recessive disorders but also those that affect complex traits such as intelligence and body growth. A large study on inbred Japanese children indicated a small but significant decrement in performance on many school tests as compared with noninbred children.

6. It has been estimated through studies of offspring of consanguineous marriages that each person has approximately one lethal equivalent per haploid genome. This means that on the average each gamete, if combined with an identical gamete, would lead to death.

REFERENCES AND SUGGESTED READING

ADAMS, M. S., AND NEEL, J. V. 1967. Children of incest. *Pediatrics* 40: 55–62.

CAVALLI-SFORZA, L. L., AND BODMER, W. F. 1971. *The Genetics of Human Populations.* San Francisco: W. H. Freeman, p. 364.

FREIRE-MAIA, N. 1957. Inbreeding in Brazil. *Am. J. Hum. Genet.* 9: 284–298.

HAMMOND, D. T., AND JACKSON, C. E. 1958. Consanguinity in a midwestern United States isolate. *Am. J. Hum. Genet.* 10: 61–63.

MORTON, N. E. 1958. Empirical risks in consanguineous marriages: birth weight, gestation time, and measurements of infants. *Am. J. Hum. Genet.* 10: 344–349.

MORTON, N. E. 1961. Morbidity of children from consanguineous marriages. *Progr. Med. Genet.* 1: 261–291.

MORTON, N. E., CROW, J. F., AND MULLER, H. J. 1956. An estimate of the mutational damage in man from data on consanguineous marriages. *Proc. Natl. Acad. Sci.* (U.S.A.) 42: 855–863.

SCHULL, W. J. 1958. Empirical risks in consanguineous marriages: sex ratio, malformation, and viability. *Am. J. Hum. Genet.* 10: 294–343.

SCHULL, W. J., AND NEEL, J. V. 1965. *The Effect of Inbreeding on Japanese Children.* New York: Harper and Row, 419pp.

SLATIS, H. M., AND HOENE, R. E. 1961. The effect of consanguinity on the distribution of continuously variable characteristics. *Am. J. Hum. Genet.* 1: 28–31.

SLATIS, H. M., REIS, R. H., AND HOENE, R. E. 1958. Consanguineous marriages in the Chicago region. *Am. J. Hum. Genet.* 10: 446–464.

REVIEW QUESTIONS

1. Define:

consanguinity   inbreeding

2. A man is affected with the autosomal recessive form of muscular dystrophy. Two of his grandchildren (first cousins) marry.

a. What is the probability that both are carriers of the disease?

b. Assuming that both are carriers, what is the probability that any given child of theirs will be affected?

c. If he had been affected with the sex-linked form, how would this have influenced your answer?

3. In a marriage involving first cousins, what is the probability that an offspring will be affected with a disease due to a rare recessive gene

a. if the common grandfather of the couple is affected?
b. if the husband's mother (the wife's aunt) is affected?
c. if the husband's paternal uncle is affected?
d. if there are no affected relatives?

4. In the pedigree below, what is the probability that the offspring of the first cousin marriage will be affected with galactosemia? What is the probability that this individual will be hemophilic? What is the probability that both conditions will be present? Both absent?

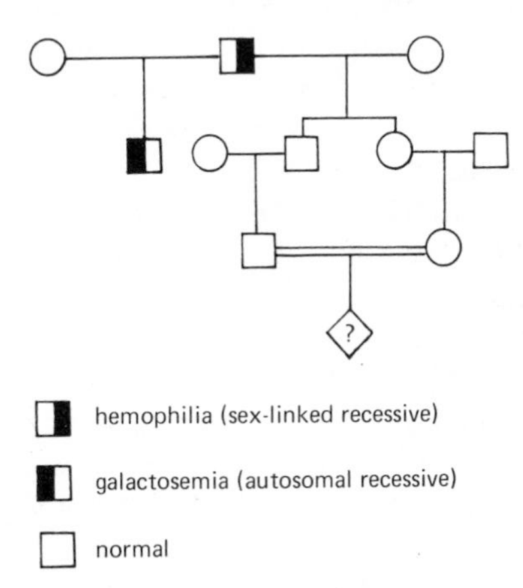

5. In a population in which 1% of the marriages are between first cousins, it is observed that 30% of the matings that produce offspring affected with a rare recessive trait are between first cousins. What is the frequency of the gene responsible for the trait?

CHAPTER TWENTY-THREE

HUMAN POPULATIONS

All human beings are classed together into a single species, *Homo sapiens*. In spite of the diversity of human types, we are, after all, similar. No one has difficulty classifying persons, however primitive the persons may seem, into human versus nonhuman groups. One criterion commonly applied to determine if two groups of organisms are members of a single species is whether they are completely fertile in cross matings. By this criterion, all human beings obviously belong to the same species.

Much diversity does exist among human beings, nevertheless, and it exists at many different levels. Each of us as an individual differs from other individuals from the same background. Each family differs somewhat from other families; groups of families differ from other groups of families; and, finally, the major ethnic groups differ from each other.

Various terms have been used to designate the major groups of mankind. The most common term is *race*. A race may be defined as a group of historically related persons who share a gene pool that differs from the gene pool of other groups. The difficulty with this definition is that it does not tell us how discriminating we should be in recognizing differences. Since there are hundreds of populations that differ from all other populations, we might therefore claim that there are hundreds of races. Common usage of the term, however, has been more restrictive.

THE ORIGIN OF RACES

This discussion will consider the manner in which races can arise from a population that initially is in equilibrium. It is assumed that there are two or more alleles at each of many loci. Each allele will have a certain frequency. Formation of races would be the formation of subpopulations with gene frequencies different from each other.

In order for races to arise, there must exist isolating mechanisms. In nearly every case, this has probably been geographic initially. One can imagine other isolating mechanisms, such as chromosome rearrangements and assortative mating, but these are more important in maintaining isolation than in giving rise to it. Geographic isolation would occur when any natural barrier, such as a river, mountain range, or ocean, separated two groups, making gene flow difficult. Complete isolation has undoubtedly been the exception rather than the rule. A common reason for separation might be migration of part of a population because of overcrowding. This could happen by mass migration of a portion of the population or by migration of individuals over a period of time. Having migrated, persons are more likely to seek mates among fellow emigrants than among the surrounding population.

Through variations on the preceding factors, isolated populations arise. It does not automatically follow, however, that the isolated population will in time be different. Differences will develop only if one or more of the following events also occur.

Genetic Drift

If two alleles, A and a, in a population have frequencies $p = 0.6$ and $q = 0.4$, respectively, the most probable values of p and q in a group of persons drawn from that population are $p = 0.6$ and $q = 0.4$. The expected deviation from those values is given by the standard error, $s = \sqrt{pq/2N}$, where N is the number of persons in the sample. Sixty-eight percent of the time, the value of p in the sample will not deviate more than one standard error; 95% of the time, p will not deviate more than two. For large values of N, s is very small and the expected deviation is correspondingly small. If the sample of persons is small, the chances that p in the sample will differ from p of the parent population are much greater. For the gene frequencies given earlier, $p = 0.6$ and $q = 0.4$, a sample of 500 persons would have $s = 0.0155$. If 500 persons contributed to the next generation, 68% of the time p would lie between 0.5845 and 0.6155 and 95% of the time p would be between 0.5690 and 0.6310. If only 50 persons contributed, $s = 0.049$. Now, $0.551 < p < 0.649$ 68% of the time, and $0.502 < p < 0.698$ 95% of the time. Formation of a new population by such a small number of persons therefore may lead purely by chance to a different gene frequency in the new population.

These chance fluctuations of genes are called *genetic drift*. The term implies that sampling error alone is responsible for changes in gene frequency. A consequence is that shifts are equally likely to occur in either direction, leading to either an increase or a decrease of p. Because Wright first explored the effects of genetic drift systematically, this phenomenon is sometimes called the *Sewall Wright* effect. For the large mobile exogamous populations of the present time, genetic drift has little chance to operate. During most of his evolution, man lived in small, relatively immobile, endogamous communities, in which the gene frequencies at any one point in time would almost certainly differ from the gene frequencies of the previous generation.

An example of genetic drift in a human population has been reported by Glass and co-workers in a study of the Dunkers. A religious isolate in Pennsylvania, they were founded in this country by a small group that immigrated from Germany. Analysis of a series of blood groups shows them to deviate considerably both from the German population and from the surrounding American population. Because of the small size of the group it is postulated that the change in gene frequencies has resulted purely from genetic drift. This example points out the possibility that a small endogamous group may be genetically isolated and subject to drift, even though it exists in the midst of a very large outbred group. The effective population size, that is, the number of persons in the breeding group at any one time, is of course smaller than the total count of persons.

Genetic drift has been proposed as the explanation for many neutral or detrimental genes that occur in frequencies higher than might be expected. Whether Tay–Sachs disease occurs at high frequency among Ashkenazi Jews because of selection or drift was discussed in Chapter 21. A variant of Nieman–Pick disease is found in Nova Scotia with a heterozygote frequency of 10 to 26% among children in certain regions. All known cases have been traced to one couple born in Nova Scotia in the late 1600s (Winsor and Welch, 1978). These and other examples point out how often such chance increases occur, even to the point of becoming a significant genetic feature of a large population.

The Founder Effect. A special case of genetic drift is sometimes referred to as the founder effect or founder principle. This occurs when a new population is formed from a very small number of individuals whose gene pool differs by chance from that of the parent population. Of the examples available, one of the more interesting and fully documented is that provided by Neel and co-workers (1964). The Xavante Indians of South America consist of a series of small villages, each with several hundred persons. When one such village divided to form two new villages, the split occurred largely along family lines. The total number of persons in the new villages was approximately 200 and 100. The effective breeding size of the new villages was much smaller, and of course the fact that families tended to stay together in the split further reduced the gene pool.

This kind of split must have occurred many times during the period when man spread from his ancestral evolutionary home in Africa to occupy the entire land mass of the earth. Most such small populations disappeared, either through merger with others or through extinction. Occasionally such a small population would flourish and leave a much larger genetic legacy to the world.

Chance deviation of gene frequencies also occurs in small populations due to unequal contributions of genes from the persons in the populations. This is illustrated by the reproductive history of the Xavantes. The chief of a village has more wives than other males and leaves more children. In very small villages he may sire a high proportion of the next generation. This could lead to very high inbreeding. The adoption of incest taboos reduces very close inbreeding, but over time the effective gene pool of a small village might become very small indeed.

Selection

The effects of selection on gene frequency were treated in Chapter 21. In order for selection to lead to different gene frequencies in different populations, it must be supposed that the fitness of various genotypes differs among the populations. This variation, in turn, would usually reflect the action of different environments. Thus *AA* might be the favored genotype in one environment whereas *aa* would have an advantage in another. In more general terms, the fitness coefficients w_1, w_2, w_3, and so on, must differ in the different populations.

Since selection is frequently a function of environment, populations that inhabit similar though widely separated environments may have similar types and frequencies of certain genes, even though the populations are otherwise different. Also, populations known to be unrelated but that occupy a common geographic area may have similar frequencies. Conversely, closely related populations in different environments may have diverged in their gene frequencies. It is thus hazardous to infer the relationships of populations on the basis of one genetic locus or only a few loci.

Even complexes of genes may be misleading. An example is found in skin pigmentation. For the most part, races that have inhabited tropical areas for thousands of years are darkly pigmented. Those that have lived in more temperate regions have less pigment. The dark pigment is thought to protect persons from the intense rays of the equatorial sun. Persons with little pigment cannot tolerate sun to the extent that highly pigmented persons can. On the other hand, in regions where there is very little sun, more of the radiation is thought to be available for conversion of ergosterol to vitamin D in persons without pigment. Several races are highly pigmented—Africans, Melanesians, Micronesians, and Australians. Selection has favored the development of pigment independently in these groups. The Africans and Australians seem to be no more closely related—indeed, perhaps they are less closely related—than are Africans and whites. Yet from the point of view of pigment, they are similar.

Another physical trait thought to be controlled by selection is the ratio of body surface area to mass. In general, animals living in cold climates are larger and rounder; in hot climates they are smaller and thinner. Heat loss is largely a function of surface area. For this reason, it is thought that stocky persons are favored in cold climates, where the problem is to conserve heat, and thin persons are favored in hot climates, where the problem is to get rid of heat. The very thin populations are all tropical, while the arctic inhabitants are heavy-set. The inheritance of body form is very complex and many factors other than selection for heat economy undoubtedly influence selection for body types.

Not all selection is related to external environment. Particularly in the case of man, social factors may give rise to and perpetuate population differences that are inherited. Each culture has its own theories on the ideal types of mates. Although the ideal may rarely exist, there are usually some persons in the culture who approach the ideal more than do others. To the extent that these persons have a favored opportunity to reproduce, the gene pool of the population will shift in their direction. If the ideal type is sufficiently stable over a long period, the effect on the gene pool could be appreciable. On the other hand, during the past century in Western civilization, there have been frequent changes in the concept of ideal types.

The influence of social forces on mate selection tends to perpetuate the separation of populations that might otherwise blend. There are many examples, including the continued existence of separate white and black populations in America. The geographic isolation that permitted separate populations to arise no longer exists. Perpetuation of two populations by social factors alone is doomed to failure in the long run. There is a small but continued exchange of genes between the two populations, which will cause the gene pools gradually to converge. The speed with which this happens is influenced by social factors, but the ultimate outcome is not.

Mutation

Mutation is not generally considered to have a major role in formation of isolates. In a freely interbreeding population, a mutant gene either would be rapidly eliminated or, if it were favorable and escaped chance elimination, would spread through the entire population, displacing other alleles until an equilibrium frequency corresponding to its relative fitness was attained. The fact that only a part of the population possessed the mutant allele at some time would not be a basis for dividing the population into groups.

Once a population is isolated from others, mutations can then become an important factor in further increasing the isolation. A particular mutation always occurs in a particular population, and the population in which it occurs has the potential to change genetically. If specific mutations were frequent, then all populations would have the same potential. But mutations—particularly favorable mutations—are exceedingly rare. A normal gene mutates to a nonfunctional form approximately once per 100,000 gametes. Mutations that lead to a specific change in amino acid sequence of the protein product, as in hemoglobin A to S, must be much rarer, possibly only one in 10^7 or 10^8 gametes. In populations of a billion persons, even these rare mutations must occur occasionally, but prehistoric populations were much smaller, and most mutations did not occur in most populations, although they must have occurred in a few populations somewhere. When a favorable mutation did occur, chances are that it occurred only once, and the population in which it occurred could therefore become different from other populations.

It is not possible to prove that single events of mutations gave rise to all of the alleles of a specific type of now existing. Perhaps the most likely possibility is hemoglobin C. This β-chain variant is found only among persons of African origin. Furthermore, its distribution in Africa is very limited and reaches a maximum concentration in the Upper Volta (Figure 21-2). Hemoglobin C seems to offer some protection against malaria in persons heterozygous for Hbs A and C (Chapter 21). The single focal point of distribution suggests that the Hb C mutation has occurred only once. If it occurred in other places and times, it was lost, either through chance (genetic drift) or because it offered no advantage in those environments.

The destiny of most new mutations, whether favorable or detrimental, is extinction. This is because chance eliminates most mutations. But chance also provides for the recurrence of mutations, and occasionally one will survive. It it is favored in all environments, it will gradually displace its alleles, given sufficient time. So long as a tenuous contact is maintained among divergent groups, a pathway is present for the flow and exchange of genes. The human populations have probably been few indeed that were so isolated as never to give or receive genes from the outside. The time necessary for a gene to spread under primitive conditions was very long.

THE EVOLUTION OF HUMAN POPULATIONS

Knowledge of prehuman and early human populations comes primarily from the skeletal remains of individuals who happened to meet their end in a place favorable for the preservation of bony material. There have been many discoveries of fossil material in recent decades, and the "missing links" are now largely known. It is still not possible to specify a precise sequence of evolutionary forms, but there are enough early hominid forms known to account for the general pattern of evolution from apelike ancestors to modern man.

A discussion of the varieties of prehistoric man can be found in books on physical anthropology. The only question that will be considered here is how the present races of man are related to each other in terms of evolution. Two extreme possibilities can be imagined (Figure 23-1). The more generally accepted is that all of the existing races of man diverged after man had evolved to *Homo sapiens*. Divergence may have occurred earlier, but only a single line survived to give rise to modern man. At no time was the line homogeneous; there was always variability, although there was also opportunity for gene exchange among all parts of the population, at least until *Homo sapiens* appeared.

Another hypothesis is that the present major groups of man diverged prior to the final evolutionary process. Evolution to *Homo sapiens* would have been by parallel evolution. This occurs when similar environments cause two separate populations to evolve in the same manner to the same final product. Many examples of parallel evolution are known in lower organisms. There is no reason to suppose it could not have occurred in man also, given the same genetic variability in the various populations.

A third possibility is a combination of the first two. Fossil evidence favors coexistence of several forms of early man. Some of these forms

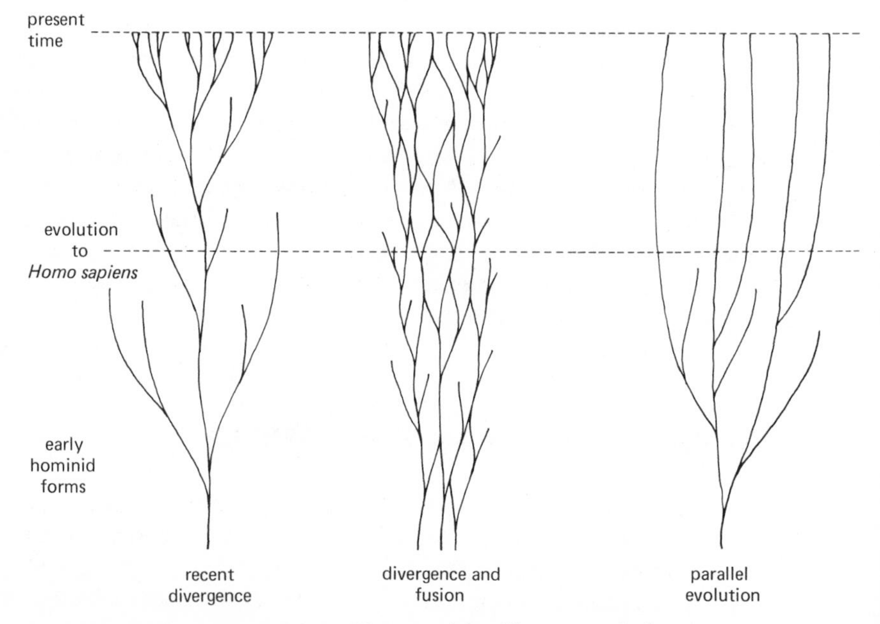

Figure 23-1 Three models of human evolution.

undoubtedly became isolated and eventually extinct. Other forms could probably interbreed and exchange genes, although the skeletal remains are different among the groups. Particularly advantageous genes could be distributed among otherwise distinct groups without leading to a loss of the major group differences. Occasionally two groups might join, forming a new population.

The third idea is similar to the first in that it assumes that the ancestors of modern man were never completely isolated from each other. It resembles the second theory in that it assumes that the progenitors of presently existing races had already formed separate populations prior to the final steps of evolution.

The exact happening of evolution may never be known. It is known though that all races now existing appear to be able to breed effectively with all other races to produce fertile offspring. This means that, unlike the case of the donkey and horse, the chromosome complements of all human races are equivalent. Indeed, the karyotypes of all existing populations are identical. If the present racial groups had been divergent for a long evolutionary time, differences would almost certainly have arisen—as they have in other closely related species—which would have led to unbalanced chromosome complements in crosses, either in the F_1 or F_2 generations. It is therefore necessary to conclude that the different lines of man were never totally isolated over a long period of time.

Anthropologists differ on the number of races that exist today. Some think it is best to recognize only six races; others recognize up to two dozen races. The disagreement is not in whether these groups exist; it is in whether each should be assigned separate race status. The present classifications of man are based primarily on morphology. For centuries, this was the only readily observable variation. The term "morphology" today includes many aspects of body structure not readily observable without modern instruments, such as x-ray machines. As new genetic loci have been recognized and differences in allele frequencies among populations have been noted, these results have generally confirmed the relationships among populations based only on physical features.

Each of the major races is composed of many "local" races or populations. Perhaps none of the latter has all the features characteristic of the major racial group to which it belongs. As a group, the local races resemble each other in many ways but also differ. In some instances, local races may show features intermediate between major racial groups, suggesting an affinity with both and testifying to the arbitrariness with which lines between races are drawn.

Table 23-1 is a list of the major races as viewed by Garn. A few local races are included, although the list is very incomplete. Even the local races frequently are composed of groups of related but distinct popula-

Table 23-1 **THE MAJOR RACES OF MAN, INCLUDING SOME LOCAL RACES***

Major race	Local races
1. Amerindian	North American Indians
	Central American Indians
	South American Indians
2. Polynesian	
3. Micronesian	
4. Melanesian-Papuan	
5. Australian	
6. Asiatic	Japanese
	Chinese
	Filipino
7. Indian	Hindu
	Dravidian
8. European	North European
	Mediterranean
9. African	Sudanese
	Forest Negro
	Bantu
	Bushmen

* From Garn, 1971.

tions. The grouping in Table 23-1 is somewhat arbitrary, in the sense that the major races may not be equivalent in terms of their separation from each other. Some have been more isolated and have evolved more independently than others. Furthermore, the list is not one with which all anthropologists agree. The differences among populations are weighed differently by some, leading to other classifications. Nevertheless, the list indicates the diversity of mankind and the population units into which man is divided. The pre-Columbian location of these races is shown in Figure 23-2.

The arbitrariness of classification of human populations into discrete groups is demonstrated by the distribution of gene frequencies in various regions of the world. If a well-studied system such as the ABO groups are plotted as in Figure 23-3 and 23-4, the frequencies are seen to change gradually from region to region, forming clines rather than abrupt differences. Where abrupt differences are detected in present-day populations, they ordinarily can be traced to recent migration. Clines, of course, may have their origins in several ways. They may result from diffusion of genes across the boundary between two originally distinct groups, the gene exchange proceeding to the point that the boundary can no longer be detected. Conversely, an originally homogeneous population may differentiate as a result of selection associated with geo-

Figure 23-2 Polar-projection map of the world showing the limits of the nine geographical races described in the text. In each case geographical barriers set off the races. (*From S. M. Garn, 1971.* Human Races, *3rd Ed. Springfield, Ill.: Charles C Thomas.*)

graphic variation. For example, if the geographic range is from far north to the tropics, a population may, over time, show the effects of selection for genes favoring heat retention on the one hand and heat loss on the other.

Genes as Indicators of Population Origins

Phylogenetic Relationships. If one assumes that the recentness of divergence of populations is associated with similarities in gene frequencies, one may use gene frequency comparisons to gain insight into the evolutionary relationships among populations. For example, the I^A allele frequencies are 0.28 in U.S. whites, 0.26 in English, and 0.10 in certain American Indians, suggesting that U.S. whites are more closely related to Englishmen than to Indians. In constructing an evolutionary tree, the branch point between U.S. whites and English would be placed very recent; that between American Indians and the other two populations would be in the more remote past. A single genetic system

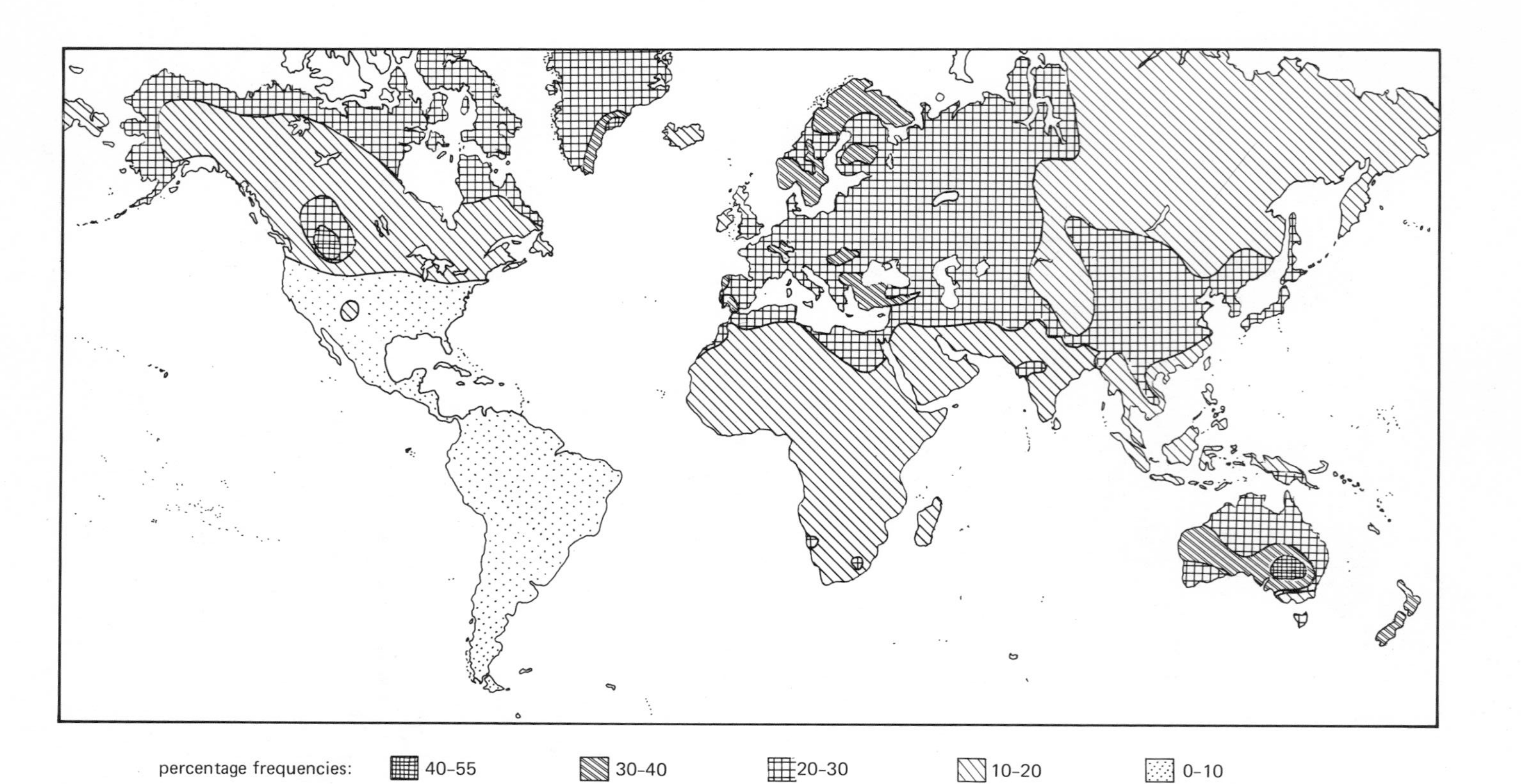

Figure 23-3 Distribution of blood group A gene in the aboriginal populations of the world. (*Redrawn from Mourant and co-workers, 1958.*)

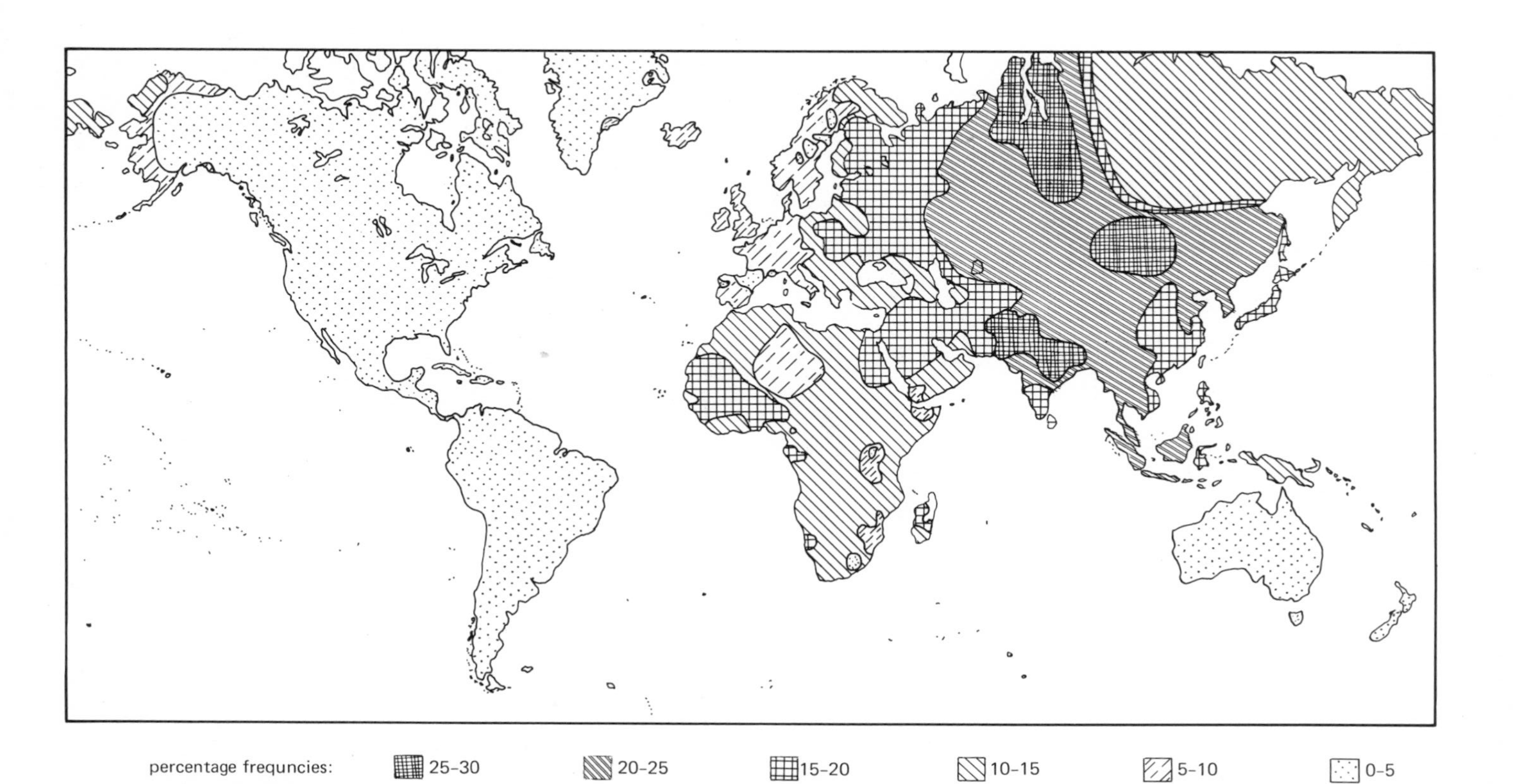

Figure 23-4 Distribution of blood group B in the aboriginal populations of the world. (*Redrawn from Mourant and co-workers, 1958.*)

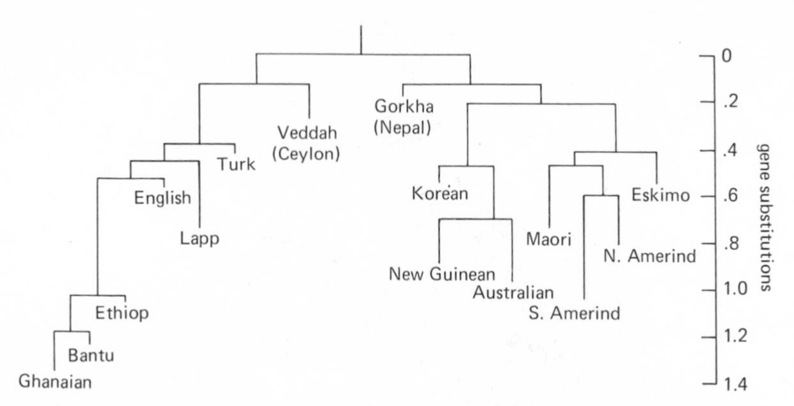

Figure 23-5 Most likely tree of descent for fifteen human populations selected to represent the world population. Method of minimum path. (*From Cavalli-Sforza, Barrai, and Edwards, 1964.*)

could be misleading, however, since Japanese, for example, have an I^A frequency of 0.27. Adding the I^B allele would help discriminate further, but a single locus can give only limited information on affinity of populations.

Various mathematical models have been used to combine information on frequencies of alleles at many loci into a single measure of genetic distance between two populations. Some of these approaches are summarized by Cavalli-Sforza and Bodmer (1971). The information from a number of populations can be combined to form a minimum evolutionary tree, such as that shown in Figure 23-5. Such a tree is generally in agreement with what is known from other sources, even though it was constructed from data on only five loci. Additional loci should improve the accuracy. A similar phylogenetic tree is shown plotted against geographic location in Figure 23-6.

Hybrid Populations. Chapter 20 outlined the method of calculating gene frequencies in mixed populations:

$$p_t = ap_a + bp_b + \ldots + np_n,$$

where p_t is the frequency of an allele in a hybrid population and a, b, . . . , n are the relative contributions of the parental populations with allele frequencies p_a, p_b, . . . , p_n. Use of this formula requires that the parental allele frequencies be known.

A number of estimates of the origin of American blacks have shown some 22% of the genes of non-Southern blacks to be of white origin, the remaining 78% being West African in origin (Table 23-2). Some loci are much more informative than others, the most useful being the Duffy blood group, since the frequency of the Fy^a allele is near zero in Africa

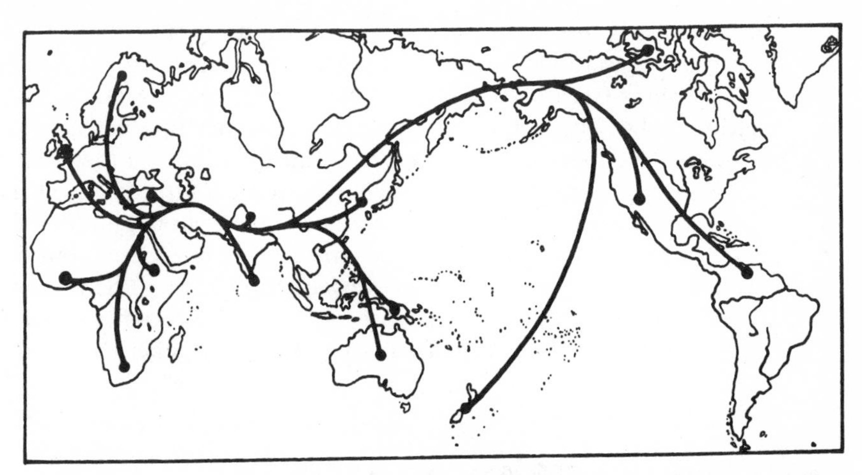

Figure 23-6 Topology of the minimum-evolution tree projected on the world map to show correspondence with the probable manner of diffusion of man. (*From Cavalli-Sforza, Barrai, and Edwards, 1964.*)

but is 0.43 in Europeans. Certain American black populations have a lower admixture of white genes. For example, studies in Georgia showed only 11% white genes, and the Charleston, South Carolina, estimate is 4%. These different estimates presumably reflect different histories of the groups.

Many other hybrid populations exist but have been less extensively studied. Various American Indian groups have been studied, especially

Table 23-2 **ESTIMATES OF THE PROPORTION OF CAUCASIAN GENES M BASED ON Fy^a FREQUENCIES IN VARIOUS AMERICAN BLACK GROUPS***

Region and Locality	Blacks N	Blacks $q \pm$ SE	Whites N	Whites $q \pm$ SE	$M \pm$ SE
Non-Southern					
New York City	179	.0809 ± .0147			.189 ± .034
Detroit	404	.1114 ± .0114			.260 ± .027
Oakland, Calif.	3146	.0941 ± .0038	5046	.4286 ± .0058	.2195 ± .0093
Southern					
Charleston, S.C.	515	.0157 ± .0039			.0366 ± .0091
Evans and Bullock counties, Ga.	304	.0454 ± .0086	322	.422 ± .0224	.106 ± .020

* The frequency of Fy^a is assumed to be zero in the ancestral African groups. N = number in sample; $q = Fy^a$ frequency; SE = standard error. (From T. E. Reed, 1969. *Science* 165:762–768. © American Association for the Advancement of Science.)

in Mexico where there is a major contribution of Indian genes to the present population. One of the genetically interesting "Indian" groups is the Black Caribs of the eastern coast of Central America. This population is thought to have originated when slave ships were wrecked during the seventeenth century on St. Vincent island in the West Indies. The several hundred African survivors joined with the Carib Indians, who were under attack by the expanding European population. The culture of the Caribs, including the language, was adopted, but disagreements between the two groups subsequently arose. The present Black Caribs are essentially pure African, with a high frequency of the African Rh allele R^o and a very high frequency of the Hb^S gene (Firschein, 1961).

GENETIC DIFFERENCES AMONG RACES

Many references have been made in previous chapters to differences in frequencies of certain genetic traits in different populations. A complete list of known differences would be very long indeed. For blood groups and certain related markers, major compendia such as that of Mourant et al. (1976) should be consulted.

There are several traits of special interest, of which the following have been selected as examples.

Lactase Persistence

The major sugar in milk is lactose, a disaccharide composed of one molecule each of glucose and galactose. Lactose is split into glucose and galactose by the enzyme lactase in the small intestine. Rare infants are homozygous for a gene that causes absence of lactase, making it impossible to break down and absorb lactose. The symptoms of gastrointestinal distress and diarrhea can be avoided by placing the infant on a milk-free diet.

In most species, including most human beings, the level of lactase decreases sharply at about the time of weaning. There is debate as to the extent that this decrease is a response to decreased exposure to milk, but in the case of many human beings, the high levels of lactase continue through adulthood. Persons can therefore be classified as positive or negative for persistence of lactase. That this persistence is hereditary and explained as a simple Mendelian trait has now been extensively documented, with high lactase levels (persistence) dominant to low levels (Lisker et al., 1975; Flatz and Rotthauwe, 1977). Adults who have low lactase levels are often called lactase deficient, but in view of the fact that the majority of the population of the world does not have persistent lactase (Table 23-3), it would perhaps be better to designate them just as nonpersistent or lactase-negative.

Table 23-3 **DISTRIBUTION OF LACTASE-POSITIVE AND -NEGATIVE PERSONS AMONG ADULTS IN VARIOUS POPULATIONS***

Population	Number examined	Lactase-positive	Lactase-negative	Frequency of gene for persistent lactase
Australian aborigines	121	116	5	.80
Thai	279	8	271	.01
India (Delhi)	70	50	20	.47
Chinese	61	7	54	.01
Lebanon (Arabs)	81	8	73	.05
Uganda (Bantu)	64	8	56	.06
Greece	700	371	329	.31
Sweden	700	679	21	.83
Finland	953	784	169	.58
U.S. whites	771	592	179	.52
U.S. blacks	349	76	273	.12
Spain	267	209	58	.53
Mexico (rural)	401	105	296	.14
Indians (Oklahoma)	36	7	29	.10

* The gene frequency of the allele for persistent lactase is calculated assuming negative persons are homozygous for a recessive allele. (Data from Flatz and Rotthauwe, 1977.)

Persistent lactase can be assessed by a lactose tolerance test (LTT), in which 2 grams lactose/kg, up to 50 grams total, is ingested. Persons with persistent lactase show a rise in blood glucose; persons with lactose intolerance show no such rise, with prompt development of diarrhea. Since lactase may persist in humans beyond the weaning period even in those who are not tolerant, the test is valid only for adults. The results of this test correspond well with history of milk tolerance in individuals. Even those without persistent lactase can tolerate small quantities of lactose.

The frequency of lactase-positive and negative persons in various populations is shown in Table 23-3. The gene frequencies are calculated on the hypothesis that lactase-negative persons are homozygous for a recessive gene. The variation among different populations is striking, with the highest frequency of the lactase-positive allele in Sweden, followed by Australian aborigines. European populations are generally high, but most others are low.

It is interesting to compare the gene frequency with milk consumption patterns. Such an analysis does not bear out the supposition that in populations with high milk consumption there has been selection for the lactase-positive allele (Flatz and Rotthauwe, 1977). Indeed, milking is thought to have originated in precisely those areas with lactose intolerance. It should be remembered, though, that fermentation of milk destroys lactase, making it acceptable for lactase-negative persons. The reason for the present distribution of gene frequencies is therefore unknown.

The presence of lactase-negative persons in many populations does have importance for nutrition programs. Milk and milk products are an important source of protein and other nutrients for many adult European populations and their descendants. Such diets are not acceptable to most persons in other parts of the world unless the milk is processed to remove lactose. Efforts to export milk products from European countries and the United States have not been highly successful, possibly due in part to this problem.

Race and Intelligence

Few topics have generated as much heat as has the question of the extent to which differences in IQ among racial groups are inherited. Evidence for genetic factors in intelligence were reviewed in Chapter 18. It was concluded that variation in intelligence within a population is due in large part to heredity. Does it follow then that differences between populations are hereditary?

The subject has been reviewed by many authors, the most extensive review being that of Loehlin, Lindzey, and Spuhler (1975). The question of whether there are racial differences in IQ is not seriously questioned. In a variety of studies, the average score for U.S. blacks has consistently been lower than the average for whites. At this point one must ask what such IQ tests measure. Also, is it an appropriate measure for all groups tested.

The IQ tests were developed to predict performance of children and others in school or in other situations. A typical test consists of a number of items (questions or problems), some of which can be solved by most of the population, others of which can be solved by only a few. The test items have been selected so that most of the population performs at intermediate levels, with the distribution of performance being roughly a normal distribution with mean of 100. In the commonly used Stanford–Binet test, the standard deviation is 15 points, that is, two-thirds of the population score between 85 and 115. The tests could be constructed to give other distributions, but this distribution correlates well with performance in real life situations.

Standardization of a test is an important step in test development. It means that a population is selected as the standard population, and test items are selected that are meaningful for that population. In the case of the Stanford–Binet test and the Wechsler Intelligence Scale for Children (WISC), the tests were standardized on U.S. white children. There is extensive evidence that the tests are valid for most children in this group. The tests are not as valid for other groups. Knowledge and skills that an "average" white child would be expected to acquire by a certain age are not necessarily the same that would be acquired by an "average" black child of the same age. So long as there are cultural differences associated with race, it will be impossible to use a single IQ test standardized on one or the other group to measure the intelligence of members of both groups. Blacks, *on the average*, do indeed perform less well on IQ tests, but the validity of IQ tests as a measure of innate intelligence in blacks cannot be accepted on present grounds. The IQ tests undoubtedly measure something related to intelligence in blacks, but a score of 100 in a black child does not necessarily equal a score of 100 in a white child.

The question of whether racial differences in IQ (not intelligence) are inherited is thus not a well formulated question. We have seen that variations in intelligence, as measured by IQ, are largely genetic in whites (Chapter 18). We would suppose that variations in intelligence among blacks also is largely genetic. But there is no reason to insist at present that average differences in IQ between the two populations is biological in nature. Until truly culture-free IQ tests are developed and standardized, one cannot know which racial groups are more intelligent or less intelligent.

THE EFFECTS OF RACE CROSSING

If populations differ in their gene pool, then a legitimate question is whether matings between persons of different races are biologically equal to matings within racial groups. One might suppose that evolution has favored complexes of genes or special chromosome arrangements within races and that mating between races breaks up these complexes, leading to a less favorable genome. The offspring might be less fit and have reduced fertility. On the other hand, the effect of race crossing is to increase the heterozygosity in the F_1, and in some species, of which corn is an example, this leads to greater fitness of the F_1 (*hybrid vigor* or *heterosis*). With both increased and decreased fitness in other species to serve as examples, conclusions about human beings must be based on direct observation.

Opportunities for careful studies of crosses between races have not been frequent, in spite of the common occurrence of such crosses. Too often, the liaisons have been of a temporary or clandestine nature, with no way to assess fertility. The offspring of such matings have often been considered outcasts, belonging to neither parental group. Thus they have not been exposed to environments that would permit comparison with the parental groups. For many of the attributes for which persons are evaluated, inequalities of environment can vitiate conclusions on the origin of differences.

One of the better opportunities for study of race crossing is found in

Hawaii. Whites, Orientals, and Polynesians are all well represented. The absence of social barriers permits persons of different races to intermarry, and it is possible therefore to compare offspring from various parental combinations. In an extensive study of such variables as fetal and neonatal mortality, growth rates, and malformations among hybrid offspring, Morton, Chung, and Mi (1967) were unable to detect any effect of race crossing in Hawaii. In a further study of cleft lip and palate, Ching and Chung (1974) were unable to detect any effects on frequency of these traits in the hybrids. Although the conclusions of such studies always are limited by the number of matings available for study, it is apparent that any possible effect, whether positive or negative, must be very small.

It was at one time suggested that the effects of the sickle cell gene are more severe in American blacks than in Africans. The consequences of homozygosity for Hb S were markedly deleterious in the United States, but comparable cases were not reported in Africa. It was thought that the Hb S genes interacted with the "European" genes of American blacks to produce a severe disease but that in Africa combinations of genes had evolved that were not so deleterious. Subsequently, this difference was shown to be a fallacy based on inadequate medical reporting from Africa. Sickle cell disease occurs in Africa as well as in the United States, and the disease is just as severe in Africa. The two genetic backgrounds appear not to make a difference.

SUMMARY

1. *Homo sapiens* is divided into many distinctive populations. The larger population divisions are sometimes referred to as races, a race being a group of historically related persons who share a common gene pool. Each race is subdivided into many local populations, and there is no sharp division of one population from the next.

2. Races arise by divergence of their gene pools. This requires some degree of breeding isolation between the diverging groups. Gene frequencies may diverge because of random drift or the founder effect. The latter occurs when populations are founded by a very small number of persons not representative of the parent population. Selection has also played an important role in the adaptation of some groups to specific environments. In some instances, mutation may also have contributed to population diversity, although this is not considered to be a major factor.

3. During human evolution, the various major isolated groups probably were never totally isolated over long periods. This is supported by the fertility of interracial crosses and by the identical karyotypes found in all human populations. An advantageous mutation arising in one would have the opportunity eventually to diffuse throughout all other populations.

4. Present day populations differ in allele frequencies at many loci, and these differences are sometimes useful in discerning the affinities among populations. Especially if a number of loci are considered simultaneously, it is possible to relate populations by their similarities in allele frequencies, forming a branching network that is consistent with other evidence on population origins.

5. Gene frequencies can also be used to study the origins of hybrid populations, such as the American blacks. It has been shown, for example, that while most American black populations have approximately 78% African ancestry and 22% European, a few isolated populations have a much greater proportion of African genes.

6. An example of a major genetic variation among populations is lactase deficiency or, its alternate, persistence of intestinal lactase into adulthood. Lactase is required to break down lactose, the principal sugar of milk. In most species and in most persons, lactase decreases after weaning, and such persons are intolerant of large amounts of milk. If lactase persists, milk is tolerated. Persistence of lactase is a simple dominant trait with highest frequency in Sweden and low frequencies in most non-European populations.

7. IQ is a measure of intelligence, but IQ tests that are standardized in one population cannot be transferred to other populations and give comparably reliable results. The average IQ scores of U.S. blacks, as measured by conventional intelligence tests standardized in whites, is lower than the average scores for whites. Although there is ample evidence that the variations in IQ in whites are strongly influenced by heredity, the question of whether the differences between races is inherited is not meaningful, since the validity of the IQ test as a comparable measure of intelligence in blacks is not established.

8. In theory, hybrid populations may have greater vigor because of increased heterozygosity, or they may have less vigor because of the breakup of coadapted gene complexes. Critical observations in human hybrid populations are limited, but the studies that have been done show no effect of hybridization, supporting the idea that all groups are basically similar genetically, even though they may differ in the frequency of alleles at a variety of loci.

REFERENCES AND SUGGESTED READING

BUETTNER-JANUSCH, J. 1973. *Physical Anthropology: A Perspective.* New York: John Wiley, 572 pp.

CAVALLI-SFORZA, L. L., BARRAI, I., AND EDWARDS, A. W. F. 1964. Analysis of human evolution under random genetic drift. *Cold Spring Harbor Symp. Quant. Biol.* 29: 9–20.

CAVALLI-SFORZA, L. L., AND BODMER, W. F. 1971. *The Genetics of Human Populations.* San Francisco: W. H. Freeman, 965 pp.

CAVALLI-SFORZA, L. L., AND EDWARDS, A. W. F. 1967. Phylogenetic analysis: models and estimation procedures. *Am. J. Hum. Genet.* 19: 233–257.

CHING, G. H. S., AND CHUNG, C. S. 1974. A genetic study of cleft lip and palate in Hawaii. I. Interracial crosses. *Am. J. Hum. Genet.* 26: 162–176.

DOBZHANSKY, T. 1970. *Genetics of the Evolutionary Process.* New York: Columbia Univ. Press, 505 pp.

ELSTON, R. 1971. The estimation of admixture in racial hybrids. *Ann. Hum. Genet.* 35: 9–17.

FIRSCHEIN, I. L. 1961. Population dynamics of the sickle-cell trait in the Black Caribs of British Honduras, Central America. *Am. J. Hum. Genet.* 13: 233–254.

FLATZ, G., AND ROTTHAUWE, H. W. 1977. The human lactase polymorphism: physiology and genetics of lactose absorption and malabsorption. *Prog. Med. Genet.* n.s. 2: 205–249.

GARN, S. M. 1971. *Human Races,* 3rd Ed. Springfield, Ill.: Charles C Thomas, 196 pp.

GLASS, B., AND LI, C. C. 1953. The dynamics of racial intermixture—an analysis based on the American Negro. *Am. J. Hum. Genet.* 5: 1–20.

GLASS, B., SACHS, M. S., JAHN, E., AND HESS, C. 1952. Genetic drift in a religious isolate: an analysis of the causes of variation in blood group and other gene frequencies in a small population. *Am. Naturalist* 86: 145–159.

GOLDSBY, R. A. 1977. *Race and Races,* 2nd Ed. New York: Macmillan, 158 pp.

HARRISON, G. A. (Ed.). 1961. *Genetical Variation in Human Populations.* New York: Pergamon Press, 115 pp.

LISKER, R., GONZALEZ, B., AND DALTABUIT, M. 1975. Recessive inheritance of the adult type of intestinal lactase deficiency. *Am. J. Human Genet.* 27: 662–664.

LOEHLIN, J. C., LINDZEY, G., AND SPUHLER, J. N. 1975. *Race Differences in Intelligence.* San Francisco: W. H. Freeman, 380 pp.

MAC LEAN, C. J., AND WORKMAN, P. L. 1972. Genetic studies of hybrid populations. I. Individual estimates of ancestry and their relation to observations on quantitative traits. *Ann. Hum. Genet.* 36: 341–351.

MORTON, N. E., CHUNG, C. S., AND MI, M.-P. 1967. *Genetics of Interracial Crosses in Hawaii.* Monographs in Human Genetics, No. 3. Basel/New York: S. Karger, 158 pp.

MOURANT, A. E., KOPEĆ, A. C., AND DOMANIEWSKA-SOBCZAK, K. 1976. *The Distribution of the Human Blood Groups and Other Polymorphisms,* 2nd Ed. New York: Oxford Univ. Press, 1100 pp.

NEEL, J. V., SALZANO, F. M., JUNQUEIRA, P. C., KEITER, F., AND MAYBURY-LEWIS, D. 1964. Studies on the Xavante Indians of the Brazilian Mato Grosso. *Am. J. Hum. Genet.* 16: 52–140.

OSBORNE, R. H. (Ed.). 1971. *The Biological and Social Meaning of Race.* San Francisco: W. H. Freeman, 182 pp.
REED, T. E. 1969. Caucasian genes in American Negroes. *Science* 165: 762–768.
WINSOR, E. J. T., AND WELCH, J. P. 1978. Genetic and demographic aspects of Nova Scotia Niemann-Pick disease (Type D). *Am. J. Hum. Genet.* 30: 530–538.

REVIEW QUESTIONS

1. Define:

race	heterosis
genetic drift	parallel evolution
founder effect	exogamous
hybrid population	endogamous
coadapted genes	

2. There are two alleles at the haptoglobin locus, Hp^1 and Hp^2. The frequency of Hp^1 in whites is 0.4 and in American blacks is 0.6. Other genetic markers indicate that the latter population is 22% white. What would be the expected frequency of Hp^1 in the African population that contributed to the American blacks?

3. Several studies of Hp frequencies in African and European populations suggest that for this locus the present U.S. black population contains 30% white genes rather than 22% as suggested by other loci. How might this disparity arise?

4. The gene for albinism occurs at much higher frequencies in many American Indian groups than in other populations of the world. This condition would seem to be distinctly disadvantageous in a society based on hunting and gathering. What explanations might account for the high frequency?

5. Cystic fibrosis occurs in Europeans approximately 1 in 2500 births. It occurs with a frequency near zero in Japanese. A large population of Europeans undergoes random mating with a population of Japanese of equal size. What would be the frequency of cystic fibrosis: (a) In the F_1 generation? (b) After many generations of random mating? Ignore the effects of selection.

6. Devise a hypothesis for the selective advantage of intestinal lactase persistence in Europeans but not in the rest of the world.

CHAPTER TWENTY-FOUR

GENETIC COUNSELING

Genetic counseling has become an important part of health care. Although the availability of genetic counseling is still very limited, most physicians and medical centers now recognize the need for counseling faced by many patients and families. Increasingly, counseling services are being developed to provide understanding of genetic risk and assist families in dealing with that risk.

Just what is meant by genetic counseling has changed with time and is still evolving. One definition arrived at by a group of experienced counselors is as follows (Committee on Genetic Counseling, 1975):

> Genetic counseling is a communication process which deals with the human problems associated with the occurrence, or the risk of occurrence, of a genetic disorder in a family. This process involves an attempt by one or more appropriately trained persons to help the individual or family to (1) comprehend the medical facts, including the diagnosis, probable course of the disorder, and the available management; (2) appreciate the way heredity contributes to the disorder, and the risk of recurrence in specified relatives; (3) understand the alternatives for dealing with the risk of recurrence; (4) choose the course of action which seems to them appropriate in view of their risk, their family goals, and their ethical and religious standards, and to act in accordance with that decision; and (5) to make the best possible adjustment to the disorder in an affected family member and/or to the risk of recurrence of that disorder.

Such a definition involves substantially more than providing genetic risks to those who may be concerned. Often, persons who come for counseling have additional needs unrelated to obtaining risk estimates, and it is important for the counselor to discern the emotional problems that have generated the apparent need for genetic counseling or that might result from counseling. A fully trained counselor must understand more than the basic genetic issues in a case. In this discussion, however, we will largely limit ourselves to the genetic questions.

WHO SEEKS COUNSELING?

By far the greatest number of problems presented to counselors are from persons or families concerned about the risk of recurrence of a disorder that has affected one or more members of the family. Typical examples would be parents who have had an affected child and want to know the risk in subsequent pregnancies. The mother may already be pregnant and want to know the risk for the fetus. In other instances the parents may have affected relatives and want the risk that the disorder will occur in their own family.

A somewhat different situation is the couple who may be contemplating having children and who are concerned about the risk of defect. Their concern may stem from being members of a group with increased risk for a specific disorder, such as Ashkenazi Jews and Tay–Sachs disease. Or they may be related to each other and are concerned about the risk faced by children of consanguineous marriages. These couples may be at low risk, but they need information on the magnitude of the risk.

PROCEDURES IN COUNSELING

Genetic counseling consists of several steps, some of which may occur simultaneously. The foundation of any counseling process is diagnosis. Patients referred for counseling need to have the diagnosis confirmed if possible. Many patients are referred to counseling centers without a definitive diagnosis, and one must be established, either by physicians in the genetics center or by referral to other medical specialists. Often the special tests available in a genetics center will assist in the diagnosis.

One of the first steps in counseling is the preparation of a pedigree of the patient or family seeking assistance. Such information may be important in distinguishing among different disorders, and the risk will be altered by knowledge that a condition is dominant, say, rather than recessive. The pedigree is drawn up on the basis of interview information, to which is added laboratory information when it becomes available. Whatever the source and nature of the information, it is brought together in the pedigree for evaluation and decision concerning the nature of the genetic problem, if one exists.

Communication of risks is the heart of genetic counseling. Families must be helped to understand the risks, including the probability that a child will *not* be affected, and the alternative courses of action. This phase has been designated *informative counseling* (Fraser, 1976). In addition to providing families with recurrence risks for a particular disorder, the counselor must provide information on courses of action. For example, prenatal diagnosis may be possible. Or the families may need information on related issues, such as sterilization or adoption.

Supportive counseling is often a responsibility of the genetic counselor, simply because families need professional help and the genetics counselor may be the person most available. It is for this reason especially that counselors need to be trained in the broad aspects of counseling as well as in the more specific areas of genetics.

In the course of providing service to families, the counselor or the counseling center acquire responsibilities for other relatives not seeking counseling. This happens when other persons are found to be at risk because of their relationship to a person who has sought help. For example, finding that the mother of a child with Down's syndrome is a balanced translocation carrier means that her sibs have a chance also of being carriers. If the mother is unwilling to have her sibs informed, the counselor is caught between the obligation to inform persons of their status and the obligation not to invade the privacy of persons. This conflict has yet to be resolved, although it can often be worked out as part of the counseling process.

ASSESSING GENETIC RISKS

Genetic Risks for Mendelian Traits

The simplest genetic issue facing a counselor is presented by those traits inherited in simple Mendelian fashion. The basic information is provided in previous chapters, but some typical problems are presented here.

Risks Among Relatives. It is frequently of interest to know the likelihood of heterozygosity for a recessive gene among the relatives of an affected person. Some simple rules will enable the calculation to be made. It two normal parents have one or more homozygous recessive offspring, then the mating must have been of the type $Aa \times Aa$. The probability that each parent is heterozygous is $P(Aa) = 1$. The same is true of any offspring of a homozygous person. Each parent received the a from one of the grandparents, but it could have come from either the grandmother or the grandfather. Hence each has a one-half risk of being heterozygous. Whichever it may have been, normal sibs of the parents have a half chance of being heterozygous. Offspring of heterozygous persons in the mating $AA \times Aa$ have a one-half chance of receiving a.

Normal sibs of aa person are of two types if the parents are both Aa. Some are AA and others Aa. The usual F_2 ratios should apply, giving $1AA:2Aa:1aa$. Since only normal sibs are under consideration, the ratio is $1AA:2Aa$; that is, $P(AA) = \frac{1}{3}$, $P(Aa) = \frac{2}{3}$.

The probability that a relative of an affected person is heterozygous for the gene is obtained by combining the probabilities of persons through whom they are related. Since a specific combination of events is required, the probabilities are multiplied together. An example is in Figure 24-1. The probability that III-5 is *Aa* is the product of the probability that II-3 is *Aa* (1) *times* the probability that II-4 is *Aa* if II-3 is *Aa* ($\frac{1}{2}$) *times* the probability that III-5 is *Aa* if II-4 is *Aa* ($\frac{1}{2}$). Algebraically,

$$P(\text{III-5 is } Aa) = P(\text{II-3 is } Aa) \times P(\text{II-4 is } Aa \text{ if II-3 is } Aa) \times P(\text{III-5 is } Aa \text{ if II-4 is } Aa)$$

or

$$P(\text{III-5 is } Aa) = (1)\left(\tfrac{1}{2}\right)\left(\tfrac{1}{2}\right) = \tfrac{1}{4}.$$

Similarly, the probability for IV-1 is

$$P(Aa) = \left(\tfrac{2}{3}\right)\left(\tfrac{1}{2}\right) = \tfrac{1}{3}$$

Occasionally, prospective marriage partners, each with an affected relative, wish to know the likelihood of their having affected children. The pedigree is drawn and the probability that each parent is heterozygous is calculated. The probability that both are heterozygous is the product of the probabilities for each. If both parents are heterozygous, each child would have a one-fourth chance of being affected and a three-fourths chance not to be. The probability that the first child would be affected is

$$P(aa) = P(Aa \text{ mother}) \times P(Aa \text{ father}) \times \tfrac{1}{4}.$$

The probability that all of three children would be affected is

$$P(3\ aa) = P(Aa \text{ mother}) \times P(Aa \text{ father}) \times \left(\tfrac{1}{4}\right)^3.$$

The probability that none would be affected is $1 - P(1$ or more affected). For n children it is

$$P(\text{no } aa) = 1 - \left[P(Aa \text{ mother}) \times P(Aa \text{ father}) \times \left(1 - \left(\tfrac{3}{4}\right)^n\right)\right].$$

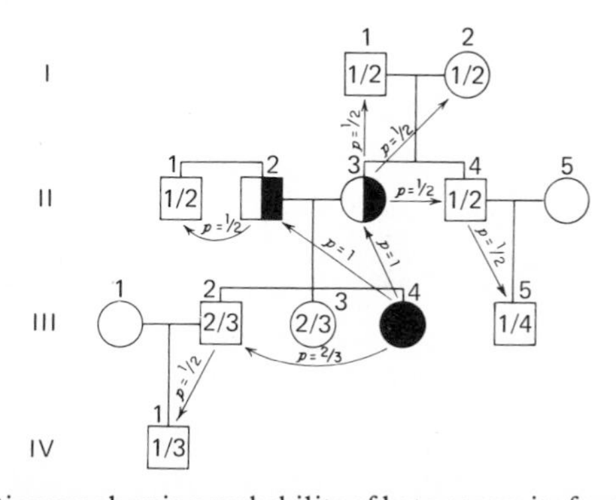

Figure 24-1 Diagram showing probability of heterozygosity for persons related to a homozygous recessive. Figures on arrows are probabilities of heterozygosity, assuming that the person the arrow is pointing from is heterozygous (except in the case of the homozygous recessive). Figures inside the symbols are the total probabilities of heterozygosity.

These calculations ignore the possibility of introduction of an *a* allele through other relatives. The likelihood that this might occur is very small unless *a* is common. They also are based on a priori considerations, that is, without knowledge that a normal (or affected) child has indeed been born. The birth of either may alter the subsequent risk very substantially.

Complex Pedigrees. Pedigrees may sometimes be subject to more than one genetic interpretation. The likelihoods of the various possible genotypes can nevertheless be calculated by setting up all possible interpretations for one observed pedigree weighted according to the probability of each. This procedure is referred to as the *Bayesian method.* Bayesian methods are used in may areas of science but are especially helpful in genetic counseling, where one often is faced with assessing the relative likelihoods of alternate explanations for past events, even though the ultimate objective is to state the likelihood of events yet to occur.

For an example, let us consider the pedigree in Figure 24-2. III-2 has a one-fourth chance of being heterozygous for albinism (ignoring the small additional chance that she might be heterozygous by virtue of inheriting an allele for albinism from her father, II-4). The probability that IV-1 would be albino is therefore one-eighth. If IV-1 turns out to be albino, then III-2 clearly is heterozygous, and all subsequent offspring would have one-half chance of being albino. However, if IV-1 is normally pigmented, this diminishes slightly the chance that III-2 is heterozygous.

To calculate the exact probability, we may consider all the possible genotypes prior to the birth of IV-1, weighted according to their relative

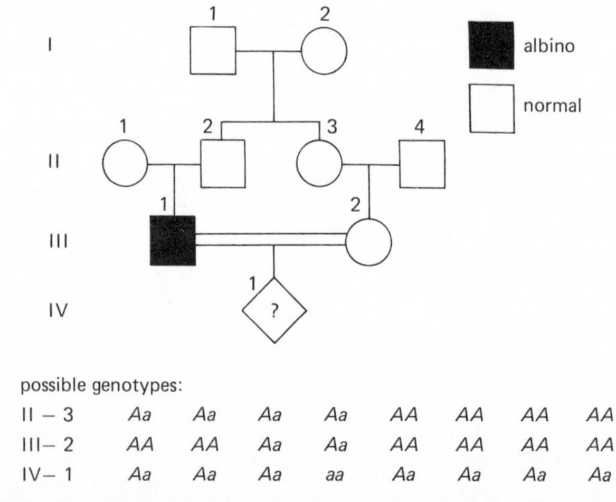

Figure 24-2 A pedigree of albinism, with marriage between first cousins, one of whom is an albino. Each column of genotypes is equally likely to occur.

frequencies, and then consider which of these is ruled out by the normal phenotype of IV-1. As diagrammed in Figure 24-2, II-3 has a one-half chance a priori of being *Aa*, III-2 has a one-fourth chance of being *Aa*, and IV-1 has a one-eighth chance of being *aa*. But if IV-1 is normal, this eliminates the column corresponding to this outcome. The new probabilities are $\frac{3}{7}$ that II-3 is *Aa*, and $\frac{1}{7}$ that III-2 is *Aa*. The risk of the second child of III-2 being albino is $\frac{1}{2} \times \frac{1}{7} = \frac{1}{14}$. If the second child is normal, this further reduces the probability that II-3 is heterozygous to $\frac{5}{13}$ and that III-2 is heterozygous to $\frac{1}{13}$. The risk for albinism for a third child would be $\frac{1}{26}$.

Expressed somewhat differently, we may define three outcomes:

1. The probability that III-2 is *AA* and both children are normal $= \frac{3}{4}$.

2. The probability that III-2 is *Aa* and both children are normal $= \frac{1}{4} \times \frac{1}{4} = \frac{1}{16}$.

3. The probability that III-2 is *Aa* and one or both children are $aa = \frac{1}{4} \times \frac{3}{4} = \frac{3}{16}$.

But the last possibility did not occur. Of the still acceptable possibilities, the probability that III-2 is *Aa* is $(2)/[(1) + (2)] = \frac{1}{16} \div (\frac{12}{16} + \frac{1}{16}) = \frac{1}{13}$.

Diseases of Late Onset. Not every genetic disease can be diagnosed at birth. This counseling problem is exemplified by Huntington's disease. If a parent has Huntington's disease, each child would have a one-half chance of receiving the Huntington gene from that parent. Since the gene is very rare, only marriages between heterozygotes and homozygous recessives will ordinarily be seen. Therefore, the risk for each child would be one-half.

Huntington's disease is a disease of late onset, however. The diagnosis is most often made when a person is 25 to 40 years old. Rarely, teenagers are diagnosed, and sometimes symptoms may not be present definitely until the affected person is in his sixties. Therefore, a parent may transmit the gene, even though he himself has not exhibited symptoms that would lead to a diagnosis. The situation may be complicated by the fact that the potential carrier parent has died of other causes prior to exhibiting symptoms.

For a disease of late onset, the estimated risk of transmitting the abnormal allele when the potential carrier parent has yet to express the trait should include an estimate of the likelihood that the parent is heterozygous. This is not simply one-half; rather, it is one-half multiplied by the proportion of heterozygotes of similar age and other relevant variables who empirically have yet to express the trait. Referring to Figure 24-3, only 25% of heterozygotes aged 40 years have failed to receive a diagnosis. Therefore, a 40-year-old person at risk (that is, with an affected parent) would have $(\frac{1}{2} \times .25)/[(\frac{1}{2} \times .25) + (.5)] = 20\%$ chance of being a heterozygote. Each of his children would have one-half this risk of receiving the gene also.

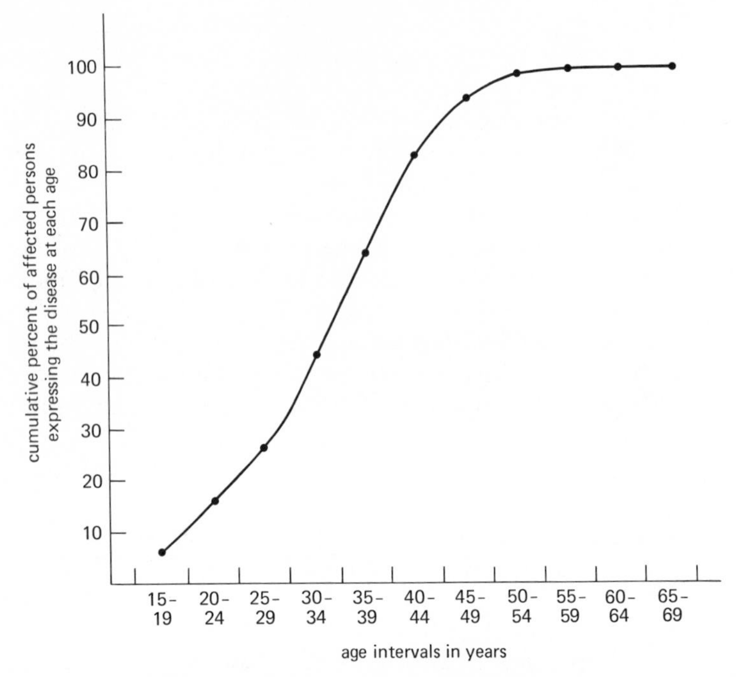

Figure 24-3 The percent of persons with Huntington's disease who express the disease (choreiform movements) at each age range. Although nearly all heterozygotes express the disease by age 50, 2 of 204 patients studied were first diagnosed above age 65. (*Data from Reed and Chandler, 1958.*)

Reduced Penetrance. Sometimes a genotype is not expressed, and a particularly difficult problem occurs in assessing the status of a person at risk for an inherited disease with reduced penetrance. In the case of retinoblastoma, the risk of developing tumors in one or both eyes if one receives the dominant gene for this condition is approximately 95%. A child whose parent was affected with inherited retinoblastoma therefore has *a priori* a 2.5% risk of carrying the gene without expressing it. As each child passes through the age of risk, the likelihoods are altered. Consider the pedigree in Figure 24-4. Although II-2 is a normal adult, prior to the birth of any children there is a 4.8% chance that she carries the gene for retinoblastoma. On the other hand, if generation III were beyond the age of risk, the likelihood that II-2 is a carrier is the sum of the likelihoods that she has the gene and did not transmit it plus the likelihood that she transmitted it but it was not expressed. This figure is divided by the sum of the two probabilities of the pedigree, that is, on the hypothesis that II-2 has the gene and the hypothesis that II-2 does not, in order to get the overall probability, 0.0256, that II-2 is a carrier.

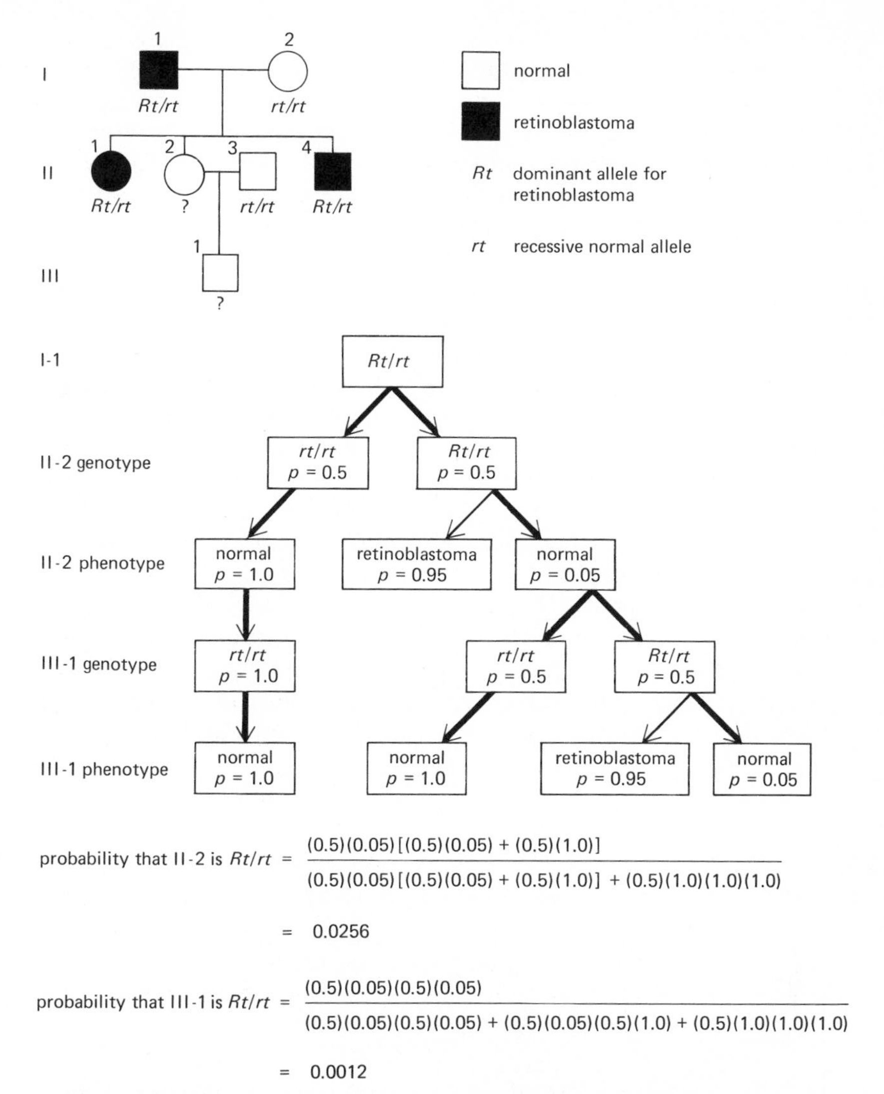

Figure 24-4 Pedigree of retinoblastoma analyzed for likelihood that normal persons at risk actually are heterozygous. The diagram below the pedigree shows the possible genotypes, those with light arrows being inconsistent with the pedigree. The calculation of probabilities consists in selecting those outcomes in generation III that are of interest and dividing by all possible outcomes consistent with the pedigree.

Empiric Risks

The precise modes of heredity for many conditions are not known. Diabetes and gout are examples of frequently occurring conditions in which there is clearly a strong genetic component, probably involving interaction with environment and other genetic systems. It is possible in such cases to predict recurrence entirely on the basis of frequency of recurrence in other families. A couple with no affected offspring will have a certain risk for a particular defect; a couple who has produced one affected child will have a higher risk; and a couple with two affected children will have perhaps a different risk for a third affected child. The amount of information available about a couple influences the predicted risk value. The more information available, the more accurate the predictions concerning future children.

Good empiric risk data are available only for a few conditions. For example, spontaneous abortions are due to many complex factors, some genetic and some environmental. In a large study, Warburton and Fraser found that the average risk of abortion is 15%. However, if a woman has already had one or more abortions, the risk increases to 25%. In spite of our lack of understanding of the etiology of abortion, the risk can be measured accurately and used in advising persons regarding their own risk.

An example of empiric risks is given in Table 24-1. Such a table is useful to estimate the likelihood that a child will be born with congenital pyloric stenosis. For example, if a family with no children came in for counseling, and if the father had been affected as an infant, the likelihood that the first child would be affected is 3.7%. If there is already one affected boy, the risk for the second child would be 10.2%. In this particular condition, the risks fall into groups, depending on whether

Table 24-1 **EMPIRIC RISKS FOR CONGENITAL PYLORIC STENOSIS**

The table has been constructed from risks among first degree relatives and in the general population in England. This table is abbreviated from a more extensive presentation in Bonaiti-Pellié and Smith (1974). Values for risk are in percent.

Index family	Neither parent affected	Father affected	Mother affected	Both parents affected
No sibs	.3	3.7	5.1	29.8
1 sib U*	.3	3.4	4.7	27.1
1 M sib A	3.2	10.2	12.1	35.3
1 F sib A	4.6	12.8	14.8	37.8
1 F sib A + 1 sib U	4.3	11.6	13.5	34.9
1 M + 1 F sib A	10.7	20.2	21.7	42.0
2 F sibs A	13.0	23.1	24.5	44.2
2 F sibs A + 1 sib U	11.9	21.0	22.5	41.2

Abbreviations: M = male; F = female; A = affected; U = unaffected.
* The sex of the unaffected sib does not alter the risk and hence is not indicated.

there has been none, one, or two affected children already born and depending on whether one or both of the parents was affected.

With empiric risks, it should be kept in mind that a particular family may have a very different risk from the population as a whole. Among close relatives of cancer patients, the increased risk is of the order of 2 to 3%. There are very rare families in which the risk appears to be higher. Proper counseling should begin with a thorough investigation of the pedigree of the family under consideration to determine whether such a family is involved.

Another problem with empiric risks is the difficulty in transferring the risks to other populations in which the gene frequencies of the (unidentified) contributing genes might differ. If the condition is genetically heterogeneous, with different causes in different populations, the risk figures derived from one population may hold only for that one population.

Linkage and Genetic Risk

The linkage of detrimental genes with genetic markers has yet to find routine application in counseling, but the possibilities are clearly great once the linkage map becomes sufficiently complete. The principal use will be in recognizing the presence of unexpressed genes. For example, the loci for G6PD structure (*Gd*) and for hemophilia A (*He*) are close together on the X chromosome. If a woman is heterozygous for hemophilia A each male fetus has a 50% chance of having received the gene for hemophilia He^- and developing into a hemophiliac. However, if the woman also is heterozygous at the *Gd* locus, say Gd^A/Gd^B, and if study of her family reveals the coupling relationship, then the predictions for the fetus may be altered. Suppose that in her genome Gd^B is coupled to He^-. Assuming a recombination of 0.05 cM, a male fetus that has red cells of G6PD type A will have only 5% risk of having also He^-. Conversely, if the fetus received Gd^B, the likelihood of receiving He^- is 95%. Such very different risks would have substantial bearing on whether or not to continue a pregnancy.

One instructive example has been reported in which prenatal diagnosis for a fetus at risk for myotonic dystrophy (MD) was approached by virtue of the close linkage of *MD*, a dominant gene, to the secretor locus, *Se* (Chapter 13). MD is a late onset disease that cannot be directly diagnosed *in utero*. The secretor locus is very close to *MD*, approximately 8 cM distant. The family represented in Figure 24-5 sought counseling when faced with the possibility that III-2 would inherit MD. It turned out that the mother, II-2, was doubly heterozygous for MD and secretor status, with the dominant allele *MD* coupled to the recessive allele *se*. Her husband was a secretor, but it could not be determined whether he was homozygous or heterozygous. The amniotic

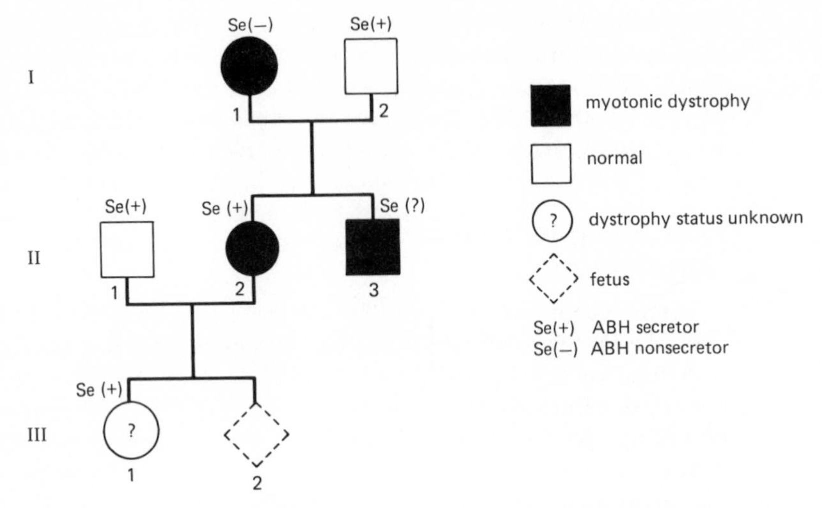

Figure 24-5 Pedigree in which myotonic dystrophy and secretor status are segregating. II-2 is heterozygous both for myotonic dystrophy and secretor status, with the dystrophy allele coupled to the recessive allele for nonsecretion. The secretor genotype of II-1 is either *Se*/*Se* or *Se*/*se*. (*Abbreviated from Schrott et al., 1973.*)

fluid was found to possess the secretor factor, indicating the fetus to be secretor-positive. Had the amniotic fluid been secretor-negative, the fetus would have inherited the *se* allele from the mother with a 92% chance that the *MD* allele would be on the same chromosome. Calculation of the probability for the *MD* allele in the face of the secretor-positive fetus is complicated by the fact that the *Se* allele could have come from either the father or the mother, and indeed the father could have been homozygous *Se*/*Se*. But the risk of myotonic dystrophy is slightly less that the 50% risk obtained by ignoring the secretor status. The parents elected to continue the pregnancy under the circumstances.

As yet there are far too few close linkages with polymorphic markers for this approach often to be of value. If one considers conditions such as Huntington's disease, it would be especially helpful to be able to say on the basis of closely linked markers that the risk of a young adult having the disorder is 3% rather than 50%. This would permit intelligent decisions on family planning while the person at risk is still young enough to find the plans useful. As the linkage map becomes more saturated, the value of linkage studies in genetic counseling will increase.

SOME ISSUES IN GENETIC COUNSELING

The growing interest in genetic counseling has raised a number of issues not previously faced. The question of who should do genetic counseling is one that will not be discussed here except to note that the persons and groups should be qualified to respond not only to the genetic questions but also to the need for follow-up and other support. A sample of other issues illustrates the extent to which counseling is tied in with areas such as law and ethics as well as all aspects of medicine.

How Much Information Should Be Provided by Counselors?

It might seem that all information acquired on a patient or family should be made available to the person concerned. Recent judicial rulings tend to support this view. Sometimes, however, the information may be detrimental, and it can be argued that some information should be exempt. In the course of studying a patient for genetic markers, the geneticist may learn of illegitimacy. Is he required to make that information known, even though it is not relevant to the purpose of the counseling and will cause emotional distress?

A related question occurs in the case of the XYY syndrome. Most evidence supports an increased risk of criminal aggression in persons with this karyotype. The magnitude of the risk may be small, but, so long as it is greater than zero, the problem remains. When a child is identified as having XYY karyotype, should his environment be structured deliberately to minimize development of aggression? Or should such structuring await an overt act of aggression that may well lead to a prison term? Our tradition of declaring a person innocent until proved guilty leads us to reject a procedure that may resemble a sentence based on potential guilt only. On the other hand, does not everyone deserve an environment that maximizes the opportunity to develop into a responsible citizen even though the environment may have to be matched to individual needs?

The issue of behavior control is one that encompasses much more than genetics. There is little agreement on whether behavior modification should be used, with or without the consent of the subject. To provide a structured environment to all XYY males, even though some would never become criminals, is offensive to many and perhaps wasteful of resources also. Clearly, much more needs to be learned about the risk in XYY males. It is possible that other forms of destructive behavior may also prove to have a genetic component. The authority and responsibility of society to deal with these situations is likely to increase as a subject of debate and perhaps as a fact.

Should Genetic Information Be Made Available to Persons Other Than Those Seeking Counseling?

The first response to this question may be that a person's genotype is his own business and that to make the information available to others is an unwarranted invasion of privacy. This may often be the case, but knowledge of genotype, especially carrier status for inherited diseases, often is first in the possession of the medical profession before it is known even to the affected persons. The release of information such as blood types may seem trivial, but the consequences may not be trivial if the blood types reveal illegitimacy. Releasing the information only to the person tested does not entirely avoid the problem. For example, the person may be too young to handle the impact of the information. In that case, what are the physician's responsibilities to withhold information? And can he withhold it legally? If withheld then, what are his responsibilities to seek out the person at a later date to transmit it? Also, what right do relatives, real or putative, have to genetic information on persons other than themselves? And what responsibility does a physician have to transmit information to other persons?

As an example, let us consider that a young woman is karyotyped and found to be heterozygous for a balanced translocation involving chromosome 21. A physician with this information would surely advise her of the risk of a child with Down's syndrome. If she were not an adult, presumably her parents or guardian could be advised. Suppose that she chooses to keep the information to herself rather than advising her normal sibs, who would have approximately a one-half chance also of carrying the translocation. If such a sib subsequently has a child with Down's syndrome, is this sufficient basis for a suit against the physician for failure to notify the sib of the potential risk? The handling of infectious disease may again serve as a precedent, but inherited diseases may present more difficult problems, since "exposure" in this instance may be through distant ancestors, and the relatives at risk may be geographically remote. The conflict between the right of the individual to privacy and the right of others to know has many arenas.

For another example, consider a man diagnosed as having Huntington's disease. His children each have a one-half chance of being affected. At what age should they be told, and whose is the responsibility for seeing that they are told? If a person known to be at risk by a physician marries and has children, only then to learn that he is affected with Huntington's disease, is the physician then subject to legal action for failing to advise him so that the social and biological responsibilities of a family could have been avoided?

Many of the above issues have been solved by individual judgment on the part of physicians, sometimes successfully, sometimes less so. But in a society that is increasingly litigation-minded, the wisdom of the individual action may not be in accord with legal requirements of the time. The public as a whole must decide where to draw the line between the rights of society and the rights of an individual. Is the gene pool public property or not? If so, what is the right of the public to manage it? With respect to information management, it may be necessary to resort to a nationwide information retrieval system so that the physician can discharge his responsibilities to relatives by entering the genetic information on a patient in the data bank. It would be the responsibility of the relatives to recall information relevant to themselves by establishing genetic relationships to other persons. Such a system would be complicated but might offer the best compromise between privacy and the need of others to know. Genetic registries have been tried—for example, in Denmark—but none has been developed to cope with the large amount of genetic information now possible and with modern social and legal attitudes.

SUMMARY

1. Genetic counseling consists in assessing the risk of occurrence or recurrence of a hereditary problem, communicating that risk to the persons seeking assistance, and helping them understand the nature of the problem and the possible choices available for dealing with it.

2. For simple Mendelian traits, the risk can be assessed from the person's pedigree. In complex pedigrees, it is often necessary to use Bayesian methods. These permit one to assign probabilities of various genotypes for persons whose genotypes are ambiguous, based on the likelihood of having produced the observed pedigree under various genetic hypotheses. Problems in counseling may be encountered because of late onset of disease and because of reduced penetrance. Each of these can be overcome if sufficient information is available from other families.

3. Even though conditions are not inherited by simple Mendelian mechanisms, estimates of risk can be given purely on the basis of empiric data. However, it should be kept in mind that the actual risk may differ if the particular family differs from the population as a whole. For example, some families have a high risk of cancer among close relatives of index cases, whereas in the general population the risk among close relatives is only slightly increased.

4. As more information becomes available on human linkage groups, these should be very helpful in providing accurate estimates of risk for diseases of late onset or in prenatal diagnosis. For example, if a genetic marker is segregating in the family and is very closely linked to a locus that is not yet expressed, it may be possible to assign a very low risk (or a very high risk) of having received a particular allele at that locus.

5. The availability of counseling raises several issues of a legal and ethical nature. These include whether and under what circumstances information should be withheld from persons seeking counseling. Also, to what extent should (or must) the counselor invade the privacy of a patient by communicating with relatives who also are at risk?

REFERENCES AND SUGGESTED READING

BERGSMA, D. (Ed.). 1973). Contemporary genetic counseling. *Birth Defects: Original Article Series* 9. The National Foundation-March of Dimes, Washington, D.C. 48 pp.

BONAITI-PELLIÉ, C., AND SMITH, C. 1974. Risk tables for genetic counselling in some common congenital malformations. *J. Med. Genet.* 11: 374–377.

COMMITTEE ON GENETIC COUNSELING. 1975. Genetic counseling. *Am. J. Hum. Genet.* 27: 240–242.

FRASER, F. C. 1974. Genetic counseling. *Am. J. Hum. Genet.* 26: 636–659.

FRASER, F. C. 1976. Genetics as a health-care service. *N. Engl. J. Med.* 295: 486–488.

FUHRMANN, W., AND VOGEL, F. 1976. *Genetic Counseling,* 2nd Ed. Springer-Verlag, New York/Heidelberg/Berlin. 138 pp.

HILTON, B., CALLAHAN, D., HARRIS, M., CONDLIFFE, P., AND BERKELEY, B., (Eds.). 1973. *Ethical Issues in Human Genetics.* Plenum Press, New York. 455 pp.

LEONARD, C. O., CHASE, G. A., AND CHILDS, B. 1972. Genetic counseling: a consumer's view. *N. Engl. J. Med.* 287: 433–439.

LUBS, H. A., AND DE LA CRUZ, F. (Eds.). 1977. *Genetic Counseling.* New York: Raven Press, 598 pp.

MAYO, O. 1970. The use of linkage in genetic counseling. *Hum. Hered.* 20: 473–485.

MILUNSKY, A., AND ANNAS, G. J. (Eds.). 1976. *Genetics and the Law.* New York: Plenum Press, 532 pp.

MURPHY, E. A., AND MUTALIK, G. S. 1969. The application of Bayesian methods in genetic counseling. *Hum. Hered.* 19: 126–151.

REED, T. E., AND CHANDLER, J. H. 1958. Huntington's chorea in Michigan. I. Demography and genetics. *Am. J. Hum. Genet.* 10: 201–225.

REILLY, P. 1977. *Genetics, Law, and Social Policy.* Cambridge, Mass.: Harvard Univ. Press, 275 pp.

RICCARDI, V. M. 1977. *The Genetic Approach to Human Disease.* New York: Oxford Univ. Press., 273 pp.

SCHROTT, H. G., KARP, L., AND OMENN, G. S. 1973. Prenatal prediction in myotonic dystrophy: guidelines for genetic counseling. *Clin. Genet.* 4: 38–45.

STEVENSON, A. C., AND DAVISON, B. C. C. 1976. *Genetic Counseling,* 2nd Ed. Philadelphia: J. B. Lippincott, 357 pp.

WARBURTON, D., AND FRASER, F. C. 1964. Spontaneous abortion risks in man: Data from reproductive histories collected in a medical genetics unit. *Am. J. Hum. Genet.* 16: 1–25.

REVIEW QUESTIONS

1. Define:

penetrance
Bayesian method
empiric risk

2. A family seeks counseling on the risk of Huntington's disease. The father is 25 years old and normal, but his mother was recently diagnosed as being affected with Huntington's disease. (a) What is the likelihood that he is heterozygous? (b) Twenty years later the same family returns. They now have a 15-year-old normal son. The father still shows no signs of Huntington's disease. What is the likelihood that the son has the gene?

3. A Frenchwoman is blood type A. (a) What is the probability that she is homozygous $I^A I^A$? (b) She marries a type O man, and their first child is type A. What is the new probability that she is homozygous? (c) How would a second type A child affect this answer?

4. A couple seeks genetic counseling because they have learned from their relatives that a form of X-linked mental retardation is in the wife's family. The counselor reconstructed the pedigree as shown. The couple already has one normal son. What is the risk that an additional son would have mental retardation?

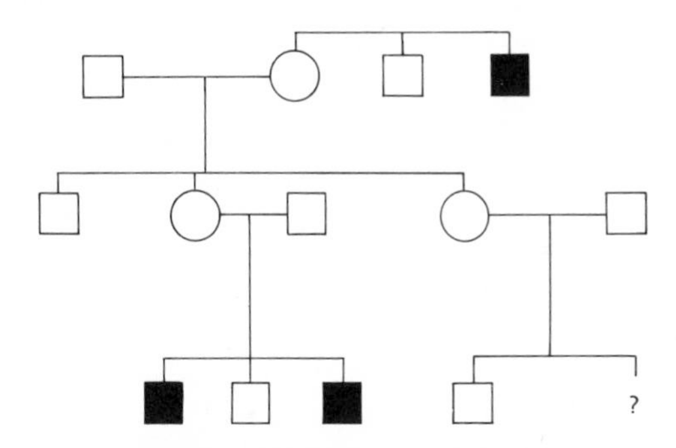

5. A couple seeks counseling because the woman's father is hemophilic and they are concerned that the child they are expecting may also be affected. Typing for various genetic markers reveals that the woman is heterozygous for G6PD, being Gd^A/Gd^B, her father has type B G6PD, and her husband also has type B G6PD. Amniocentesis shows the fetus to be a male with type B G6PD. What is the likelihood that the fetus has the gene for hemophilia? (Assume the *Gd* and hemophilia loci are 4 cM apart.)

CHAPTER TWENTY-FIVE

GENETIC SCREENING

Much of the tragedy of inherited disease could be prevented if parents at risk for producing an affected child could be identified beforehand or if affected newborns could be diagnosed and treated before degenerative changes occur. To accomplish this requires the availability of testing procedures for large numbers of persons, with the ability to provide treatment or other assistance as needed.

Genetic screening has raised a number of questions concerning the effects on the gene pool and the rights of individuals to have screening or *not* to have it. Some of these issues apply also to genetic counseling, the difference being largely whether genetic intervention is sought by individuals or provided (or required) by public agencies. One issue, especially in an ethnically heterogeneous population such as the U.S., is the limiting of screening for a particular trait only to that racial or ethnic group at high risk. For example, homozygotes for sickle cell anemia are virtually limited to blacks, and very few instances of Tay–Sachs disease are found except in Ashkenazi Jews. This biological reality is sometimes difficult to accommodate in politically acceptable ways.

PRINCIPLES OF GENETIC SCREENING

The procedures for screening can conveniently be grouped into those that recognize matings at risk, those that identify affected fetuses, and those that identify affected newborns.

Screening of Parents

Virtually all recessive disease could be eliminated if heterozygotes were identified and avoided mating with each other. Alternatively, if two heterozygotes married, they could avoid reproducing or they could have pregnancies monitored to detect homozygotes. The problem is not to prevent all heterozygotes from having children, since all of us are probably heterozygous for several undesirable recessive traits. There are estimates that we each have, on the average, two recessive alleles that, if homozygous, would be lethal. But our children have no risk unless both parents are heterozygous for the same allele.

Limited programs of this type have been introduced among Ashkenazi Jews for Tay–Sachs disease. The detection of heterozygotes is simple and inexpensive, and, if at least one parent is homozygous normal, it is not necessary to attempt prenatal diagnosis. This greatly reduces the effort required to manage this disorder. Among whites, cystic fibrosis could be prevented if all couples could be screened for heterozygosity. Unfortunately, at present the tests for heterozygosity are not very reliable. When better tests are available, identification of the one couple in 625 at risk will almost certainly be less expensive than long-term treatment of the child with cystic fibrosis.

While identification of heterozygotes appears in many ways to be the most acceptable means of preventing certain genetic diseases, objections have been raised. These objections have tended to center around two issues: invasion of privacy and misunderstanding or misuse of information. Sickle cell anemia is an example of a disease for which inexpensive screening of heterozygotes is available, and a number of voluntary screening programs have been set up.

Eight percent of U.S. blacks are heterozygous for sickle cell hemoglobin. Such heterozygotes are normal, the only serious question of risk being associated with extreme deprivation of oxygen, such as might occur in high-altitude flight. Even here most heterozygotes do not appear to have a problem, as witness the excellent performance of several in the Olympic games in Mexico City. If heterozygotes did not marry each other, or, when they do marry each other, if they would refrain from having children, the tragedy of sickle cell anemia would be prevented. These facts have been the basis for large-scale screening programs to identify heterozygotes. The cost per blood sample is trivial.

A number of problems, nongenetic to be sure, have resulted from these well-meant programs. Some of the problems might occur with any genetic screening program. Perhaps foremost is the difficulty in convincing people that a heterozygote is truly healthy. The heterozygote himself may not understand this, and, at least in the case of sickle cell anemia, advertisements designed to reach the black population have been misleading on this point. Employers have often not distinguished between sickle cell trait (as in the heterozygote) and sickle cell anemia. And in some instances persons with sickle cell trait have been charged higher insurance rates, although there is no actuarial basis for such a practice. Finally, any genetic trait has the potential of revealing illegitimacy. For the most part these problems can be ameliorated by education, both of the persons at risk and the persons with whom they deal.

But it is small wonder that the screening programs have sometimes been viewed as mixed blessings.

Prenatal Screening

Many biochemical disorders and most chromosomal defects can be detected by amniocentesis. Certain malformations, whether genetic or nongenetic, can also be detected by the technique of ultra high-frequency sound reflection. Such techniques are of little use unless parents are willing to consider abortion if the fetus proves to have a defect. Since this option is becoming more generally accepted, the question is sometimes raised as to the possibility and feasibility of amniocentesis on all pregnancies or, if not all, on those pregnancies considered to be high risk. Complications associated with amniocentesis appear to be small.

From the purely biological point of view, prenatal diagnosis and abortion has little impact. For the most part, those fetuses that would be aborted would not have the potential of developing into fertile adults. A "genetic death" is the same whether it involves a zygote that cannot implant or an adult who, for whatever reason, does not reproduce. A difference is that prenatal losses are likely to be replaced with additional pregnancies. Such compensation tends to retard the selection against recessive genes, but the effect is quite small. It should have no effect on dominant genes or chromosomal disorders.

While the thrust of prenatal diagnosis is the detection of defective fetuses, it also has the potential of demonstrating that a fetus is free of a particular defect. A pregnant woman heterozygous for a balanced translocation involving chromosome 21 may learn that the fetus will not have Down's syndrome and that it has a completely normal karyotype or a balanced translocation such as she has herself. She can thus continue the pregnancy free of that particular concern. Or parents who have already lost a child with Tay–Sachs disease may learn that their next child will not be affected. Such emotional benefits may be substantial.

One question that should be asked of every service, medical or otherwise, is the cost versus the benefit. Amniocentesis and prenatal diagnosis is expensive. Even though it is technically feasible, would the resources required for a program contribute more to public welfare if devoted to other problems? It is impossible to add the emotional and humanitarian factors into such equations, especially since each of us would weigh them differently. The exercise of making such an economic analysis is nevertheless useful in gaining perspective.

The current annual number of births in the United States is on the order of 3.15 million. Of these, some 16,000 will have chromosome abnormalities (Table 25-1), including 3000 with Down's syndrome. Virtually all of these could be recognized by prenatal diagnosis. The cost of

Table 25-1 **FREQUENCY OF CHROMOSOMAL DISORDERS IN NEWBORNS POTENTIALLY DETECTABLE PRENATALLY***

	United States	World
Population	210 million	3.86 billion
Number annual births	3.15 million	130 million
Births per year:		
Total chromosome abnormalities	15,600	643,000
Sex chromosome aneuploidies	5200	255,000
Autosomal trisomies	4100	170,000
Autosomal rearrangements	5300	218,000

* From Shaw (1972) modified with more recent estimates of population size and growth and newer estimates of aneuploidy (Jacobs, 1972).

prenatal diagnosis (amniocentesis plus karyotyping) is on the order of \$300. The cost of maintaining a mentally retarded person in an institution is approximately \$7500 per year (1974 dollars), exclusive of special medical care. For a person with Down's syndrome, the average length of stay may be on the order of 30 years. Thus the lifetime cost of maintaining such patients may average \$225,000. If this figure is multiplied by the number of such patients, then each year the number of Down's patients born represents an ultimate expenditure of \$675 million. On the other hand, it would cost \$1,080 million to monitor all pregnancies. To be sure, the procedure would identify other chromosome abnormalities as well, but these would generally not lead to long-term institutionalization.

If screening were limited to high-risk pregnancies, the economic picture changes sharply. As noted in Chapter 5, the risk of trisomy 21 among offspring of mothers in the age range 40 to 44 years is approximately 1%. Monitoring this group would cost \$30,000 per Down's fetus detected, as opposed to \$225,000 for maintaining the undetected patient in an institution. One may argue about exact values and approaches to be used in these calculations, but the conclusions are not altered.

The proportion of all Down's patients born that are from mothers in the older age range varies with the population structure and reproductive patterns. A study in Denmark covering the years 1953 to 1970 showed 19.5% of the Down's patients to be born to mothers 40 or more years of age (Mikkelsen and Stene, 1972), and similar results have been obtained elsewhere. Other high-risk pregnancies would be those in which a child with Down's syndrome had been previously born to the parents and, of course, those in which one of the parents has been identified as a balanced translocation carrier.

Similar considerations have been given Tay–Sachs disease by

O'Brien (1970) and by Nelson et al. (1978). Children with this disorder require hospitalization for approximately two years before the inevitable fatal termination of their disease. The 1970 cost of such care in a state hospital in California was $35,000. Heterozygotes can be detected at approximately $2 per test, and homozygous fetuses can be detected by amniocentesis at a cost of approximately $100. Of the 204 million persons in the United States (1970 census), 6 million are Jewish. The gene frequencies in the Jewish and non-Jewish groups are 0.013 and 0.0013, respectively (Myrianthopoulos, 1962). Based on a birth rate of 17.4 per 1000 for both groups, the annual incidence of affected children should be 18 new cases among Jews and 6 among non-Jews. (Some of the latter cases are in fact Sandhoff's disease, a related but genetically distinct disease.) This is 1 per 6000 Jewish births compared with 1 per 600,000 non-Jewish births. There are some 100,000 Jewish births per year and 3.5 million non-Jewish births.

Clearly, one would not want to do prenatal diagnosis for Tay–Sachs disease on all births or all Jewish births. Rather, prenatal diagnosis would be indicated only if both parents were heterozygous, a fact that might be established either from testing them directly or from birth of a previous child who was affected. Of the 100,000 Jewish births, 72 would be from parents both of whom were heterozygous. The cost of screening 200,000 parents would be $400,000, and the cost of prenatal testing of 72 pregnancies would be $7200. The cost of not identifying the 18 affected fetuses would be $630,000.

These kinds of estimates are very crude, since they usually are based on the most efficient testing systems and do not include the nonlaboratory costs of the program. On the other hand, parents would need to be tested only once, and if one parent were homozygous for the normal allele, as will usually be the case, the other parent would not require testing. Therefore, it clearly is economically feasible to test all Jewish couples for heterozygosity and do prenatal testing when indicated. The situation is quite different for non-Jewish couples, however, because of the much lower frequency of Tay–Sachs disease.

Screening of Newborns

Detection of genetic disease in newborns has as its goal the prevention of defect by therapeutic intervention. Usually therapy must be instituted before irreversible physiological changes have occurred. An important justification for screening is that a diagnosis made after clinical symptoms appear is too late for maximum benefit of therapy. Therefore, detection of the disease must occur before the patient attracts medical attention. A partial list of diseases that can be screened in blood and urine samples from newborns is given in Table 25-2.

Phenylketonuria (PKU) is the inherited disease for which screening

Table 25-2 **A PARTIAL LIST OF METABOLIC DISORDERS THAT CAN BE DETECTED BY MASS SCREENING OF BLOOD AND URINE SAMPLES FROM NEWBORNS. A NUMBER OF OTHER AMINOACIDURIAS, INDIVIDUALLY RARE, WOULD BE DETECTED BY THE SAME SCREENING PROCEDURES***

Defect	Material assayed	Abnormality detected	Incidence per 10^6	Treatment
Galactosemia	Blood	Low galactose-1-phosphate uridyl transferase		Avoid galactose in diet
G6PD deficiency	Blood	Low glucose-6-phosphate dehydrogenase	100,000 in certain groups	Avoid certain toxic agents
PK deficiency	Blood	Low pyruvate kinase		Splenectomy
Argininosuccinic aciduria	Blood	Low argininosuccinase	14	Low protein diet
Orotic aciduria	Blood		Very low	Uridine in diet
6PGD deficiency	Blood	Low 6-phosphogluconate dehydrogenase		
Cystinuria	Urine	Cystine excretion	139	
Hartnup disease	Urine	Aminoaciduria	38	
Histidinemia	Urine	High histidine in blood and urine	41	
Homocystinuria	Blood	High methionine	4	Low methionine diet
Maple syrup urine disease	Urine	Branched chain amino acids in urine	5	Diets low in Val, Leu, Ile
Phenylketonuria	Blood	High phenylalanine	88	Low phenyl-alanine diet
Hereditary angioneurotic edema	Blood	Low C′1-esterase inhibitor		
Sickle cell disease	Blood	Sickle cell hemoglobin	1600 in blacks, low in whites	

* Based primarily on information in Levy, 1973.

is most extensively practiced, and PKU will be a useful example to discuss some of the issues that arise in screening. As described in Chapter 11, children homozygous for the PKU gene are normal at birth, being protected by the normal metabolism of their mothers. When they are on their own metabolism, however, phenylalanine accumulates in the blood, leading to interference with function of the central nervous system. The interference is irreversible, so that diagnosis of PKU and institution of dietary treatment must occur within the first few weeks of life. Direct assays of phenylalanine hydroxylase are not possible except with a liver biopsy, and diagnosis therefore depends on detection of increased blood phenylalanine (Figure 25-1) or increased metabolites of phenylalanine in urine.

There has been much debate on the appropriate time to test for PKU. On the one hand, diagnosis should be as early as possible in order to minimize the risk of brain damage. If it is too early, some children with PKU will have failed to accumulate sufficient phenylalanine to give a positive test and will be missed. If the test is delayed to avoid any possibility of false negatives, some children will have suffered irreversible brain damage. Further, the greater the delay in testing after birth, the more difficult it is to assure that all children are tested. Any testing protocol is therefore a compromise between the errors associated with early and late testing.

A further complication arises with PKU. High levels of phenylalanine may occur without clinical problems. Possibly these are due to low

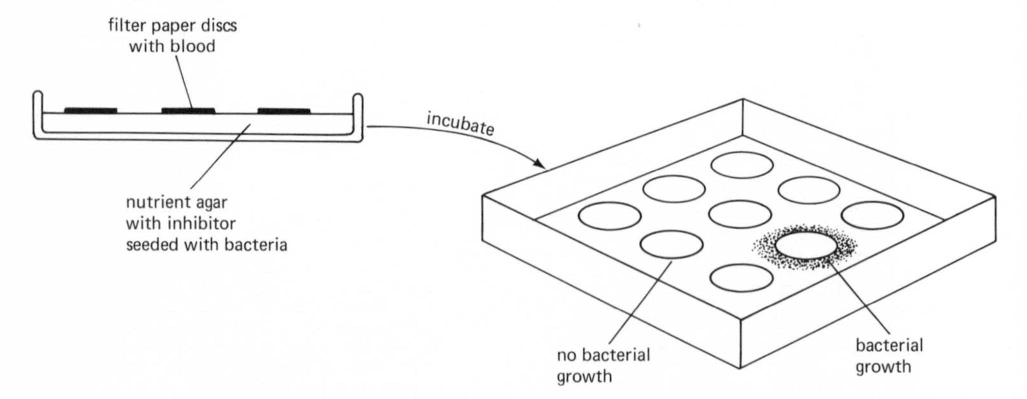

Figure 25-1 Diagram of the Guthrie test for phenylketonuria. The nutrient medium contains all the factors for growth of the bacterial strain used plus an inhibitor specific for phenylalanine. Normal blood on the filter paper disc does not contain sufficient phenylalanine to overcome the inhibition; hence, there is no bacterial growth. If the level of phenylalanine is very high in the blood, as in PKU, sufficient quantities diffuse into the medium surrounding the disc to overcome the inhibition and permit bacterial growth. In the diagram, there is growth around one disc, which would be from a person with hyperphenylalaninemia, perhaps PKU. All the other tests are negative.

(rather than zero) levels of phenylalanine hydroxylase, but dietary management is not required for normal development. Such children are difficult to distinguish from those with PKU without careful study. Yet they may be detected as "positives" in the screen for PKU. In the early screening for PKU, many such children were thought to have typical PKU and were placed on the PKU diet. Now that the condition is recognizable, more careful evaluation and follow-up of positive children can prevent this error.

As with prenatal diagnosis, the cost of detecting versus not detecting affected children is useful information to consider in decisions on whether or not to set up screening programs. It has been estimated that in Massachusetts, the screening program for inborn errors of metabolism costs approximately $250,000 per year (Levy and co-workers, 1970). This program makes use of blood samples from four-day-old infants and is designed to detect a number of defects in addition to PKU. Among the approximately 100,000 births each year in Massachusetts, eight to ten children with classical PKU are detected and treated. If these had not been treated, most would have become mentally retarded and would have been maintained in state institutions, at a total lifetime cost exceeding $1 million. Thus it is much less expensive to detect and treat persons with PKU than to support them as mental retardates. Such a simple economic analysis ignores the humanitarian issues associated with a life of normal intelligence as compared with a life of mental deficiency. Also, the entire cost of the program is compared to the benefits derived from detecting PKU, when in fact other treatable disorders also are detected. Table 25-3 lists the metabolic disorders that have been detected in the Massachusetts screening program for the period 1962 to 1979.

Not all disorders for which screening has been recommended require immediate action in the newborn period. In the case of glucose-6-phosphate dehydrogenase (G6PD) deficiency, no therapeutic intervention is necessary. Affected persons should be aware of the disorder as early as possible so that exposure to certain drugs can be avoided. For diseases such as sickle cell anemia, early screening helps identify affected persons so that they can receive better medical care, but the outcome of the disease is not affected.

COUNSELING, SCREENING, AND THE GENE POOL

Medicine has often been accused of improving the lot of the individual at the expense of the population as a whole. There has especially been concern that medicine has rescued many detrimental genes from extinction, adding them to the burden of future generations. Such is the

Table 25-3 **METABOLIC DISORDERS DETECTED IN THE MASSACHUSETTS SCREENING PROGRAM, 1962–1979***

Disorder	Total screened	Number found	Estimated frequency
Hypothyroidism	414,733	103	1:3800
Phenylketonuria	1,406,221	95	1:15,000
Hyperphenylalaninemia	1,406,221	86	1:16,000
Phenylketonuria (DHPR)	1,406,221	1	<1:1,000,000
Galactosemia	973,224	15	1:65,000
Maple syrup urine disease (MSUD)	1,266,864	3	1:400,000
MSUD (intermediate variant)	1,266,864	2	1:600,000
MSUD (intermittent variant)	1,266,864	1	<1:1,000,000
Homocystinuria	974,475	3	1:300,000
Histidinemia	666,736	37	1:18,000
Histidinemia (atypical)	666,736	2	1:300,000
Cystinuria	666,736	53	1:13,000
Hartnup disorder	666,736	36	1:18,000
Iminoglycinuria	666,736	55	1:12,000
Argininosuccinic acidemia	666,736	9	1:75,000
Cystathioninemia	666,736	11	1:60,000
Hyperglycinemia (nonketotic)	666,736	4	1:150,000
Methylmalonic acidemia	666,736	7	1:95,000
Propionic acidemia	666,736	2	1:300,000
Hyperprolinemia (type I)	666,736	2	1:300,000
Hyperprolinemia (type II)	666,736	2	1:300,000
Hyperlysinemia	666,736	2	1:300,000
Sarcosinemia	666,736	3	1:200,000
Hyperornithinemia	666,736	1	<1:600,000
Carnosinemia	666,736	1	<1:600,000
Urocanic aciduria	666,736	1	<1:600,000
Hyperglutamic aciduria	666,736	1	<1:600,000
α-Aminoadipic aciduria	666,736	1	<1:600,000
Fanconi syndrome	666,736	1	<1:600,000
Rickets (vitamin D dependant)	666,736	1	<1:600,000
Hereditary tyrosinemia	666,736	1	<1:600,000
Hereditary fructosemia	666,736	1	<1:600,000

* Provided by Dr. Harvey L. Levy.

case only if the genes are transmitted. Increased survival of persons with genetic defects or increased quality of life does not affect the gene pool unless reproduction is also increased.

The introduction of counseling and screening does raise the question of what the long-range effects of altered reproductive patterns might be on the gene pool. Several mathematical models have been developed, of which those of Crow (1973) for rare recessive alleles are presented in Table 25-4. In this table, the allele a is deleterious in homozy-

Table 25-4 **CHANGES IN THE FREQUENCY q OF A RARE RECESSIVE GENE a WITH VARIOUS REPRODUCTIVE PRACTICES. THE GENE FREQUENCY IS COUNTED IN THE ADULT STAGE***

Initial genotype frequency	Marriage frequencies
Aa: $2q$	$Aa \times Aa$: $4q^2$
AA: $1 - 2q$	$Aa \times AA$: $4q(1 - 2q)$
	$AA \times AA$: $(1 - 2q)^2$

	Adult gene frequency			Numerical example $q = 0.01$	
	After 1 generation		After n generations	After 1 generation	After 10 generations
1. Complete cure, or avoidance of marriage between heterozygotes	$2q^2 + q(1 - 2q)$	$= q$	q	0.0100	0.0100
2. Abortion or preadult death of affected homozygotes	$\dfrac{q^2 + q(1 - 2q)}{3q^2 + 4q(1 - 2q) + (1 - 2q)^2}$	$= \dfrac{q}{1 + q}$	$\dfrac{q}{1 + nq}$	0.0099	0.0091
3. No children from $Aa \times Aa$ marriages	$\dfrac{q(1 - 2q)}{4q(1 - 2q) + (1 - 2q)^2}$	$= \dfrac{q}{1 + 2q}$	$\dfrac{q}{1 + 2nq}$	0.0098	0.0083
4. Artificial insemination in $Aa \times Aa$ marriages	$q^2 + q(1 - 2q)$	$= q(1 - q)$	$\sim q(1 - nq)$	0.0099	0.0090
5. Abortion with complete compensation	$\frac{4}{3} q^2 + q(1 - 2q)$	$= q\left(1 - \frac{2q}{3}\right)$	$\sim q\left(1 - \frac{2nq}{3}\right)$	0.0099	0.0093

* From Crow, 1973.

gous combination. Its frequency among adults is q. The addition of newly mutant alleles is ignored. Several interesting conclusions can be drawn:

1. *Avoidance of marriages between heterozygotes does not change the gene frequency.* If heterozygotes for cystic fibrosis, Tay–Sachs disease, or sickle cell anemia mated only with homozygous normals, the frequency of affected children would drop to zero immediately except for the rare offspring who happened to receive a newly mutant allele from the homozygous normal parent. If $AA \times Aa$ marriages have the same average number of children as $AA \times AA$, there is no change in the frequency of a. If they have fewer or greater, the frequency of a will change accordingly.

2. *Nonreproduction by affected homozygotes slowly reduces the gene frequency.* Nonreproduction may be by abortion, early death, sterility, or voluntary action. The model assumes that such aa fetuses or individuals are not replaced in the family.

3. *If marriages between heterozygotes produce no children, the decline in gene frequency is faster than by elimination of homozygotes only.* In the usual marriage of $Aa \times Aa$, two a alleles would be eliminated for each four children produced. If no children were produced, the two a alleles associated with heterozygous children would also be eliminated from the gene pool. As the frequency of a becomes very small, the frequency of such marriages also becomes very small.

4. *Artificial insemination in* Aa × Aa *marriages would prevent homozygous affected offspring.* Since one partner would not contribute a alleles, the frequency of a would slowly decrease.

5. *Abortion with complete compensation would lead to slower reduction of* a *alleles than options 2 to 4.* Reproductive compensation means that all aa fetuses would be aborted and replaced with AA or Aa fetuses. This is different from situation 2, where aa fetuses are not replaced.

Mutation, although rare, would introduce new a alleles. Crow has further considered the effects of mutation and of selection favoring the heterozygote, as shown in Table 25-5. Again, the effect on the gene frequency is very small, but some practices do lead to an increase in a.

SOME LEGAL AND SOCIAL ISSUES RELATED TO SCREENING

Despite the possible benefits, screening has not been welcomed by everyone. The greatest objections can be traced to the conflict between the rights of individuals versus the rights of society as a whole. When may society decide that its interests are more important than the freedom of the individual to make his own decisions? There are many precedents for such actions. The requirements for immunization against in-

Table 25-5 **SAME AS TABLE 25-4 EXCEPT FOR THE ADDITIONAL ASSUMPTION OF MUTATION AT RATE m, OR EQUIVALENT SELECTION FAVORING HETEROZYGOTES***

	Gene frequency		Numerical examples			
			$m = 10^{-4}$ or het. adv. = 0.01		$m = 10^{-5}$ or het. adv. = 0.003	
	After 1 gen	After n gens	After 1 gen	After 10 gens	After 1 gen	After 10 gens
1.	$q + m$	$q + nm$	0.0101	0.0110	0.0100	0.0101
2.	$\frac{q}{1+q} + m$	$\sim \frac{q}{1+nq} + nm$	0.0100	0.0100	0.0099	0.0092
3.	$\frac{q}{1+2q} + m$	$\sim \frac{q}{1+2nq} + nm$	0.0099	0.0093	0.0098	0.0084
4.	$q(1-q) + m$	$\sim q(1-nq) + nm$	0.0100	0.0100	0.0099	0.0091
5.	$q\left(1 - \frac{2q}{3}\right) + m$	$\sim q\left(1 - \frac{2nq}{3}\right) + nm$	0.01003	0.0103	0.0099	0.0094

* From Crow, 1973.

fectious disease and for the reporting of veneral disease contacts are examples in which the public interest takes precedence over individual freedom of choice and right of privacy. In this regard, inherited disease has been compared to infectious disease by noting that transmission of the former is vertical (that is, lineally by descent); transmission of infectious disease is horizontal (that is, to persons coexisting in the same time). Many of the same legal principles may apply, however.

Compulsory screening of potential marriage partners or parents carries with it the threat that persons with "unmatched" genotypes will not be allowed to marry or to have children. Many persons concerned with individual liberties see such screening as a first step toward government intervention in the area of reproduction. Involuntary sterilization brings back bad memories of programs that were based on limited scientific knowledge at best. The issue remains though whether parents should be allowed to have children likely to be defective and likely to become the responsibility of the state. Many of the same arguments apply to prenatal screening, with the additional problem that amniocentesis is a much more invasive procedure than collecting a blood sample and abortion a much more drastic intervention than avoidance of pregnancy. Aside from the enormous effort that would be required to perform amniocentesis on a high proportion of pregnancies, many other issues are involved, including some important ethical and legal issues. The rights of fetuses to be born, to be born free of detectable genetic defect (that is, to be aborted if defect is present), the rights of parents, and the rights of society are questions yet to be fully answered in the courts and in public opinion.

The attitudes on screening of newborns are generally much more favorable, and most states of the U.S. have compulsory screening laws for phenylketonuria (Reilly, 1977). In the case of PKU, the justification is the ability to treat those affected if they are identified soon enough, and such a program is both humanitarian and cost effective. Other treatable genetic disorders, such as galactosemia and homocystinuria, have been added to the screening programs in many states.

Not all newborn screening is without controversy, however. The identification of XYY males has especially been attacked. In part, this is due to differences in opinion as to how much risk such children have for antisocial behavior. Should they be placed deliberately in a more structured environment to reduce the risk of aggressive actions? The opponents of XYY screening point out that such special treatment might well lead to a self-fulfilling prophecy, with the risk of antisocial behavior being a product of the label rather than the karyotype. This presupposes the inability to define an environment that would alter behavior, an attitude with which many would disagree.

Screening for sickle cell anemia has been strongly advocated and, in a few states, required. The main purpose is to identify parents at risk for producing children with sickle cell anemia, with the expectation that proper counseling will reduce the number of affected children. As noted earlier, such programs have had many problems. The follow-up counseling has often been inadequate. The distinction between persons homozygous and heterozygous for sickle cell hemoglobin is not always made clear. And the revelation of illegitimacy has created problems for many families. Many black leaders now strongly oppose mass screening programs for sickle cell anemia, since the problems are large compared to the benefits.

In spite of the difficulties encountered by some forms of screening, it seems likely that the programs will grow as new information becomes available and ways are found to solve the legal and social problems. Newborn screening programs are especially likely to expand and to include not only new genetic tests but also tests for nongenetic conditions, such as hypothyroidism, that can be detected with blood or urine samples and for which effective therapy is available.

SUMMARY

1. Screening of parents or potential parents is designed to provide information on those matings at risk for producing defective offspring. It is assumed that counseling provided to such persons will lead to fewer affected offspring, either through lowered reproduction or alternative choices of mates. Tay–Sachs disease is an example of an inherited disorder where screening of adults has been largely successful. In the case of sickle cell anemia, a number of problems have occurred as a result of screening programs.

2. Prenatal screening (diagnosis) has become widespread and is an effective way to prevent birth of children with a variety of biochemical and chromosomal disorders. Especially in the case of older mothers, prenatal diagnosis would substantially reduce the incidence of trisomy syndromes. Similarly, the prenatal diagnosis of Tay–Sachs disease where both parents are heterozygotes has been effective in preventing many cases of this disorder.

3. Screening of newborns has led to identification and treatment of many children affected with inherited metabolic diseases. Phenylketonuria is the disease most widely screened, and cost–benefit analyses have shown that the cost of the screening is substantially less than the cost of a lifetime of care for untreated victims.

4. Changes in reproductive patterns or fertility have very little effect on the frequencies of alleles associated with inherited disease. If mutation is considered, some will increase in frequency, but the increase is exceedingly slow.

5. The introduction of screening, especially compulsory screening, raises questions about the conflict of individual privacy versus public good. Many states have compulsory screening for phenylketonuria and other metabolic diseases, and these have generally been successful. However, screening programs for sickle cell anemia have generated major problems and illustrate the difficulties that can result from a program that fails to consider all the dimensions of the situation.

REFERENCES AND SUGGESTED READING

COMMITTEE FOR THE STUDY OF INBORN ERRORS OF METABOLISM. 1975. *Genetic Screening: Programs, Principles and Research.* National Academy of Sciences, Washington, D.C.

CROW, J. F. 1973. Population perspective. In *Ethical Issues in Human Genetics* (Hilton, B., Callahan, D., Harris, M., Condliffe, P., and Berkley, B., Eds.). New York: Plenum Press, pp. 73–80.

GUTHRIE, R. 1968. Screening for "inborn errors of metabolism" in the newborn infant—a multiple test program. *Birth Defects: Original Article Series* 4: 92.

HARRIS, M. (Ed.). 1970. *Early Diagnosis of Human Genetic Defects.* Fogarty International Center Proceedings No. 6. U.S. Government Printing Office, Washington, D.C. 229 pp.

HOLLOWAY, S. M., AND SMITH, C. 1975. Effects of various medical and social practices on the frequency of genetic disorders. *Am. J. Hum. Genet.* 27: 614–627.

JACOBS, P. A. 1972. Chromosome mutations: frequency at birth in humans. *Humangenetik* 16: 137–140.

LEVY, H. L. 1973. Genetic screening. *Adv. Hum. Genet.* 4: 1–104, 389–394.

LEVY, H. L., SHIH, V. E., AND MAC CREADY, R. A. 1970. Massachusetts metabolic disorders screening program. In Harris (1970), pp. 47–66.

LITTLEFIELD, J. W. 1972. Genetic screening (editorial). *N. Engl. J. Med.* 286: 1155–1156.

MIKKELSEN, M., AND STENE, J. 1972. The effect of maternal age on the incidence of Down's syndrome. *Humangenetik* 16: 141–146.

MILUNSKY, A., AND ANNAS, G. J. (Eds.). 1976. *Genetics and the Law.* New York: Plenum Press, 532 pp.

MURPHY, W. H., PATCHEN, L., AND GUTHRIE, R. 1972. Screening tests for argininosuccinic aciduria, orotic aciduria, and other inherited enzyme deficiencies using dried blood specimens. *Biochem. Genet.* 6: 51–59.

MYRIANTHOPOULOS, N. C. 1962. Some epidemiologic and genetic aspects of Tay–Sachs' disease. In *Cerebral Sphingolipidoses: A Symposium on Tay–Sachs' Disease and Allied Disorders* (Aronson, S. M., and Volk, B. W., Eds.). New York: Academic Press, pp. 359–374.

NELSON, W. B., SWINT, J. M., AND CASKEY, C. T. 1978. An economic evaluation of a genetic screening program for Tay–Sachs disease. *Am. J. Hum. Genet.* 30: 160–166.

NITOWSKY, H. M. 1973. The significance of screening for inborn errors of metabolism. In *Heredity and Society* (Porter, I. H., and Skalko, R. G., Eds.). New York: Academic Press, pp. 225–261.

O'BRIEN, J. 1970. Discussion in Harris (1970), pp. 62–65.

REILLY, P. 1977. *Genetics, Law, and Social Policy*. Cambridge, Mass.: Harvard Univ. Press, 275 pp.

RESEARCH GROUP ON ETHICAL, SOCIAL AND LEGAL ISSUES IN GENETIC COUNSELING AND GENETIC ENGINEERING OF THE INSTITUTE OF SOCIETY, ETHICS AND THE LIFE SCIENCES. 1972. Ethical and social issues in screening for genetic disease. *N. Eng. J. Med.* 286: 1129–1132.

SHAW, M. W. 1972. Chromosome mutations in man. In *Mutagenic Effects of Environmental Contaminants* (Sutton, H. E., and Harris, M. I., Eds.). New York: Academic Press, pp. 81–97.

THOMAS, G. H., AND HOWELL, R. R. 1973. *Selected Screening Tests for Genetic Metabolic Diseases*. Chicago: Year Book Medical Publishers, 101 pp.

REVIEW QUESTIONS

1. Cystic fibrosis has a frequency of 1 in 2500 births among U.S. whites. What would be the gene frequency after three generations
(a) if heterozygotes avoided marriage with each other?
(b) if there is no restriction on marriages of heterozygotes with each other, but no children are produced by such marriages?
(c) if all affected fetuses could be aborted?

2. The following questions have no "answers." List as many arguments as you can on both sides of the question.
(a) Does a fetus have a right to be born, whether that fetus is defective or normal?
(b) Do parents have a right to impose a preventable burden on the state?
(c) Does society have the right to prevent persons from marrying or reproducing if the persons are known to have a high risk of defective offspring?
(d) Should persons who know that they carry detrimental genes be required by law to so inform a potential spouse? Should the information be a matter of public record?

APPENDIX A

SOME ELEMENTS OF PROBABILITY AND STATISTICS

When a chemist makes measurements such as temperature, the observations are limited in accuracy only by the instruments used. If better instruments were available, more accurate observations could be made. Temperature is a measure of the mean kinetic energy of a population of molecules. Very few of the molecules actually possess energy equal to the mean, however, for the majority move faster or slower than the mean value. If one were to select ten molecules at random, they would very likely not include a molecule with the mean energy of the parent population. Furthermore, the mean of the ten molecules would almost certainly vary slightly from the population mean.

Chemists commonly deal with 10^{20} molecules at a time rather than 10. The chance deviations that occur with ten molecules do not happen when observations are made on such immense numbers, and sampling error can therefore be safely ignored. Biologists, however, commonly make observations on small numbers where chance deviation can occur. It is necessary to recognize the possibility of chance deviation and to estimate the reliability of observations. Geneticists in particular have been faced with this problem. Mendel recognized the chance nature of gene transmission and the possibility of deviation from ideal ratios, although the statistical procedures for evaluating the significance of deviations were not available to him.

This appendix will be concerned with some of the basic laws of probability and how they are used to solve some genetic problems. Many of the statistical techniques commonly in use require more background than can be assumed or provided in this treatment. The material given here should provide an appreciation of the nature of statistical inference as well as a means of solving many of the simpler problems encountered in genetics.

SOME BASIC CONCEPTS OF PROBABILITY

The Meaning of Probability Statements

A statement of probability is the likelihood of a "favorable" event among all possible events. Favorable, as used in this context, merely means the event or events of interest, whether or not they might be rated desirable by other criteria. Probability can vary from 0 to 1, 0 indicating no possibility of the favorable event, and 1 indicating no alternate possibility; that is, the event is the only one possible. If $P = \frac{1}{2}$, then half of the time a favorable event will occur.

Probability can be expressed symbolically as

$$P(e_f) = \frac{\Sigma e_f}{\Sigma e_i},$$

where $P(e_f)$ is the probability of a favorable event (e_f), Σe_f is the sum of all events that are favorable, and Σe_i is the sum of all events that are possible. The values e_f and e_i may be in any units that express the relative frequency of each event. Since the units appear both in numerator and denominator, they cancel out.

Consider a die with six sides, each equally likely to appear when the die is thrown. If we ask what the probability is that a 3 will be thrown, the answer is $P(3) = \frac{1}{6}$. The probability of any specific side is $\frac{1}{6}$ and there are six sides. The total number of choices is expressed by $6 \times \frac{1}{6} = 1$. Therefore

$$P(3) = \frac{\frac{1}{6}}{6 \times \frac{1}{6}} = \frac{1}{6}.$$

To calculate the probability of obtaining an even number, one must sum the favorable outcomes, in this case 2, 4, or 6. Then

$$P(\text{even}) = \frac{3 \times \frac{1}{6}}{6 \times \frac{1}{6}} = \frac{1}{2}.$$

The various possibilities need not be equally likely. Suppose we place 100 marbles into a jar. Ten are yellow, 30 are red, 15 are white, 20 are black, and 25 are green. The probability of drawing a red marble is

$$P(\text{red}) = \frac{30}{10 + 30 + 15 + 20 + 25} = 0.30.$$

If the black marbles are removed from the jar, the probability of drawing a red marble becomes

$$P(\text{red}) = \frac{30}{10 + 30 + 15 + 25} = 0.375.$$

Thus reducing the number of unfavorable events increases the likelihood of a favorable event.

Suppose that in the original series of 100 marbles, we wish to know the probability of drawing either a white or black marble. In this case, two possible outcomes are favorable, and

$$P(\text{black or white}) = \frac{15 + 20}{10 + 30 + 15 + 20 + 25} = 0.35.$$

The same answer is obtained by calculating the separate probabilities for black and white, followed by addition; for example, $P(\text{black}) = 0.15$, $P(\text{white}) = 0.20$, $\therefore P(\text{black or white}) = 0.35$.

The last example illustrates the important principle that if there are two or more favorable and mutually exclusive events, the total probability can be obtained by summing the probabilities of the separate events.

Another important principle is the independence of chance events from previous trials. When a coin is flipped, the probability of a head or tail is $\frac{1}{2}$ for each. If a given trial yields a head, the next flip still has equal chances for a head or tail. Even if a sequence of five heads is thrown, the sixth still has a half chance of being a head.

This fact permits the computation of probabilities for complex sequences and combinations of events. The probability of a sequence of events is simply the combined product of each event taken separately. The probability of throwing three heads is

$$\begin{aligned} P(3\text{H}) &= P(\text{H}) \cdot P(\text{H}) \cdot P(\text{H}) \\ &= \tfrac{1}{2} \cdot \tfrac{1}{2} \cdot \tfrac{1}{2} = \tfrac{1}{8}. \end{aligned}$$

This procedure is the same whether one coin is tossed three times or three coins are tossed together. In either case each component event contributes its probability.

The Binomial Distribution

Statements that the toss of a coin has two equally probable outcomes or that there are five colors of marbles, each with a certain likelihood of being drawn, are simple forms of probability distributions. It is useful to express probability distributions in mathematical form, both to aid in thinking more clearly and because certain operations can be applied to mathematical statements that would not be possible otherwise.

A type of problem frequently arises in genetics that can be readily handled by a binomial distribution of the type $(x + y)^n$. In the preceding example, the probability of three heads was shown to be $\frac{1}{8}$. This is also the probability for three tails or for any specific sequence, such as a head followed by two tails. Frequently the interest is in the combination—in this case, one head and two tails—rather than in the sequence by which it was obtained. The same totals could have been obtained by two other sequences: tail, head, tail; and tail, tail, head. Each sequence has the probability $\frac{1}{8}$, and since each is a favorable outcome, the total probability of two tails and a head can be obtained by adding them to get $\frac{3}{8}$.

Such enumeration is simple in the example but may become complex for larger numbers. Fortunately, the binomial expansion permits calculation of the terms appropriate for any combination of events. The two probabilities associated with two alternate events, for example, heads versus tails, are designated p and q. The binomial is then expressed as

$$(p + q)^n,$$

where n is the number of events in the combination. In the above example, n was 3, corresponding to three coins or three tosses. The expansion of $(p + q)^3$ is

$$p^3 + 3p^2q + 3pq^2 + q^3.$$

For $p = q = \frac{1}{2}$, the expansion becomes

$$\tfrac{1}{8} + \tfrac{3}{8} + \tfrac{3}{8} + \tfrac{1}{8} = 1.$$

If p is the probability of a head, then the four terms are the probabilities of three heads, two heads and a tail, one head and two tails, and three tails.

The binomial expansion for any value of n can always be obtained by multiplying the quantity $(p + q)$ by itself n times. This is not necessary, however. Each term in the expansion consists of two parts—the algebraic terms p and q with their exponents, and the coefficient. The exponents correspond to the events of probability p and q respectively. For example, a combination of 10 heads and 20 tails would give $p^{10}q^{20}$. This expression, however, is the probability of a specific sequence of 10 heads and 20 tails. There are many possible sequences, the exact number being given by the coefficient.

There are two convenient ways to arrive at coefficients for individual terms in the expansion. For small values of n, and particularly if the entire expansion is desired, the Pascal triangle is helpful. Consider the expansion for small values of n.

$$\begin{array}{rc} n = 0 & 1 \\ 1 & 1p + 1q \\ 2 & 1p^2 + 2pq + 1q^2 \\ 3 & 1p^3 + 3p^2q + 3pq^2 + 1q^3 \\ 4 & 1p^4 + 4p^3q + 6p^2q^2 + 4pq^3 + 1q^4 \end{array}$$

Regularities in the exponents of p and q are obvious and familiar. The coefficients also fit into a regular pattern somewhat less obviously. Each is the sum of the two nearest coefficients in the line above. With this information, one can construct a Pascal triangle that yields any desired coefficient.

n =														
0							1							
1						1		1						
2					1		2		1					
3				1		3		3		1				
4			1		4		6		4		1			
5		1		5		10		10		5		1		
6	1		6		15		20		15		6		1	

etc.

The Pascal triangle is very useful for small values of n. But n may be very large, and there are $(n + 1)$ terms in the expansion. Furthermore, interest may be limited to only a few terms. The general formula for any term in the expansion is

$$\frac{n!}{x!(n - x)!} p^x q^{n-x},$$

also written

$$\binom{n}{x} p^x q^{n-x},$$

where x is the exponent appropriate for p. In a complete expansion, x varies from n to 0. Suppose we wish to calculate the probability of tossing four heads and two tails. In this case, $x = 4$ and $n = 6$. The appropriate term in the expansion of $(p + q)^6$ is

$$\begin{aligned} P(4\text{H},2\text{T}) &= \frac{6!}{4!2!} p^4 q^2 \\ &= \frac{1 \cdot 2 \cdot 3 \cdot 4 \cdot 5 \cdot 6}{1 \cdot 2 \cdot 3 \cdot 4 \cdot 1 \cdot 2} p^4 q^2 \\ &= 15 p^4 q^2, \end{aligned}$$

which is the same answer given in the Pascal triangle for $n = 6$. Substituting for p and q gives

$$\begin{aligned} P(4\text{H},2\text{T}) &= 15(\tfrac{1}{2})^4(\tfrac{1}{2})^2 \\ &= \tfrac{15}{64}. \end{aligned}$$

Similarly, the probability of throwing any combination of 50 heads and 50 tails is

$$\begin{aligned} P(50\text{H},50\text{T}) &= \frac{100!}{50!50!} p^{50} q^{50} \\ &= 0.08. \end{aligned}$$

This is a small number, which can be calculated readily even though the calculations involve very large numbers.

The Multinomial Distribution

Using the same reasoning as for the binomial distribution, one can set up a multinomial distribution with any numbers of terms, p, q, r, s . . . , where the sum of the terms is equal to one. The general formula for the distribution is

$$(p + q + r \ldots)^n = \sum \frac{n!}{x!y!z! \ldots} p^x q^y r^z \ldots,$$

where $x, y, z, \ldots$ are the exponents associated with $p, q, r, \ldots$ and add up to n. For example, for a three-allele system, the distribution of genotypes in the population would be

$$(p + q + r)^2 = p^2 + q^2 + r^2 + 2pq + 2pr + 2qr.$$

The distributions of genotypes for a three-allele locus on chromosome 21 in a population with Down's syndrome (assuming independent origin of the alleles) would be

$$(p + q + r)^3 = p^3 + q^3 + r^3 + 3p^2q + 3pq^2 + 3p^2r + 3pr^2 + 3q^2r + 3qr^2 + 6pqr.$$

The Normal Distribution

For small values of n, one can graph the probability for obtaining any combination of choices without great effort. For example, consider ten tosses of a coin (or ten children from a backcross mating). The likelihood of heads or tails (or dominant versus recessive phenotypes) is equal to $\frac{1}{2}$. The likelihood of all possible outcomes is shown in Figure A-1. The distribution is centered around the most likely outcome, 5, but this outcome is expected only $\frac{1}{4}$ of the time. However, 89% of the time the outcome will be no more deviant than 3 : 7 or 7 : 3.

As one extends the number of trials, the curve becomes smoother until it can be approximated mathematically by a continuous curve that assumes an infinite number of trials. Such a curve is shown in Figure A-2. It is the so-called *normal* or Gaussian distribution that describes the chance deviation about some mean value that characterizes many biological and other systems and is the limit of the binomial distribution as n approaches infinity. The distribution can be used in very much the same way as the discontinuous distribution of Figure A-1. If one selects any two points along the horizontal axis, the area under the curve defined by these boundaries relative to the entire area under the

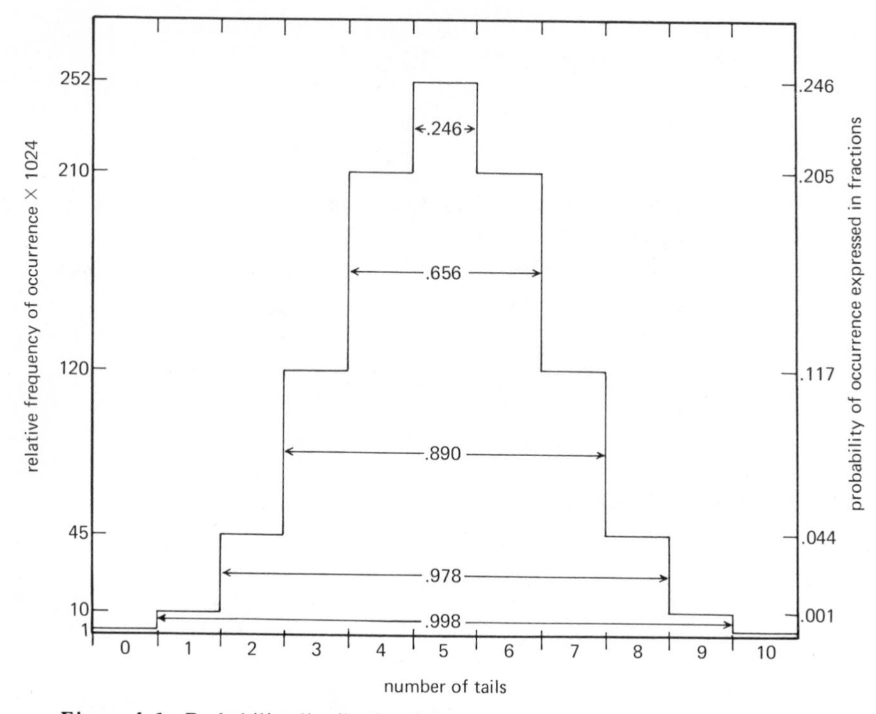

Figure A-1 Probability distribution for a toss of ten coins. There are $(\frac{1}{2})^{10} = 1024$ possible sequences, giving 11 combinations in the relative frequencies shown. The left ordinate is the number of sequences corresponding to a given combination of heads and tails. The right ordinate is the same divided by 1024 to convert to fractions. The figures inside the distribution are sums of areas as indicated. These values represent the total probability of any combination within the area.

curve is the probability that a particular event will fall within the boundaries. This assumes that the normal distribution is an appropriate model and has the mean and deviation from the mean that are characteristic of the biological situation.

Variability of a set of observations about the mean is ordinarily expressed as the *standard deviation.* This is calculated by the formula

$$s = \sqrt{\frac{\Sigma_i^n (x_i - \bar{x})^2}{n - 1}},$$

where s is the standard deviation, x_i are the observations, $\bar{x}$ is the mean, of the observations, and n is the number of observations. If the observations are proportions such as heads versus tails rather than measured quantities, the standard deviation is calculated by

$$s = \sqrt{pq/n},$$

where p and q are the probabilities of the two alternative events.

Distributions may differ either in their mean or in their variation about the mean (or both). Figure A-3 illustrates both of these possibilities, where the abscissa is plotted in terms of some units of measurement. The normal curve, as shown in Figure A-2, is plotted against s. This has the effect of removing the scale dimensions, an operation that is useful for many statistical procedures.

The likelihood of an observed deviation from an expected mean, in the case of proportions, is given by z, the normal deviate:

$$z = \frac{p_0 - p}{s},$$

where p_0 is an observed proportion of events, such as heads, and p the hypothetical.

Evaluation of z can be illustrated by the following. A coin is tossed 100 times. Heads turn up 40 times and tails 60. Can we conclude that the chances of heads versus tails are not equal? Let us set up the hypothesis that the chances are equal, that is, $p = 0.5$. The observed $p_0 = 0.4$.

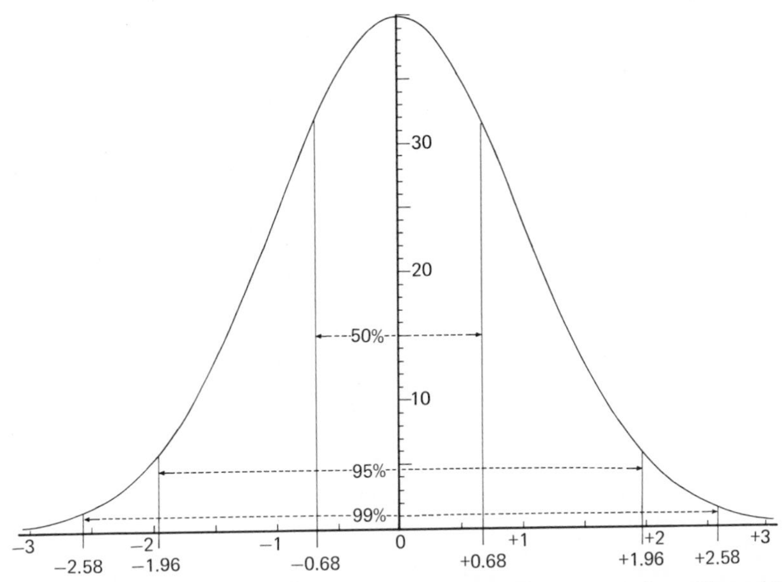

Figure A-2 Normal distribution plotted against standard deviation. The total area under the curve approaches 1 as the curve is extended on both sides, although the height of the curve approaches zero. The area under a particular portion of the curve equals the probability that a deviation from the mean will fall within that portion. For example, the area bounded by −0.68 SD and +0.68 SD = 0.50. Therefore, 50% of the time, the mean value of a sample will fall within ±0.68 SD of the true population mean. The boundaries for 95% and 99% of the area also are indicated.

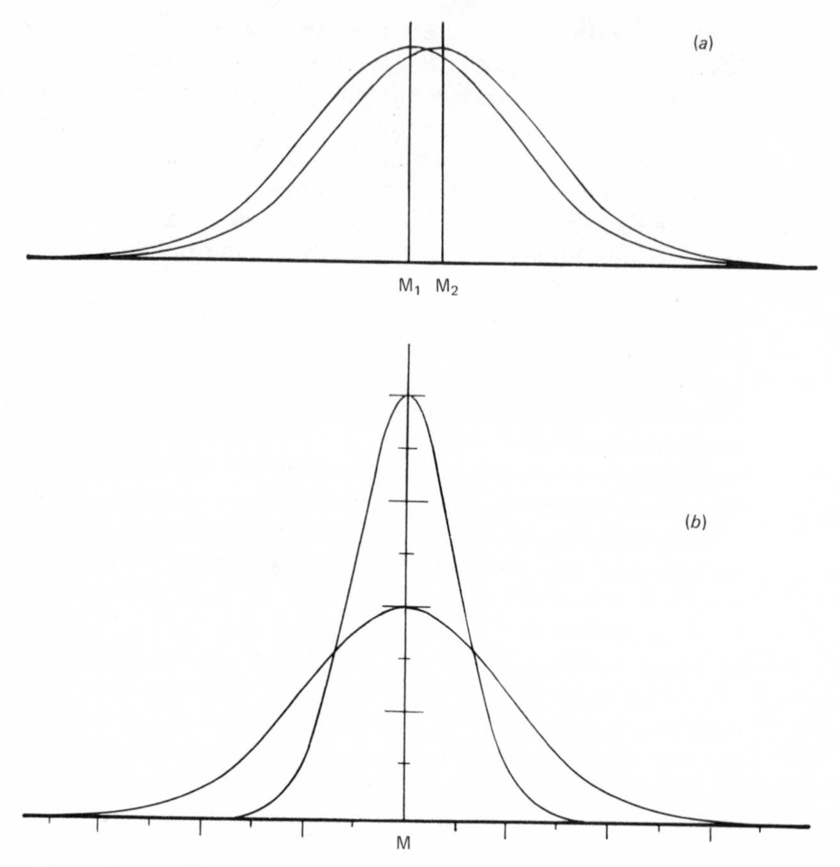

Figure A-3 (*a*) Two normal distributions with the same areas and the same variation about the mean but with slightly different means. (*b*) Two normal distributions with the same mean and same areas but with different variation about the mean.

The observed deviation, expressed as units of standard deviation, is then

$$z = \frac{0.4 - 0.5}{\sqrt{(0.5)(0.5)/100}}$$
$$= -2.00$$

Referring to A-2, we see that a deviation of 2.00 along the abscissa includes just over 95% of the area of the curve. Therefore, a 40:60 deviation will occur less than 5% of the time by chance when $p = 0.5$. We may therefore reject the hypothesis that $p = 0.5$, with only a 5% risk of making an error if $p = 0.5$ in fact. Had the deviation been large enough to yield $z > 2.58$, the risk would have been only 1%.

TESTS OF SIGNIFICANCE

In genetics, as in many other sciences, the question frequently arises as to whether a set of observations conforms to a certain hypothesis. This often takes the form of asking whether the classification of persons as affected or nonaffected yields a ratio compatible with simple Mendelian inheritance. This problem is the same as asking whether the distribution of heads and tails in a number of coin tosses is compatible with the hypothesis of equal chances for each to turn up.

In order to answer the question, one must formulate a null hypothesis; that is, there is no difference between the theoretical value of p and the value in the population studied. In terms of a coin, we would say the ratio of heads and tails is not different from the expected if heads and tails are equally likely. If we can prove the hypothesis unlikely, then we may wish to conclude that heads and tails are not equally likely and that the coin lands on one side more often than the other.

For example, if a coin were tossed ten times, a 10:0 distribution would occur only 0.002 of the time when heads and tails are in fact equally likely. If this distribution were tossed, one could conclude that the coin results were biased and expect to be wrong only 0.2% of the time if in fact the coin is not biased. A 9:1 distribution would also lead one to discard the null hypothesis, but in this case, and error would be made 2.2% of the time. An 8:2 distribution would lead to error 11% of the time if the hypothesis were rejected. For most purposes this error is too large, and the hypothesis would be allowed to stand pending more data.

At what point should the null hypothesis be rejected? The answer will depend on the likelihood of alternate explanations and the consequences of a wrong decision. For most investigations, a result that will happen only 5% of the time by chance under a hypothesis is considered adequate to reject the hypothesis. If the results could happen only 1% of the time by chance, the hypothesis can be rejected with greater confidence. On occasions, erroneous rejection of a hypothesis may have serious consequences. A more conservative level, such as 0.1%, might be used.

There are many tests for determining whether or not a particular set of observations deviates from an expected distribution (and therefore from a particular hypothesis or model), the various tests being appropriate for particular kinds of observations and models. In the above discussion, exact probabilities were calculated for binomial distributions for small values of n. In the case of normally distributed variables, the normal deviate z is sometimes appropriate, and indeed most statistical tests ultimately are based on the normal deviate. Of the various tests, the following are most often used in problems encountered in elementary genetics.

Chi-square Analysis

The χ^2 (chi-square) analysis is related in theory to the normal deviate analysis outlined above. However, it has some special applications that justify separate consideration.

Goodness of Fit. A measure of deviation from theory of a set of observations can be measured by χ^2, given by the formula

$$\chi^2 = \sum \frac{(E - O)^2}{E},$$

where E = expected value, O = observed value, and Σ indicates summation for all categories. In the problem of the previous section, the expected values (on the hypothesis of $p = 0.5$) would be 50 heads and 50 tails. The observed values were 40 and 60. Therefore,

$$\chi^2 = \frac{(50 - 40)^2}{50} + \frac{(50 - 60)^2}{50}$$
$$= 4.00.$$

This value of 4.00 is the square of 2.00 obtained by the earlier method, since, for the special case of two alternatives (heads or tails), $\chi^2 = z^2$.

In order to decide whether the value of χ^2 is too large to represent chance occurrence, it is necessary to consult a table of χ^2, such as is given in Table A-1. To use the table, one must know the degrees of freedom. The degrees of freedom (df) are the number of categories that can be varied independently without changing the total number of observations. In the above problem, there is one degree of freedom, since only one category can be varied, either heads or tails. If the proportion of heads is arbitrarily specified to be 0.45, then the proportion of tails must

Table A-1 **DISTRIBUTION OF χ^2**

The values of P are the probability that χ^2 will exceed by chance the value given in the table.*

Degrees of freedom	0.99	0.95	0.50	P 0.10	0.05	0.02	0.01
1	0.00016	0.0039	0.45	2.71	3.84	5.41	6.64
2	0.0201	0.103	1.39	4.61	5.99	7.82	9.21
3	0.115	0.35	2.37	6.25	7.82	9.84	11.35
4	0.297	0.71	3.36	7.78	9.49	11.67	13.28
5	0.554	1.15	4.35	9.24	11.07	13.39	15.09
10	2.56	3.94	9.34	15.99	18.31	21.16	23.21

* Abridged from Table III, R. A. Fisher, *Statistical Methods for Research Workers*. Edinburgh: Oliver and Boyd, Ltd., by permission of the author and publishers.

be 0.55. If one were studying the distribution of dice scores, there would be five degrees of freedom since a die has six sides.

For degree of freedom, a χ^2 of 3.84 is larger than 95% of the values obtained by chance. The χ^2 of 4.00 exceeds this value, indicating that a 40:60 distribution would occur less than 5% of the time by chance, the same conclusion reached by the earlier method.

Chi square can readily be applied to data consisting of more than one category. In the cross *AaBb* × *aabb*, four types of offspring are expected: *AaBb, aaBb, Aabb,* and *aabb*. In the absence of linkage or other factors, they should be in equal proportions. Actual observation gave 45 *AaBb,* 40 *aaBb,* 60 *Aabb,* and 55 *aabb*. Are these values compatible with a 1:1:1:1 ratio? The expected value for each category is 50; hence,

$$\chi^2 = \frac{(50-45)^2}{50} + \frac{(50-40)^2}{50} + \frac{(50-60)^2}{50} + \frac{(50-55)^2}{50}$$

$$= 5.00, \text{ df} = 3.$$

For three degrees of freedom, χ^2 must exceed 7.8 to invalidate the hypothesis.

Contingency. Another important use of χ^2 is in testing for contingency or association. The example just cited is a test of association between the *A* locus and the *B* locus. Since a specific cross was tested and simple Mendelian inheritance was involved, it was possible to arrive at the "expected" values on theoretical grounds and independently of the data to be tested. In other cases, it is desirable to test for association between two traits in the absence of external hypotheses from which to derive expected values.

An example would be the association between hair color and eye color. Do blonds tend to have blue eyes more often than brown eyes? Or perhaps they tend to have brown eyes. Observation of 100 persons gave 35 blue-eyed blonds, 10 brown-eyed blonds, 15 blue-eyed brunettes, and 40 brown-eyed brunettes. A contingency table is set up as follows:

Observed	Blue eyes	Brown eyes	Totals
Blond	35	10	45
Brunette	15	40	55
Totals	50	50	100

In order to calculate χ^2, it is necessary to compute a table of expected values. From the marginal totals, it is seen that $\frac{50}{100}$ of the persons have blue eyes and $\frac{45}{100}$ are blonds. We wish to test the null hypothesis of no association. Absence of association would lead us to expect that $\frac{50}{100}$ of the 45 blonds would have blue eyes. Similarly, $\frac{50}{100}$ of the 55 brunettes should have blue eyes, and so on. A table expected values can be constructed by these procedures, for example,

Expected	Blue eyes	Brown eyes	Totals
Blond	22.5	22.5	45
Brunette	27.5	27.5	55
Totals	50	50	100

Note that the marginal totals must agree with those in the table of observed data. It is necessary to calculate only one of the entries by multiplication. The remaining three can be obtained by subtraction from the marginal totals. For this reason, there is only 1 df in a 2 × 2 contingency table.

Calculation of the χ^2 is as before:

$$\chi^2 = \frac{(22.5 - 35)^2}{22.5} + \frac{(22.5 - 10)^2}{22.5} + \frac{(27.5 - 15)^2}{27.5} + \frac{(27.5 - 40)^2}{27.5}$$

$$= 25.25.$$

This value clearly exceeds that required for 1% level of significance for 1 df. We must conclude that the null hypothesis is incorrect and that there is indeed association between hair color and eye color, there being an excess of blue-eyed blonds and brown-eyed brunettes, and a deficiencey of blue-eyed brunettes and brown-eyed blonds.

Contingency tables can be constructed larger than the 2 × 2 illustrated. The degrees of freedom in an $m \times n$ table is $(m - 1) \cdot (n - 1)$.

Yates Correction for χ^2. The χ^2 table is derived from a continuous curve, although the distribution of χ^2 itself is discontinuous. This approximation is of little significance unless the numbers of observations are small. If the numbers are small, χ^2 is likely to be erroneously large, and a hypothesis may be incorrectly rejected.

Yates has suggested a means of correction. It consists in reducing the $(E - O)$ values by $\frac{1}{2}$. Thus in a goodness of fit test, the observed values might be 4, 7, 6, 3 with expected values 5, 5, 5, 5. Chi-square would be calculated

$$\chi^2 = \frac{(4.5 - 5)^2}{5} + \frac{(6.5 - 5)^2}{5} + \frac{(5.5 - 5)^2}{5} + \frac{(3.5 - 5)^2}{5}$$

$$= 1.00, \text{ 3 df.}$$

Without the correction, $\chi^2 = 2.00$.

There is no fast rule to decide when to use the Yates correction. If the expected values in each cell are under 10, then the correction probably will make a difference.

CORRELATION

It is often useful to have a means of expressing quantitatively the resemblance between relatives. In other instances it may be desirable to express the relationship between two measures made on a series of individuals. This can be done with the *correlation coefficient, r*. We will be concerned only with the simplest case of linear correlation, that is, when the relationship between two variables X and Y is such that for any change in X there is a proportional change in Y. If Y is plotted against X, a straightline relationship is observed.

For the situation in which there are two series of measurements to be compared, for example IQs of wives and husbands or height and weight of a series of individuals, r can be calculated by the formula

$$r = \frac{\Sigma(X - \overline{X})(Y - \overline{Y})}{\sqrt{\Sigma(X - \overline{X})^2\Sigma(Y - \overline{Y})^2}},$$

where X is an observed value of one trait, $\overline{X}$ is the mean value, Y is a corresponding observation on the second trait, and $\overline{Y}$ is the mean value. (The derivation of r and its relationship to other statistical parameters are given in textbooks of statistics.)

The coefficient r can assume any value from -1.0 to $+1.0$. The correlation coefficient may best be thought of as a measure of the variation shared in common by two variables. If none of the variation is shared, then r is zero and a plot of the two variables shows a random scatter. If the correlation is significantly positive (or negative), then the two variables are somewhat related, and knowledge of one is useful in predicting the other. There still is unshared variation unless $r = +1.0$ or -1.0.

Reliability of a correlation coefficient is dependent on the number of observations on which it is based. Tests of significance are available for deciding whether a correlation coefficient is significantly different from zero or from another calculated correlation coefficient.

REFERENCES AND SUGGESTED READING

CAVALLI-SFORZA, L. L., AND BODMER, W. F. 1971. *The Genetics of Human Populations*. San Francisco: W. H. Freeman, pp. 805–849.

DIXON, W. J., AND MASSEY, F. J. Jr. 1957. *Introduction to Statistical Analysis,* 2nd Ed. New York: McGraw-Hill, 488 pp.
EMERY, A. E. H. 1976. *Methodology in Medical Genetics.* Edinburgh: Churchill Livingstone, 157 pp.
FISHER, R. A. 1958. *Statistical Methods for Research Workers,* 13th Ed. Edinburgh: Oliver and Boyd, 356 pp.
KEMPTHORNE, O. 1957. *An Introduction to Genetic Statistics.* New York: Wiley, 545 pp.
MOSIMANN, J. E. 1968. *Elementary Probability for the Biological Sciences.* New York: Appleton-Century Crofts, 255 pp.
SNEDECOR, G. W., AND COCHRAN, W. G. 1967. *Statistical Methods,* 6th Ed. Ames: Iowa State Univ. Press.

APPENDIX B

HUMAN LINKAGE GROUPS

The assignment of human loci to specific chromosomes continues at such a high rate that any tabulation is out of date. The list of chromosomal assignments in Table B-1 is intended to include only those assignments that are based on good evidence, although some of these loci will undoubtedly move even before this is published. Most of the assignments and references can be found in McKusick (1978) and Human Gene Mapping 4 (1977). Only the autosomal assignments are shown. A catalog of X-linked loci is included in McKusick (1978).

In addition to the loci in Table B-1, several groups of loci have been shown to be linked by pedigree studies, but no assignment to a specific chromosome has been made. A list of these is given in Table B-2.

Table B-1 **LIST OF HUMAN LOCI ASSIGNED TO SPECIFIC AUTOSOMES**

Those loci polymorphic in one or more major populations are in **boldface.**

Symbol	McKusick number	Trait	Remarks
Chromosome 1			
AdV 12-CMS-1p	10292	adenovirus 12 site 1p	1p36
AdV 12-CMS-1q	10293	adenovirus 12 site 1q	1q42
AK-2	10302	adenylate kinase-2	1p31-1pter
Amy-1	**10470**	**α-amylase, salivary**	**1p**
Amy-2	**10465**	**α-amylase, pancreatic**	**1p**
Cae	11620	cataract, zonular pulverulent	1p
Fy	**11070**	**Duffy blood group**	**1q2**
El-1	13050	elliptocytosis-1	1p, 3 cM from *Rh*
ENO-1	17243	enolase-1	1p34-1p36
αFUC	**23000**	**α-L-fucosidase**	**1p32-1p34**
FH-M	13686	fumarate hydratase, mitochondrial	
FH-S	13685	fumarate hydratase, soluble	1q42-1qter
GDH	13809	glucose dehydrogenase	1p21-1pter
GUK-1	13927	guanylate kinase-1	1q31-1q42
GUK-2	13928	guanylate kinase-2	
Mtr	15658	5-methyltetrahydrofolate:L-homocysteine S-methyltransferase	
PepC	**17000**	**peptidase C**	**1q41-1q43**
PKU	26160	phenylalanine hydroxylase	near *Amy*
PGM-1	**17190**	**phosphoglucomutase-1**	
6PGD	**17220**	**6-phosphogluconate dehydrogenase**	
Rh	**11170**	**Rhesus blood group**	**1p32-1pter**
RN5S	18042	5S RNA	1q42-1q43
Sc	**11175**	**Scianna blood group**	**1p**
UMPK	**19171**	**uridine monophosphate kinase**	**1p32**
UGPP-1	19175	uridyl diphosphate glucose pyrophosphorylase-1	1q21-1q23

Chromosome 2			
ACP-1	**17150**	acid phosphatase-1	2p23
AHH	10833	aryl hydrocarbon hydroxylase	2p
GAL+ ACT	13703	galactose enzyme activator	2p11-2p22
If-1	14757	interferon-1	
IDH-S	14770	isocitrate dehydrogenase, soluble	
MDH-S	15420	malate dehydrogenase, soluble	2p23-2pter
UGPP-2	19176	uridyl diphosphate glucose pyrophosphorylase-2	
Chromosome 3			
ACY-1	10462	aminoacylase-1	
GPx-1	**23170**	**glutathione peroxidase-1**	
HVS	14245	herpes virus sensitivity	
TSAF8	11695	cell cycle controller-G1	
Chromosome 4			
Alb	**10360**	**albumin**	
Gc	**13920**	**group-specific component (vitamin D binding protein)**	**4q11-4q13**
PepS	17025	peptidase S	4pter-4q21
PGM-2	**17200**	**phosphoglucomutase-2**	**4p14-4q21**
PRPP-AT	17245	phosphoribosylpyrophosphate amidotransferase	4pter-4q21
Chromosome 5			
AVr	10745	antiviral regulator	5p
ARS-B	25320	arylsulfatase B	
DTS	12615	diphtheria toxin sensitivity	5q15-5qter
HexB	14265	hexosaminidase β subunit	5pter-5q13
If-2	14758	interferon-2	
Leu-RS	15135	leucyl-tRNA synthetase	
Chromosome 6			
ADRH	20191	adrenal hyperplasia (21-hydroxylase deficiency)	near *HLA-B*
ASD-2	10880	atrial septal defect	
GOT-M	**13815**	**glutamate-oxaloacetate transaminase, mitochondrial**	
GLO-I	**13875**	**glyoxylase I**	**6p21-6p22**
Hch	14160	hemochromatosis	part of *MHC*

Table B-1 **(Continued)**

Symbol	McKusick number	Trait	Remarks
Chromosome 6 (*Continued*)			
MHC		**major histocompatibility complex**	**6p2100-6p22**
ME-1	15425	malic enzyme, soluble	6p21-6q16
MRBC	15805	monkey RBC receptor	
NDF	20270	neutrophil differentiation factor	part of *MHC*
OPCA-1	16440	olivopontocerebellar atrophy I	
PGM-3	**17210**	**phosphoglucomutase-3**	**6p21-6qter**
PA	17337	plasminogen activator	
SOD-2	14746	superoxide dismutase-2 (mitochondrial)	6q16-6q21
SA6	18551	surface antigen (SA 6)	
Chromosome 7			
ASL	20790	argininosuccinate lyase	7pter-7q22
Col-1	12015	collagen-1	
Co	**11045**	**Colton blood group**	
GUS	25322	β-glucuronidase	7pter-7q22
HaF	23400	Hageman factor	7q
H4	14275	histone H4	
HADH	14345	hydroxy-acyl CoA dehydrogenase	
Km	**14720**	**Ig kappa L chain**	
Jk	**11100**	**Kidd blood group**	**7q**
MDH-M	**15410**	**malate dehydrogenase, mitochondrial**	**7p22-7q22**
NCR	16282	neutrophil chemotactic response	
SA7	18552	surface antigen 7	7p12-7pter
UP	19174	uridine phosphorylase	
Chromosome 8			
CF7	22750	clotting factor VII	

GSR	**13830**	**glutathione reductase**	**8p21**
LETS	15128	large external transformation-sensitive protein	
Chromosome 9			
ABO	**11030**	**ABO blood group**	
ACON-S	**10088**	**aconitase, soluble**	
AK-1	**10300**	**adenylate kinase-1**	**close to *ABO***
AK-3	10303	adenylate kinase-3	9p12-9q33
ASS	21570	argininosuccinate synthetase	
GALT	**23040**	**galactose-1-phosphate uridyl transferase**	
NPa	16120	nail-patella syndrome	close to *ABO*
Chromosome 10			
ADK	10275	adenosine kinase	
EMP-130	13371	external membrane protein-130	
GSS	13825	glutamate-gamma-semialdehyde synthetase	
GOT-S	13818	glutamate oxaloacetate transaminase, soluble	10q24-10q26
HK-1	14260	hexokinase-1	10pter-10q24
PP	17903	inorganic pyrophosphatase	10pter-10q24
FUSE	17475	polykaryocytosis inducer	
Chromosome 11			
ACP-2	17165	acid phosphatase-2	11p12-11cen
EsA4	13322	esterase-A4	11cen-11q22
Hb BETA	**14190**	**β-globin complex**	
LDH-A	15000	lactate dehydrogenase A	11p12.08-11p12.03
LEA-1	15125	lethal antigen a1	11p13-11pter
LEA-2	15126	lethal antigen a2	11q13-11qter
LEA-3	15127	lethal antigen a3	11p13-11pter
SA11	18555	surface antigen 11	11p
W-AGR	27780	Wilms tumor-AGR triad	11p3
Chromosome 12			
CS	11895	citrate synthase, mitochondrial	
ENO-2	13136	enolase-2	
GAPD	13840	glyceraldehyde-3-phosphate dehydrogenase	12p12.2-12pter

Table B-1 **(*Continued*)**

Symbol	McKusick number	Trait	Remarks
Chromosome 12 (*Continued*)			
LDH-B	15010	lactate dehydrogenase B	12p12.1-12p12.2
PepB	16990	peptidase B	12q21
SHMT	13845	serine hydroxymethyltransferase	12pter-12q14
SA12	18556	surface antigen 12	
TPI-1	19045	triose phosphate isomerase-1	12p12.2-12pter
TPI-2	19046	triose phosphate isomerase-2	
Chromosome 13			
EsD	**13328**	**esterase D**	**13q3**
RB-1	18020	retinoblastoma-1	
rRNA	18045	ribosomal RNA	13p12
Chromosome 14			
EBV	13285	Epstein-Barr virus expression	
EMP-195	13374	external membrane protein-195	
NP	16405	nucleoside phosphorylase	14q12-14q20
rRNA	18045	ribosomal RNA	14p12
Trp-RS	19105	tryptophanyl-tRNA synthetase	14q21-14qter
Chromosome 15			
HexA	27280	hexosaminidase α subunit	15q22-15qter
IDH-M	14765	isocitrate dehydrogenase, mitochondrial	15q21-15qter
MPI	15455	mannose phosphate isomerase	15q22-15qter
αMAN-A	15458	α-D-mannosidase, cytoplasmic	15q11-15qter
β2M	10970	β2-microglobulin	15q14-15q21
PK3	17905	pyruvate kinase-3	15q22-15qter
rRNA	18045	ribosomal RNA	15p12
Chromosome 16			
APRT	**10260**	**adenine phosphoribosyltransferase**	**16q**
Hb ALPHA	14180	α-globin complex	

Hpα	**14010**	**haptoglobin α subunit**	**16q22**
LCAT	24590	lecithin-cholesterol acyltransferase	very near *Hpα*
TK-M	18829	thymidine kinase, mitochondrial	
Chromosome 17			
AdV12-CMS-17	10297	adenovirus-12 site	17q21-17q22
Col-1	12015	collagen-1	17q
GALK	**23020**	**galactokinase**	**17q21-17q22**
αGLU	**23230**	**α-glucosidase, acid**	**17q**
TK-S	18830	thymidine kinase, soluble	17q21-17q22
Chromosome 18			
PepA	**16980**	**peptidase A**	**18q23-18qter**
Chromosome 19			
GPI	17240	glucosephosphate isomerase	19pter-19q13
αMAN-B	24850	α-D-mannosidase, lysosomal	19pter-19q13
PepD	**17010**	**peptidase D**	
PVS	17385	poliovirus sensitivity	19q
Chromosome 20			
ADA	**10270**	**adenosine deaminase**	**20p11-20qter**
DCE	12565	desmosterol-to-cholesterol enzyme	
ITP	14752	inosine triphosphatase	
Chromosome 21			
AVP	10745	antiviral protein; interferon receptor	21q21-21qter
PGS	13844	phosphoribosyl glycineamide synthetase	
rRNA	18045	ribosomal RNA	21p12
SOD-1	14745	superoxide dismutase-1	21q22.1
Chromosome 22			
ACON-M	10084	aconitase, mitochondrial	
ARS-A	25010	aryl sulfatase A	
CML	15141	chronic myeloid leukemia	22q12
DIA-1	25080	cytochrome b5 reductase (diaphorase)	
GALB	10417	α-galactosidase B	22q13-22qter
rRNA	18045	ribosomal RNA	22p12

Table B-2 LINKAGE GROUPS NOT YET ASSIGNED TO CHROMOSOMES

Loci and intervals (cM)	Traits (McKusick number)
Lu. . .13. . .*Se*. . .4. . .*Dm*	*Lu*, Lutheran blood group (11120) *Se*, secretor type (18210) *Dm*, myotonic dystrophy (16090)
MNSs. . .*Scl*	*MNSs*, blood group (11130) *Scl*, sclerotylosis (18160)
Tf. . .E_1	*Tf*, transferrin (19000) E_1, pseudocholinesterase-1 (17740)
PTC. . .*K*	*PTC*, phenylthiocarbamide taste sensitivity (17120) *K*, Kell blood group (11090)
GPT. . .*EB*	*GPT*, glutamate-pyruvate transaminase (13815) *EB*, epidermolysis bullosa, Ogna type (13195)
Le. . .*C3*	*Le*, Lewis blood group (11110) *C3*, complement component 3 (12070)
Gm. . .*Pi*	*Gm*, immunoglobulin H chain region (14700) *Pi*, protease inhibitor; α_1-antitrypsin (10740)

REFERENCES

MCKUSICK, V. A. 1978. *Mendelian Inheritance in Man*, 5th Ed. Baltimore: Johns Hopkins Univ. Press, 976 pp.

HUMAN GENE MAPPING 4 (1977): Fourth International Workshop on Human Gene Mapping. *Birth Defects: Original Article Series* 14 (1978), New York: The National Foundation; also published in *Cytogenet. Cell Genet.* 22 (1–6), 1978.

ANSWERS TO REVIEW QUESTIONS

CHAPTER TWO

2. 1:3.
3. 1:1, all normal.
4. $\frac{3}{4}$ normal, $\frac{1}{4}$ albino.
6. (a) *ddEE* × *ddEE* → *ddEE* or *DDee* × *DDee* → *DDee*.
(b) *DDee* × *ddEE* → *DdEe*.
(c) *Ddee* × *ddEE* → *DdEe* + *ddEe* or *DDee* × ddEe → *DdEe* + *Ddee*.
7. *RrTt* × *RrTt* → *rrtt*. Subsequent children: $\frac{1}{16}$ *RRTT*, $\frac{1}{16}$ *RRtt*, $\frac{1}{16}$ *rrTT*, $\frac{1}{16}$ *rrtt*, $\frac{1}{8}$ *RRTt*, $\frac{1}{8}$ *RrTT*, $\frac{1}{8}$ *rrTt*, $\frac{1}{8}$ *Rrtt*, $\frac{1}{4}$ *RrTt*.
10. (a) Autosomal dominant.
(b) Autosomal dominant and recessive.
(c) Autosomal recessive and X-linked recessive.
(d) Autosomal recessive.
(e) None.
11. (b) $\frac{1}{2}$ for each child, (c) $\frac{1}{4}$ for each child, (d) $\frac{1}{2}$ for males, zero for females.
12. Rule out: (a) X-linked dominant or recessive, (b) X-linked recessive, (c) autosomal recessive, X-linked dominant or recessive, (d) X-linked dominant or recessive, autosomal dominant, (e) dominant, (f) dominant. None of the examples is consistent with Y linkage. Most likely: (a) autosomal dominant, (b) autosomal dominant, (c) autosomal dominant, (d) autosomal recessive, (e) X-linked recessive, (f) X-linked recessive.

CHAPTER THREE

2. (a) 5.8×10^9.
(b) 2.9×10^6.
3. 30%, 20%, 20%.
4. GC-rich regions due to three hydrogen bonds per base pair.
7. No crossover: *ABC, abc;* one crossover: *ABc, abC, Abc, aBC;* two crossovers: *AbC, aBc*.

CHAPTER FIVE

2. MI nondisjunction in males: *XY, O;* MII: *XX, XY, O;* MI + MII: *XY, XXY, Y, XYY, X, XXYY, O*. MI nondisjunction in females: *XX, O;* MII: *XX, O;* MI + MII: *X, XX, XXX, XXXX, O*.
4. (a) Father.
   (b) Mother.
   (c) Either.
   (d) Father.
   (e) Mother.
5. 46,XY Xg(a+) × 47,XXX Xg(a−) → 47,XXY Xg(a+).

CHAPTER SIX

2. (a) 46,−22,+t(21q22q)
   (b) 46,X,r(X).
   (c) 46,5p−.
   (d) 46,XY,rcp(1p:7p).
   (e) 46,XX,del(8)(p21).
5. 45,−21,−22,+t(21q22q).

CHAPTER SEVEN

3. In case (b), many cases would arise by nondisjunction at MII, increasing the homozygosity at the *Xg* locus and hence the frequency of *Xg*/*Xg*. By contrast, in case (a), the frequencies should be the same as in normal females.
4. (a) Female with Turner's syndrome.
   (b) Female with Turner's syndrome.
   (c) Female with streak gonads, normal height.
   (d) Female with streak gonads.
   (e) Normal male.
   (f) Normal male (?).
   (g) Female, streak gonads.
   (h) Normal female.
6. None.

CHAPTER EIGHT

3. (1) Substitution of amino acid with same charge; (2) mutation to another codon that codes for the same amino acid.

5. (a) The mRNA would be "constitutive."
   (b) The mRNA would be "constitutive."
7. (1) XO females are not identical to XX females. (2) Both alleles at specific loci (*Xg*, X-linked ichthyosis) are active in the same cells.

CHAPTER NINE

4. The frequency of homozygotes should equal the mutation rate, ca. $10^{-5} - 10^{-6}$.

CHAPTER TEN

2. Parent/offspring combinations where both traits are present in both persons.
5. (1) 7. (2) Gln, Lys, Glu, Leu, Ser, Tyr.

CHAPTER ELEVEN

3. One has a mutation in the structure of the apoenzyme that affects coenzyme (vitamin) binding. The other mutation affects some other part of the apoenzyme or causes absence of the apoenzyme.
4. (a) Recessive. Half the level of enzyme is generally adequate to perform the metabolic conversions.
   (b) Dominant. The uninhibited enzyme will be expressed, regardless of the presence of normal enzyme.
   (c) Dominant. Half the level of inactivating enzyme is generally adequate or at least will be strongly expressed.
   (d) Recessive. Half normal transport activity will ordinarily be adequate, although exceptions are known.

CHAPTER TWELVE

2. $\gamma 1$, a(+)z(+); $\gamma 3$, g(+).
3. A, G, M, D, E.
4. Subclasses refer to Ig molecules within a class but that have small structural differences in the constant regions of the heavy chains. The different related subclasses are produced by different loci.
5. Three.

6. Four: L and H chain variable and constant genes.
7. A number of loci, such as those producing adenosine deaminase and nucleoside phosphorylase, IR genes, and genes of unknown function, mutants of which produce some of the agammaglobulinemia diseases.

CHAPTER THIRTEEN

2. Type A_2 persons have anti-B in their plasma, which can be used to identify B(+) persons (types B and AB). Of the B-positive persons, those whose plasma reacts with A_2 cells would be type B, the remainder would be type AB. Type O cells will react neither with A plasma nor B plasma. Absorption of anti-A plasma with A_2 cells will leave activity only against A_1 cells.
3. R_1r, R_1R_2, R_0R_0, and rr''. The highest risks would be those with factor D: R_1r, R_1R_2, and R_0R_0.
4. (c), (e) = (f), (a), (b), (d).
5.

Genotype	Antigens on red cells					Antigens in secretions				
	A	B	H	Le^a	Le^b	A	B	H	Le^a	Le^b
AB, H, Se, Le	+	+	±	−	+	+	+	±	+	+
AB, H, sese, Le	+	+	±	+	−	−	−	−	+	−
AB, H, Se, lele	+	+	±	−	−	+	+	±	−	−
AB, H, sese, lele	+	+	±	−	−	−	−	−	−	−
AB, hh, Se, Le	−	−	−	+	−	−	−	−	+	−
AB, hh, Se, lele	−	−	−	−	−	−	−	−	−	−

6. (a) rr.
   (b) R^1R^2 (R^zr, R^zR^o, R^1r'', R^or^y, R^2r').
   (c) R^1r, R^1R^o (R^or').
   (d) R^or, R^oR^o.
   (e) $r'r$.
   (f) R^1R^1 (R^1r').

CHAPTER FOURTEEN

7. Translocation of the *TK*+ gene to another chromosome; reverse mutation of mouse *TK*− gene; mutation of suppressor genes.
8. The enzyme has two or more subunits produced by different loci.
9. (a) Contamination of culture with maternal cells. Karyotype should be compared to mother to assure nonidentity.
   (b) The fetus is heterozygous for classical Tay–Sachs disease and

a rare variant of hexosaminidase A that is active under *in vitro* test conditions but not in vivo. Parents could both be tested to assure heterozygous expression *in vitro*.

CHAPTER FIFTEEN

2. Expected: 100 affected, 100 nonaffected; observed: 85 affected, 115 nonaffected. $\chi^2(1\text{ df}) = 4.50$. $P < 0.05$. Yes.
3. Expected: 50 brown-eyed, 50 blue-eyed; observed: 45 brown-eyed, 55 blue-eyed. $\chi^2(1\text{ df}) = 1.00$. $P > 0.05$. Yes.
4. Expected: 105.742 affected, 119.258 nonaffected; observed: 122 affected, 103 nonaffected. $\chi^2(1\text{ df}) = 4.716$. $P < 0.05$. No.
5. Expected: 27 affected, 81 nonaffected; observed: 30 affected, 78 nonaffected. $\chi^2(1\text{ df}) = 0.44$. $P > 0.05$. Yes.
6. Expected: 50 affected, 150 nonaffected; observed: 65 affected, 135 nonaffected. $\chi^2(1\text{ df}) = 6.00$. $\mathrm{P} < 0.02$. No.
9. $q' = 0.288$.

CHAPTER NINETEEN

2. Expected: 100 recombinants, 100 nonrecombinants; observed: 174 recombinants, 226 nonrecombinants. $\chi^2(1\text{ df}) = 13.52$. $P << 0.01$. Yes.
3. Expected: 300 albinos, 100 pigmented; observed: 288 albinos, 112 pigmented. $\chi^2(1\text{ df}) = 1.92$. $P > 0.05$. No.
4. 3.25% affected and blood type B, 3.25% aff. type AB, 21.75% aff. type O, 21.75% aff. type A, 3.25% nonaff. type O, 3.25% nonaff. type A, 21.75% nonaff. type B, 21.75% nonaff. type AB.
5. Recombinant: III-2, -7; nonrecombinant: III-1, -4; uncertain: III-3, -5, -6.
6. All children of his sons would be normal for both traits, as would all daughters of his daughters. His daughters' sons would be at risk as follows: 2.5% colorblind, 2.5% G6PD deficiency, 47.5% both traits, 47.5% neither trait.
7. 87%.

CHAPTER TWENTY

2. (a) 0.5714.
   (b) 0.0816.

3. (a) 0.01.
   (b) 1.98%.
   (c) 0.000 392.
4. 0.04. 0.0768.
5. (a) 0.0004.
   (b) 0.0392.
   (c) 0.0753.
   (d) 0.00154.
6. $L^M = L^N = 0.50$. $\chi^2(1 \text{ df}) = 4.0; P < 0.05$; no. 0.401 M, 0.464 MN, 0.134 N.
7. 0.0098, 0.126.

CHAPTER TWENTY-ONE

2. (a) $q_{10} = 0.05$.
   (b) 40 generations.
3. $q_1 = 0.909$. $q_{10} = 0.05$. $q_{100} = 0.00909$. 990 generations.
4. 0.167. 0.200.
5. 0.25.

CHAPTER TWENTY-TWO

2. (a) $\frac{1}{4}$.
   (b) $\frac{1}{4}$.
   (c) It would depend on sex of children in F_1. If both were sons, neither grandchild could receive the X-linked allele. If both were daughters, they would be heterozygous. The great-grandsons would not have an increased risk due to consanguinity, since they cannot be homozygous for a sex-linked gene. The great-granddaughters could have increased risk only if the grandson entering into the consanguineous marriage were affected.
3. (a) $\frac{1}{16}$
   (b) $\frac{1}{12}$
   (c) $\frac{1}{6}q$, where q is the frequency of the rare allele.
   (d) $q^2 + \frac{1}{16}pq$.
4. $\frac{1}{64}$. $\frac{1}{4}$ if male, zero if female, overall probability $= \frac{1}{8}$. $\frac{1}{512}$. $\frac{441}{512}$.
5. 0.0015.

CHAPTER TWENTY-THREE

2. 0.656.
5. (a) 0.0001, (b) 0.0001.

CHAPTER TWENTY-FOUR

2. (a) 0.44.
(b) 0.045.
3. (a) 0.182.
(b) 0.308.
(c) 0.471.
4. $\frac{1}{6}$.
5. 0.96.

CHAPTER TWENTY-FIVE

1. (a) 0.020.
(b) 0.0179.
(c) 0.0189.

GLOSSARY

Acentric a chromosome fragment without a centromere.

Acrocentric a chromosome with the centromere near one end.

Allele an alternate form of a gene.

Allograft a tissue graft between genetically nonidentical members of the same species.

Allosteric site a site on an enzyme, other than the catalytic site, that combines with metabolites to regulate activity of the enzyme.

Allotype a term used especially in immunology to express the phenotype in a system involving genetic variation.

Amino acid protein building blocks of the general formula H_2NCHR-$COOH$.

Amniocentesis sampling the amniotic fluid by direct tap.

Amorph an allele that produces a completely inactive product or no product.

Anaphase the period in mitosis and meiosis when the chromosomes move to opposite poles.

Aneuploidy having a loss or gain of one or more chromosomes as compared to the basic complement.

Antibody an immunoglobulin produced in response to exposure to an antigen and capable of combining very specifically with the antigen.

Anticodon a sequence of three nucleotides in tRNA that pairs with the codon in mRNA.

Antigen any substance capable of inducing antibodies.

Ascertainment the identification or location of families and individuals for study.

Assortative mating mating between persons with similar traits.

Autosome a chromosome other than the sex chromosomes.

Backcross the cross of an F_1 hybrid with one of the parental lines.

Bacteriophage a bacterial virus.

Bands, chromosome areas of light or dark staining produced by a number of techniques. C-bands, G-bands, and Q-bands are so designated because of the technique and stain used.

Barr body sex chromatin.

Base a substance alkaline in nature. Of particular interest in genetics are the purine and pyrimidine bases in nucleic acid.

Bayesian method the assessment of relative probabilities of prior events on the basis of their likelihoods of having produced a particular result.

Bivalent the figure produced by pairing of homologous chromosomes in meiosis.

C-bands bands observed when chromosomes are treated and stained to show regions of constitutive heterochromatin.

Centimorgan a measure of the crossover frequency between linked genes. One centimorgan equals 1% recombination.

Centriole a small body in animal cells that lies next to the nucleus. Usually there are two. They serve as centers for spindle fiber formation.

Centromere the region of the chromosome that becomes associated with spindle fibers during mitosis and meiosis and is important in movement of chromosomes to the poles.

Centromeric index the ratio of the short arm of a chromosome to the total length, ordinarily expressed as percent.

Chiasma (pl. chiasmata) points at which homologous chromosomes remain attached after pairing has ceased in meiosis.

Chimera in human genetics, the term refers to a person with two different genotypic complements arising from more than one zygote.

Chromatid either of two structures that comprise the duplicated chromosome arms prior to division of the centromere to form daughter chromosomes. Each chromatid becomes a chromosome after division of the centromere.

Chromatin nuclear substance with characteristic staining properties, generally equated with chromosomal material.

Chromomere heavily staining concentrations of chromatin often occurring in regular patterns along the chromosomes at certain stages of the cell cycle. Chromomeres probably represent regions of extensive coiling or folding of the chromosome.

Chromosome the structures in the cell nucleus that store and transmit genetic information.

Chromosome lag failure of a chromosome to remain in its complement during anaphase.

Cis equivalent to *coupled.* The alleles at two linked loci are in the *cis* configuration if they are on the same chromosome.

Cistron a functional genetic unit, as defined by a *cis-trans* test. Two mutations are said to be in the same cistron if, in a diploid cell or organism, they cannot complement each other in the *trans* configuration.

Clastogen an agent that breaks chromosomes.

Cline a gradient in gene frequencies over space.

Clone a population of cells originally derived from a single cell by mitosis.

Coadaptation two or more genes that interact and that are subject to natural selection as a single unit.

Codominant alleles are codominant if each is expressed independent of the presence of the other.

Codon a sequence of three nucleotides (in mRNA) or deoxyribonucleotides (in DNA) that specify a particular amino acid in a polypeptide chain.

Coenzyme a small molecular weight substance necessary for the catalytic activity of an enzyme. The B vitamins are coenzymes.

Compensation the replacement of nonsurviving offspring with additional offspring.

Complement a complex mixture of plasma components that are bound in antigen-antibody reactions and that are necessary for cell lysis.

Complementation functional independence of two mutations as tested in the *trans* configuration or in heterokaryons. Two such mutants that cannot produce normal function when in the *trans* configuration are said to be noncomplementary and therefore in the same complementation group.

Congenital present at birth. A congenital trait may be inherited or environmental in origin.

Consanguinity being related through a common ancestor.

Constitutive enzymes are constitutive if they are produced at fixed rates regardless of need.

Constriction an attenuated region of a chromosome. The *primary* constriction is associated with the centromere. Other constrictions are called *secondary*.

Contact inhibition inhibition of further growth of cells in culture when a confluent monolayer has formed.

Co-repressor a metabolite that combines with a repressor to inhibit function of an operon.

Coupling on the same chromosome; in the *cis* configuration.

***Cri-du-chat* syndrome** a syndrome involving mental retardation and other defects due to partial deletion of the short arm of chromosome 5.

Crossing over a process in which homologous chromosomes (chromatids) exchange distal segments. Crossing over involves breakage and physical exchange of segments, followed by repair of the breaks. Crossing over is a regular event in meiosis but occurs only rarely in mitosis.

Cross reacting material (CRM) a protein that has lost its biological activity but that still can be detected immunologically.

C-terminus the end of a polypeptide chain that has a free carboxyl group.

Cytogenetics the field of investigation concerned with studies of chromosomes.

Cytokinesis cell division.

Cytoplasmic inheritance extranuclear inheritance.

Deletion loss of a segment of a chromosome.

Denaturation loss of biological activity by treatment with heat, acid, and so on. Denaturation occurs with large molecules such as proteins or nucleic acid because of loss of the three -dimensional structure.

Deoxyribonucleic acid (DNA) a polymer composed of units of deoxyribonucleotides. DNA serves as the primary storage site of genetic information.

Derivative chromosome a structurally rearranged chromosome generated by a single rearrangement involving two or more chromosomes.

Dermatoglyphics the study of the pattern of dermal ridges on the surface of the fingers, toes, palms, and soles.

Diakinesis a stage in meiosis in which chromosomes are maximally condensed.

Dicentric having two centromeres on one chromosome.

Dihybrid heterozygous at two loci.

Diploid having two sets of chromosomes.

Diplotene a stage of meiotic prophase in which the chromatids become visibly separate.

Disjunction the separation of chromosomes in meiosis or mitosis.

Dispermy fertilization by two sperm.

Dizygotic dizygotic (DZ) twins arise from two separate ova fertilized by separate sperm. DZ twins are also called fraternal twins.

DNA deoxyribonucleic acid.

Domain in immunoglobulins, a region of intrachain homology. There are 4 domains in heavy chains and 2 in light chains.

Dominant an allele is dominant if it is expressed when only one copy is present.

Duplication a term used when a chromosomal segment is present twice, often in tandem.

Dysgenic detrimental to the genetic quality of a species.

Effector a small metabolite that combines with repressor to influence the activity of an operon.

Electromorph a variant of a protein detected by a difference in electrophoretic mobility.

Empiric risk the risk of occurrence of a trait based on empiric observation rather than genetic theory.

Endogamous mating within the family or related group.

Endonuclease an enzyme that breaks the bonds between adjoining nucleotides located in the interior of a polynucleotide chain.

Endoreduplication duplication of chromosomes without cell division, giving a tetraploid complement of chromosomes.

Enzyme a biological catalyst, consisting of a protein, called the *apoenzyme*, and, on occasion, a nonprotein coenzyme.

Epistasis interaction between loci such that the genotype at one locus prevents expression of the genotype at the second locus. The term is also used in a more general sense to mean gene interaction.

Equational division the separation of chromosomes into daughter cells with complements similar to the parent cell. The term is used especially for the second division of meiosis.

Equatorial plane the plane on which the chromosomes align at metaphase in mitosis and meiosis.

Euchromatin regions of chromosomes characterized by certain patterns of condensation and staining. Most known genes appear to be located in euchromatic regions.

Eugenics a philosophy that is concerned with improving the genetic quality of a population.

Eukaryote an organism characterized by true chromosomes and nuclei. Eukaryotes include all organisms except viruses, bacteria, and blue-green algae.

Euploid having the basic haploid complement of chromosomes or a multiple of the basic complement.

Exogamous mating outside the family or related group.

Exon a sequence of DNA that is translated into protein.

Exonuclease an enzyme that breaks bonds between adjoining nucleotides only when one of the nucleotides is in the terminal position of the polynucleotide chain.

Expressivity the variation in phenotype associated with a particular genotype.

Feedback inhibition inhibition of an enzyme by a product of the pathway catalyzed by the enzyme.

Fertilization the fusion of an egg and sperm to form a zygote.

Fibroblast one of the predominant cell types in connective tissue.

Fitness the ability of an organism to survive and reproduce.

Founder principle changes in gene frequency in small populations as a result of their origin from only a few founders from the parental population.

Frameshift mutation a mutation in a structural gene caused by addition or deletion of one or more deoxyribonucleotides, causing the translation of the corresponding mRNA to be shifted so that the nucleotide sequences are read in a different register.

G-bands bands observed when chromosomes are stained with one of the modified Giemsa methods.

Gamete a mature germ cell, either ovum or spermatozoon.

Gene a unit of heredity. At present, genes are usually equated with units of function, that is, the sequence of DNA required to code for one polypeptide chain or one RNA molecule (other than mRNA).

Gene pool the total pool of genes in a population.

Genetic code the sequences of three nucleotides in DNA or RNA that specify amino acids when translated into polypeptide chains.

Genetic drift changes in allele frequencies due to chance sampling of the parental population.

Genetic marker any single gene trait that can be used to follow the transmission of chromosomal regions in a mating.

Genetic load the decrease in fitness of a population as a result of detrimental alleles.

Genetotrophic disease a nutritional deficiency disease caused by inherited requirement for an unusually high level of a nutrient.

Genome a haploid set of genes. The term also is often used to refer to the complete gene complement, without regard to ploidy.

Genotype the genetic constitution of an organism.

Gonad the reproductive organs in which the germ cells are found and gametes produced. The ovary in females and testes in males.

Gonosome a sex chromosome.

Gynandromorph an individual who is a sexual mosaic, with certain tissues of male phenotype and others of female.

Haploid having a single set of chromosomes, that is, only one of each type of chromosome.

Haplotype a term used to indicate the haploid genetic composition of a complex locus, for example, the HL-A locus.

Hardy-Weinberg law a law stating the expected combinations of alleles under conditions of random mating.

Hemizygous refers to loci for which a single copy is present, as in the case of X-linked genes in males.

Hermaphrodite organisms in which both male and female gonads are present in the same individual.

Heterochromatin portions of chromosomes characterized by heavy staining due to a more compact structure during certain parts of the cell cycle

Heterogametic producing two types of gametes with respect to sex chromosomes. In mammals, males are heterogametic, producing X-bearing and Y-bearing sperm.

Heterograft a tissue graft between different species.

Heterokaryon a cell having two or more genetically nonidentical nuclei.

Heteromorphic describes homologous chromosomes that differ in appearance.

Heteronuclear a cell culture that no longer has the original complement of chromosomes.

Heteroploid heteronuclear.

Heteropycnotic densely staining regions of chromosomes resulting from condensed packing of heterochromatin.

Heterosis the increased vigor of hybrids compared with that of the parental lines.

Heterozygous having different alleles at a genetic locus.

Hinge region in immunoglobulins, the segment between the first and second constant region domains, characterized by interchain disulfide bonds.

Histocompatibility the acceptance of tissue grafts.

Histone a group of basic proteins associated with chromosomes.

Holandric refers to inheritance

due to genes on the Y chromosome.

Holoenzyme an enzyme consisting of an apoenzyme and a coenzyme.

Homogametic producing gametes of a single type. In mammals, females are homogametic.

Homograft a tissue graft between members of the same species.

Homologous chromosomes chromosomes that are identical in content of gene loci.

Homonuclear a cell culture that has retained the original chromosomal complement.

Homozygous having indistinguishable alleles at a particular locus on both chromosomes.

Hybrid a cross between two distinct races or species. A cell hybrid is a cell derived from two different cultured cell lines that have fused.

Idiogram a diagram of the chromosomal composition (karyotype) of a cell.

Inbreeding mating between related persons.

Inbreeding coefficient the coefficient F that expresses the likelihood that two alleles at a locus will be identical by descent.

Incidence the frequency of a trait at birth.

Index case the person through whom a family is located. Same as *proband* or *propositus*.

Initiator codon the codon that serves to initiate translation of mRNA into a polypeptide chain.

Interference the phenomenon of reduced crossing over in the immediate vicinity of another crossover.

Interphase the part of the cell cycle between divisions.

Interstitial region the chromosome segment between the centromere and the point of a translocation.

Intron an untranslated sequence of DNA located within a structural gene. Also known as *intervening sequence*.

Inversion a chromosome rearrangement in which a central segment produced by two breaks is inverted prior to repair of the breaks. See *paracentric* and *pericentric*.

Isoalleles alleles that appear to be identical under many circumstances but that can be distinguished in one or more test systems.

Isochromosome a chromosome with two identical arms.

Isogenic having identical genotypes.

Isozyme enzymes that can be distinguished by some property, usually electrical charge, but that act enzymatically on the same substance.

Karyotype the chromosome complement.

Kinetochore centromere.

Landmark a feature used in identifying chromosomes, such as the ends of the arms, the centromere, certain prominent bands.

Late-labeling refers to incorporation of tritium-labeled thymidine late in the period of DNA synthesis.

Lateral asymmetry differential incorporation of BrdU into homologous positions of sister chromatids due to greater concentration of adenine in one of two complementary strands, producing differences in staining or fluorescence of certain dyes.

Leptotene the early stage of meiotic prophase before the chromatids become visible as separate structures.

Linkage loci are linked if they are located on the same chromosome, even though they may be too remote to demonstrate linkage directly in a breeding test.

Locus the position of a gene on a chromosome.

Lymphocyte one of the major groups of white blood cells.

Lyon hypothesis the hypothesis that all the X chromosomes in a cell but one are inactivated early in development.

Lysosome a cell organelle that contains many of the hydrolytic enzymes. The function seems to be primarily to break down metabolites.

Map unit one map unit is equal to 1% recombination between linked genes. A map unit is also known as a *centimorgan*.

Meiosis division of germ cells leading to formation of haploid gametes.

Meiotic drive deviation from a 1:1 ratio of segregating alleles in gametes.

Messenger RNA (mRNA) the ribonucleic transcription product of the nucleus that is translated by ribosomes into specific polypeptide sequences.

Metacentric chromosomes having a centromere in the middle.

Metafemale an individual whose constitution of sex chromosomes consists of three X chromosomes. A triplo-X person.

Metaphase the stage of mitosis and meiosis at which the maximally condensed chromosomes are arrayed in a plane between the poles of the cell.

Metaphase plate the plane on which metaphase chromosomes are aligned. Equatorial plane.

Missense mutation a substitution in a codon that causes a different amino acid to be incorporated at the corresponding position in the polypeptide chain.

Mitochondria the cytoplasmic organelles associated with aerobic oxidation and energy production.

Mitogen a substance that stimulates cell division.

Mitosis cell division in which chromosome replication leads to daughter cells identical in chromosome content to the parent cell.

Monosomy having only one copy of a particular chromosome rather than two.

Monozygotic monozygotic (MZ) twins arise from a single zygote. MZ twins also are called identical.

Morgan a unit of recombination equal to 100%. Ordinarily recombination distances are measured in centimorgans (1 cM = 1% recombination).

Mosaic an individual with two or more genetically different cell lines derived from a single zygote.

Multifactorial a trait influenced by variation at several loci. Equivalent terms are *multigenic* and *polygenic*.

Mutagen any substance that increases the mutation rate.

Mutation any change in the genetic material that is heritable.

Nondisjunction the failure of chromosomes to separate properly in cell division.

Nonpenetrance failure of a phenotype to express the genotype.

Nonsense mutation mutation of a codon to a chain-terminating codon.

NOR nucleolar organizer region.

Nuclease an enzyme that breaks down nucleic acids.

Nucleic acid a polymer composed of chains of nucleotides.

Nucleolus a nuclear body that is associated with specific regions of chromosomes and that is involved in the synthesis of ribosomal RNA.

Nucleoside a purine or pyrimidine base to which is attached a sugar (ribose in ribonucleosides and deoxyribose in deoxyribonucleosides).

Nucleosome a histone-DNA complex observed in the dissociation of chromosomes. Also known as *nu* body.

Nucleotide a nucleoside to which a phosphate group is attached via the 5′-hydroxyl on the sugar.

Nullisomic being without a particular chromosome.

Oöcyte the female germ cell during the stages of meiosis. The term *primary oöcyte* refers to stages of the first meiotic division, producing thereby a secondary oöcyte.

Oögenesis formation of the ova.

Oögonium a primordial female germ cell that divides by mitosis, eventually giving rise to oöcytes.

Operator the regulator site in a Jacob–Monod operon. Specific repressor molecules bind to the operator, shutting off transcription of the adjacent structural genes.

Operon a system of gene regulation studied in bacteria by Jacob and Monod. The operon consists of a segment of DNA containing an operator to which repressor molecules can bind, a promoter to which RNA polymerase binds, and one or more structural genes that are transcribed as a unit.

Ovarian follicle a fluid-filled sac in the ovary in which is located a primary oöcyte.

Overdominant greater fitness of a heterozygote in comparison to the two homozygotes.

Ovotestis a gonad that contains both ovarian and testicular tissue.

Pachytene the stage in prophase of meiosis I during which homologous chromosomes are completely paired and some shortening and coiling are apparent.

Pairing the side-by-side alignment of homologous chromosomes during first meiotic prophase. Pairing is highly specific.

Panmixis random mating.

Paracentric refers to chromosomal events, such as inversions, that do not include the centromere.

Parasexual genetic recombination that occurs in some organisms, such as fungi, by other than regular sexual processes. In fungi and recently in mammalian cell hybrids, the parasexual cycle occurs by cell fusion with formation of heterokaryons, nuclear fusion, and eventual loss of chromosomes to return to a haploid (in fungi) or diploid (in mammals) state.

Parthenogenesis formation of an embryo from an unfertilized ovum.

Partial sex linkage refers to loci in a homologous region of the X and Y chromosomes. Such homology has not been demonstrated in man.

Pedigree a diagram showing phenotypes and biological relationships among members of a family, often for several generations.

Penetrance the frequency with which a particular genotype is expressed, regardless of the variation in expression.

Pericentric refers to an event that involves both arms of a chromosome.

Pharmacogenetics the study of genetic variability in response to and metabolism of drugs.

Phenocopy an environmentally induced phenotype that resembles an inherited trait.

Phenotype the observable organism, produced by interaction of the genotype with the environment.

Phytohemagglutinin a plant extract that agglutinates red cells and that is used in cell cultures because of its mitogenic action.

Pleiotropy the production of apparently independent phenotypic effects by a single allele.

Point mutation a mutation affecting a single nucleotide pair. The term also is applied to any mutation that is no larger than the smallest deletion that can be resolved.

Polar body one of the products of meiosis in oöcytes, extruded as a small body from the major part of the cytoplasm. The first polar body contains one complement of chromosomes from meiosis I. The second polar body contains a complement from meiosis II.

Polygenic a trait influenced by variation at several loci. Equivalent terms are *multigenic* and *multifactorial*.

Polymerase an enzyme that catalyzes the formation of a polymer from its constituent building blocks. Examples are DNA polymerase and RNA polymerase.

Polymorphism existing in two or more forms within a species. In genetic polymorphisms, the alleles are in frequencies greater than can be maintained by mutation.

Polypeptide a polymer of amino

acids joined together by peptide bonds.

Polyploid having three or more sets of the basic haploid complement of chromosomes.

Polysome a complex consisting of several ribosomes held together by mRNA.

Polysomic a cell or individual with one or more chromosomes in addition to the normal complement.

Polytene chromosomes, as in certain cells of *Diptera,* in which replication without separation has led to giant multistranded chromosomes.

Prevalence the frequency of a trait in the living population at any one time.

Proband the person through whom a pedigree is ascertained. Same as *index case* or *propositus.*

Prophase the early phase of mitosis or meiosis from the time that the chromosomes first become visible to the beginning of metaphase when they are maximally condensed.

Prokaryote an organism that lacks true chromosomes. Prokaryotes include bacteria and blue-green algae.

Promoter in the operon, the region between the operator and the structural genes to which RNA polymerase binds.

Propositus same as *proband* or *index case.*

Protamine a very basic polypeptide of small molecular weight associated with the chromosomes of sperm of many animals.

Protein a polymer of amino acids joined by peptide bonds. In their biologically active states, proteins are folded into specific three-dimensional structures and function as catalysts in metabolism and to some extent as structural elements of cells and tissues. All gene action is expressed in terms of the kinds and amounts of proteins.

Pseudohermaphrodite a person with gonads of one sex only but with sexual organs that have some features of the other sex.

Q-bands fluorescent bands observed when chromosomes are stained with quinacrine or a derivative.

Quadrivalent the pairing of four chromosomes in a single figure observed in meiosis of organisms heterozygous for a translocation.

R-bands bands observed when chromosomes are treated and stained in a special way. The R-bands stain the reverse of the G-bands.

Race a major population having a gene pool distinct from other populations of the same species. Often used in the same sense as *subspecies.*

Rad a unit of absorbed ionizing radiation equal to 0.01 joules per kilogram of any medium.

Reading frame the reading of an undifferentiated RNA sequence as a series of codons of three nucleotides each. The reading frame depends on the point at which translation begins.

Recessive a trait that is expressed only if the individual is homozygous for the corresponding allele.

Recombinant chromosome a structurally rearranged chromosome resulting from meiotic crossing over between a displaced segment and its normally located homolog.

Recombination the formation of new combinations of alleles following segregation during meiosis.

Reduction division first meiotic division in which the number of chromosomes is reduced by one-half.

Region in karyotyping, the space between two adjacent landmarks.

Regulator gene a gene whose product is a repressor molecule that regulates the activity of an operon.

Rem the quantity of ionizing radiation that is equivalent in biological damage to one rad.

Repressor the substance (protein) produced by a regular gene that combines with the operator to regulate activity of an operon.

Repulsion alleles at linked loci are said to be in repulsion if they are on different chromosomes.

Reverse mutation mutation from a previously mutant form to wild type.

Reverse transcriptase an enzyme that catalyzes the formation of DNA using RNA as a template.

Ribonucleic acid (RNA) a polymer of ribonucleotides. The principal ribonucleotides in RNA are guanylic acid, cytidylic acid, uridylic acid, and adenylic acid.

Ribonucleotides nucleotides in which the pentose is D-ribose.

Ribosome particles that occur both in cytoplasm and nucleus. They are RNA-protein complexes and are the sites of protein synthesis.

Ring chromosome a chromosome in which breaks have occurred on either side of the centromere, with the two proximal ends joining together in the repair process to form a ring.

Robertsonian translocation translocation between acrocentric chromosomes such that both long arms are attached to the same centromere. The reciprocal centric fragment usually becomes lost in time.

Roentgen a unit of ionizing radiation exposure equal to 2.58×10^{-4} coulomb per kilogram air.

Satellite a chromosome segment attached to the main part of the arm by means of a thin strand.

Secondary constriction a constricted region of a chromosome other than the primary constriction (centromeric region).

Segregation the separation of pairs of alleles at meiosis.

Segregation distortion the recovery of alleles at other than the expected 1 : 1 ratio from a heterozygote.

Segregation ratio the ratio of alleles recovered from a particular mating type.

Sex chromatin a densely staining body in the nuclei of cells having more than one X chromosome. The sex chromatin body is the inactive X chromosome, and the number is equal to the number of X chromosomes less one.

Sex chromosome an X or Y chromosome.

Sex influenced a trait the expression of which is modified by the sex of the person in whom it occurs.

Sex limited a trait that is expressed only in one sex.

Sex linkage refers to genes on the X chromosome.

Sex ratio the ratio of males to females (or to the total) in a population. The primary sex ratio is the ratio at conception. The secondary sex ratio is the ratio at birth.

Sib (sibling) a brother or sister from the same parents.

Sibship brothers and sisters from the same family.

Simian crease a single flexion crease across the palm of the hand.

Somatic refers to all cells of the body other than germ cells.

Sperm (spermatozoön) the mature male gamete.

Spermatid one of the haploid products of meiosis in males prior to maturation into a spermatozoon.

Spermatocyte a male germ cell during the period of meiosis. A primary spermatocyte is in meiosis I and a secondary spermatocyte is in meiosis II.

Spermatogonium a cell in the stem line of male germ cells that divides by mitosis.

Spindle the structure formed during nuclear division that aligns the chromosomes and pulls them to opposite poles.

Stratification the coexistence of two or more subpopulations within a larger population.

Streak gonad a streak of fibrous tissue that replaces the normal gonad.

Structural gene a gene that codes for the amino acid sequence of a polypeptide chain.

Submetacentric describing a chromosome with a centromere between the center and one end.

Suppressor mutation a mutation that restores a wild-type phenotype to a mutant organism by a mechanism other than back mutation to the original gene structure.

Synapsis pairing of homologous chromosomes in meiosis.

Synatinemal complex a structure visible by electron microscopy that occurs between paired meiotic chromosomes.

Syngamy the fusion of male and female gametes.

Syntenic loci on the same chromosome.

Telocentric a chromosome having a terminal centromere.

Telomere the terminal structure of the chromosome arms. The telomere is not ordinarily visible as a distinct entity, but its existence is inferred from the stability of a normal chromosome as opposed to the instability of a broken chromosome arm.

Telophase the stage in nuclear division in which the new complements of daughter chromosomes become uncoiled and disperse, and new nuclear membranes are formed.

Teratogen any substance that causes malformations in embryonic development.

Terminalization movement of chiasmata to the chromosome ends in late meiotic prophase.

Terminating codon a codon that terminates synthesis of a polypeptide chain.

Test cross a cross of a double heterozygote with the double recessive parent, used to study the assortment of two loci.

Tetrad a bivalent in meiosis consisting of the four chromatids.

Transcription the replication of genetic information as coded in the nucleotide sequences of RNA and DNA. Either RNA or DNA may serve as the template depending on the enzymes and nucleotides present, and the product in either case is RNA or DNA.

Transduction the transfer of host genetic information from one cell to another by means of a virus. The phenomenon has been unambiguously demonstrated only in bacteria.

Transfer RNA (tRNA) small molecular weight RNA that functions in protein synthesis. Each amino acid has one or more tRNA molecules to which it binds. Each tRNA is specific for one amino acid and has a sequence of three nucleotides that constitute the anticodon.

Transformation (1) transfer of genetic information from one cell to another by means of free DNA; and (2) loss of contact inhibition by cultured cells.

Transition a mutation involving substitution of a purine for a purine or a pyrimidine for a pyrimidine in the DNA structure.

Translation the synthesis of a polypeptide chain using mRNA to direct the amino acid sequence. The mRNA is translated into protein.

Translocation change in location of a chromosome segment to another chromosome.

Transversion a mutation involving substitution of a purine for a pyrimidine or a pyrimidine for a purine in the DNA structure.

Triploid having three sets of the basic haploid chromosome complement.

Trisomic having three copies of a particular chromosome.

Trivalent in meiosis, three chromosomes paired into a single figure.

Wild type a normal phenotype of an organism. Also a normal allele as compared to a mutant allele.

X-chromosome inactivation the inactivation of all but one X chromosome in cells having more than one. According to the Lyon hypothesis, this occurs early in embryogenesis.

Zygote the diploid cell formed from fusion of sperm and ovum.

Zygotene the stage in meiosis in which pairing of homologous chromosomes occurs.

INDEX